The Warsaw Pact, 1969-1985:
The Pinnacle and Path to Dissolution

by Matěj Bílý

The Warsaw Pact, 1969-1985:
The Pinnacle and Path to Dissolution

by Matěj Bílý

Academica Press
Washington – London

Library of Congress Cataloging-in-Publication Data
Names: Bílý, Matěj, author.
Title: The Warsaw pact, 1969-1985 : the pinnacle and path to dissolution / Matěj Bílý.
Description: Washington : Academica Press, [2020] | Includes bibliographical references and index.
Identifiers: LCCN 2019954543
| ISBN 978-1680531947 (hardcover : alk. paper) | ISBN 9781680532173 (pbk. : alk. paper)
LC record available at https://lccn.loc.gov/2019954543
Cover image: „Shield-84" maneuvers in Czechoslovakia.
Source: ČTK (Czech Press Agency)

Contents

Introduction

Since its outbreak in the second half of the 1940s, the Cold War has been a phenomenon examined by a countless number of scholars, from political scientists and historians to security analysts and international relations experts. This fact is not surprising. Even during the Cold War era, the issue was very hot - in many ways, a completely new process that, for the first time in history, carried with it the very real threat of total annihilation of human civilization in a global nuclear war. From this point of view, it seems essential to examine the situation in detail as well as to analyze retrospectively the reasons why the power competition of two world superpowers never reached such extremes or even any significant military confrontation. More than a quarter-century after the Cold War's end, we can say with certainty that the basic outlines and course of this very specific conflict have been satisfactorily mapped through many scientific fields. Many interpretative perspectives have been established as well. These facts, however, do not mean that all aspects and factors of the four decades long and very dynamical historical process have been analyzed in necessary detail, or that enough time had passed, an essential element for any unbiased and thorough research.

One of the integral features of the bipolar nature of the Cold War world was the existence of two power blocs framed by large military-political pacts: the North Atlantic Treaty Organization (NATO) in the West and the Warsaw Treaty Organization in the East. The above-mentioned words are especially true for the latter, although not exclusively: Even more than twenty years after the organization's dissolution, there was no comprehensive historical monograph adequately analyzing the 36 years of its history. Actually, the alliance of the Soviet Union and its European satellites not only influenced the Cold War's development, but it was also an important part of the then-existing reality of the Soviet sphere of influence in Europe. Not only was the organization a formal umbrella for allied ties in the geopolitical region, but many significant processes in the so-called Eastern Bloc were controlled through it. Foreign and national security policy of the Soviet Union and its satellites was often presented on behalf of the Warsaw Pact. Therefore, it is an urgent task for the historiographer dealing with the Cold War - especially with the Soviet sphere of influence on the Old Continent - to thoroughly examine and assess the Warsaw Treaty Organization and its actual role. The urgency of the task partially comes from the fact that at least the Czech public mostly recalls the Warsaw Pact in association with the suppression of the Prague Spring by five armies of the organization's member-states in 1968. And yet, many other aspects of the alliance's existence remain relatively unknown or are limited to medially attractive, sensational issues, such as the dislocation of Soviet missiles from member-states' territories or the plans of military campaigns through Western Europe, all of which are lacking the context of the Cold War's reality.

This monograph seeks to deepen significantly the current scientific knowledge of the Warsaw Treaty Organization. It is based on my Ph.D. thesis defended in June 2015 at Charles University's Institute of World History in Prague.[1] Using the available primary sources, along with editions of documents and secondary literature, the present aim is to analyze the complexity of the Organization's history in the period of Soviet leader Leonid Iliych Brezhnev's pinnacle and the relatively short intermezzo of his successors, Yuri Vladimirovich Andropov and Konstantin Ustinovich Chernenko. Subsequently, the timeframe of this book is from March 1969 to May 1985.

These dates are not picked randomly. The former was a moment when the alliance reformed its military structures and thus finished its first stage of existence and its initial shaping. I focused on that period (1955-1969) in my master thesis, *The Warsaw Pact. Beginning and Crises*, defended at the Institute of World History, Charles University in Prague in 2010. At the same institute one year later, I defended my rigorous thesis, *The Warsaw Pact in 1950s and 1960s*, based on that work. This book follows up on my previous research of the Warsaw Treaty Organization. The latter date, May 1985, is related to a prolongation of the alliance, as

---

[1] Bílý, Matěj, "*Varšavská smlouva 1969–1985. Vrchol a cesta k zániku,*" (PhD diss., Charles University in Prague, 2015).

its thirty-year validity had expired on that date. Simultaneously, Mikhail Sergeyevich Gorbachev was establishing his power in the Soviet Union. After careful consideration, I dropped my initial intention to also cover the era of his rule. Gorbachev's changes in the Eastern Bloc are undoubtedly a very specific as well as important historical period. The ongoing shifts in the Soviet sphere of influence in Europe affected the Warsaw Treaty Organization significantly. Thus, a proper analysis of that process would result in a fundamental extension of a text that is quite large already. As it stands, the Warsaw Pact in the era of Gorbachev and its following break-up deserve the attention of historiography and future research will undoubtedly focus on such an issue in an adequate way.

Structure of the book and methodology

The general research aim was to analyze the Warsaw Treaty Organization's internal development so as to define its real role in the mechanisms of the Soviet sphere of influence in Europe as well as its influence on international affairs. The book never had the ambition to be military history in nature. However, if we want to explain the processes within the alliance comprehensively, it is clear that beside its political structures we must also deal with their military counterparts to a large extent. Therefore, the work follows two parallel lines which inevitably intersect at certain times. Such an approach corresponds with the fact that the Warsaw Pact's military and political frameworks operated mostly separately after 1969, as the following text shows. In the chapters focused on military development, I tried to put the issue into the context of the time. The above-mentioned dual-track mechanism in the alliance should by no means overshadow the basic fact that in the Eastern Bloc reality, a political power remained in charge of shaping the basic contours of national security strategy. As to be explained, the military elites' recommendations were taken into account in particular decisions, but were not ultimate. In that regard, it is good to point out that a detailed analysis of the relation between political authorities and military leadership – that probably differed in the Warsaw Pact's individual member-states – goes far beyond the scope of this book. Nonetheless, it was necessary to outline, at least generally, what the situation was like in the Soviet Union. As the

Eastern superpower was a key to processes within the alliance, it was also essential to clarify who set the agenda in the examined period. I point out in advance that the presented conclusions in that respect are not categorical, due to the unavailability of primary Soviet sources. It is possible that future research will revise them.

Although the book primarily focuses on events within the alliance and the development of the member-states' cooperation features in its framework, the Warsaw Pact's functioning nature makes it necessary to put these processes into the broader context. The issues that must be explained are the background of the Cold War's development, events in the Eastern Bloc, and the Soviet Union as the dominant power within the Pact. The work has absolutely no ambition to be a synthesis of the history of the Soviet sphere of influence from 1969-1985. Therefore, some issues are simply outlined in order to allow for a better understanding of the matter of processes within the Warsaw Treaty Organization itself. The chapter examining the alliance's role in the crisis of the Polish communist regime in the early 1980s is an exception, as the text deals with some aspects of the events in more detail. Even in this case, I tried to keep those factors which could have affected the Warsaw Pact's possible actions, both potentially and actually.

At first sight, the topic could seem quite attractive. However, this is misleading. We are reminded that the book covers a long period of sixteen years when more dramatic and dynamic moments alternated with duller times, certainly less interesting from the reader's perspective. But a scientific historiography should not focus on ground-breaking and crucial events only. If we want to understand any historical process comprehensively, we must inevitably cover the events fully, including the calmer, less important periods. That is the reason why the following text can seem too large and detailed in some segments. I suppose, however, that even these passages make sense in respect to the above-mentioned aims of the book. Based on a broad heuristic, they can be helpful to other scholars examining the Eastern Bloc and the Cold War in the future.

Each period and issue is given different attention in the text. Such an approach reflects the fact that in the Warsaw Pact's history can be found few relatively short periods – especially during big crises – rich in events.

These periods contrast sharply with the much poorer periods, and also times when the alliance remained passive. Chiefly, I pay great attention to two phenomenons – the Polish crisis in early 1980s and the Romanian attitude to cooperation within the organization. I saw the Polish events and the subsequent move away from the so-called Brezhnev doctrine as a significant shift in Soviet policy as well as a crossroads for the Warsaw Pact, such that I felt a need to analyze the dynamics of the ongoing processes in both the alliance and the Soviet leadership very thoroughly. The same applies to the Romanian question. It was given great attention for two reasons: First, I regard it appropriate to explain the issue and its genesis comprehensively. Otherwise, it would be necessary to deal with the Romanian stances at almost every single Warsaw Pact political and military meeting at the expense of the fluency of reading. Second, the "balancing" policy of Ceausescu's regime on the Warsaw Treaty Organization reveals an important finding – the fact that the Eastern Bloc's erosion began much sooner than usually claimed. I demonstrate that the USSR virtually lost its levers for making its satellites obedient as far back as the first half of the 1970s. Such a lack of effective mechanisms was fully exposed during the course of the Polish crisis.

In the absence of a comprehensive historiographic book on the Warsaw Pact's history, I focused on the functioning of the organization itself. Considering the scope of the issue, I was unable to significantly compare the Eastern alliance's mechanisms with those working in its Western counterpart at the time. It is no question that such research would be very beneficial and most assuredly will be done in the future. The scope of the issue also prevented a deep analysis of one of the very important Cold War aspects: the real balance of military forces between two blocs. The same holds for the volumes of defense expenditures. After all, many analysts have dealt with both issues for several decades. So, the book examines these factors on the basis of existing literature. I did no significant archival research primarily focused on such fields.

The time frame of the book suggests attending to the approach that largely connects the development stages in the Warsaw Pact's existence with the individual Soviet leaders' rule. Despite the fact that the reform of the alliance's military structures is taken as the beginning of the text, the framing essentially corresponds to a moment when the CPSU General Secretary Leonid Brezhnev, after suppression of the Prague Spring, consolidated his position within the Soviet political leadership. It was at that moment when the pinnacle of Brezhnevism began. Such an approach was proven right during the research itself, which clearly confirms that the top Soviet leadership defined features of both political and military cooperation in the Warsaw Pact's structures.

The text is basically chronological. The chapters are split into three main periods reflecting the book's subtitle *Pinnacle and Path to Dissolution*. The first is the period before 1975, when the Helsinki Final Act was signed. That moment can be seen as a political pinnacle of the Warsaw Treaty Organization. The following period ends with the outbreak of the so-called Euromissile crisis and the Soviet invasion of Afghanistan, when the détente process virtually failed. As the book shows, the alliance and its activity mostly stagnated during those years. The last period of 1980-1985 was characterized by the so-called Second Cold War. In that stage, despite some attempts to adjust its mechanisms, the alliance's rigid structures proved to be absolutely unable to react to the dynamical challenges of the time. So, the Warsaw Pact finally turned to the path leading to its dissolution.

I did not deem it necessary to over-stretch readers with a final summary of the individual chapters' conclusions. Instead, the book contains a short epilogue about the general conclusions and a brief outline of the Warsaw Pact's fate. I also tried to explain in short why the Pact could not survive the end of the Cold War.

The book is primarily based on broad archival research. Therefore, the main method used is a contained analysis of the documents. As explained later, these are materials from Czech, Polish, German, and British archives. For obvious reasons the largest part of the heuristic base consists of the documents of the Czechoslovak provenience. This applies particularly to military sources. And so, many processes are inevitably demonstrated and explained on the Czechoslovak example. Nonetheless, I deliberately tried to avoid Czechoslovak or any other national perspectives and I sought not to make general conclusions in that respect.

As far as the secondary literature is concerned, I decided to ignore Sovietological titles published before the Warsaw Pact's dissolution. In most cases, these are obsolete analyses which would not benefit my declared aim: to explain the development of the Warsaw Treaty Organization on the basis of primary sources – all while avoiding speculations. Moreover, some more relevant observations are often reflected in historiographical works of a later date which the text refers to. To a large extent, this is related to the fact that the heuristic base includes only a few memoirs. Not only are there a very limited number of memoirs dealing with the Warsaw Pact's history directly, but such a source also forms one of the pillars of the secondary literature. Unfortunately, during my research I was unable to collaborate with those still-living witnesses whose testimony would be relevant to the analyzed processes and who could provide another interesting perspective that would enliven the often sterile, stark talk of the archival documentation. I compensate for this deficiency by using some already published testimonies, especially those of former East-European generals.

The study of the secondary literature proved the need for a cautious approach towards the term Warsaw Pact; not only when researching the literature, but also during the work with the documentation. Namely, the term is used not only regarding the alliance itself and regarding its structures, but it also has served in the Western discourse as a general denomination of the states in the Soviet sphere of influence in Europe. In that respect, the following text uses the term only in those moments when it is actually related to the organization. Otherwise, I use different names, such as the Warsaw Pact's member-states, the Eastern Bloc, or the Soviet sphere of influence in Europe.

In the text, relatively vague terms, such as "state leadership" or "military leadership", are used often. The former means a garniture in power in a particular country at the time, and the latter means specific military institutions and their members with decisive powers. I am aware that such a denomination of power structures – often not unanimous, with different actors, elements, and factions – is simplified. However, it is an almost inevitable approach to a text dealing with such a broad issue. The same applies for the names of the individual states' capitals, which frequently stand as an abbreviation for all the above-mentioned groups.

Dealing with weapons and military equipment, the text primarily keeps the original names rather than code-names used by NATO at the time. In that regard, I am aware that the latter names are more familiar these days. Therefore, the Western code-names will be given in brackets following the first mention of the original names.

Current state of historical knowledge

Many security analysts, political scientists, and pundits dealt with the Warsaw Pact in the time of its existence. Their works, however, were mostly of a *Sovietological* nature, based on sources publicly available, such as media coverage, declarations, both political and military officials' statements, or defector's testimonies. This research would not reflect the primary sources kept secret at the time, especially the internal documentation, both political and military, of the states of the Soviet sphere of influence in Europe, which is absolutely essential for any detailed analysis. Relevant historical research of the Warsaw Pact really started after its dissolution in 1991. In the first half of the 1990s, the former Soviet satellites' as well as Russian archives slowly began to open revealing more details about the processes whose nature had merely been a subject of speculation before. As is typical for a scientific examination of any phenomenon, and also in this case, partial papers emerged and relevant documents were published first. In that respect, an effort by the American historian of Czech origin, Vojtech Mastny, and his American colleague, Mark Kramer, must be noted. They both stand behind many crucial texts related to the Warsaw Pact's history.[2] A majority of these

---

[2] Namely Kramer, Mark, "Warsaw Pact Military Planning in Central Europe: Revelations From the East German Archives," *Cold War International Project Bulletin,* no. 2 (1992): 1, 13–19; Kramer, Mark, "The Lesson of the Cuban Missile Crisis for Warsaw Pact Nuclear Operations," *Cold War International Project Bulletin,* no. 5 (1995):. 59, 110–115; Kramer, Mark, "Comentary on Soyuz-75 and Shchit-88 Military Excercise Documents," *Cold War International Project e-Dossier,* no. 20 (2011); Mastny, Vojtech, "We are in a Bind: Polish and Czechoslovak Attempts at Reforming the Warsaw Pact, 1956–1969," *Cold War International Project Bulletin,* no. 11 (1998): 230–235; Mastny, Vojtech, "*The*

were published at the website *Cold War International History Project Bulletin,* issued by the Wilson Center (a Washington-based institute). Another important step was taken in 1999, when the institute *Parallel History Project on Cooperative Security* was established. The institute's focus was not only on the history of the Warsaw Pact, but of NATO as well. On the project's website, a large archive of available documents of both alliances was built; in many cases supplemented by partial studies. Many testimonies by former high-profile military officials can be found there as well.

Despite all the resources mentioned above, the fact is that all the historical texts related to the Warsaw Pact published so far consist of basic general introductions, not-so- comprehensive volumes, and collective monographs; or mostly focus on the military aspects of the Pact's existence. At the moment, the most significant work is *A Cardboard Castle? An Inside History of the Warsaw Pact* (New York 2005) by Malcolm Byrne and Vojtech Mastny. The book is foremost a large edition of documents. Therefore, the introduction, which basically explained the alliance's internal development put it into the Cold War and Eastern Bloc context, remained limited for obvious reasons. To the period my book examines are related no more than approximately 25 pages, which naturally cannot cover all the aspects of an almost two decades-long existence of the military-political alliance, let alone describe and analyze them in detail. Although Mastny's work should by no means be ignored – and the following text points to it frequently – a more critical reading, and especially a confrontation with the heuristic base, reveal that some of author's conclusions are highly speculative.

Similar deficiencies appear as well in the most significant Czech work so far, *Plánování nemyslitelného* by Petr Luňák (Prague 2007), which is weighed down by its primary content, namely, an edition of the Czechoslovak People's Army war plans. Therefore, its introduction, which seeks to explain the Warsaw Pact's history in general, is very often missing some important political aspects. Luňák's main conclusion that the alliance of the Eastern Bloc nations was an offensive military

Soviet Union and the Origins of the Warsaw Pact," Accessed October 16, 2019. http://www.php.isn.ethz.ch/lory1.ethz.ch/collections/coll_pcc/into_VM.html.

formation planning an occupation of Western Europe is presented in a scandalous manner and particularly fails to reflect the Cold War context as well as actual relations between political and military power in the Soviet Union and its European sphere of influence.

The Polish historiography also deals with the Warsaw Pact significantly. In that respect, an extensive study and edition *PRL w politycznych strukturach Układu Warszawskiego w latach 1955–1980* by Wanda Jarząbek (Warsaw 2008) must be mentioned first and foremost. Although the author focuses primarily on the Polish activities within the alliance's framework, the initial study (approximately two hundred pages) is one of the best general analyses of the Warsaw Treaty Organization thus far. Unfortunately, it is limited to the period of the Polish leaders Władysław Gomułka and Edward Gierek. Jarząbek avoided the Sovietological methods while also keeping an ideological neutrality, something that earlier studies on the issue lacked quite often. She makes very few speculations. However, the parts dealing with the development after 1970 are much flatter and heavily thesis-driven, in comparison to the previous era, and are frequently limited to a list of issues that the Polish representatives actively engaged in at the alliance's meetings. Nevertheless, the work was a valuable source for my book and its quality sharply contrasts with an earlier text *Układ Warszawski: Działalność polityczna 1955–1991* by another Polish author, Jaczek Ślusarczyk (Warsaw 1992). This title proves that the works dealing with the Warsaw Pact written before the Eastern Bloc's collapse, and in the first half of the 1990s, are not very relevant today because of their restricted heuristic bases.

Valuable, and often much more detailed and precise findings can be found in some volumes and collective monographs. These are represented first and foremost by *NATO and the Warsaw Pact. Intrabloc Conflicts*, edited by Marry Ann Heiss and S. Victor Papacosma (Kent 2008). The volume contains a number of high-quality papers by reputable historians cooperating with the Parallel History Project, such as Douglas Selvage (United States), Jordan Baev (Bulgaria), Bernard Scheafer (Germany), or Csába Békés (Hungary), all of whom are quoted frequently in this book's individual chapters. Some important military aspects of the Warsaw Pact are revealed in the volume *War Plans and Alliances in the*

Cold War, edited by Vojtech Mastny, Sven Holtsmark, and Andreas Wenger (London 2006). Unfortunately, it mostly focuses on the 1960s. Therefore, the most beneficial source for the period I examined was Frade P. Jensen's paper *The Warsaw Pact's Special Target. Planning the Seizure of Denmark*, which presents a relatively unique interpretation of the Warsaw Pact's offensive doctrine. The other papers in the volume just show that, for obvious reasons, the historiography dealing with the Warsaw Treaty Organization has primarily focused on the 1950s and 1960s thus far, and the literature related to the Pact's development after 1975 is not very extensive.

The German historiography examines the Warsaw Pact as well. In this case, unfortunately, military aspects stand in the limelight even more. The collective monographs *Der Warschauer Pakt: von der Gründung bis zum Zusammenbruch: 1955 bis 1991/ im Auftrag des Militärgeschichtlichen Forschungsamtes*, edited by Torsten Diedrich, Winfried Neinmann, and Christian Ostermann (Berlin 2009), focused on the situation of the individual member-states' armed forces. This text helped me compensate for – to some extent – the absence of military documents from the former Eastern Bloc countries' archives beside Czechoslovakia. The same applies to the earlier synthesis *Die Rotte Bündnis. Entwicklung und Zerfall des Warschauer Paktes 1955–1991* (Bonn 1996) by Frank Umbach, a security analyst, which is based mostly on the East-German military documentation. Similar sources also reference an excellent paper by Beatrice Heuser whose analysis of the Warsaw Pact's military doctrine, written in 1993, was outstanding at the time.[3]

After my work was finished, two significant monographs dealing with the Warsaw Treaty Organization were published: the synthesis *The Warsaw Pact Reconsidered. International Relations in Eastern Europe, 1955–69*[4] by Laurien Crump, a Dutch historian, and *Romania and the Warsaw Pact, 1955-1968. From Obedience to Defiance*[5] by Laurentiu-

---

[3] Heuser, Beatrice, "Warsaw Pact Military Doctrine in the 1970s and 1980s: Findings in the East German Archives," *Comparative Strategy*, no. 12 (1993): 437–451.

[4] Crump, Laurien, *The Warsaw Pact Reconsidered. International Relations in Eastern Europe, 1955–69* (London/New York: Routledge, 2015).

[5] Dumitru, Laurentiu Cristian, *Romania nad the Warsaw Pact, 1955–1968. From Obedience to Defiance* (Rome: Italian Academic Publishing, 2014).

Cristiana Dumitru, a Romanian historian. Both texts, however, focus on the period before 1969 and thus would have been little benefit for this book.

Sources

The monograph is mostly based on extensive research in Czech, Polish, and German archival collections. Some British documents are also reflected in the sidelines. First of all, it must be admitted that we have still not gotten an opportunity to study the original Soviet materials regarding the Warsaw Pact's activities. Materials for the period examined in this book remain in inaccessible Russian archives. This applies to documentation both military and political. Obviously, the Soviet politburo records of meeting, where the Warsaw Pact's summits were certainly discussed along with documents of the alliance's Unified Command based in Moscow, would enhance the work's scope significantly. However, the possibility of such an extensive research program in the near future seems almost utopian considering the current political situation in Russia.

As already suggested, the main part of the heuristic base consists of the documentation from Czech archives, due to their accessibility. It includes materials of various former Czechoslovak government, diplomatic, and military institutions as well as the CPCz leadership. During my many years of research, I studied almost all the documents related to the Warsaw Treaty Organization in the Czech archives. First of all, the collections of the National Archive of the Czech Republic must be mentioned, containing the CPCz CC Presidium documents. Accessibility of these collections for the entire examined period was a great advantage. There are located not only detailed reports on the Warsaw Pact's political bodies' meetings, but also many sources on the Soviet foreign policy and the international situation, which reveal the Eastern Bloc countries' objectives at the time. Many of these documents shed light on the relations between the individual alliance's member-states. A limiting factor is that these are almost solely informational reports. We do not have minutes of the meetings where the party leadership members would have most likely discussed the situation within the Warsaw Pact much more candidly, a likely contrast to the stark language of the mentioned documents.

This is a problem with the alliance's meetings in general: for the examined period, there is only a very limited number of stenographic protocols of those sessions. In the vast majority of cases, we are left with reports about those meetings worked out by the individual member-states. Nonetheless, for most of them we have at least a full transcript of speeches delivered by each delegation. It is not ruled out that the complete records of discussions can be found in Russian archives.

Of great value to my research were documents from the Archive of the Ministry of Foreign Affairs of the Czech Republic. Especially useful were the collections *Teritoriální odbory – tajné* (Territorial Departments – Secret), *Dokumentace teritoriálních odborů* (Documentation of Territorial Departments), and *Porady kolegia* (Collegium Briefings. Although these diplomatic cables about the member-states' interaction alone do not allow a reconstruction of the situation within the Warsaw Treaty Organization, they often reveal the real intentions of actions taken by the alliance. As its political structures served to a large extent as the formal auspices, which the book demonstrates, many issues related to the Warsaw Pact were discussed in diplomatic side-channels rather than the official alliance's framework. Unfortunately, the former collection is opened for researchers to 1974 only. After that date, my research was limited to the Documentation of Territorial Departments' collection that allows studying records up to 1989, but without the materials related to the Soviet Union. Undoubtedly, access to these sources would have made some conclusions for the second half of the 1970s and the first half of the 1980s more precise.

After many administrative complications, I was given access to the documentation of the Military Historical Archive of the Czech Republic, namely the collection *Federální ministerstvo obrany/Operační správa* (Federal Ministry of Defense/Operational Department), that is often imprecisely referred to as the Warsaw Pact collections by some historians. As the first researcher ever, I was able to study the entire collection except for the original documents created directly in the alliance's Unified Command. During the Warsaw Pact's dissolution, on February 21, 1991, Czechoslovakia signed a Protocol for the termination of validity of the military agreements within the Warsaw Treaty Organization and dissolution of its military bodies and structures. Article

III/b of the agreement committed the signatories *"not to publish the Warsaw Pact's documents and not to provide them to any third sides"*. This factor was, to some extent, eliminated by the fact that the provision does not apply to Czech translations of these Russian-language materials as well as copies made by the Czechoslovak side. The book reflects mostly reports on preparations and course of the alliance's military bodies' meetings, military agreements, and regular communication between the Czechoslovak military institutions and the Unified Command. Protocols on assignment of the Czechoslovak troops to the Pact's formation and military analyses of actual tasks, served as a useful base for analysis of the main developmental trends in that field. Let me point out in advance that in the examined period the individual Warsaw Pact's member states did not have any complex information about the overall military strategy of the Pact. Therefore, unlike many earlier works, this book tries to avoid enormous generalizations based on the Czechoslovak documentation in that matter.

There were a number of important documents I also obtained in Archiwum Akt Nowych in Poland, mostly from the collection *Polska Zjednoczona Partia Robotnicza – Komitet Centralny w Warszawie* (Polish United Workers' Party – Central Committee in Warsaw). The volume of these documents corresponds to the fact that Poland was generally very active within the alliance. Moreover, these documents are accessible for the entire examined period. However, like the Czechoslovak materials, except for few moments, they do not contain records of the Polish leadership discussions about the alliance's meetings. As already mentioned Wanda Jarząbek correctly notes, there is a relatively huge difference between the evidential value of documents from Gomułka's era and Gierek's era. Since the 1970s, the Polish materials related to the Warsaw Pact's meeting typically move away from stenographic protocols to brief general reports. That obviously entails a significant subjectivity as well as limited evidential value. However, one can agree with the Polish historian that the amount of space given to Soviet officials' speeches in these records about the alliance's meetings only confirms that these speeches were taken by the Soviet satellites as a setting of the Kremlin's actual objectives and thus were also the guidelines for their diplomatic corps and general foreign policy. Fortunately, the PUWP CC politburo stenographic protocols exist for the critical period of 1980-1981 and were an important source for the chapter dealing with the Warsaw Pact's role in those events.

My archival research also included a complete analysis of the East-German SED politburo collection located in Bundesarchiv in Berlin. This documentation, too, is accessible without any time restrictions. As the GDR was one of the most loyal Soviet satellites and closely coordinated its steps within the alliance's framework with the Kremlin for most of the examined period, the East-German documents – in comparison to the Czechoslovak and Polish ones – often provide a better insight into the background of the Warsaw Pact's meeting. Just as in the case of all the other above-mentioned archival collections, I did not focus solely on the documentation related to the organization itself, but taking into account its mechanisms, I also analyzed the reports on some bilateral meetings – particularly those involving the top Soviet officials – where the alliance's strategy was frequently shaped.

To a limited extent, the text refers to the materials from The National Archives in Great Britain, mostly to the British Foreign Office documents. These include either analyses of some of the Warsaw Pact's actions worked out within NATO and for its needs, or British diplomats' reports on talks with their Warsaw Pact member-states' counterparts. It must be pointed out that these sources are not essential for the book and they are primarily used as explanations of some Western reactions to the steps of the Warsaw Treaty Organization.

Besides archival research, the book is based on various editions of documents as well as online archives. For the development within the Warsaw Pact itself, the above-mentioned publications edited by Mastny, Luňák, or Jarząbek are essential. The absence of research in some former alliance's member-states' archives – Bulgarian, Hungarian, and Romanian – is compensated, to some extent, by Mastny's edition and the already-mentioned online archive *Parallel History Project on Cooperative Security*, which collects copies of both political and military materials of all the Warsaw Pact's countries, including the three named. Moreover, some of them were translated to English or German. That fact is crucial for me, because my absent Bulgarian, Hungarian, and Romanian language skills prevented me from working with the originals. In all the mentioned cases, one has to remember that these are just selections, often very narrow, rather than a complete documentation.

Once again, the absence of a comprehensive set of the Soviet documents must be pointed out. In recent years, an effort by Russian historians and archivists resulted in publishing extensive editions of the Soviet leadership's materials. Unfortunately, these only go up to the 1960s and so they do not correspond to the period examined in this book. The Soviet documents noted in the text are individually revealed pieces. The exemption is the CPSU CC politburo documentation on the Polish crisis in 1980-1981 disclosed by the Russian Federation.

As suggested already, the text also had to frequently reflect the general development of the Cold War and the events in the Soviet sphere of influence. Besides the summarizing literature, references to the editions of documents are made often. This is the case of the Polish events of December 1970 based on materials published in *Tajne dokumenty Biura Politycznego. Grudzień 1970,* edited by Paweł Domański (London 1991), or the Euromissile crisis based on *The Euromissiles Crisis and the End of the Cold War: 1977–1987,* collected by Timothy McDonell (Rome 2009). The analysis of the crisis of the Polish communist regime in the early 1980s stands as a separate chapter. In this regard, I could use the extensive editions *From Solidarity to Martial Law: The Polish Crisis of 1980–1981: A Documentary History* by Andrzej Paczkowski and Malcolm Byrne (New York 2007), and *Przed i po 13 grudnia. Państwa bloku wschodniego wobec kryzysu w PRL 1980–1982,* edited by Łukasz Kamiński, Iskra Baeva, and Petr Blažek (Warsaw 2006, 2007). Both publications include not only the Eastern Bloc states' and the Soviet documents, but also some Western materials - all translated to English or Polish, respectively.

In certain parts, the text points to selected writings of the Soviet military officials and theorists published in the military press at the time. Undoubtedly, a very cautious approach and critical evaluation is necessary when such a source is analyzed. These articles were tendentious, to a large extent, and written with an extreme emphasis on ideological starting points. Moreover, any far-reaching conclusions about their overall nature could have been made after a wider analysis of the Soviet military journals only, which goes beyond the scope of this book. Regarding the first half of the 1970s, I quote the texts from *Selected Soviet Military Writings 1970–1975*, edited by W. F. Scott (Washington 1976). Obviously, the publication is just an arbitrary selection of articles from the Soviet military press. Therefore, the source is used in this book only to complete the picture, rather than as a ground for any ultimate conclusions. I consider speeches and materials by the top Soviet military officials presented at the alliance's closed meetings to be a much more relevant source for understanding the Soviet military thinking about the Warsaw Treaty Organization. These are noted in the text frequently.

The book relies on press to a very limited extent. Such a source is used mostly in the chapter dealing with the Polish crisis in the early 1980s. The Warsaw Pact played an important propagandistic role in those events and was deliberately used to threaten the Polish opposition. Therefore, it was necessary to reflect on, at least basically, some stories published in the press at the time. In that respect, I did not research the journals and newspapers from the period completely. Instead, I used the press cutting collection in the National Archive of the Czech Republic. Its undisputed advantage is the fact that it includes a very broad compilation of stories related to the Warsaw Treaty Organization published in main newspapers, both Western and Eastern, in 1980-1982. Moreover, it seems that these materials were subjected to neither ideological nor instrumental selection.

I do not consider it necessary to go into the numerous secondary literature that is noted in this book in the passages dealing with the context of the Eastern Bloc and the Cold War. Nonetheless, it is necessary to look deeper, at least generally, at the literature focusing on the Polish crisis. The issue of a potential military invasion of Poland in 1980 and 1981 is the subject of many historical studies. Most of them, however, are based on either selected or not very relevant sources. Therefore, some conclusions drawn so far are not convincing or, worse still, even purely speculative. This book, based on the primary sources, makes some of these conclusions more precise or questions their validity. Moreover, since a comprehensive monograph on the Warsaw Pact's history is still absent, the alliance's role in the Polish events has still not been examined properly yet. During my research, I could not leave aside Vojtech Mastny's and Mark Kramer's papers[6] as well as the well-written introductions in the above-mentioned editions by Andrzej Paczkowski and Łukasz Kamiński. I also note numerous papers and documents published in the *Cold War International History Project Bulletin*, mostly by Polish and German historians. It was impossible to ignore a record of the meeting of witnesses of the events, who served as the Polish and Soviet high-rank officials, both military and political, at the time. The conference took place in the fall of 1997 in Jarchanka, Poland, and the record was published as *Wejdą nie wejdą: Polska 1980–1982 konferencja w Jachrance listopad 1997* (London 1997).

In terms of outlining the Soviet policy, I was left to rely on the secondary literature. In that respect, the findings by the reputable Russian historians, Dmitry Volkogonov and Vladislav Zubok, must be pointed out. The former's work *The Rise and Fall of the Soviet Empire* (New York 1998) is a result of his intensive research in the Russian archives in the first half of the 1990s when he managed to collect a sizeable volume of key-documents, including the CPSU CC politburo materials. Unfortunately, I was not given an opportunity to study directly that documentation whose copies Volkogonov handed to the United States (now stored in the Library of Congress). It is very likely that particular documents from the collection called "Volkogonov Archive" are relevant

---

[6] Kramer, Mark, "The Warsaw Pact and the Polish Crisis of 1980–81: Honecker's Call for Military Intervention," *Cold War International History Project Bulletin*, no. 5 (1995): 124; Kramer, Mark, "Poland 1980–81: Soviet Policy during the Polish Cisis," *Cold War International History Project Bulletin*, no. 5 (1995): 1, 116–126; Kramer, Mark, "In Case Military Assistance is Provided to Poland: Soviet Preparation for Military Contingencies, August 1980,". *Cold War International History Project Bulletin*, no. 11 (1998): 102–109; Kramer, Mark, "Jaruzelski, the Soviet Union, and the Imposition of Martial Law in Poland: New Light on the Mystery of December 1981," *Cold War International History Project Bulletin*, no. 11 (1998): 5–14; Kramer, Mark, "Colonel Kuklinski and Polish Crisis, 1980–81," *Cold War International History Project Bulletin*, no. 11 (1998): 48–59; Kramer, Mark, "Die Sowjetunion, der Warschauer Pakt und blockinterne Krisen während der Brežněv-Ära," in *Der Warschauer Pakt. Von der Gründung bis zum Zusammenbruch 1955 bis 1991*, eds. Torsten Diedrich, Winfried Heinemann, and Christian Ostermann (Berlin: Christoph Links Verlag, 2009): 273–336; Mastny, Vojtech, "The Soviet Non-Invasion of Poland in 1980–1981 and the End of the Cold War," *Europe-Asia Studies*, no. 51 (1999): 189–211.

for the issues examined in this work. Regrettably, it is necessary to critically acknowledge that Volkogonov's book includes few important findings on the Polish crisis. Unlike the other parts, this period makes only rare references to the sources. Instead, it consists of personal observations of the author who maintained personal ties with some members of the Soviet leadership at the time. These are, however, full of inaccuracies and numerous events are dated either vaguely or dubiously. Zubok's book *The Failed Empire* (University of North Carolina 2007) reflects on not only documents, but also various memoirs by the top Soviet power structures' members and their aides. As far as the Czech historiography dealing with the history of the Soviet Union and its sphere of influence in Europe is concerned, works by Karel Durman, Bohuslav Litera, Miroslav Tejchman, or Jan Wanner were particularly useful for this book.[7]

***

An opportunity to study in Poland, Germany, and the United Kingdom - allowed by a number of grant and scholarship projects between 2011 and 2015 - made an important contribution to writing this book.[8] The text includes some slightly modified and amended passages already

---

[7] Durman, Karel, *Útěk od praporů. Kreml a krize impéria 1964–1991* (Praha: Karolinum, 1998); Litera, Bohuslav, "Unifikace sovětského bloku v 70. letech," *Slovanský přehled,* no. 1 (1994): 81–96; Litera, Bohuslav: "Sovětský svaz, 'Eurorakety' a eroze sovětského bloku," *Slovanský přehle,* no. 3 (1994): 259–276; Tejchman, Miroslav and Litera, Bohuslav, *Moskva a socialistické země na Balkáně 1964–1989* (Praha: Historický ústav, 2009); Tejchman, Miroslav, *Nicolae Ceausescu* (Praha: Nakladatelství Lidové noviny, 2004); Vykoukal, Jiří and Litera, Bohuslav and Tejchman, Miroslav, *Východ. Vznik, vývoj a rozpad sovětského bloku 1944–1989* (Praha: Libri, 2000); Wanner, Jan, "Sovětská hierarchie a politika SSSR 1968–1982," *Slovanský přehled,* no. 2 (1994): 165–186; Wanner, Jan, *Brežněv a východní Evropa* (Praha: Karolinum, 1995).

[8] Namely Projektové účelové stipendium FF UK č. 1746/A, *Varšavská smlouva a polská krize 1980–1981* (2011); Projekt rozvoje vědních oborů na Univerzitě Karlově č. 12/205605, *Profilace – asimilace – koexistence – integrace – reflexe (vývoj jazykových, konfesních, etnických a národních identit v areálu východní a jihovýchodní Evropy* (2012); Vnitřní grant FF UK č. 009/2013, *Varšavská smlouva mezi pražským jarem a helsinskou konferencí* (2013); the stay abroad under the Erasmus programme (2013); the stay abroad under The Anglo-Czech Educational Fund scholarship (2014–2015).

published in a form of particular papers during my doctoral studies.[9] In this case, I do not refer to my own previously published works in this book.

In writing the following text, the intensive consultations with my tutor, doc. PhDr. Jan Pelikán, CSc, were a big help for me. He gave me valuable advice all the time and subjected the text to a needful constructive criticism. The same applies to PhDr. Ondřej Vojtěchovský, Ph.D. and the other members and participants of a doctoral seminary at the Institute of World History, Faculty of Arts, Charles University in Prague, who read the incipient text regularly, discussed it, and gave their suggestions. Those are namely Mgr. Kathleen Geaney, Ph.D., PhDr. Ondřej Žíla, Ph.D., Mgr. Milan Sovilj, Ph.D., Mgr. Uroše Lazareviče, Mgr. Daniel Slavík, Ph.D. and Mgr. Přemysl Vinš. Finally, I must appreciate my consultations with Vladislav M. Zubok of the Department of International History, London School of Economics and Political Science, who devoted his time to me during my stay in Great Britain. I would like to express my special

---

[9] Bílý, Matěj, "Bratrská pomoc Polsku? Role Varšavské smlouvy v první fázi polské krize: červenec–prosinec 1980," *The Twentieth Century – Dvacáté století*, no. 1 (2012): 159–182; Bílý, Matěj, "Prosinec 1980: účast ČSSR v pokusu o multilaterální řešení polské krize," in *České, slovenské a československé dějiny 20. století VII*, eds. Tomáš Hradečný and Pavel Horák (Ústí nad Orlicí: Oftis, 2012), 207–217; Bílý, Matěj, "Rumunsko a Varšavská smlouva v první polovině 70. let," in *VI. ročník studentské konference o Balkáně – sborník vystoupení*, ed. Hana Suchardová (Praha: Rada pro mezinárodní vztahy, 2012) 18–25; Bílý, Matěj, "1981: Role Varšavské smlouvy ve druhé fázi polské krize," *Historie a vojenství*, no. 2 (2013): 20–40; Bílý, Matěj, "ČSSR v organizaci Varšavské smlouvy na přelomu 60. a 70. let," in *České, slovenské a československé dějiny 20. století VIII*, eds. Tomáš Hradečný and Pavel Horák (Ústí nad Orlicí: Oftis, 2013) 273–283; Bílý, Matěj, "Gierekovo vedení a prosazování polských zájmů v organizaci Varšavské smlouvy," *Historie a vojenství*, no, 3 (2014): 4–16; Bílý, Matěj, "Romania in the political structures of the Warsaw Treaty Organization at the turn of 1960s and 1970," *Oriens Aliter*, no 2 (2014): 44–69; Bílý, Matěj, "Od pražského jara k helsinskému procesu. Vývoj politické spolupráce v organizaci Varšavské smlouvy na přelomu 60. a 70. let," *Východočeské listy historické*, no. 33 (2015): 89–124; Bílý, Matěj, "Statut Varšavské smlouvy pro období války. Kontext – schválení – dopady," *Historie a vojenství*, no. 2 (2015): 15-27; Bílý, Matěj, "ČSSR a krach détente. Nástin československé politiky v rámci Varšavské smlouvy v druhé polovině 70. let," in *České, slovenské a československé dějiny 20. století IX*, eds. Tomáš Hradecký, Pavel Horák and Pavel Boštík (Ústí nad Orlicí: Oftis, 2015): 417-426; Bílý, Matěj, "Počátky pokusu o reformu Varšavské smlouvy," *The Twentieth Century – Dvacáté století*, no. 1 (2011): 159–172.

gratitude to everybody mentioned above. The same holds for PhDr. Tomáš Klusoň of the Archive of Ministry of Foreign Affairs of the Czech Republic, who helped me with recherché of the collections there. I also thank Mgr. Josef Zikeš, the director of the Central Military Archive in Prague, who gave me permission to study yet to be sorted documentation, and numerous staff of the archive who were patient with my time-consuming research in a nonstandard regime. Special thanks go to Mgr. Ondřej Matějka, the deputy-director of the Institute for the Study of Totalitarian Regimes (Prague, Czech Republic), for his support in publishing this work in English,[10] and Patrick Keenan, M.A. for a proofreading of my English manuscript.

---

[10] Originally Bílý, Matěj, *Varšavská smlouva 1969-1985. Vrchol a cesta k zániku* (Praha: Ústav pro studium totalitních režimů, 2016).

Outline of the Warsaw Pact development in 1955-1969[11]

In 1969, the Warsaw Treaty Organization completed the first stage of its existence. During its first fourteen years, it evolved dynamically. The Pact's foundation in 1955 did not alter the existing international situation at all, and had only formally terminated the Soviet European satellites' armed forces. At the time the documents reforming the alliance's military structures were approved at the Political Consultative Committee session in Budapest, the Warsaw Pact represented a real collective military power and an organization putting forward some significant international initiatives. Given the general functioning of the Eastern Bloc, however, all the processes within the Pact remained reliant primarily on the will of the Soviet Union.

Between 1955 and 1969, the situation within the Warsaw Pact fully depended on development of relations in the Soviet sphere of influence and, namely, on Moscow's actual military and political course. The Soviet leadership adapted its policy towards the Pact completely to meet its imminent needs. The Soviet Union managed to use the Warsaw Pact's establishment – originally a politically motivated move seen as a temporary, short-term solution – also in the long-term. Despite the failure of Khrushchev's plan to significantly strengthen the Soviet Union's position in Europe through dissolution of both military blocs followed by withdrawal of all foreign troops from the European states' territories, the alliance's establishment affected relations between Moscow and its satellites in the sphere of military cooperation.

Until the second Berlin crisis, the Warsaw Pact actually did not exceed the scope of an umbrella organization used for defining and implementing armament programs along with developing the Eastern Bloc

---

[11] Based on Bílý, Matěj, "*Varšavská smlouva v 50. a 60. letech,*" (PhDr diss., Charles University in Prague, 2011): 221–222.

countries' armed forces. Moreover, its realization often proceeded on a bilateral basis, outside the alliance's structures, which inevitably created problems in coordination of the individual member-states' activities. The alliance's political framework was ignored by the Kremlin except for its single instrumental activation in 1958 designed to give support to Khrushchev's international initiatives. In talks with the Western powers' officials, however, Moscow was able to use the Warsaw Pact's existence as a tool preventing their potential interference with the Soviet sphere of influence.

Given Khrushchev's decision to put aside the signing of a separate peace treaty with the GDR, the second Berlin crisis ended without military confrontation. Nonetheless, ever since then, the Warsaw Pact represented a real military power and the Soviet military command presumed that a potential war would have been fought on the alliance's basis. This shift, however, did not change the fact that the Pact's command structures remained strictly in the hands of the Soviet generality. In addition, Moscow did not deem it necessary to reform in any way – formally or virtually – the existing bodies of the Pact, both military and political.

The key impetus for the Warsaw Pact's reform did not occur before the power shock in the Kremlin in 1964. In its attempts to stabilize the situation in the Eastern Bloc, the new Moscow leadership decided to strengthen the Warsaw Pact's status as well. However, the fundamental principle of Soviet dominance would still have been preserved. In that time, it became very clear that the original plan of Khrushchev's to dissolve NATO and the Warsaw Pact simultaneously was going to be impossible to implement. The project, seen initially as temporary, soon became a long-term organization. Therefore, the essence of Brezhnev's leadership reflected on the alliance's reforms, focusing on expansion and specification of its vaguely described military and political bodies' structure. The reforms – in praxis just a formal strengthening of collective principles of cooperation – should have also silenced some member-states' objections to the existing state of things.

Most of the Soviet satellites welcomed the Kremlin's initiative to adjust the Warsaw Pact's political and military framework. But the alliance's reform met opposition from Romania. Ceausescu's leadership fully understood the real situation in the Eastern Bloc. Given their experience with previous practice within the organization, Bucharest decided to improve its position in relation to Moscow by an effort to disintegrate the Warsaw Pact rather than to strengthen its framework. Nonetheless, the Romanian attitude to the alliance's reform in the 1960s represented just one of many means used by Ceausescu's regime to get at least limited freedom of action within the Soviet sphere of influence. This process contained some risks; however, the basic fact was that despite all the moves to a more independent foreign policy taken by Bucharest, the domestic situation in Romania did not deviate significantly from the model of a Leninist-Stalinist regime. That was the reason why Czechoslovakia, fully loyal to the Warsaw Pact, rather than Romania, became a victim of Soviet aggression in 1968. The intervention demonstrated that Moscow prioritized preserving the Leninist-Stalinist regimes' nature in its sphere of influence over an ultimate loyalty to the Eastern Bloc's organization like the Warsaw Pact and the Comecon. Otherwise, either Romania or Albania would become a victim of Soviet military intervention. The latter, after all, used the military suppression of the Prague Spring as a pretext to withdraw officially from the alliance's structures, which the country effectively had left back in 1961, at the time of its rift with Moscow.

At the turn of the 1960s and 1970s, the Warsaw Pact's reaction to the international development was overdue. The alliance's scheme after its reform in 1969 placed emphasis on the Eastern Bloc countries' military cooperation in peacetime. In the context of the Cold War and ongoing détente, the scope of military cooperation inexplicably overshadowed the Pact's political importance. Despite this fact, the Warsaw Pact's significance on the international stage was much bigger than in 1955, when the organization had been founded.

Cooperation within the
Warsaw Pact political structures
during the first half of the 1970s

The Warsaw Pact's reform, approved in March 1969 by the Political Consultative Committee meeting in Budapest, concerned the military framework of the alliance only. Its political structures continuously adhered formally to a few vague articles of the founding charter. The Political Consultative Committee remained nominally the highest as well as the single official political body of the organization. However, new international challenges at the turn of the 1960s and 1970s, particularly regarding tendencies towards easing West-East tensions, also resulted in a significant rise of political cooperation in the Warsaw Treaty Organization.

Once Leonid Ilyich Brezhnev became head of the Soviet leadership, the Warsaw Pact went through few changes. Nevertheless, the Pact still did not constitute the key allied ties between the Eastern Bloc countries. Those remained secured by the network of bilateral treaties whose roots went back to Stalin's era.[12] In the second half of the 1960s,

---

[12] The Central and Eastern-European countries were militarily interconnected during 1943-1949. A total of 23 bilateral allied treaties concerning eight states of the emerging Eastern Bloc were conducted in that period, generally on the basis of 'everyone with everybody'. The only exception was Albania, which signed the agreements with Bulgaria and Yugoslavia only. The key point of those documents was the signatories' proclamation of a joint defense against a potential aggressor. The network of the bilateral treaties initially did not include the GDR. Despite friendly relations between East Berlin and the other Soviet satellites, no official alliance was set up. Absence of a peace agreement with Germany played a key role. On the contrary, Yugoslavia actively took part in establishing that security system. However, in the wake of the beginning of the Soviet-Yugoslavian rift, Moscow repudiated an allied treaty with Belgrade, explicitly noting the political process with László Rajk. The Soviet satellites followed that move soon. Therefore, Yugoslavia remained outside the military ties of the Eastern Bloc.

the twenty-year validity of these documents expired. Although there was a multilateral alliance coming from the Warsaw Treaty, Moscow considered it necessary to preserve bilateral ties within its sphere of influence. Signatures of the new allied agreements between the Eastern Bloc states at the turn of the 1960s and 1970s, however, largely concerned the Warsaw Pact as well. For example, Moscow implemented an article into a Soviet-Czechoslovak document that explicitly bound the pledge of "strengthening unity of the socialist countries" in compliance with the provisions of the alliance.[13] This way, Brezhnev's leadership probably sought to ensure a unified foreign policy of the Soviet sphere of influence in Europe. Simultaneously, the Kremlin interconnected the bilateral allied ties with the multilateral mechanisms. Alongside the existing alliance, the USSR considered the conclusions of those new agreements as an integral part of the complex allied system of the entire Bloc.[14] Links to the Warsaw Pact significantly differed in the individual treaties. It was the most obvious in the Soviet-Romanian document. By request of Bucharest, the alliance was labeled as an answer to the establishment of NATO. Unlike the other Soviet satellites, Romania also enforced a statement that the country would comply with the charter of the Warsaw Pact until the end of its validity only.[15] Moreover, there were some remarkable differences

---

Kořalková, Květa, *Vytváření systému dvoustranných spojeneckých smluv mezi evropskými socialistickými zeměmi (1943-1949)* (Praha: Academia, 1966).

[13] Comparison of Czechoslovak and Soviet drafts on a new allied treaty between CSSR and USSR, undated, 1970, TO(t) 1970-1974, box 1, i. č. 89, sign. 020/112, AMZV, Prague.

[14] *Záznam o návštěvě rady sov. Velvyslanectví v Praze s V.V. Astafjeva u vedoucího 1. t. o. 26. června 1970*, TO(t) 1970-1974, box 4, sign. 020/117, č. j. 023.757/70-1, AMZV, Prague.

[15] In February 1968, Romanian-Soviet talks were held in Bucharest in order to work out the new bilateral *Treaty on Friendship, Cooperation, and Mutual Assistance* with a validity of twenty years. Finalization of the text was extremely complicated. Among other things, the Romanian diplomats refused an explicit commitment to adhere to the duties of membership in the Warsaw Pact. Bucharest correctly objected that validity of the alliance would expire in 1975. In fact, such a formulation would prolong its existence to 1988. A perspective of simultaneous dissolution of the military blocs - which Romania at least publicly strove for - would also recede. Moscow stayed stubborn. The Kremlin considered a pledge of adherence to the provisions of the alliance as a key aspect of the new treaty. The

in the pledge of mutual military assistance. For instance, a condition for such help to the GDR was an action of the entire Warsaw Treaty Organization. On the other hand, the agreement between the Soviet Union and Poland supposed mutual military assistance solely in case of attack by "*West-German militarism and its allies*".[16] Those contrasts most likely reflected the capabilities and willingness of individual Eastern Bloc countries' leaderships to defend their interests during the negotiations with Moscow.

At the time, the West overestimated the treaties between the Eastern Bloc states. They were considered an additional mechanism to strengthen the Warsaw Pact. Some of their articles were also construed as a codification of the so-called Brezhnev doctrine.[17] In fact, the entire situation proved that, regardless of partial modifications in the Warsaw Treaty Organization, the USSR did not mean it as a keystone of its European sphere of influence's cohesion; even after the 15-year existence of the Pact, the Kremlin insisted on bilateral allied ties.

Transformation of the Warsaw Pact in the begging of the CSCE process

The Eastern Bloc faced many challenges at the end of the 1960s, both internal and international. In the wake of military intervention in Czechoslovakia, maintaining stability within the Soviet sphere of influence was necessary. Considering this, the main issue was an approach towards the Romanian regime - on the one hand rebellious, on the other hand fully devoted to the principles of Leninism-Stalinism. After a series

---

Romanian side therefore tried to assert an amendment stating that the obligation is valid till the end of the Warsaw Pact's existence only. Retagan, Mihai, *Ve stínu pražského jara* (Praha: Argo, 2002): 45, 83.

[16] *Porovnání spojeneckých smluv ČSSR a NDR, PLR, BLR, RSR, a SSSR s PLR, NDR, BLR a MLR s čs. návrhem nové čs.-sovětské spojenecké smlouvy podle hlavních zásad,* undated, 1970, TO(t) 1970-1974, box. 1, i. č. 89, sign. 020/112, AMZV, Prague.

[17] *Ohlas na podepsání Smlouvy o přátelství, spolupráci a vzájemné pomoci ze dne 6. května 1970 v socialistických, kapitalistických a rozvojových zemích,* 16. 5. 1970, ibid.

of border clashes, the USSR found itself at the edge of war with China.[18] Such a situation only increased the Kremlin's need for internal cohesion of the Eastern Bloc.[19] At the same time, favorable conditions arose to resolve two long-term key issues - the German question, and the convening of the all-European security conference. The Warsaw Treaty Organization had been involved in both issues in the past. Actually, their re-opening provided grounds for activation of the alliance's political structures. In March 1969, the Political Consultative Committee declared the Warsaw Pact's readiness to start talks on the potential of an all-European security conference without any preconditions, for the first time.[20] This paved the way for new, intensive, international negotiations. Moscow authorized individual members of the alliance to take part in them. From the beginning, with regard to the dissimilar diplomatic abilities of the Soviet satellites, their roles in preparation of the conference should have been different.[21] Multilateral meetings under the auspices of the Warsaw Treaty Organization should have provided an appropriate coordination forum. As the Pact represented practically a single, official, collective political mechanism of the Soviet sphere of influence in Europe, it seemed to be a

---

[18] A number of clashes took place at the Soviet-China border between March 2 and March 15, 1969. More followed until August 13, 1969. Schaefer, Bernd, "The Sino-Soviet Conflict and the Warsaw Pact, 1969–1980," in *NATO and the Warsaw Pact. Intrabloc Conflicts*, eds. Mary Ann Heiss and S. Victor Papacosma (Kent: Kent State University Press, 2008): 208.

[19] Antonín Benčík, for example, points to the Chinese factor in the Soviet efforts to stabilize the Warsaw Treaty Organization. The beginning of the idea of reforming the alliance in the 1960s actually coincided with the eruption of an open rift between Moscow and Beijing. At the time, the USSR sought to ensure calm in Europe in order to have potentially free hands in the Far East. Benčík, Antonín, *Vojenské otázky československé reformy 1967-1970*; vol. 1, (Praha: Doplněk, 1996): 12.

[20] The previous appeals of the Warsaw Treaty Organization for holding the all-European security conference stipulated conditions of a clear guarantee of the borders in Europe, solution of the German question, or rejected participation of the US at the summit.

[21] *Mimořádná politická zpráva k ohlasům na budapešťskou Výzvu členských států Varšavské smlouvy*, 30. 5. 1969, DTO 1945-1989, i. č. 34, e. č. 42, č. j. 023.010/69, AMZV, Prague.

proper center for the formulation of the Bloc's foreign strategy.[22] It soon became clear that technical changes of cooperation within the alliance framework were needed to make this conception work. In context of its existence so far, the Warsaw Pact therefore showed extraordinary activity at the turn of the 1960s and 1970s.[23]

The events of 1969 suggested that in the beginning the USSR apparently had no coherent plan about the course of the intensified alliance meetings, as they differed from both previous and future experiences. Even the meeting of deputy foreign ministers in East Berlin, May 21-22, had some atypical features. The Warsaw Pact member states were supposed to report about their latest bilateral talks with Western countries regarding holding the all-European security conference and coordinating their next strategy. The Soviet instructions for future steps were unusually general. The Pact members were given relatively free reign with their own diplomatic initiatives. That inspired false hopes in some Soviet satellites' leadership that political cooperation within the alliance framework could really attain a new dimension.[24]

As it soon became clear, individual Warsaw Pact member-states sought to pursue their own interests in the dynamical international

---

[22] Wanner, *Brežněv a východní Evropa*, 38.

[23] During 1969 and 1970, three Political Consultative Committee sessions and one informal meeting of the Warsaw Pact member states' leaders were held. The foreign affairs ministers negotiated under the auspices of the alliance twice. During the same period, their deputies met four times. In addition, recently established collective military bodies, especially the Committee of Defense Ministers and the Military Council of the Unified Armed Forces, began to work in parallel. The Warsaw Pact's structures had showed a similar activity in the first half of 1966 only, when Brezhnev instigated the intensive talks on potential reform of the alliance.

[24] Hungary was a typical example. However, considering previous experience, Kadar's leadership questioned the real Kremlin's intentions. Budapest was not sure whether the extensive freedom and absence of strict directives just meant that Moscow either had no clear vision of the outline of the all-European security conference or hid its real priorities from the Warsaw Pact members. *Note on the Meeting of the Deputy Foreign Ministers*, May 21, 1969, Accessed October 16, 2019. http://www.php.isn.ethz.ch/kms2.isn.ethz.ch/serviceengine/Files/PHP/172 53/ipublicationdocument_singledocument/9b7b446c-693b-42d6-92a5-80729582 0045/en/690521_Note.pdf.

situation. They intended to use the process of convening the all-European security conference to enforce their specific, previously unfeasible, priorities.[25] At the turn of the 1960s and 1970s, Czechoslovakia and Bulgaria were the only members of the alliance which did not significantly mobilize their diplomatic efforts. After suppression of the Prague Spring, Czechoslovak foreign policy remained largely paralyzed.[26] Bulgaria, the most loyal Soviet satellite, resigned on their own stances in advance. During internal discussions, Sofia clearly admitted its intention to fully support the proposals of Moscow.[27] The interests of the rest of the Warsaw Pact member-states were varied and their methods to achieve those aims were often antagonistic. The GDR sought to convince its allies to focus on the main East-German priority - full international recognition of a German communist state. Ulbricht's leadership hoped that it would be able to enforce this as an additional precondition for holding the all-European security conference. Poland was very active as well. During 1969, it prepared an ambitious draft of the *European Security Treaty*. The main objective of that proposal was to set up a new, collective security system on the Continent, which would later lead to wide disarmament talks with not only superpowers, but also every interested country as a participant. On the contrary, Romania strove to ensure that the proposed security conference would take place on the basis of the equal and sovereign states. Bucharest considered this as an opportunity to continue weakening the

---

[25] Mastny, Vojtech and Byrne, Malcolm, eds., *A Cardboard Castle? An Inside History of the Warsaw Pact*, (New York: CEU Press, 2005), 40.

[26] In fully keeping with the official instructions from the Warsaw Pact meetings, Czechoslovak diplomacy strove for holding the all-European security conference in bilateral talks with Western countries. However, Prague soon realized that that the effort was pointless. In 1970, Vasil Biľak had to admit that many of the Western statesmen showed an ostentatious lack of concern about developing relations with the incoming normalization regime. *Zpráva o průběhu a výsledcích budapešťské porady ministrů zahraničních věcí členských států Varšavské smlouvy*, 14. 7. 1970, 1261/0/5, sv. 134, a. j. 211/info4, NA, Prague.

[27] Hungarian Foreign Ministry Memorandum of Soviet-Hungarian Consultations on the European Security Conference, October 10, 1969, in *A Cardboard Castle?* (doc. 67), 347; *Charakteristika postojů BLR k problematice evropské bezpečnosti*, 18. 2. 1971, TO(t) 1970-1974, box 7, i. č. 12, sign. 013/311, č. j. 021.200/71-2, AMZV, Prague.

Kremlin's influence on its own decision-making process.[28] Finally, Hungary was interested in fast and overall improvement of East-West relations, which would result in strengthening economic cooperation between the blocs.

This position of Hungarian leader Janos Kadar was the closest to the Kremlin's vision. Hungary definitely did not strive for such a level of autonomy like Romania. Unlike Poland, East Germany, or Czechoslovakia, Hungary had no unresolved disputes with Western Germany. In that sense, Hungarian leadership watched with great concern the above-mentioned states' efforts to review the Bucharest declaration of the Political Consultative Committee and to stipulate the preconditions of the Warsaw Pact on resolving the German question before the all-European security conference. This could have potentially disrupted or hindered the entire process.[29] Setting aside a passive Bulgaria, Hungary became the main Soviet partner within the Warsaw Pact, at least in the initial stages of the Conference on Security and Co-operation in Europe (CSCE) preparations. Budapest did not resign on its own initiatives completely. However, the HSWP CC First Secretary Janos Kadar acted more loyally and flexibly than his Polish and East-German counterparts, Władysław Gomułka and Walter Ulbricht. As Hungary had no serious discords with West, it could focus on cooperative measures of economic development, as the Kremlin also preferred.[30]

Polish activities posed the main threat to a smooth realization of the Soviet vision. Gomułka was afraid that the alliance was leaving the so-called 1967 "Warsaw Declaration", which had made a guarantee of Polish western post-war borders a precondition for establishing diplomatic relations with the FRG.[31] In May 1969, he offered Bonn separate talks

---

[28] Mastny, *A Cardboard Castle?*, 40.

[29] Békés, Csába, "Studená válka, détente a sovětský blok. Vývoj koordinace zahraniční politiky sovětského bloku (1953– 1975)," *Soudobé dějiny,* no. 1-2 (2011): 78.

[30] Békés, Csába, "*Hungary and the Warsaw Pact, 1954-1989: Documents on the Impact of a Small State within the Eastern Bloc,*" Accessed October 16, 2019. http://www.php.isn.ethz.ch/lory1.ethz.ch/collections/coll_hun/intro07f3.html?navinfo=15 711.

[31] In February 1967, the meeting of the Warsaw Pact member-states' foreign affairs ministers in the Polish capital defined conditions of the Eastern Bloc

about mutual reconciliation. Not only did he act publicly, but also without previous consultation with the Kremlin.[32]

It is not hard to grasp that the East-German regime officials soon protested in Moscow against Gomulka's moves.[33] Considering that, Polish actions brought the risk of a new dispute within the Eastern Bloc. Moreover, Warsaw became involved in the all-European security conference issue much deeper than the Kremlin would appreciate. In fact, the Polish draft of the *European Security Treaty* aimed to establish the new collective security system followed-up by the start of a debate on the dissolution of military-political blocs. An idea of setting-up nuclear-free zones in Europe, as the Rapacki Plan had outlined before,[34] was revived

---

countries for reconciliation with West Germany. Bonn was supposed to confirm all the borderlines in Europe, pronounce the Munich agreement null *ex tunc*, accept the existence of two German states, and surrender all the nuclear weapons ambitions as well as its claims on West Berlin. The resolution was not signed by Romania, which had established diplomatic relations with the FRG separately only recently.

[32] The Warsaw Treaty Organization's willingness to start the negotiations on the all-European security conference without any preconditions scared the Polish leadership. There was no mention of the fundamental principle of confirmation of the post-war frontiers by West Germany and eventually by other Western powers. However, such a step was one of the main foreign policy priorities of Warsaw. Moreover, the bilateral talks between Moscow and Bonn began, also concerning the issue of territorial guarantees. Gomulka soon realized he had to start a separate dialogue with West Germany in order to achieve confirmation of the Oder-Neisse borderline. Selvage, Douglas, "The Warsaw Pact and the European security conference, 1964–69," in *Origins of the European security system: the Helsinky process revisited, 1965–75*, eds. Andreas Wenger, Vojtech Mastny and Christian Nuenlist (Abingdon: Routledge, 2008): 94.

[33] Selvage, Douglas, "The Warsaw Pact and the German Question 1955–1970: Conflict and Consensus," in *NATO and the Warsaw Pact. Intrabloc Conflicts*, eds. Mary Ann Heiss and S. Victor Papacosma (Kent: Kent State University Press, 2008): 186.

[34] In October 1957, Adam Rapacki, Polish minister of foreign affairs, in UN General Assembly proposed establishing a nuclear weapons-free zone in Central Europe. An agreement under the auspices of not only the Soviet Union but also the Western powers, was supposed to concern Poland, Czechoslovakia, and both German states. The initiative had not been consulted with Moscow in advance. Thus, it was the first real independent foreign policy move by one of the Soviet satellites.

as well.[35] We must point out that Gomułka's leadership acted on these issues very cautiously. Their proposals tactically and consistently referred to selective paragraphs of the Warsaw Pact's highest body, the 1966 and 1969 resolutions, and the communist and workers parties' conference held in Karlovy vary in 1967.[36]

The USSR decided to unify such various stances of its satellites through the Warsaw Treaty Organization. Moscow chose a relatively uncommon consultative model - the ministers of foreign affairs' meeting.[37] Soviet leadership came to conclusion that the Pact could be used as a formal auspice of the multilateral talks in the Eastern Bloc more often. Preparation of the all-European security conference was the sole agenda of the summit that took place late October in Czechoslovakia.[38] Multilateral consultation within the Warsaw Pact's political framework was welcomed by all the Eastern Bloc countries interested in international

---

[35] Polish Proposals for the Conference on Security and Disarmament, October, 24, 1969, in *A Cardboard Castle?* (doc. 68): 350.

[36] Collections of Polish documents concerning preparation of the all-European security conference, undated 1969, PZPR KC, XIA/246, AAN, Warsaw.

[37] According to a 1956 Political Consultative Committee resolution, a committee for foreign policy affairs should have been established. However, that never happened. Before 1969, the foreign affairs ministers met collectively for three times only. In 1959, they just declared their support of Khrushchev's approach towards the issue of West Berlin. In 1966, they failed to agree on a potential reform of the alliance's political structures. One year later, after Romania established diplomatic ties with West Germany, disregarding the rest of the Eastern Bloc countries, the ministers redefined conditions of the Warsaw Pact for reconciliation with Bonn.

[38] The place was chosen by Moscow, basically on Czechoslovak request. In this manner, the incoming Husák's normalization regime sought to prove its loyalty to the Warsaw Pact and the Eastern Bloc, in general. The Kremlin, of course, appreciated such activity and complied to the Czechoslovak request despite the fact that Bucharest also strove for the hosting. *K poradě ministrů zahraničních věcí členských států Varšavské smlouvy v Praze*, 29. 10. 1969, 1261/0/5, sv. 107, a. j. 177/3, NA, Prague; *Zpráva o průběhu a výsledcích pražské porady ministrů zahraničních věcí členských států Varšavské smlouvy a návrhy na další postup v oblasti evropské bezpečnosti*, 14. 11. 1969, 1261/0/5, sv. 110, a. j. 181/info2, NA, Prague; *Vývoj vztahů mezi ČSSR a Sovětským svazem po dubnovém plénu ÚV KSČ v roce 1969*, 17. 6. 1971, TO(t) 1970-1974, box 2, i. č. 89, sign. 020/112, č. j. 021.226/71-1, AMZV, Prague.

development at the time. For example, even the Polish leadership did not see the upcoming foreign ministers' meeting as a threat to its interests. On the contrary, Poles considered using the summit to put forward an official proposal for review of the previously mentioned Warsaw Declaration from 1967. That would provide additional permissions of the entire Pact to bilateral talks, which Poland started with Bonn at the time.[39] Therefore, we can say that in 1969, there existed a short period when some Eastern Bloc countries estimated a new and more straightforward character of political cooperation within the Warsaw Treaty Organization.[40] For instance, this recalls a conversation between Polish deputy foreign minister Józef Winiewicz and his Soviet counterpart Vladimir Semyonov. The Soviet diplomat appreciated some aspects of the PRP's proposals. As he stated, potential Soviet support, however, should have depended on the immediate strategy of Moscow. Winiewicz opposed him: The fate of the Polish initiatives should have relied on the decision of the upcoming official Warsaw Pact meeting and not Moscow alone. In the wake of activation of the Pact, Poland tried this way to test the formal decision-making mechanism of the alliance that the USSR had often bypassed in the past.[41] It was part of a wider strategy in Gomulka's leadership: Essential strengthening of foreign policy coordination within the Warsaw Pact should have prevented the Kremlin from putting its own global priorities above the interests of the rest of the alliance's member-states. An internal report by the Polish foreign ministry from April 1969 warned

---

[39] The signatories of the 'Warsaw package' - the Warsaw Pact members without Romania - should have met the day after the meetings of foreign affairs ministers in Prague to discuss the issue collectively.

[40] *NOTATKA w sprawie naszego stanowiska wobec europejskej konferencji bezpieczeństwa i współpracy*, 30. 9. 1969, PZPR KC, XIA/246, AAN, Waraw; Report by Polish ministry of foreign affairs to the PWUP CC First Secretary W. Gomulka on Soviet stances on holding the all-European security conference, 8. 10. 1969, PZPR KC, XIA/87, AAN, Warsaw.

[41] Jarząbek, Wanda, "*Poland in the Warsaw Pact 1955–1991: An Appraisal of the Role of Poland in the Political Structures of the Warsaw Pact*," Accessed October 16, 2019. http: // www.php.isn.ethz.ch / lory1.ethz.ch / collections / coll_poland/ Introductionb85a.html?navinfo=111216.In this regard, Jarząbek wrongly notes that the Political Consultative Committee session was held in Prague in October 1969. In fact, it was just a meeting of foreign affairs ministers.

against such tendencies. Poland therefore appealed for frequent multilateral political talks within the Warsaw Pact. As the stances of the *entire* Bloc should have been defined there, this measure would at least reduce Moscow's ability to only assert their own aims during the preparation of the all-European security conference. Soviet goals were definitely less ambitious than Polish ones.[42] At that stage, the Soviet Union sought nothing but confirmation of the existing borders - and in fact, even a political *status quo* - as well as a pledge of nonviolence during potential review of the post-war situation in Europe.[43]

Moscow left nothing to chance and selectively sought support of its satellites on a bilateral basis. The USSR mainly focused on Hungary, as its positions were closest to Soviet ones. The Kremlin tactically tried to give the false impression that it really needed Hungarian assistance to prevent Poland, Eastern Germany, and Romania from putting forward their proposals.[44] The USSR labeled them all as too "ambitious"; their complexity could potentially delay the preparation of the all-European security conference undesirably. Hungarian leadership probably succumbed to the illusion of becoming a truly close partner of the Kremlin. Therefore, Budapest promised to support Soviet stances at the upcoming alliance meeting as it expected that Moscow would take into account some of the Hungarian priorities regarding inter-bloc cooperation and would help to put those into a unified foreign policy tack of the Warsaw Pact.[45]

---

[42] Jarząbek, Wanda, *PRL w politycznych strukturach Układu Warszawskiego w latach 1955–1980* (Warszawa: PAN, 2008), 72-73.

[43] Mastny, *A Cardboard Castle?*, 40.

[44] On this purpose, in the second half of October, the series of bilateral consultation between Soviet deputy foreign minister Vladimir Semyonov and Hungarian, Czechoslovak, Polish, and Romanian representatives took place in Moscow. In this stage, Romania intended to assert also dissolution of the blocs, withdrawal of foreign troops, the foreign military bases ban and reduction of 'demonstrations of force'. Moscow therefore instructed the rest of the Warsaw Pact member-states to refuse Ceausescu's proposals. The USSR argued that NATO opposed such principles for long time, so asserting those could potentially lead to a deadlock in the conference.

[45] Hungarian Foreign Ministry Memorandum of Soviet-Hungarian Consultations on the European Security Conference, October 18, 1969, in *A Cardboard Castle?* (doc. 67): 347.

At the turn of the 1960s and 1970s, there was an obvious misunderstanding of purposes surrounding the intensified political cooperation within the Warsaw Treaty Organization between the Soviet Union and its satellites. Some of the Eastern bloc countries considered more frequent consultations as an opportunity to at least partially assert their own interests on the global stage through a common-formulated strategy of the entire Bloc. In fact, Moscow used the phrase about "unified foreign policy tack of the Warsaw Pact" as both a comfortable and effective tool of elimination of the inappropriate individual initiatives of its European satellites.

The great naiveté of the consideration of substantially improving the way the Warsaw Pact worked was illustrated by the actual run of the foreign ministers' meeting which took place in Prague and Lány in late October, 1969. It was still impossible to assert anything within the Pact against the will of Moscow. However, the features of the summit obviously went through some small changes. For example, the chairmanship rotated between member-states' delegations for the first time.[46] Compared to the past, the agenda was better prepared for since the deputy foreign ministers had discussed it twice previously.[47] The delegation of Poland was also allowed to present its draft of the *European security treaty* without any complication. On the other hand, Moscow had verified that the rest of the Pact members would not support it. Moscow's previously pleaded assistance from Hungary appeared to be unnecessary. Discussion on Polish initiatives was blocked by East-German and Romanian delegations. Both declared that their party leaderships needed more time to study the documents in detail. In fact, this was just an

---

[46] Czechoslovak, Bulgarian, and Hungarian foreign affairs ministers alternated in the chairmanship during the meeting. Henceforward, similar rotations took place at the Warsaw Pact political gatherings, till the end of Brezhnev's era.

[47] *Zpráva o průběhu a výsledcích pražské porady ministrů zahraničních věcí členských států Varšavské smlouvy a návrhy na další postup v oblasti evropské bezpečnosti*, 14. 11. 1969, 1261/0/5, sv. 110, a. j. 181/info2, NA, Prague; Preparation of the meeting of foreign affairs ministers was the only reason why the talks of deputies under the auspices of the alliance took place in Moscow on 24th October. *Politbüro des ZK Reinschriftenprotokoll nr. 39*, 21. 10. 1969, DY 30/ J IV 2/2/1247, BArch, Berlin; *Konferenz der Außenminister der Mitliedstaaten des Warschauer Vertrages*, 28. 10. 1969, DY 30/ J IV 2/2/1248, BArch, Berlin.

obstructive tactic.[48] Although SED leadership had received the material in advance, it made a purpose-built decision that it would not release an official statement until mid-November - after the Warsaw Pact meeting. Poland would have been informed *ex-post* only.[49] This practically meant that the GDR wanted to consult the issue with Soviet comrades and generally accept their stances.[50]

Considering the above-mentioned, the Polish minister of foreign affairs, Stefan Jędrychowski, assured the allies that his country would not put forward the proposal in the meantime and would loyally wait for the response of every Warsaw Pact member. The Soviet delegation, with massive support by the GDR's representatives, thereafter 'explained' to the Polish diplomats that their ambitious conception would allow the West to impede the all-European security conference process. The USSR also criticized the possible unilateral weakening of Soviet military force and potential Western interference with relations of the socialist countries.[51] Ministers of foreign affairs of the Warsaw Pact member-states therefore passed the documents reflecting mostly the Soviet positions. A Moscow-initiated appeal for the non-use of force and acceptance of all existing state frontiers in Europe was less complex than the Polish proposals. However, Warsaw considered it a meaningful minimum.[52] The main Polish interest was to gain a guarantee of the Oder-Neisse borderline being as wide as

---

[48] In the mid-1960s, Romania began to use that phrase at the Warsaw Pact meetings as a suitable excuse for avoiding clearly disapproving positions.

[49] *Konferenz der Außenminister der Mitliedstaaten des Warschauer Vertrages*, 28. 10. 1969, DY 30/ J IV 2/2/1248, BArch, Berlin.

[50] In this stage of preparation of the all-European security conference, the SED leadership declared its intention to consult all its steps with the Kremlin and only then take clear positions within the Warsaw Pact. *Maßnahmen in Auswetrung der Konferenz der Minister für Auswärtige Angelegenheiten der Warschauer Vertragestaaten zun Problem der europşaischen Sicherheit am 30. un 31. Oktober 1969 in Prag*, 18. 11. 1969, DY 30 / J IV 2/2/1253, BArch, Berlin; *Direktive für das Auftreten des Vertreters der DDR auf der Konferenz der stellvertretendes Außenminister der Mitgliedstaaten des Warschauer Vertrages*, 20. 1. 1970, DY 30/ J IV 2/2/1264, BArch, Berlin.

[51] East German Evaluation of Polish Proposal for a European Security Treaty, November 13, 1969, in *A Cardboard Castle?* (doc. 69): 354

[52] Selvage, "The Warsaw Pact and the European security conference," 98.

possible. Gomulka's leadership therefore joined the Kremlin-outlined strategy of the Warsaw Pact, because it aimed for clear recognition of all the borders in Europe. Even the usually rebellious Romania was satisfied, as the final declaration emphasized state sovereignty, territorial integrity, and non-intervention with the internal affairs of the other countries. Those principles represented the basis of Ceausescu's regime rhetoric.[53] In fact, with such ingenious compromises Moscow was able to gain support for its policy within the Warsaw Pact.[54]

After a long time, the Prague meeting proceeded in a less ostensible and more diplomatic atmosphere. No serious conflicts arose.[55] That strengthened the willingness of the Eastern Bloc countries to hold talks on international issues under the auspices of the Warsaw Pact more frequently. In late October 1969, a decision on holding regular multilateral meetings of deputy foreign ministers was made. The meetings were supposed to be a flexible reaction to international events and a preliminary coordination of joint answers of the Pact.[56] Nevertheless, the main purpose

---

[53] *Zpráva o průběhu a výsledcích pražské porady ministrů zahraničních věcí členských států Varšavské smlouvy a návrhy na další postup v oblasti evropské bezpečnosti*, 14. 11. 1969, 1261/0/5, sv. 110, a. j. 181/info2, NA, Prague.

[54] The final documents should have avoided any controversial stances. Internally approved strategy of the Warsaw Pact suggested maximal softening of the matters considered sensitive by the West. The intention was to undermine positions of the Western politicians who opposed the idea of the all-European security conference. In terms of this tactic, the resolution emphasized two points: to rule out the use or threat of force; and the development of trade, economic, and scientific-technological cooperation. The East-West relation was outlined as a win-win game. Such a strategy aimed to comprehensively accelerate preparations of the conference. The first half of 1970 and the Finnish capital were proposed as a potential term and a hosting place of the summit. *Oficiální ohlas v západních zemích na poradu ministrů zahraničních věcí členských států Varšavské smlouvy v Praze; Prohlášení porady ministrů zahraničních věcí členských států Varšavské smlouvy*, 31. 10. 1969, 1261/0/5, sv. 110, a. j. 181/info2, NA, Prague.

[55] *Zpráva o průběhu a výsledcích pražské porady ministrů zahraničních věcí členských států Varšavské smlouvy a návrhy na další postup v oblasti evropské bezpečnosti*, 14. 11. 1969, ibid.

[56] *Maßnahmen in Auswetrung der Konferenz der Minister für Auswärtige Angelegenheiten der Warschauer Vertragestaaten zun Problem der europ§aischen Sicherheit am 30. un 31. Oktober 1969 in Prag*, 18. 11. 1969, J IV 2/2/1253, BArch, Berlin; In the past, the deputy foreign ministers' meetings were

of the activation of the alliance's political structures was mostly preparation of the all-European security conference. Its relatively wide and initially unclear agenda provided to individual member states an opportunity to strive for their own priorities.[57] However, the majority of the Soviet satellites supposed that the only way to succeed was to coordinate the efforts of the entire Bloc. Therefore, they sought wider support for their proposals within the alliance's framework. With the exception of Romania, there was consensus on more frequent meetings and their more effective form. An effort to strengthen internal cooperation was documented by setting up a commission authorized to work out joint economic proposals of the Warsaw Pact for the intended security conference. The fact that Hungary, instead of the Soviet Union, chaired that body signaled certain, but cosmetic shifts only.[58] After an internally unstable period in the second half of the 1960s, the willingness of Soviet satellites to cooperate under the auspices of the Warsaw Treaty Organization on preparation for the all-European security conference became a source of the Pact's cohesion.

This positive experience contributed to the fact that the Warsaw Pact member-states' meetings were held in December 1969 in Moscow, formally initiated by Eastern Germany.[59] In this situation, when intensive

---

held under the auspices of the Warsaw Pact sporadically. Henceforward, their talks should have become more operative and practical. For the first time, the deputies were supposed to meet immediately after the North Atlantic Council summit in December 1969. They were also entrusted with discussing a unified approach of the Warsaw Treaty Organization towards Albania. *Report to the [Hungarian] Revolutionary Workers' and Peasants' Government on the Meeting of the Deputy Foreign Ministers*, October 30, 1969, Accessed October 16, 2019. http://www.php.isn.ethz.ch/kms2.isn.ethz.ch/serviceengine/Files/PHP/17255/ipu blicationdocument_singledocument/ae87f3dc-ba08-4980-8f6b-493c98e1ebcb/en /691030_Report_E.pdf.

[57] Mastny, *A Cardboard Castle?*, 40.

[58] It was probably just not a very significant Soviet sop. Moscow rewarded Budapest for its above-described collaboration as the Hungarian priorities actually focused on inter-bloc trade cooperation.

[59] The East-German politburo protocol reveals that Ulbricht initiated the meeting on direct instruction from Moscow. *Politbüro des ZK Reinschriftenprotokoll nr.46*, 25. 11. 1969, DY 30/ J IV 2/2/1254, BArch, Berlin; *Zpráva o*

talks between the Soviet Union and Western Germany were starting, Brezhnev wanted to coordinate the future strategy of the Eastern Bloc regarding the German question.[60] We must remember that "the militaristic threat of the FRG" served, from the establishing of the Warsaw Pact, as one of its main cohesive elements. *Ostpolitik* of the new West-German social-democratic Chancellor Willi Brandt brought such a challenge that Brezhnev considered it necessary to hold the alliance summit before the USSR began negotiations with Bonn. It was a substantial shift, as the Kremlin in previous periods usually informed the satellites about its intentions *ex-post* only.[61]

        Aware of the different priorities of individual Warsaw Pact member-states, Brezhnev let all the present party leaders speak first. Then he acted as a constructive arbiter. He summed-up the speeches and tried to find a compromise.[62] The CPSU CC General Secretary assured that in talks with West Germany, the Soviet Union would also take into account the basic demands of the allies - confirmation of all existing borderlines in Europe, *de jure* recognition of the GDR, and abrogation of the Munich

---

připravovaném setkání představitelů států Varšavské smlouvy, 10. 11. 1969, 1261/0/5, sv. 108, a. j. 179/1, NA, Prague.
[60] Mastny, Vojtech, "Superpower Détente: US-Soviet Relations, 1969-1972," *Bulletin of the German Historical Institute,* no 1 (2004): 22.
[61] Mastny, *A Cardboard Castle?*, 42, 358.
[62] The meeting revealed the different attitudes of individual Warsaw Pact member-states towards further strategy in the German question. The GDR appealed for a cautious approach; Ulbricht warned that Brandt fully followed the previous course of Konrad Adenauer. He asked for ultimate fulfillment of the 'Warsaw package' from 1967. On the contrary, Poland and Romania asserted quick reconciliation with Bonn. The offended East-German leader also noted that there was no reason for separate negotiations on confirmation of the frontiers between Warsaw and Bonn, as the Oder-Neisse line defined the Polish borders with East Germany. Gomulka tried to defuse East-German criticism with the same rhetoric used by Moscow to suppress the Polish initiatives: He warned that increased duress on West Germany to recognize the East-German state could have led to undesirable delays in holding the all-European security conference. Gomluka refused to discuss the issue of beginning the separate talks of Poland and West Germany. He only laconically stated that those negotiations would be beneficial for other socialist countries as well.

agreement since its signature.[63] The Soviet leader used the ongoing process of reconciliation between the Eastern Bloc and the FRG as a pretext to request absolute unity of the Warsaw Pact. Regarding Polish-East-German disputes, he warned against Brandt's alleged attempts to split its member states. The Soviet representatives often used similar, usually very vague, phrases with no real content in the following years, as they pushed for a unified foreign policy of the alliance. On this count, Brezhnev was strongly backed, especially by Ulbricht. Because the GDR was unrecognized by the West, defending East-German interests largely depended on their promotion to the joint course of the Warsaw Pact.[64] Gomulka also declaratively adopted those principles. However, he did not want to draw attention away from separate Polish activities. As mentioned above, Poland was actually interested in strengthening alliance cooperation. After all, the communiqué of the meeting was again compiled in a way that reflected the key demands of all Warsaw Pact members. The document repeated an appeal for international recognition of the GDR and confirmation of all the frontiers in Europe, including Polish ones.[65]

---

[63] Hungarian Report of Warsaw Pact Summit on Policy toward West Germany, January 7, 1970, in *A Cardboard Castle?* (doc. 71): 358.

[64] Ulbricht knew that international recognition of the GDR fully depended on support of the USSR and the Warsaw Treaty Organization. However, the SED CC First Secretary also praised the accomplishments of a joint strategy of the alliance because of fears of East-German isolation in the German question, as a potential result of separate moves by the Soviet Union and Poland. On the other hand, those were neither strictly ideological nor instrumental outcries. At the time, the GDR actually firmly adhered to the official policy of the Warsaw Pact during its bilateral talks with West Germany. Immediately after Brandt's government came to power, East Berlin declared its willingness to normalize relations of the German states. That was no separate move by Ulbricht, but part of the strategy previously approved within the Warsaw Pact. *Záznam k současným vztahům NDR-NSR a k vývoji v NSR*, 1. 4. 1970, TO(t) 1970-1974, box 1, i. č. 68, sign. 0344/111, č. j. 021.983/70-1, AMZV, Prague; *Zpráva o jednání předsedy vlády NDR s W. Stopha s bonnským kancléřem W. Brandtem v Kesselu 21. 5. 1970*, 28. 5. 1970, č. j. 023122, ibid; *Zpráva o výsledcích jednání stranických a vládních delegací ČSSR a NDR v říjnu 1970*, 6. 11. 1970, sign. 0344/112, kr. 3, č. j. 025.922/70-4, ibid.

[65] *Dokumenty spotkania przywódców prtyjnych i państwowych siedmiu krajów socjalistycznych*, 10. 12. 1969, PZPR KC, V/89 (2899), AAN, Warsaw.

Final public appeals of the above-mentioned Warsaw Pact's meetings created the illusion of absolute unity. In fact, disagreements between member-states were still prevalent. Poland, which ostensibly submitted at the alliance summits, gives a good example. The Poles assured the other member-states they would respect the unified foreign policy tack and would not publicize their much more complex collective security proposals. However, they took part in backroom negotiations with the West. In December 1969, the North Atlantic Council dealt with both the official Warsaw Pact initiatives as well as the wider separate Polish draft of the *European Security Treaty*.[66]

Perspective of reform of the Warsaw Pact political structures

As outlined before, the turn of the 1960s and 1970s saw all the Warsaw Pact member-states, except for Romania, interested in intensified mutual political cooperation. It is also good to add that the formal scope was considered a minor issue. The previously mentioned meeting of the party leaders in Moscow gives an example. Ulbricht, who formally initiated it, called for a composition of the delegations, which did not differ from the Political Consultative Committee session. Nevertheless, it was labeled as an officially unattached meeting of the Warsaw Pact member-states' top officials.[67]

Tendencies appeared in late 1969 to use more frequent talks, under the auspices of the alliance, to uphold formal reform of its political

---

[66] Selvage, "The Warsaw Pact and the European security conference", 99-100; During the above- mentioned foreign affairs ministers meeting in Prague, Poland tried to get the Warsaw Pact's permission for this strategy. However, the other delegations remained silent on Polish proposal. *Zpráva o průběhu a výsledcích pražské porady ministrů zahraničních věcí členských států Varšavské smlouvy a návrhy na další postup v oblasti evropské bezpečnosti*, 14. 11. 1969, 1261/0/5, sv. 110, a. j. 181/info2, NA, Prague.

[67] *Zpráva o připravovaném setkání představitelů států Varšavské smlouvy*, 10. 11. 1969, sv. 108, a. j. 179/1, ibid; One year later, an internal East-German documentation labeled the Political Consultative Committee session as 'the meeting of leading officials of the communist and workers parties and governments of the Warsaw Treaty Organization member states' only. *Politbüro des ZK Reinschriftenprotokoll nr. 53,* 30. 11. 1970, DY 30/ J IV 2/2/1313, BArch, Berlin.

structures. Hungary took the initiative as it deemed that only the Warsaw Pact's unity would help maintain its active role in the process of convening the all-European security conference. In order to improve mutual political cooperation in the upcoming key period, the head of Hungarian diplomacy, Janos Peter, revived the idea of establishing the Committee of Ministers of Foreign Affairs. In December, he put forward that proposal during talks with his Soviet counterpart Andrei Gromyko in Moscow. In early 1970, the Kremlin blessed the idea and authorized Hungarian leadership to start bilateral negotiations with the other Pact's members on the issue.[68] We can make three conclusions from that approach: In spite of sub-changes in cooperation within the alliance, Moscow's position remained dominant at the turn of the 1960s and '70s. It was still unimaginable even for the loyal Pact members to initiate anything without previous Soviet permission, disregarding the good will. The Kremlin's willingness to leave activity on this issue to Hungarian leadership suggests that the USSR considered the formalization of changes within the Warsaw Pact's political structures either unrealistic or unimportant. However, the Soviet Union began to treat alliance member states more as partners, not only as pure satellites.

Janos Kadar, who sent letters to his counterparts, made an official proposal on reform of the Warsaw Pact's political framework in mid-January 1970. As he argued, the need for new statutes of political meetings arose from the simple fact that the alliance had permanent military bodies only. He correctly mentioned that even within NATO a system of political bodies existed. The absence of a consistently working secretariat stood in the limelight of Hungarian criticism. The Warsaw Pact's secretary general, deputy foreign minister of the USSR Nikolai Firyubin,[69] was unreservedly accused of not paying attention to the alliance. Mutual consultations

---

[68] Békés, Csába, "*Records of the Meetings of the Warsaw Pact Deputy Foreign Ministers,*" Accessed October 16, 2019. http://www.php.isn.ethz.ch/lory1. ethz. ch /collections/coll_defomin/intro_bekes72dc.html?navinfo=15700.

[69] According to initial 1956 intentions of Moscow, a joint secretariat with the secretary general as a head should have been established within the Warsaw Pact. In fact, this never happened. Thus, the secretary general remained a strictly formal position with no real powers or significant duties. Firyubin held that post from 1960 till his death in 1983.

shaping the unified foreign policy course of the Pact were supposed to become more systematic. The Political Consultative Committee should hold two obligatory sessions per year. There was also a call for establishing a committee of ministers of foreign affairs and a Moscow-based permanent secretariat.[70] Hungarian initiatives gained strong support, especially from Polish leadership. The foreign affairs ministry of Poland immediately worked out their own analysis of the situation within the Pact. It essentially corresponded with Budapest's stances; according to Warsaw, an absence of permanent political bodies was the reason for the entire alliance's weakness.[71] Polish consideration of the alteration of the Warsaw Pact went even further: Similar to the 1950s, the internal documents talked again about the need for revision of the founding charter.[72]

In January 1970, Hungary made a significant effort to get support for its proposals. They were put forward at the alliance meeting of deputy foreign ministers in Sofia, where no one but Romanian representative Nicolae Ecobescu opposed such intentions. He used the same counterarguments as Ceausescu's leadership had operated with back in the 1960s when it was striving to prevent reform of the Warsaw Pact.[73] The rest of the alliance member-states were supportive of the Hungarian proposals. Budapest therefore tried to convince lone Romania to alter its rigid position. In case Romanian opposition persisted, Hungary prepared an alternative plan: the Warsaw Pact member-states' foreign ministers should, at the very least, be officially authorized to formulate the alliance's foreign policy together. Hungarian diplomat Frigyes Puja, later that

---

[70] *Notatka w sprawie organizacyjnego zinstytucjonalizowania systemu Układu Warszawskiego*, 20. 1. 1970, PZPR KC, XIA/106, AAN, Warsaw.

[71] Jarząbek, *PRL w politycznych strukturach*, 37-38.

[72] Nevertheless, the Polish materials also suggested measures which would hardly improve the Warsaw Pact's practical mechanisms. For instance, Poland considered proposing a new Institute for International Affairs to be established within the alliance. This mostly academic institution should have put forward various conceptions of a further foreign policy strategy of the Pact and analyzed the international situation. *Notatka dot. systemu konsultacji i koordynacji polityki zagranicznej państw Układu Warszawskiego*, undated, probably 1970, PZPR KC, XIB 171, AAN, Warsaw.

[73] *Informace o poradě náměstků k Evropské bezpečnosti*, 27. 1. 1970, TO(t) 1970-1974, box. 1, i. č. 82, sign. 017/111, č. j. 020.534/70-2, AMZV, Prague.

month, secretly visited Bucharest in order to negotiate Romanian support for one of the proposals. However, his mission failed.[74]

Despite the intensified activities of the Warsaw Pact's political structures, formal reform seemed to be impassable in the early 1970s. Hungarian historian Csába Bekes speculates that behind the initiatives of Hungary could have stood the Kremlin's directives. Taking into account the next development, this thesis seems to be implausible. Soviet leadership did not reckon it worthy to confront Bucharest on that issue again, as the USSR perhaps considered formalization of the shifts in the Pact's mechanisms needless. Leaving the door open for future activities of its satellites, the Kremlin did not want to engage itself significantly. Hungary therefore suggested continual non-binding bilateral talks on the issue. The USSR just requested the Soviet government be informed about the development. The Soviet Union did not intend to put reform of the Warsaw Pact's political structures in front of Romania as this approach definitely failed back in the 1960s. Moscow therefore appealed rather for frequent meetings under the auspices of the Pact. There should have been no statutory documents defining those meetings until the moment they became routine praxis.[75]

After the failure of talks on reform of the Warsaw Pact's political structures in 1966, the USSR was content with establishing statutory-free consultative mechanisms, which enabled them to activate the organization whenever the Kremlin considered it necessary. Furthermore, as the Prague Spring experience revealed, Moscow was able to comfortably bypass the alliance structures, if needed. Brezhnev also organized many various informal and bilateral talks. This praxis climaxed in the early 1970s in an attempt to establish a tradition of multilateral Crimean meetings.[76] Those

---

[74] Békés, "*Records of the Meetings*".

[75] *Podkladový materiál pro kolegium MZV ČSSR o československo-sovětských vztazích zaslaný velvyslancem ČSSR v Moskvě*, 1. 9. 1970, TO(t) 1970-1974, box. 1, i. č. 89, sign. 020/112, č. j. 0236/70, AMZV, Prague; *Zpráva o sovětsko-rumunských vztazích*, 13. 9. 1971, 1261/0/6, sv. 12, a. j. 11/info1, NA, Prague.

[76] Jarząbek, *PRL w politycznych strukturach*, 39. The party leaders spent their holidays in Crimea in the 1960s already. However, nothing suggests that the Soviet leadership used their stays for systematic consultation at the time, although there were undoubtedly always some political talks.

were intended as some less formal form of contact between the Warsaw Pact member-states' leaders. Unlike the official alliance gatherings, where the delegations usually gave their speeches and then signed more or less pre-discussed documents, meetings in the summer resorts of Yalta and Sochi were much more spontaneous and open. Since 1971, Brezhnev tried to shift the annual talks with the Soviet European satellites' leaders during their vacations to become a tradition.[77] According to Moscow's idea, the Eastern Bloc countries were definitely not supposed to cooperate solely within the official organizations, Comecon and the Warsaw Pact. There should have been numerous extra talks between party and state representatives on many levels.[78] Through the coordination of foreign policy within the scope of the Warsaw Treaty Organization, Moscow pursued their main goals: It created an illusion for the outside world that the relations of Pact member-states really respected the principles of so-called 'socialist internationalism'. In addition, more frequent alliance sessions should have provided more stable ground for foreign policy cooperation within the Eastern Bloc.[79]

The previously mentioned meetings of deputy foreign ministers substantially improved the way the Warsaw Pact worked. On the initiative of Hungary, they started being held regularly. In January 1970 during sessions in Sofia, a decision was made that those meetings would follow every significant NATO summit.[80] Although no detailed analysis of this forum had been worked out yet, those frequent meetings most likely

---

[77] Due to disputes with Romania, only the first three Crimean meetings took place on multilateral basis. Baev, Jordan, "*The "Crimean Meetings" of the Warsaw Pact Countries' Leaders,*" Accessed October 16, 2019. 2003. http://www.php.isn.ethz. ch/lory1.ethz.ch/collections/crimea_meetings4a8c.html?navinfo=16037.

[78] Tejchman and Litera, *Moskva a socialistické země na Balkáně*, 84-86.

[79] Litera, "Unifikace sovětského bloku"*, 85.

[80] *Report to the Hungarian Socialist Workers' Party's Political Committee and to the Government on the Meeting of the Deputy Foreign Ministers*, January 27, 1970, Accessed October 16, 2019. http://www.php.isn.ethz.ch/kms2.isn.ethz.ch/ serviceengine/Files/PHP/17267/ipublicationdocument_singledocument/c8bc03d b-dd9f-419a-b324-42208a13879f/en/700126_Report_E.pdf.

became a key political mechanism of the Warsaw Pact in the 1970s.[81] Their success was bolstered by the fact that they lacked any formal rules. Therefore, even rebellious Romania did not oppose their holding. According to Bucharest, consultations without any binding scope were the only useful and proper form of political cooperation within the Warsaw Pact. Furthermore, the representative level of deputy foreign ministers was relatively low. Thus, even contentious and controversial issues could have been discussed and, unlike at the high and hyped meetings, individual member-states were not obligated to take an official stance. Only non-binding exchanges of information and preparation of upcoming declarations took place there. Recommendations only could emerge from the talks of deputies to be further assessed by the state and party leaderships. Later in the 1970s, that forum provided to the member-states a significant insight into actual Soviet intentions concerning the Warsaw Pact.[82] That was very important because relatively frequent alliance talks on higher levels were just temporary, lasting until 1972 only. The meetings of deputy foreign ministers therefore "filtered" problematic issues. The USSR and other member states verified acceptability of their proposals before they put forward those at a higher level. If some deeper discord appeared, Moscow prevented such an item from being discussed in main alliance bodies. This was one of the factors why the Warsaw Treaty Organization did not fail to perform relatively unanimously. However, postponing the solution of some problematic issues inevitably created long-term latent tensions.[83]

Paradoxically, the absence of formal rules for deputy foreign ministers' meetings strengthened their importance as they could have been held operatively *ad hoc*. We can even say that in this particular period in the 1970s, to some extent, they played the role of a non-existing permanent secretariat of the alliance. However, the competences of that forum soon

---

[81] During 1970-1985, only nine Political Consultative Committee sessions and fourteen foreign affairs ministers' meetings took place. However, there is evidence of twenty-three talks of the deputies at least.

[82] Békés, "*Records of the Meetings*".

[83] Tejchman and Litera, *Moskva a socialistické země na Balkáně*, 89.

became a sticking point. In 1971, at the meeting in Budapest, Soviet foreign vice minister, Leonid Ilyichev, conveyed the opinion of Moscow that those talks should not have in praxis shaped the documents to be later approved by the ministers of foreign affairs on behalf of the Warsaw Pact. The Polish delegation even opposed the principle that deputies set up the agenda for talks of their superiors.[84] Warsaw insisted the deputy foreign ministers' meetings to be a strictly advisory forum. The USSR soon decided to specify the role of those talks. Moscow rejected the claim of the head of Hungarian diplomacy, Janos Peter, that unconfined meetings could bring many interesting suggestions. Since 1971, the Kremlin clearly defined their agenda in advance.[85] In this manner, the Soviet Union actually restrained, from its point of view, undesirable and unnecessary multilateral political mechanisms, which in the previous two years crystallized within the Eastern Bloc through the Warsaw Treaty Organization.

Purpose and scenario of the Warsaw Pact political meetings

The activity of the Warsaw Pact political structures continued even in 1970, disregarding Romanian sub-attempts to scale down the alliance meetings. The talks were still mostly a side-product of the Eastern Bloc's effort to convene the all-European security conference and to maintain the initiative role in this process. At the time, particularly some Soviet satellites showed interest in closer cooperation in the scope of the alliance. In January 1970, party bodies of SED and CPCz discussed further possible steps. The internationally unrecognized GDR and Czechoslovak normalization regime with

---

[84] *Minutes of the Meeting of the Deputy Foreign Ministers*, June 19, 1970, Accessed October 16, 2019. http://www.php.isn.ethz.ch/kms2.isn.ethz.ch/ service engine/Files/PHP/17297/ipublicationdocument_singledocument/bf221ca2-5d8b-4ce6-add5-7e7389786508/en/700619_Minutes_E.pdf.

[85] *Záznam o schozdke s. ministra Ing. J. Marka s námestníkom ministra zahraničných vecí ZSSR s. Rodionovom dňa 12.5. 1971,* TO(t) 1970-1974, box 4, i. č. 89, sign. 020/117, č. j. 010.777/71-KM, AMZV, Prague.

low foreign credentials sought in the Warsaw Treaty Organization
backing a chance for engagement in preparation of the all-European
security conference. Thus, both countries intended to maximize
the importance of alliance cooperation. They suggested negotiations
at least on the level of ministers of foreign affairs, and it would
be even better to hold a straight Political Consultative Committee
session.[86]

The next meeting of ministers of foreign affairs was held 21-
22 of June 1970 in the Hungarian capital. Compared with previous
years, however, it brought very little results in all aspects. The
North Atlantic Council, just a few days before, officially declared its
willingness to negotiate regularly on the all-European security conference.
NATO also agreed on a potential location - Helsinki.[87] Nevertheless,
widely differing opinions on the scheme and agenda of the potential
summit still existed between the West and the East. The Warsaw
Pact member-states therefore had to admit that their previous
short-term proposal of the conference was unrealistic.[88] Because
of Romania, there was also no shift in the issue of potentially establishing
regular alliance consultations of the ministers of foreign affairs.
However, the temporary moderation of Ceausescu's regime was
illustrated by the fact that unlike in the second half of the 1960s,
Romanian delegations repressed all of their protests. Representatives of

---

[86] *Informace o některých názorech vedení SED na další postup ke svolání
celoevropské konference o bezpečnosti v Evropě*, 19. 1. 1970, box 8, i. č. 68, sign.
0344/118, č. j. 020.410/71-4, ibid.

[87] Vojtech Mastny speculates that the Finnish proposal on holding the all-
European conference in Helsinki from April 1969 was actually pre-discussed with
Moscow, as Urho Kekkonen, president of Finland, kept close ties with the Soviet
secret service. Mastny, "Superpower Détente", 20.

[88] In autumn 1969, at the meeting in Moscow, the deputy foreign ministers still
believed that the socialist countries should have striven for holding the all-
European security conference before the end of upcoming year. *Konferenz der
Außenminister der Mitliedstaaten des Warschauer Vertrages*, 28. 10. 1969, DY
30/ J IV 2/2/1248, BArch, Berlin.

Romania only took reserved attitudes towards some issues and their consent often had to be obtained through lengthy negotiation.[89]

The Budapest meeting brought in the scheme that became typical for most of the Warsaw Pact political gatherings. As the preserved records suggest, unlike in 1969, there was no longer open discussion. The speeches of the delegations had mostly a reporting and phrasal structure instead.[90] In fact, everything important had been discussed in advance at the deputy foreign ministers' meeting, which had taken place shortly before the session of their superiors,[91] proven by the reaction of Poland. As Warsaw still strove for shifting cooperation within the alliance framework to a higher level, it proposed to postpone the main foreign affairs ministers' meeting by two days. That move

---

[89] *Zpráva o průběhu a výsledcích budapešťské porady ministrů zahraničních věcí členských států Varšavské smlouvy*, 14. 7. 1970, 1261/0/5, sv. 134, a. j. 211/info4, NA, Prague; That was obvious even at the preliminary meeting of deputy foreign ministers. Romanian representative Macovescu swore blindly that his country considered it useful to cooperate with the Warsaw Pact members in the CSCE negotiations in order to maintain activity on the Eastern side. He even used a term "*we, the Warsaw Pact members*". Macovescu also appreciated draft communiqué prepared by the Soviet and Hungarian delegations as a proper basis for shaping a joint position. A unified strategy only should not have limited individual states in their bilateral talks. However, during approval of the resolution, the Romanian delegation tended towards extreme wordplay. This resulted in sharp arguments with the Hungarian representatives which was undoubtedly also fueled by Bucharest's unwillingness to accept the proposed reforms of the Warsaw Pact's political structures. *Minutes of the Meeting of the Deputy Foreign Ministers*, June 19, 1970, Accessed October 16, 2019. http://www.php.isn.ethz.ch/kms2.isn.ethz.ch/serviceengine/Files/PHP/17297/ipublicationdocument_singledocument/bf221 ca2-5d8b-4ce6-add5-7e7389786508/en/700619_Minutes_E.pdf.

[90] *Materiały i dokumenty z narady ministrow spraw zagranicznych panst-stron Układu Warszawskiego, odbytej v budapeszcie w dniach 21-22 czerwca 1970r.*, PZPR KC, XIA/248, AAN, Warsaw.

[91] The preliminary meeting of deputies related to the Budapest talks took place on the 19th of June 1970 in the Hungarian capital. Moreover, Moscow also discussed the agenda of foreign affairs ministers' gathering bilaterally. *Report to the Hungarian Socialist Workers' Party's Political Committee and to the Government on the Meeting of the Deputy Foreign Ministers*, January 27, 1970, Accessed October 16, 2019. http://www.php.isn.ethz.ch/kms2.isn.ethz.ch/serviceengine/Files/PHP/17267/ipublicationdocument_singledocument/c8bc03db-dd9f-419a-b 324-42208a13879f/en/700126_Report_E.pdf.

would allow the leaderships of the Warsaw Pact member-states to familiarize themselves in detail with the outcomes of recent deputies' talks. However, the USSR refused the Polish request.[92]

According to Soviet strategy, the Warsaw Treaty Organization's political meetings should have been given a new, very limited purpose: They were nothing more than a counteraction to the North Atlantic Council sessions, providing to the Eastern Bloc states the opportunity to jointly react on NATO stances.[93] In that stage of the Cold War, the USSR probably came to the conclusion that the outward presentation of the Warsaw Pact as a clear NATO counterpart, to the greatest extent possible, was favorable. Ministers of foreign affairs' meetings, under the auspices of the alliance, were supposed to give the impression they were analogous to similar talks within NATO. In a situation when two *equal* military-political pacts existed, it was naturally much easier to negotiate in the early stages of the détente process on the basis of two leading powers as the representatives of the current blocs.[94] This conception could have significantly helped the USSR to maintain sufficient control over its sphere of influence during alternating international situations. In this regard, Bohuslav Litera considers the Eastern Bloc's cohesion to be one of the key aims of Soviet policy towards its

---

[92] *Informace V.Biľaka o nadcházející poradě ministrů zahraničních věcí v Budapešti*, 19. 6. 1970, 1261/0/5, sv. 131, a. j. 206/25, NA, Prague.

[93] *Draft of the Speech by Hungarian Deputy Foreign Minister Frigyes Puja for the Meeting of the Deputy Foreign Ministers*, June 19, 1970, Accesed October 16, 2019. http://www.php.isn.ethz.ch/kms2.isn.ethz.ch/serviceengine/Files/PHP/ 173 00 / ipublicationdocument_singledocument/da490bf2-98aa-46fc-ab20- 123b86e7 0032/en/700619_Draft_Speech_ENG.pdf.

[94] Both superpowers preferred détente to be rather a bilateral issue. They saw easing tensions as a process respecting the hierarchy within the blocs. However, such an approach collided with political priorities of their allies. NATO, as well as the Warsaw Treaty Organization members, considered détente more dynamical, as a potential opportunity for altering the existing *status quo*. Those countries therefore emphasized the multilateral feature of easing tensions. Wenger, Andreas and Mastny, Vojtech, "New perspectives of the origin of the CSCE process," in *Origins of the European security system: the Helsinky process revisited, 1965–75*, eds. Andreas Wenger, Vojtech Mastny and Christian Nuenlist (Abingdon: Routledge, 2008): 5.

European satellites in the 1970s. East was supposed to act as obvious counterbalance of West.[95]

Collective talks of foreign affairs ministers were used by Moscow mostly to present some of its initiatives on behalf of the entire Warsaw Treaty Organization. For instance, the Pact officially offered to USA and Canada participation in the proposed all-European security conference, as well as agreeing to put cultural issues on the agenda.[96] As previously mentioned Gromyko's considerations, of establishing some sort of international security body in Europe, was moved to the position of an official demand of the alliance. By Soviet request, the Warsaw Pact also approved the appeal for reducing the numbers of foreign troops dislocated on European countries' territory. The move was actually a straight directive of the Kremlin.[97]

The unification of foreign policy of the Warsaw Pact was not asserted by the USSR only. However, it is obvious that Moscow had a decisive influence on the issue. Specifically, the Polish minister of foreign affairs, Stefan Jędrychowski, saw consistent cooperation within the alliance as a key factor for the success of the Helsinki process.[98] As we previously suggested, the Soviet satellites also showed concerns for potential apparent disunity that could facilitate Western instigation of disputes among them. Hungary, for instance, warned against such a scenario. However, Budapest suggested using the same tactics: The Warsaw Pact member states should strive to deepen the existing discords within NATO. The described approach also did not mean that all the Soviet satellites, especially Poland and Hungary, completely resigned to

---

[95] Litera, "Unifikace sovětského bloku," 84.

[96] Rey, Marie-Pierre, "The USSR and the Helsinki process, 1969-75," in *Origins of the European security system: the Helsinky process revisited, 1965–75*, eds. Andreas Wenger, Vojtech Mastny and Christian Nuenlist (Abingdon: Routledge, 2008): 67.

[97] *Memorandum států Varšavské smlouvy*, 22. 7. 1970, 1261/0/5, sv. 134, a. j. 211/info4, NA, Prague; *Zpráva o nejdůležitějších opatřeních sovětské zahraniční politiky v období ledna – června 1970*, 27. 5. 1970, TO(t) 1970-1974, box. 8, i. č. 89, sign. 020/311, č. j. 0122/70, AMZV, Prague.

[98] *Materiały i dokumenty z narady ministrow spraw zagranicznych panst-stron Układu Warszawskiego, odbytej w budapeszcie w dniach 21-22 czerwca 1970r.*, PZPR KC, XIA/248, AAN, Warsaw.

their own diplomatic activities, and in the Helsinki process, spoke solely through the Warsaw Treaty Organization. However, cooperation in the scope of the alliance should have been intensified. That would have resulted in a joint presentation of key initiatives and declarations determining positions of the Eastern Bloc. Individual member states were supposed to actively enter in bilateral talks with Western countries and to act in accordance with collectively defined stances of the alliance.[99] The Kremlin considered that tactic very useful. At the time, Soviet diplomacy was able to use the formal existence of the Warsaw Pact's multilateral mechanisms skillfully during negotiations with third-party countries.[100]

Negotiations on the German question

As explained before, the political talks within the Warsaw Pact at the turn of the 1960s and '70s were activated not only because of the all-European security conference, but also because of the German question. Brezhnev rightfully considered the signing of Soviet-West-German treaty on friendship in August 1970 to be a major shift and a critical moment for upcoming political development in Europe. He wanted to discuss the situation with the Soviet satellites' leaders. Therefore, he decided to hold the Political Consultative Committee meeting in Moscow.[101] The main

---

[99] *Minutes of the Meeting of the Deputy Foreign Ministers*, June 19, 1970, Accessed October 16, 2019. http://www.php.isn.ethz.ch/kms2.isn.ethz.ch/service engine/Files/PHP/17297/ipublicationdocument_singledocument/bf221ca2-5d8b-4ce6-add5-7e7389786508/en/700619_Minutes_E.pdf; *Draft of the Speech by Hungarian Deputy Foreign Minister Frigyes Puja for the Meeting of the Deputy Foreign Ministers*, June 19, 1970, Accesed October 16, 2019. http://www.php.isn. ethz.ch/kms2.isn.ethz.ch/serviceengine/Files/PHP/17300/ipublicationdocument_ singledocument / da490bf2-98aa-46fc-ab20-123b86e70032 / en / 700619_Draft_ Speech_ENG.pdf.

[100] According to the Soviet diplomat G. I. Solovyev, the prime minister of Sweden, Olof Palme, during his visit to Moscow in June 1970, probed what was the USSR's attitude towards mutual reduction of armed forces in Europe proposed by NATO. The Soviet officials claimed they were not authorized to answer the question because the Warsaw Pact had not taken any joint position yet. *Záznam o návštěvě s. Z. Picka u rady odboru Skandinávských zemí MID SSSR s. G. I. Solovjeva*, 29. 6. 1970, TO(t) 1970-1974, box. 4, i. č. 89, sign. 020/117, č. j. 0161/70, AMZV, Prague.

[101] Brezhnev probably just seized the opportunity. Because of signing the Soviet-West-German treaty, the Warsaw Pact member-states' officials gathered in the

pillars of the concluded agreement actually significantly concerned the other Warsaw Pact countries, as they contained confirmation of a border *status quo* in Europe, including the Oder-Neisse line. It was to be the first of a series of documents which Willy Brandt's government intended to use for reconciliation between West Germany and the Eastern Bloc. The treaties with Poland, East Germany, and Czechoslovakia would have followed. Bonn also committed to support holding the all-European conference as well as accepting the conception of both German states' representation in the United Nations.[102] Regarding this, the Soviet leader - perhaps clearly instrumentally - talked about the direct effect of the Warsaw Pact's initiatives as well as military intervention in Czechoslovakia on conclusion of the treaty.[103] With reference to the continual existence of NATO, he warned against overblown optimism. However, he admitted that the Eastern Bloc states succeeded in causing some disruption among NATO members, especially in Washington-Bonn ties.[104] As already mentioned, that strategy was previously suggested at the Warsaw Pact gatherings, after all.

The burning issue of the reform of the alliance's political structures was discussed as well. The Political Consultative Committee

---

Soviet capital on August 12. *Minutes of the Hungarian Party Politburo Meeting on the August 1970 PCC Meeting*, August 25, 1970, Accessed October 16, 2019. http://www.php.isn.ethz.ch/kms2.isn.ethz.ch/serviceengine/Files/PHP/18034/ipu blicationdocument_singledocument/89875ee9-6b04-46a4-b06b-9295e3d36f45/ en/Minutes_Hungarian_Party_1970_Eng.pdf.

[102] Those commitments were included in the amendment 'Agreement on signatories' intentions' which, however, was not published by request of West-German diplomacy.

[103] Gustáv Husák naturally fully supported such evaluations made by Brezhnev. *Vystoupení s. L. I. Brežněva na poradě vedoucích představitelů zemí Varšavské smlouvy v Moskvě dne 20. srpna 1970*; *Informace o průběhu diskuse na zasedání Politického poradního výboru Varšavské smlouvy v Moskvě dne 20. srpna 1970*, 1261/0/5, sv. 135, a. j. 213/1, NA, Prague.

[104] However, Brezhnev's notes on the alleged worsening of U.S.-West-German relations proved to be wrong. In the early months of 1971, even Gromyko had to admit this fact. *Zpráva o poradě ministrů zahraničních věcí členských států Varšavské smlouvy konané ve dnech 18.-19. 2. 1971 v Bukurešti*, 26. 2. 1971, sv. 154, a. j. 238/7, ibid.

dealt with the question for the first time after two and half years. Brezhnev noted that there had been a significant step up in political cooperation within the Warsaw Pact since 1969. Therefore, he called to reconsider the possibility of a statutory anchoring of the ongoing meetings and finally establishing advisory bodies to the Political Consultative Committee.[105] The words of the Soviet leader contained nothing imperative. That proved of such marginal importance that the Kremlin requested formalization of changes within the Warsaw Pact. Gomułka reacted widely on Brezhnev's notice. He provocatively stated that in comparison to the Warsaw Treaty Organization, NATO still had better mechanisms and structure; for example, its supreme body meetings were held twice a year. Despite recent and separate Polish international moves, the PUWP CC First Secretary again called for establishing better conditions for mutual consultations and looking for new forms of alliance cooperation. Regarding the tactic of deepening disputes among NATO members, he warned against a similar Western approach towards the socialist countries.[106] Hungarian and Czechoslovak delegations were less eager. They were unanimous in a stance that to continue holding *ad hoc* meetings was a plausible alternative to the creation of new bodies. However, Romania blocked further discussion on the issue once again.[107]

In 1970, an unprecedented *second* session of the Political Consultative Committee was held in the same year.[108] In November, only

---

[105] *Informace o průběhu zasedání politického poradního výboru Varšavské smlouvy v Moskvě dne 20. srpna 1970*, 22. 8. 1970; *Vystoupení s. L. I. Brežněva na poradě vedoucích představitelů zemí Varšavské smlouvy v Moskvě dne 20. srpna 1970*, sv. 135, a. j. 213/1, ibid,

[106] *Posiedzenie Politycznego Komitetu Doradczego Państw Układu Warszawskiego Moskwa, 20 sierpnia 1970 r., Wystąpienie I sekretarza KC PZPR tow. Władysława Gomułki*, PZPR KC, XIA/248, AAN, Warsaw.

[107] *Minutes of the Hungarian Party Politburo Meeting on the August 1970 PCC Meeting*, August 25, 1970, Accessed October 16, 2019. http://www.php. isn. ethz. ch/kms2.isn.ethz.ch/serviceengine/Files/PHP/18034/ipublicationdocument_singl edocument / 89875ee9 - 6b04-46a4-b06b-9295e3d36f45/en/Minutes_Hungarian_ Party_1970_Eng.pdf.

[108] Two Political Consultative Committee sessions in one year were held neither before nor after.

three months after their last meeting, Wladyslaw Gomułka initiated such a step in his conversation with Brezhnev. This way, he wanted to use the Warsaw Pact to affect, from his point of view, the undesirable positions of the East-German leader Walter Ulbricht.[109] His policy went through significant alteration and began to differ from the unified policy tack of the alliance during 1970.[110] In August, the offended SED CC First

---

[109] Personal relations between Ulbricht and Gomulka were rarely bad. The tensions climaxed after November 1967, when both leaders argued in Moscow on East-German unwillingness to extensively buy the Polish industrial goods, pointing to a poor quality of the products. Their mutual animosity then never decreased till the political fall of Gomulka in December 1970. Selvage, Douglas, "The Treaty of Warsaw (1970): The Warsaw Pact Context," *Bulletin of the German Historical Institute*, no. 1 (2004): 70-71.

[110] During 1969-1971, the East-German attitude towards the European security conference made a 180-degree change. Before the Prague Spring, Ulbricht believed that the internal rifts in West Germany would create an opportunity for the spreading of socialism. He did not abandon the idea of establishing some form of German confederation on the socialist basis. In the wake of suppression of the Prague Spring, he also expected that Moscow would label Brandt's *Ostpolitik* as *"ideological branch of imperialism"*. After Brandt won the 1969 election, Ulbricht still insisted on the condition that Bonn had to fully recognize East Germany according to international law. The Kremlin began to fear that the East-German leader could become an obstacle to establishing correct relations with the new West-German government. Since late 1969, Moscow therefore started to closely cooperate with Erich Honecker, the SED CC secretary, in order to deprive Ulbricht of his influence on foreign policy. Henceforward, the power of Honecker's political-line supporters grew in the East-German politburo. In July 1970, Brezhnev informed Honecker that the Kremlin was also strongly disturbed by Ulbricht's foreign policy. On behalf of the Soviet politburo, he appealed for replacement of the SED CC First Secretary and asked for daily reports on the situation. Weakening of Ulbricht's influence was proven in October 1970 by the East-German memorandum on "the Warsaw Treaty Organization positions and approach towards convening the CSCE". Despite it still containing the ideological phrases about spreading of socialism in the West through the left-wing orientated groups as well as the trade unions, the document emphasized mostly the development of mutual economic and scientific-technological cooperation of both blocs. Since autumn 1971, Moscow strongly urged the beginning of negotiations between two German states. Rhetoric of the GDR took a strictly defensive position and stressed maintaining peace in Europe, confirmation of the frontiers, and economic collaboration. In late 1971, Honecker strengthened his positions insofar that after a discussion within the party leadership, he decided to send a letter to Moscow signed by 21 politburo members which proposed to replace the

Secretary demonstratively did not take part in the Warsaw Pact's supreme body meeting.[111] Brezhnev ignored Gomulka's appeal initially.[112] However, he later decided to hold the Political Consultative Committee. The meeting took place on December 2 in East Berlin, probably due to demonstrative reasons. After signing the Polish-West-German treaty, the CPSU CC General Secretary considered it useful to warn against Brandt's *Ostpolitik* once more. Through that policy, Bonn allegedly only attempted to initiate discord among the Warsaw Pact countries. Ulbricht's move away from the unified course of the alliance, which he had previously upheld loudly, proved that the words of the Soviet leader in fact did not lack content.[113] The meeting therefore served mainly to set the next

---

SED CC First Secretary regarding his poor economic decisions and high age. Bange, Oliver and Kieninger, Stephan, "Negotiating one's own demise? The GDR's Foreign Ministry and the CSCE negotiations: Plans, preparations, tactics and presumptions," *Cold War International History Project e-Dossier*, no. 17 (2007): 3-6; Kopstein, Jeffrey, "Ulbricht Embattled. The Quest for Socialist Modernity in the Light of New Sources," *Europe-Asia Studies*, no. 4 (1994): 611; Caciagli, Federica, "The GDR's target in the early CSCE process. Another missed opportunity to freeze the division of German, 1969-73," *Origins of the European security system: the Helsinky process revisited, 1965–75*, eds. Andreas Wenger, Vojtech Mastny and Christian Nuenlist (Abingdon: Routledge, 2008): 111.

[111] According to official explanation, Ulbricht did not take part due to his poor health and physicians' advice. Animosity in East-German leadership fully showed-up. The chairman of the state council, Willi Stoph, became the head of the GDR's delegation, whereas he totally failed to consult his speech with the SED CC First Secretary. Such a situation, when the factions within the East-German leadership frequently put forward incoherent proposals, bothered the Kremlin. Therefore, just one day after the Political Consultative Committee session, the Soviet diplomats separately discussed a possible solution with the SED representatives and urged the unity of the party leadership of all the Eastern Bloc states to be outwardly presented. *Minutes of the Hungarian Party Politburo Meeting on the August 1970 PCC Meeting*, August 25, 1970, Accessed October 16, 2019. http://www.php.isn.ethz.ch/kms2.isn.ethz.ch/serviceengine/Files/PHP /18034/ipublicationdocument_singledocument/89875ee9-6b04-46a4-b06b-9295e3d36f45/en/Minutes_Hungarian_Party_1970_Eng.pdf.

[112] *Notatka z rozmowy tow. Gomułky z tow. Breżniewem /Budapeszt, 23. XI. 1970 r./*, PZPR KC, XIA/88, AAN, Warsaw.

[113] For long time, the SED CC First Secretary was one of the main protagonists of a joint approach of the Warsaw Pact towards the German question as well as the all-European security conference. Keep in mind that until spring 1970, his positions reflected primarily the stances of Moscow. Regarding the separate talks

engagement of the Warsaw Treaty Organization in the German question. By conclusion of the treaty confirming the western borders of Poland, Gomulka achieved his goal. Henceforward, he pragmatically advocated the alliance's course of aiming towards strengthening the international position of the GDR.[114] Aware of its low international credibility, the Czechoslovak normalization regime also strove for future involvement of the Warsaw Pact in the issue of reconciliation between the West-German state and the Eastern Bloc. However, Husák did not present that as a purely Czechoslovak interest. He instrumentally related it to the other priorities like international recognition of the GDR or defining the status of West Berlin. Despite the recent signing of bilateral agreements of the Soviet Union and Poland with West Germany, according to Husák's view, the Warsaw Pact should have acted together in the future establishment of diplomatic relations with Bonn.[115]

The Berlin meeting of the alliance's supreme body had again mostly a perfunctory character that was proven by the fact that the duration of such a top multilateral summit was not supposed to exceed four and a half hours.[116] Its single real outcome was a traditionally common

---

of the Soviet Union and Poland with Bonn, however, the GDR gradually increased its demands for more involvement of the Warsaw Pact in the international recognition of East Germany and its admission to international organizations. *Direktive für das Auftreten des Ministers für Auswärtige Angelegenheiten der DDR auf der Konferenz der Außenminister der Mitgliedstaaten des Warschauer Vertrages (21.-22. Jini 1970 in Budapest)*, 8. 6. 1970, DY 30/ J IV 2/2/1287, BArch, Berlin; *Stanovisko NDR k přípravě konference o evropské bezpečnosti*, 31. 3. 1970, TO(t) 1970-1974, box 1, i. č. 68, sign. 0344/111, č. j. 021980, AMZV, Prague.

[114] The East-German government officially appreciated the conclusion of the Polish-West-German treaty, of course. Berlin called it a significant contribution to European security as well as respecting *status quo* on the continent, and therefore also to future recognition of the GDR. It was also emphasized that East Germany already confirmed Polish western borders in 1950. *Anlage Nr. 2 zum Protokoll Nr. 54/70 vom 8. 12. 1970*, DY 30/ J IV 2/2/1314, BArch, Berlin.

[115] *Informace o průběhu zasedání politického poradního výboru Varšavské smlouvy v Berlíně dne 2. prosince 1970*, 5. 12. 1970, 1261/0/5, sv. 146, a. j. 225/1, NA, Praha.

[116] *Maßnahmeplan zur Vorbereitung des Treffens der Partei- und Staatsführungen der Warschauer Vertragsstaaten am Mittwoch, dem 2. 12. 1970*, DY 30/J IV 2/2J/3250, BArch, Berlin.

declaration defining the Warsaw Pact's position on actual world events. From the Soviet point of view, this was the only desirable purpose. The support for wider international recognition of the GDR was declared officially as one of the Pact's priorities.[117] The Political Consultative Committee resolution served as authorization for individual Warsaw Treaty Organization members to start particular diplomatic activity on the issue. During the 1970s, that strategy became increasingly typical. However, maximal efforts to present unity resulted in searching for compromised positions that had no potential for solving the real problems, as the East-German case shows. During internal conversations of SED leadership, even after the Political Consultative Committee session, Ulbricht outlined maintaining his different attitude, especially towards the status of West Berlin. He had supported the common declaration only because he did not want to reveal existing discords within the Warsaw Pact.[118] After all, during previous months, Poland under Gomulka's leadership had acted in the same way.

Therefore, Romania remained the only country that tried to modify declarations of the alliance substantially. Romanian attitude finely reflected the real political role of the Warsaw Treaty Organization at the beginning of the 1970s. After 1969, the Pact increasingly tended to become a forum covering unified foreign policy tack of the Eastern Bloc states. Despite the fact that there were official negotiations among the

---

[117] For example, Gomulka formally proposed joint action of the Warsaw Pact in Third World countries in order to get support for establishing diplomatic ties with the East-German state.

[118] At the Political Consultative Committee session, Ulbricht failed in his appeals for the Warsaw Pact to support his efforts to prevent any political involvement of the FRG in West Berlin. According to Ulbricht's vision, the western sectors of the divided city should have been neither part of a West-German state nor an independent political unit. West Berlin ought to get the status of an autonomous territory within the GDR. The Soviet Union as well as the rest of its satellites, e.g. Poland, were aware that such a radical conception would never be accepted by the Western powers. *Zpráva ZÚ o 14. zasedání SED, konaném ve dnech 9.-11.12. 1970*, TO(t) 1970-1974, box 1, i. č. 68, sign. 0344/111, č. j. 026.853/70-4, AMZV, Prague; *Informace o jednání ministra zahraničních věcí PLR s. Jendrychowského v NDR ve dnech 6.-8. 1. 1971*, 13. 1. 1971, č. j. 020.190/71-4, ibid; *Návštěva ministra zahraničních věcí NDR O. Winzera v ČSSR*, 20. 1. 1971, box 4, sign. 0344/112, č. j. 026.779/70-4, ibid.

equals, the USSR acted in a hegemonic manner. On the contrary, Bucharest considered the talks within the Warsaw Treaty Organization, which were not based on any statutory document, a non-exclusive consultative forum consisting of some socialist countries. Romania reserved the right to shape its foreign policy course outside the alliance structures. Unlike in the 1960s, the rest of the Warsaw Pact members saw different Romanian positions only as a specific issue that did not pose a threat to the implementation of key priorities.

Gradual talk's attenuation

The CSCE preparation officially took place on the basis of individual sovereign states. However, the conception of two existing military-political blocs persisted. Decisive stances were in fact declared by top gatherings of NATO and the Warsaw Treaty Organization.[119] The USSR considered such a state favorable. In order to react to the North Atlantic Council summit, which defined Western conditions for holding the conference, in February 1971, the Kremlin initiated the next Warsaw Pact meeting of foreign affairs ministers in Bucharest.[120] There, Gromyko had to finally admit that the holding of CSCE in the immediate future was not realistic and success in the issue would require long-term efforts and patient negotiation.[121] Moscow therefore used the alliance meeting to

---

[119] *Zestawienie stanowisk Układu Warszawskiego i NATO w sprawach związanych z Europejską Konferencją Bezpieczeństwa i Współpracy*, undated, after June 11, 1971, PZPR KC, XIB 171, AAN, Warsaw.

[120] Even the Western diplomats were surprised by the fact that the meeting took place only two months after the last Political Consultative Committee session. *Ohlas bukurešťské porady ministrů zahraničních věcí členských zemí VS*, 26. 2. 1971, DTO 1945-1989, i. č. 34, e. č. 52, č. j. 021.324/71-2, AMZV, Prague. Arrangement of the meeting was fully directed by the Kremlin. In January, during his conversation with Winzer, Gromyko clearly declared that the Soviet leadership would solely decide on the holding and usefulness of the foreign affairs ministers' gathering. *Vermerk über die Konsultation des Ministers für Auswärtige Angelegenheiten der DDR, Genossen Otto Winzer, mit dem Minister für Auswärtige Angelegenheiten der UdSSR, Genossen A.A. Gromyko, am 11. 1. 1971 in Moskau*, 12. 1. 1971, DY 30/J IV 2/2J/3289, BArch, Berlin.

[121] *Zpráva o poradě ministrů zahraničních věcí členských států Varšavské smlouvy konané ve dnech 18.-19. 2. 1971 v Bukurešti*, 26. 2. 1971, 1261/0/5, sv. 154, a. j. 238/7, NA, Prague.

instruct its satellites how to act in the international arena regarding ongoing dynamical events; the Eastern Bloc's two priorities should have been talks on holding the CSCE as well as emancipation of the GDR by all available means, both bilateral and multilateral.[122] The Warsaw Treaty Organization was to play no exclusive role in those talks. It provided a scope for unification of its members' approach only; individual Warsaw Pact countries committed to intense diplomatic activity. Under the auspices of the joint declaration of the alliance, the Warsaw Pact countries would have the backing and support for their diplomatic efforts.[123] Nevertheless, the Pact itself never gained a completely privileged position within the consultative mechanism of the Eastern Bloc. That is illustrated also by the then-appeal of the East-German minister of foreign affairs Otto Winzer, for frequent bilateral talks to be held alongside the multilateral Warsaw Pact meetings.[124] In addition, all the international moves of the alliance member-states were consulted in praxis by the Soviet foreign affairs ministry through its embassies in Moscow. That state fully met the Kremlin's vision.[125]

---

[122] The GDR strove for support of the Warsaw Pact not only in the issue of membership of both German states in the UN, but also admission of East Germany to the World Health Organization or invitation to the European Economic Commission meeting. In this regard, the alliance unanimously admitted a need for coordinated action. The communiqué on behalf of the Warsaw Treaty Organization urged international equalization of the GDR, after all. *Komuniké z porady ministrů zahraničních věcí členských států Varšavské smlouvy*, 19. 2. 1971, 1261/0/5, sv. 154, a. j. 238/7, NA, Prague.

[123] *Protokolární zápis z porady ministrů zahraničních věcí členských států Varšavské smlouvy*, 19. 2. 1971; *Komuniké z porady ministrů zahraničních věcí členských států Varšavské smlouvy*, 19. 2. 1971, 1261/0/5, sv. 154, a. j. 238/7, NA, Prague; *Informace o vývoji situace v otázkách evropské bezpečnosti a spolupráce*, 6. 10. 1971, PK 1953-1989, kat. č. 621, k. č. 144, AMZV, Prague.

[124] *Návštěva ministra zahraničních věcí NDR O. Winzera v ČSSR*, 20. 1. 1971, TO(t) 1970-1974, box 4, i. č. 68, sign. 0344/112, č. j. 026.779/70-4, AMZV, Prague.

[125] The briefings of the ambassadors and other embassies' staff of individual satellites in the Soviet capital remained an important instrument. However, some shift in conception of foreign policy cooperation between the USSR and its satellites was indicated, for example, by the fact that the Czechoslovak embassy in Moscow at the time began to receive instructions by Soviet diplomats which were later handed over to other Czechoslovak embassies, mostly in the developing

In political dimensions, the USSR most likely considered the Warsaw Treaty Organization to be primarily a tool for internationally promoting the CSCE.[126] Therefore, the next alliance meeting was held after more than ten months, in the moment when the Kremlin found that new challenges in the Helsinki process occurred and the Eastern Bloc had to face them through close cooperation. In November 1971, in the wake of favorable international development, Moscow attempted to accelerate the preparation of the security conference.[127] For this purpose, the meeting of foreign affairs ministers took place. The Warsaw Pact member-states should have pushed for holding the CSCE in 1972 as well as collectively preventing Western initiatives which were undesirable from their point of view. The key problem was the NATO countries' condition: to put on the conference agenda talks on the mutual reduction of the armed forces.[128] On Moscow's directive, the Warsaw Treaty Organization instrumentally pointed to the complexity of that burning issue and the need for cohesive consultations on that measure within the alliance framework in advance. Also, the proposal on the free exchange of individuals, ideas, and information to be guaranteed by the conference naturally raised fears of the Leninist-Stalinist regimes ruling in the Eastern Bloc countries. However, all those stances were formulated again in advance, not at the Warsaw Pact meeting itself. Unlike the past, the session was held prior to the North Atlantic Council summit - this time in order to affect its

---

countries. Unlike the previous era, those embassies were directly authorized to make particular foreign policy moves. Thus, the Soviet satellites increasingly played the roles of liaison and tool of the Kremlin's diplomacy. *Rozbor zpravodajské činnosti ZÚ Moskva k otázkám zahraniční politiky a návrhy na zdokonalení činnosti v této oblasti*, 8. 2. 1972, PK 1953-1989, kat. č. 632, k. č. 145, č. j. 026.550/71-1, AMZV, Prague; *Zpráva o plnění hlavních úkolů vyplývajících ze spojenecké smlouvy se SSSR*, 17. 6. 1971, TO(t) 1970-1974, box 2, i. č. 89, sign. 020/112, č. j. 021.226/71-1, AMZV, Prague.

[126] The Eastern Bloc countries' approach towards the CSCE issue was definitely not coordinated at the Warsaw Pact meetings only, but also during many bilateral talks. Jarząbek, PRL w politycznych strukturach, 75.

[127] The USSR appreciated especially the agreement on status of West Berlin signed by four powers of the former anti-Nazi coalition and Brezhnev's successful talks with Brandt and Georges Pompidou, the president of France.

[128] NATO was unanimous on that issue, except for France.

outcomes. In fact, top political gatherings in the scope of the Warsaw Pact clearly returned to the non-operative scheme known from Khruschev's era. Instead of really dealing with actual problems, the meetings were limited to presentations of previously arranged initiatives and statements. Moscow did not oppose multilateral cooperation within the Eastern Bloc completely. However, it was transferred to lower levels. For example, the alliance members were supposed to intensify their mutual interaction in the scope of newly established expert groups on particular international issues.[129]

In January 1972, the Warsaw Treaty Organization's political structures' efficiency was stagnating; this was demonstrated by the following Political Consultative Committee session in Prague. We have a stenographic transcript of the meeting, which was not very typical for that time. Therefore, we can well analyze the real character of the gathering. The Agenda was again defined in advance, at the party leaders' talks in Warsaw in December of the previous year. In fact, that meant it was merely a formality for the authorization of the USSR to prepare all the documentation that was afterwards to be sent to the alliance member-states for formal approval.[130] At least in the case of the GDR, the whole politburo dealt with the materials prepared by Moscow and then instructed East-German delegates which changes in the text they should have striven for.[131] However, those were usually limited to some small stylistic

---

[129] *Informace o nadcházející poradě ministrů zahraničních věcí členských států Varšavské smlouvy ve Varšavě ve dnech 30.11.-1.12.1971*, 25. 11. 1971, 1261/0/6, sv. 20, a. j. 20/5, NA, Prague; *Zpráva o průběhu porady ministrů zahraničních věcí členských států Varšavské smlouvy ve Varšavě ve dnech 30. listopadu – 1. prosince 1971*, 13. 12. 1971, sv. 25, a. j. 26/info1, ibid. However, those expert groups were not largely activated until the rule of Brezhnev's successor, Yuri Andropov.

[130] *Meeting of the Political Consultative Committee of the Warsaw Treaty Member-States, Prague, 25-26 January 1972*, Accessed October 16, 2019. http://www.php.isn.ethz.ch/kms2.isn.ethz.ch/serviceengine/Files/PHP/18108/ipu blicationdocument_singledocument/d411ef76-8567-41fb-ab97-cce6a5d62d87/en/Stenographic_Rec_1972_Eng.pdf.

[131] *Anlage Nr. 1 zum Protokoll Nr. 2/72 vom 18. 1. 1972*, DY 30/ J IV 2/2/1375, BArch, Berlin; *Příprava zasedání politického poradního výboru států Varšavské smlouvy*, 21. 1. 1972, 1261/0/6, sv. 27, a. j. 29/1, NA, Prague.

modifications and either the sharpening or relaxing of diction related to the actual priorities of individual Warsaw Pact members. Therefore, there were often petty disputes among the delegations.[132] Such praxis was typical in the following era.[133] At the meeting itself, present foreign affairs ministers were charged with the last modification of the final declarations. The alliance secretariat lacked any significant activity. In fact, every delegation set its representative within that work group. The Warsaw Pact general secretary Nikolai Firyubin then formally supervised their work. The whole meeting apparently showed some features of the real discussion. There was a voting on every drafted proposal, however rudimentary. Approval of all the member-states was necessary. Similar to the meetings of foreign affairs ministers, rotation of chairmanship took place at the Political Consultative Committee as well.[134]

The Prague summit proves that Moscow at the time did not intend to use the Warsaw Treaty Organization as a platform for solving the disputes among the countries of the Soviet sphere of influence. Significant division remained, especially on the German question. Ratification of the treaties signed in 1970 by the Soviet Union, Poland, and West Germany was postponed distinctly in Bonn. Czechoslovakia, alongside with East Germany, tried to take advantage of this situation. Both countries considered urging Poland to wait for reconciliation with West Germany for the fulfillment of their own requirements: abrogation of the Munich agreement and admission of the GDR to the United Nations as a full member.[135] The Polish foreign affairs ministry feared that Moscow would succumb to the duress of Prague. In this scenario, the Warsaw Pact meeting would be used to adopt a directive ordering the government of

---

[132] Nowak, Jerzy, *Od hegemonii do agonii: upadek Układu Warszawskiego – polska perspektywa* (Warszawa: Bellona, 2011), 62.

[133] *Komuniqué der Tagung des Politischen Beratenden Ausschusses der Teilnehmerstaaten des Warchauer Vertrages*, 15. 4. 1974, DY 30/ J IV 2/2/1499, BArch, Berlin.

[134] *Meeting of the Political Consultative Committee of the Warsaw Treaty Member-States, Prague, 25-26 January 1972*, Accessed October 16, 2019. http://www.php.isn.ethz.ch/kms2.isn.ethz.ch/serviceengine/Files/PHP/18108/ipu blicationdocument_singledocument / d411ef76-8567-41fb-ab97-cce6a5d62d87 / en/Stenographic_Rec_1972_Eng.pdf.

[135] Békés, "Studená válka, détente a sovětský blok," 80.

Poland to postpone the establishment of diplomatic relations with Bonn.[136] However, this did not happen. Bringing the ongoing ratification process of the treaties with West Germany to an end was top priority for the Kremlin. The USSR could not allow anything to jeopardize it.[137] Disputes among the Warsaw Pact members Moscow also assigned led to potentially damaging factors. Although the German question was the main item of the summit in Prague,[138] Brezhnev characteristically decided to postpone the solution of the discord. As he did not prefer to deal with the situation within the official alliance structures, he rather waited for the informal, less watched party leaders' meeting, which took place in Crimea in July.[139]

The Warsaw Treaty Organization's supreme body session in Prague therefore mostly served as favorable political aegis for the ongoing massive armaments programs of the Unified Armed Forces. Those were approved without any serious debate. That scheme was barely unique. In general, we can conclude that military issues were discussed at the

---

[136] Jarząbek, *PRL w politycznych strukturach*, 69-70.

[137] The Kremlin, therefore, opposed any links between ratification of both documents and the start of the multilateral negotiations on holding the all-European conference, for example. *Zpráva o průběhu porady ministrů zahraničních věcí členských států Varšavské smlouvy ve Varšavě ve dnech 30. listopadu – 1. prosince 1971*, 13. 12. 1971, 1261/0/6, sv. 25, a. j. 26/info1, NA, Prague.

[138] Contradictions appeared mostly in stances of Gierek and Honecker. The PUWP CC First Secretary suggested Bulgaria and Hungary - the countries which had no unresolved disputes with West Germany - establish diplomatic relations as well, right after ratification of Polish and Soviet treaties with Bonn. The Warsaw Treaty Organization should have provided only additional support for Czechoslovakia and East Germany during their negotiations with the FRG. On the contrary, Honecker urged the entire Warsaw Pact to wait for abrogation of the Munich agreement and admission of the GDR to the UN. *Meeting of the Political Consultative Committee of the Warsaw Treaty Member-States, Prague, 25-26 January 1972*, Accessed October 16, 2019. http://www.php.isn.ethz.ch/kms2.isn. ethz.ch/serviceengine/Files/PHP/18108/ipublicationdocument_singledocument/d 411ef76-8567-41fb-ab97-cce6a5d62d87/en/Stenographic_Rec_1972_Eng.pdf.

[139] *Notatka z przebiegu Spotkania I-szych sekretarzy bratnich partii na Krymie /31 lipca 1972/*, PZPR KC, XIA/612, AAN, Warsaw. Brezhnev succeeded in persuading Erich Honecker to stay silent on establishing diplomatic relations between Poland and West Germany. In exchange, he promised to ensure that the rest of the Warsaw Pact would wait for admission of both German states to UN. Thus, Poland secured its previous exception and actually established ties with Bonn in autumn 1972.

Political Consultative Committee meetings in a very common, even vague manner. Mainly, it was alliance military structures that dealt with the particular aspects.[140]

Hitherto, the praxis clearly showed that the Warsaw Treaty Organization's supreme body meetings were not very operative. In 1972, at the summit in Prague, however, calls for holding them more frequently appeared again. In this regard, the new PUWP CC First Secretary Edward Gierek followed up on the long-term efforts of his predecessor. He urged convening two regular sessions of the body per year. In his reaction, Brezhnev blindly swore that all the member-states definitely wished more political talks to be held within the alliance. The Soviet leader encouraged them not to be afraid of initiating such meetings; he promised Moscow would always gladly support them.[141] In this connection, Brezhnev once more presented his visions of the Pact's future - it should have been more about political alliance, rather than military organization.[142] The following developments show that those words lacked content. In fact, the Political Consultative Committee session in Prague ended the period of higher activity of the Warsaw Pact political structures at the turn of the 1960s and '70s. At least some Soviet satellites were undoubtedly still interested in the continuity of such cooperation. However, they lacked any means to achieve that. The vague alliance founding charter missed a more precise definition of conditions for convening the meetings. In March 1972, for instance, Jędrychowski and his East-German counterpart Oskar Fischer agreed that holding the session of foreign affairs ministers would be useful. Shortly after West Germany ratified the treaties with the Soviet Union and Poland, the meeting was supposed to coordinate the next

---

[140] *Překlad usnesení států-účastníků Varšavské smlouvy, přijatého na zasedání Politického poradního výboru 26. ledna 1972; Spolupráce s armádami Varšavské smlouvy – usnesení Politického poradního výboru z 26.1. 1972*, 27. 10. 1973, FMO GŠ/OS, box. 8, p. č. 41, č. j. 39033, VÚA, Prague.

[141] *Meeting of the Political Consultative Committee of the Warsaw Treaty Member-States, Prague, 25-26 January 1972*, Accessed October 16, 2019. http://www.php.isn.ethz.ch/kms2.isn.ethz.ch/serviceengine/Files/PHP/18108/ipu blicationdocument_singledocument / d411ef76-8567-41fb-ab97-cce6a5d62d87 / en/ Stenographic_Rec_1972_Eng.pdf.

[142] Mastny, *A Cardboard Castle?*, 42.

approach of the Pact towards the German question.[143] Nevertheless, Moscow approved the gathering before early 1973, 14 months after the last similar consultation within the alliance framework had taken place. That was, in addition, the final Warsaw Pact gathering of foreign affairs ministers for the next three years.[144]

With the absence of key Soviet documents, the question of what reasons were behind Moscow's decision to attenuate political cooperation within the Warsaw Treaty Organization again can hardly be answered. It cannot be said there was lack of proper agenda to be consulted. On the contrary, both Helsinki's process and détente accelerated at the time. It is important to point out that there was no complete freezing of talks; the Kremlin, however, preferred different forums, like the already mentioned summer meetings in Crimea, instead of the Warsaw Pact structures. The heart of political consultation within the alliance framework was transferred to lower levels, most notably to the deputy foreign ministers' meetings. The Kremlin took an accommodative approach towards those. In certain cases, the USSR actually heeded the appeals of its satellites for holding such consultations. In May 1973, the meeting of deputy foreign ministers took place in Moscow on Polish initiative, for instance. However, it was just a coincidence. The Kremlin only used the proposal made by Poland. In fact, at the time, the USSR itself considered it necessary to instruct the Warsaw Pact members that the alliance had to prevent the disarmament issue from being put on the CSCE agenda at all costs. Moscow deemed it a potential obstacle to the conference. Thus, holding all the Warsaw Pact meetings, which provided some hope to the Soviet satellites that they would be able to assert some of their interests

---

[143] *Zpráva o stavu přípravy a svolání celoevropské konference o bezpečnosti a spolupráci a o úkolech FMZV*, 18. 4. 1972, PK 1953-1989, kat. č. 639, k. č. 146, č. j. 010.441/72, AMZV, Prague.

[144] Thus, Anna Locher is wrong in her claim that the foreign affairs ministers met in the scope of the Warsaw Pact almost annually during 1969-1976. Locher, Anna, "*Shaping the Policies of the Alliance – The Committee of Ministers of Foreign Affairs of the Warsaw Pact, 1976-1990*," Accessed October 16, 2019. http://www.php.isn.ethz.ch/lory1.ethz.ch/collections/coll_cmfa/cmfa_intro7e2f.html?navinfo=15699.

into the unified course of the alliance,[145] depended strictly on the will of Moscow.

Regress of systematic consultations on the higher representative level, under the auspices of the Warsaw Pact, was probably related to an alteration of the Kremlin's general tactics in détente. Both the USA and USSR increasingly tended to negotiate on key issues bilaterally, rather than on the basis of the blocs. This approach was also proven by the Agreement on the Prevention of Nuclear War initiated by Moscow. In June 1972, this document was signed in the United States during Brezhnev's visit by the two superpowers only, and not on the level of NATO or the Warsaw Treaty Organization. One year later, the Soviet General Secretary wanted to instruct the Eastern Bloc party leaders which direction their following foreign policy efforts should take. However, instead of convening the Warsaw Treaty Organization summit for this purpose, he held rather informal talks in Crimea again.[146] At the time, Gromyko assured the alliance members that Moscow still counted on multilateral consultations. Nevertheless, their scheme was not clear.[147]

Before the Helsinki Conference, only one more Political Consultative Committee session was held. The meeting in Warsaw, April 17-18, 1974, finally defined the body's working method, which remained largely constant till the death of Brezhnev. First of all, it is important to say that before 1974, there was no systematic holding of the Warsaw

---

[145] The GDR, for example, urged more frequent talks of deputy foreign ministers. In fact, the Warsaw Treaty Organization meetings were one of the few opportunities for East Germany to actively participate in international politics. *Report on the Meeting of the Deputy Foreign Ministers*, May 21, 1973, Accessed October 16, 2019. http://www.php.isn.ethz.ch/kms2.isn.ethz.ch/serviceengine/Files/PHP/17337/ipublicationdocument_singledocument/4ae29316-b523-45ab-9 2ea-917f86dc216d/en/730521_Report_E.pdf.

[146] In comparison to the official Warsaw Treaty Organization gatherings, actual international and economic issues were discussed in a less formal atmosphere at this meeting. *Stenogram narady aktywu partyjnego poświęconej spotkaniu Krymskiemu przywódców krajów socjalistycznych*, 3. 7. 1973, PZPR KC, XIA/ 404, AAN, Warsaw.

[147] *Informace o návštěvě gen.taj. ÚV KSSS L.Brežněva v USA a Francii, přednesená čl. PB ÚV KSSS a min.ZV SSSR A.A. Gromykem*, 26. 7. 1973, TO(t) 1970-1974, box 1, i. č. 89, sign. 020/111, č. j. 024006, AMZV, Prague.

Pact's supreme body sessions. In the first half of the 1970s, Brezhnev simply decided that such gatherings would be useful. Regarding this, he reflected the appeals of other party leaders very sporadically.[148] However, the meeting in Warsaw was the first to be held almost regularly at once every two years. It no longer provided grounds for key international issues to be openly discussed. In fact, Brezhnev himself determined the agenda as well as the spirit of the final documents.[149] The situation reflected his dominant position in shaping Soviet foreign policy at the time.[150] The invitations, consisting also of the items of the meeting, were sent to the Warsaw Pact members by the hosting country. However, that was nothing more than a formality.[151]

The agenda of the Political Consultative Committee session in Warsaw consisted of following steps in the European security issues and détente, in preparation for celebration of the 20th anniversary of the Pact's founding and a report by the Supreme Commander, Marshal Yakubovski, on the work of Unified Command.[152] The last item again was discussed

---

[148] During his visit in Moscow in late 1973, Edward Gierek, who intended to assert Polish priorities within the Warsaw Pact during frequent collective talks, proposed also the Political Consultative Committee session to be held. He reiterated that Janos Kadar had expressed the same interest as well. The Soviet leader briefly said that a consultation on the actual international situation was really necessary. However, he promised no concrete meeting. *Informacja o przebiegu i rezultatach rozmów tow. Gierka z tow. Leonidem Breżniewm v Moskwie*, 29. 12. 1973, PZPR, XIV/591, AAN, Warsaw; *Tezy do rozmów z towarzyszami radzieckimi*, undated 1973, XIB/126, ibid.

[149] *Postanowienie Biura Politycznego KC PZPR*, 9. 3. 1974, XIA/587, ibid.

[150] At the time, Brezhnev already shaped the main course of Soviet policies. However, he neither held absolute power nor strove for a dictatorial or despotic position within the politburo. In praxis, the top CPSU leadership remained organized and ran on a collective basis. Unlike Khrushchev, Brezhnev did not make himself an expert in many fields. He listened to the advice of his colleagues and allowed them to oppose him and disagree with his opinions. Bacon, Erwin, "Reconsidering Brezhnev," in *Brezhnev Reconsidered*, eds. Erwin Bacon and Mark Sandle (New York: Palgrave, 2002): 14-15.

[151] Latter by the PUWP CC First Secretary, E. Gierek, to the SED CC First Secretary, E. Honecker, March 23, 1974, DY 30/2351, BArch, Berlin.

[152] *Zpráva o průběhu a výsledcích zasedání Politického poradního výboru států Varšavské smlouvy*, 24. 4. 1974, 1261/0/6, sv. 115, a. j. 117/1, NA, Prague.

just superficially, as was typical.[153] The trend established two years before, when all the delegations factually agreed with Brezhnev's evaluation of the international situation and proposals on following moves in the Helsinki process and Vienna disarmament talks, continued. The speeches of the other leaders gave neither key information nor new suggestions.[154] Therefore, the only "vitalization" of the previously written meeting's scenario was an unexpected verbal assault from Erich Honecker and Edward Gierek on Romanian policy, which resulted in an argument with Nicolae Ceausescu. As we explain in more detail in the chapter dealing with the Romanian issue, this event had no serious impact.[155]

In the limelight was Brezhnev's speech, of course. In fact, his thesis became a directive for upcoming international activities of the Soviet satellites.[156] The strengthening of the Warsaw Treaty

---

[153] In fact, Yakubovski referred in general terms to the realization of military directives approved by the Political Consultative Committee in 1969 and 1972. He stated that the bilateral protocols on the assignment of member states' troops to the Unified Armed Forces' formation were fulfilled successfully. *Informace členů Kolegia ministra národní obrany o zasedání Politického poradního výboru členských států Varšavské smlouvy*, undated 1974, FMO GŠ/OS, k. 8, p. č. 41, č. j. 39033, VÚA, Prague.

[154] Husák appreciated the diplomatic activities of the Warsaw Pact allies during the convening treaty between Czechoslovakia and West Germany. He was also thankful for the role the alliance played in defending socialism in his country. The CPCz CC General Secretary supported all the plans furthering the Pact's development, even arguing with a cliché that the Warsaw Pact meant a guarantee of peace in Europe. The content of Kadar's speech was similarly poor. He ascribed a decisive role in the reversion of the Cold War to the Warsaw Treaty Organization and the initiate Soviet peace policy. Bulgarian dictator Zhivkov talked in the same way. *Vystoupení s. G. Husáka na poradě politického poradního výboru Varšavské smlouvy 17. dubna 1974; Stručná charakteristika vystoupení vedoucích jednotlivých delegací na zasedání politického poradního výboru*, 24. 4. 1974, 1261/0/6, sv. 115, a. j. 117/1, NA, Prague.

[155] *Zpráva o průběhu a výsledcích zasedání Politického poradního výboru států Varšavské smlouvy*, 24. 4. 1974, ibid.

[156] Brezhnev spoke about the Soviet views on the actual international situation at length. In the center of his speech was naturally the Helsinki process. In this regard, he declared that the priority was a confirmation of the frontiers in Europe. He did not intend to allow the Final Act to set the basis for Western propagandistic activities aiming to undermine the socialist ideology in the Eastern Bloc countries. In the terms of strengthening trust between the East and West, the Soviet leader

Organization's role in shaping and coordination of unified foreign policy tack finally gained support of all the members, including Romania. This especially concerned a formalization of consultative mechanisms. Janos Kadar made those visions more specific. According to the Hungarian stance, the Political Consultative Committee should have been held regularly once a year, supplemented by systematic talks of foreign affairs ministers and their deputies. However, Brezhnev definitely did not intend to limit cooperation between the socialist states to the Warsaw Pact members only. Therefore, he emphasized deepening the existing interaction with Cuba, North Vietnam, North Korea, Mongolia, and potentially even Yugoslavia.[157]

In regard to the upcoming 20th anniversary of the Warsaw Pact's existence, the Soviet leader did not miss the opportunity to summarize. He very much appreciated the changes the alliance had gone through since he had become the head of the CPSU. Brezhnev stressed significance of its resolutions for the alteration of the international situation. That was more than just a propagandistic statement; the alliance's political structures really became an important tool of the Kremlin for asserting Soviet foreign policy initiatives on the international stage. The Soviet General Secretary marked prevention of establishing the unified nuclear forces of NATO, changes in West-German policy, and strengthening the GDR position as an independent and sovereign country as the main successes of the Warsaw Treaty Organization. Post-war settlement in Europe was in fact completed, and all the issues initially related to the peace treaty with Germany were actually solved even without its actual signing, Brezhnev stated. He therefore appreciated that almost everything that the Political

---

also admitted the possibility of mutual notification of big military maneuvers and exchange of the observers there. However, he refused such an extension which would provide NATO detailed information on dislocation and combat preparation of the Warsaw Pact troops. This position was understood as coming out of the then-strategic conception that a nuclear attack by the enemy was expected to target huge groups of the Eastern military units.

[157] *Zpráva o průběhu a výsledcích zasedání Politického poradního výboru států Varšavské smlouvy*, 24. 4. 1974; *Stručná charakteristika vystoupení vedoucích jednotlivých delegací na zasedání politického poradního výboru*, 24. 4. 1974; *Vystoupení s. L.I. Brežněva na poradě politického poradního výboru Varšavské smlouvy 17. dubna 1974*, 1261/0/6, sv. 115, a. j. 117/1, NA, Prague.

Consultative Committee declarations had strove for in previous years was accomplished.[158] On one hand, we must acknowledge that the Warsaw Treaty Organization was really involved in those issues. On the other hand, it is important to emphasize that the role of the alliance was limited to covering Soviet policy. The fact that in various cases the unified course of the Warsaw Pact at least partially reflected priorities of the other member-states, could not change the central point. Those were mostly the Kremlin's efforts to keep stable the Soviet sphere of influence.

The relatively moderate tone of the Brezhnev speech that focused mostly on successes in détente was reflected very little by the quite sharp final communiqué of the summit in Warsaw. Using phrases and ideological language, the document warned against opponents of détente, "the imperialists" and "the militarists striving for activation of NATO" as well as talked about "the Cold War supporters" in the West. Paradoxically, the diction of many passages of the text brought back the Cold War rhetoric of the Warsaw Pact known from Khrushchev's era. However, it also contained an appeal to Western politicians for joining efforts to hold the upcoming all-European security conference successfully. In this connection, it did not neglect to stress the fact that since its establishment, the Warsaw Treaty Organization alone strove to hold such a summit. The document also vaguely announced the upcoming strengthening of the alliance's political framework.[159] As before, despite its strictly European range, the Warsaw Pact's supreme body also replied to events in other world regions. The Political Consultative Committee published separate resolutions on situations in Vietnam, the Middle East, and Chile.[160] In addition, the Soviet leadership intended to assault China harshly in final documents. Beijing should have been accused of a secret deal with NATO. However, such an anti-Chinese declaration failed because of a Romanian veto.[161]

---

[158] *Vystoupení s. L.I. Brežněva na poradě politického poradního výboru Varšavské smlouvy 17. dubna 1974,* ibid.

[159] *Komuniké ze zasedání politického poradního výboru států Varšavské smlouvy,* undated, ibid.

[160] *Anlage Nr. 2 zum Protokoll Nr. 14/74 vom 15. 4. 1974,* DY 30/ J IV 2/2/1499, BArch, Berlin.

[161] Mastny, *A Cardboard Castle?,* 43.

On the contrary, Ceausescu was not successful in asserting an appeal for the simultaneous dissolution of NATO *and* the Warsaw Treaty Organization to be repeated in the document. Brezhnev refused such a perspective; during the toast, he rather stressed that considering the absence of a disarmament agreement, the future improvement of the defense capabilities of the Warsaw Pact was necessary. Neither RCP Executive Committee First Secretary actually opposed that explanation.[162]

Establishing of Committee of Ministers of Foreign Affairs and Joint Secretariat

The real utility of the Political Consultative Committee sessions faded. However, their importance was increasingly exaggerated in meretricious ways; adoration of their significance became a norm for the Eastern Bloc states' top officials. In the period between the Warsaw Pact's supreme body meetings, those were stressed as a proof of the socialist countries' unity. The USSR assured its satellites that in the talks with the United States, it is shown in the documents collectively approved by the Political Consultative Committee.[163] Nevertheless, that was usually just a phrase lacking any content.

On the frontlines of the efforts to raise the cooperation within the Warsaw Pact to a higher level, Poland stood above all under the leadership of the new PUWP CC First Secretary Edward Gierek. He came to power after widespread social unrests which had erupted on the Baltic coast in late 1970. The situation resulted in significant personal changes in party leadership, including the forced resignation of Wladyslaw Gomulka. It

---

[162] In Romania, the Executive Committee was an analogous body to the politburo or central committee presidium in other Eastern Bloc countries. *Zpráva o průběhu a výsledcích zasedání Politického poradního výboru států Varšavské smlouvy*, 24. 4. 1974, 1261/0/6, sv. 115, a. j. 117/1, NA, Prague.

[163] Speech by PUWP CC First Secretary E. Gierek at Polish embassy in Moscow in the presence of CPSU CC General Secretary L. Brezhnev and the prime minister of USSR A. Kosygin, December 17, 1974, PZPR KC, XIV/592, AAN, Warsaw; *ИНФОРМАЦИЯ о переговорах Й.И. Брежнева, Н.В. Подгорного, А.Н. Косыгина и А.А. Громыко с президентом США Р. Никсоном и государственным секретарем США Г. Киссинджером, находившимися в Советском Союзе с 27 июня по июлю 1974 г.*, XIA/600, ibid.

must be added that Moscow naturally carefully monitored the events in Poland. However, the riots aimed at internal and mostly economic issues. There was neither a significant anti-Soviet campaign nor the challenging of the country's membership in the Warsaw Pact.[164] After all, Brezhnev tried at all costs to pre-empt such escalation that would require sending the armed forces of the USSR, or even the other Warsaw Pact member-states. Remember, no more than two years had elapsed since the intervention in Czechoslovakia. Repeating military action in the Eastern Bloc could have potentially negative international impacts on the Soviet Union as well as the entire communist movement.[165]

---

[164] Czechoslovak diplomats considered an absence of the anti-Soviet acts as a key difference from the student unrests which had erupted in Poland in March 1968. *Pričiny vzniku politickej krízy v PĽR a hodnotenie súčasnej politickej situácie*, 22. 2. 1971, TO(t) 1970-1974, box 9, i. č. 79, sign. 016/311, , č. j. 021181, AMZV, Prague. The fact that protests aimed neither at socialism in general nor orientation of Polish foreign policy was stressed even by the PUWP politburo member Stanisław Kociołek at the party plenum. Stenogram VII plenum KC PZPR z 20 grudnia 1970 r. (fragmenty), in *Tajne dokumenty Biura Politycznego. Grudzień 1970*, (doc. 12), ed. Paweł Domański, (London: Aneks, 1991): 107.

[165] The Soviet directives clearly urged for calm in the situation and to reestablish order by Polish means. It is important to add that the Soviet leadership fully respected Gomulka's order to use force. Nevertheless, the CPSU politburo called for political measures and stabilization without extreme bloodshed. The Polish politburo discussed the possibility of foreign military intervention on December 19. The PUWP CC secretary Zenon Klizsko said that the country faced a situation in which asking the Soviet army for help was necessary. However, the Polish leadership did not consider repeating the Czechoslovak scenario at all. Referring to his conversation with Brezhnev, the prime minister Józef Czyrankiewicz rejected such an option. He emphasized that no Soviet soldier would cross Polish borders without agreement between the Polish and Soviet leadership. No solution on the basis of the Warsaw Treaty Organization was discussed. The defense minister Wojciech Jaruzelski conveyed the content of his recent conversation with his Soviet counterpart. Grechko sent a message that the Kremlin definitely did not insist on deployment of the army but on the contrary preferred political measures. The Soviet defense minister actually did not appreciate the fact that the second largest army of the Warsaw Treaty Organization should have to shoot into workers. In this regard, he expressed concerns about the possible weakening of the alliance, most likely meaning its propagandistic image. Jaruzelski tried to persuade the hesitating Polish politburo members for a nonviolent approach - which in fact meant significant changes in the party leadership - claiming that the condition of the armed forces did not allow maintaining order in the capital in

Nevertheless, shortly after Gierek came to power, he tried to immediately convince the Kremlin about absolute loyalty of new PUWP leadership for any case. For this reason, he naturally stressed the need for Polish orientation to the Soviet Union.[166] Therefore, in January 1971, the new leader of the Polish communists assured Leonid Brezhnev that he had no ambitions for independent foreign policy.[167] He swore Poland would shape its course primarily according to Soviet priorities. Polish leadership intended to consult its diplomatic effort both bilaterally with the Kremlin as well as with the other Eastern Bloc countries within the Warsaw Treaty Organization structures.[168] The Soviet General Secretary in turn

---

case the situation went worse. Domański, *Tajne dokumenty*, iv-v; Protokół nr 19 z posiedzenia Biura Politycznego KC PZPR w dniu 19 grudnia 1970 r. - godz. 14.20, in *Tajne dokumenty Biura Politycznego* (doc. 11): 86; Kosman, Marceli, *Los Generała. Wokoł medialnego wizerunku Wojciecha Jaruzelskiego* (Torun: Adam Marszałek, 2008), 95.

[166] *Sprava o poľsko-sovietskych vzťahoch za rok 1971*, 7. 2. 1972, TO(t) 1970-1974, box 9, i. č. 79, sign. 016/311, č. j. 020.947/72-2, AMZV, Prague.

[167] Friszke, Andrzej, *Polska Gierka* (Warszawa: WSiP, 1995), 25.

[168] During his first visits in the Soviet satellites, Gierek presented practically the same phrases. He emphasized Polish membership and willingness to cooperate in the Warsaw Treaty Organization; he assured them that the situation in Poland significantly differed from the events of the Prague Spring. The PUWP CC First Secretary stressed that stances of the party leadership were totally incomparable to recent developments in Czechoslovakia and the Polish army played well during the crisis. Moreover, the altered Polish leadership still informed the ambassadors and military attachés of the Warsaw Pact member-states about the situation in the country. They also tried to send the message that their attitude towards capitalist countries would not differ from Polish allies; Warsaw declared its intention to abide by a joint course of the alliance and that Poland was still interested in frequent meetings of the Pact. *Protokoł ze spotkania I sekretarza KC PZPR tow.Edwarda Gierka i prezesa Rady Ministrów PRL tow. Piotra Jaroszewicza z I sekretarzem KC SED tow. Waltrem Ulbrichtem i przewodniczącym Rady Ministrów NRD tow. Willi Stophem*, 11. 1. 1971, PZPR KC, XIA/48, AAN, Warsaw; *Protokoł ze spotkania I sekretarza KC PZPR tow.Edwarda Gierka i prezesa Rady Ministrów PRL tow. Piotra Jaroszewicza z I sekretarzem KC KPCz tow.Gustavem Husakiem i przewodniczącym rządu CSRS Lubomirem Sztrougalem*, 16. 1. 1971, XIA/38, ibid; Report by the deputy chief of Main Political Administarion of PPA, O. Korzenicki, to the military attachés of the Warsaw Pact member-states on deployment of the Polish army during the demonstrations in December 1970, March 8, 1971, DY 30/J IV 2/2J/3370, BArch,

appreciated that Polish diplomacy comes out of the unified course of the alliance. For the recent international moves which bothered Moscow he blamed Gomulka and his choleric trait.[169] Replacement of the PUWP CC First Secretary thus did not bring any significant shift in Polish foreign policy priorities. In the early 1970s, there was also a clear continual line from the previous decade in the approach of Poland towards the Warsaw Treaty Organization. Polish officials still showed their interest in reforming the alliance's political structures.[170] Gierek's leadership did not leave the efforts to protect national priorities completely; however, it modified the tactics used.[171]

In the Spring of 1971, the 24th CPSU congress explicitly marked the Warsaw Treaty Organization as the main coordinative foreign policy body of the socialist commonwealth. In this regard, Moscow appreciated the reform of the alliance's military framework and stressed its further strengthening. However, the Soviet communists declared that not only a military, but a mostly political dimension of the Pact, had to be seen.[172] In June 1971, taking into account that statement, the Polish foreign affairs minister Jędrychowski, during his visit to Hungary, talked again about the

---

Berlin; Report by the PUWP leadership on the actual political and economic situation in Poland, March 8, 1971, DY 30/J IV 2/2J/3369, BArch, Berlin.

[169] In fact, this universal phrase meant that despite Gomulka's more independent foreign policy moves Moscow was not seriously bothered by a Polish course of action. Most likely, Brezhnev especially defaulted to the Soviet view over much of the Polish involvement in the German question and preparation of the all-European security conference. Protokół z rozmowy Edwarda Gierka z Leonidem Breźniewem w Moskwie, dnia 5 stycznia 1971 r., in *Tajne dokumenty Biura Politycznego* (doc. 13): 126.

[170] Jarząbek, "*Poland in the Warsaw Pact*".

[171] The most immediate priority of Gierek's leadership's policy remained successfully finishing the reconciliation with West Germany. That required a smooth ratification of the mutual treaty. Report by the PUWP leadership on the actual political and economic situation in Poland, March 8, 1971, DY 30/J IV 2/2J/3369, BArch, Berlin; *Správa o vývoji vnútropolitickej situácii v PĽR v roku 1969-1970, zahraničnej politiky PĽR a rozvoja čsl.-polskej spolupráce vo všetkých oblastech*, 28. 10. 1971, TO(t) 1970-1974, box 9, i. č. 79, sign. 016/311, č. j. 025998, AMZV, Prague.

[172] *Politická zpráva ZÚ Moskva č. 10 o výsledcích XXIV. sjezdu KSSS*, 22. 7. 1971, TO(t) 1970-1974, box 8, sign. 020/311, , č. j. 023.940/71-1, AMZV, Prague.

need for additional changes in the alliance.[173] Thereby he stated the eminent interest of the new Polish leadership to engage in the issue. In this manner, Poland probably tried to face a further deepening of Moscow's dominant position within the Warsaw Pact. After all, at the time, the Polish representatives cautiously pointed to such a possibility during the informal diplomatic talks with their Eastern Bloc states' counterparts.[174] The top Polish leadership was fully aware how the real mechanisms of the Warsaw Pact worked. Therefore, it is not likely that Poland saw some reform of the alliance's political structures as an opportunity to breach Soviet hegemony significantly. It strove for partial adjustments aiming to slight the strengthening of Polish sovereignty only. From the Polish point of view, the establishment of regular meetings could at least subtly improve conditions for more systematic, better-organized efforts to assert the country's priorities into an officially approved unified course of the Pact. In addition, this strategy could theoretically bring some, however feeble, changes to modify some of Moscow's intentions which were not in favor of Poland. In the Warsaw Treaty Organization, all the resolutions and documents had to be approved strictly unanimously. If there was a lack of consent of a single delegation, the declaration could not have been officially presented on behalf of the alliance.[175] That provided certain, but strictly limited, opportunities for the member-states to maneuver and assert their own interests. Unlike the bilateral talks, the resolutions of either the Political Consultative Committee or the other alliance political bodies were largely obligatory.[176] If the Soviet satellites were able to push their priorities there, it gave them permission to strive for those, as well as ensuring at least basic support of the other countries of the Bloc.

As we explain in the following chapter, the above-mentioned Polish strategy had to inevitably clash with Romanian attitudes. Both

---

[173] *Výsledky návštěvy polské delegace na nejvyšší úrovni a ministra zahraničních věcí PLR a MLR*, 28. 6. 1971, TO(t) 1970-1974, box 5, i. č. 58, sign. 015/311, č. j. 023483, AMZV, Prague.

[174] *Informácia o výhľadovom zameraní poľskej zahraničnej politiky*, 11. 10. 1971, TO(t) 1970-1974, box 1, i. č. 79, sign. 016/111, č.j 025.103/71-2, AMZV, Prague.

[175] Ślusarczyk, Jaczek, *Układ Warszawski: Działalność polityczna 1955-1991* (Warszawa: Instytut Studiów Politycznych Polskiej Akademii Nauk, 1992), 48.

[176] Jarząbek, *PRL w politycznych strukturach*, 89.

Gierek's and Ceausescu's leaderships actually aimed for essentially the same goal: to improve their position in the international arena. However, they differed in the extent of their claims as well as preferred tactics. Poland attempted to widen opportunities only to uphold their positions into the joint course of the Warsaw Pact and thus supported reform of the alliance's political structures. The regime of Nicolae Ceausescu was much more ambitious. Its long-term priority was to minimize the influence of Moscow on Romanian policy. Poland never went so far. The idea of a unified foreign policy of the Bloc coordinated through the Warsaw Treaty Organization was not attractive to Bucharest, as it provided definitively less opportunities than Romania demanded. Therefore, Romania strove for disintegration of cooperation within the alliance while it considered it a burden.

In the summer of 1973, at the Crimean meeting of Eastern Bloc states' party leaders, Brezhnev outlined CPSU's opinions on the future role of the Warsaw Treaty Organization; besides further improvements in military cooperation, the time had allegedly come to deepen the coordination of the foreign policies of the Pact members. Those words were apparently interpreted by Gierek as Soviet consent to long-postponed reforms. He immediately supported the idea, however, he had not presented any specific proposals yet.[177] Gierek's optimism was probably also fueled by Brezhnev's consideration of the potential creation of a certain - for the Eastern Bloc -favorable sort of collective security system in Europe.[178] Remember, that it was a cornerstone of the initial Polish draft of the CSCE agenda that was blocked by the Soviet Union with the assistance of other alliance members in 1969. The Crimean meeting thus encouraged Polish leadership to action. Late 1973, the PUWP CC First Secretary unequivocally supported prolongation of the Warsaw Treaty

---

[177] *Wystąpienie tow. Edwarda Gierka na spotkaniu I Sektretarzy Komitétow Centralnych partii komunistycznych i robotnicznych krajów socjalistycznych na Krymie /1973 r./*, undated, prlobably July 1973, PZPR KC, XIA/613, AAN, Warsaw.

[178] *Stenogram narady aktywu partyjnego poświęconej spotkaniu Krymskiemu przywódców krajów socjalistycznych*, 3. 7. 1973, PZPR KC, XIA/404, AAN, Warsaw.

Organization, as its validity was to end in 1975. In this regard, he openly urged the realization of reforming the alliance's political structures on its 20th anniversary.[179] However, Brezhnev did not react to the proposal. He most likely still did not attach much importance to the matter.[180] Unlike Poland, Moscow considered the appeals for improving political cooperation within the Pact as empty phrases.[181]

The Polish leader tirelessly stressed deepening the political cooperation within the Warsaw Pact framework, as already mentioned in the Political Consultative Committee session in Warsaw in April 1974.[182] He finally achieved some significant break-through. The Romanian delegation preliminarily accepted the establishing of the Committee of Foreign Affairs Ministers, which was supposed to improve existing alliance consultations. After almost ten years, Romania stopped blocking reform of the Warsaw Pact's political structures.[183] Nevertheless, Moscow stayed lax. Since June, the USSR tried to prevent the issue being put on the agenda of the official talks of the Eastern Bloc countries.[184] The Soviet approach was well-illustrated by the conversation of Polish diplomat

---

[179] *Tezy do rozmów w Moskwie. II. Sprawy międzynarodowe*, undated, probably December 1973, PZPR KC, XIV/591, AAN, Warsaw; *Tezy do rozmów z towarzyszami radzieckimi*, undated 1973, XIB/126, ibid.

[180] In this regard, it is good to add that according to testimony of Jerzy Nowak, former Polish diplomat, the Crimean meetings were largely affected by huge alcohol consumption. Therefore, some ideas presented there cannot be taken so seriously. Nowak, *Od hegemonii do agonii*, 50.

[181] Phrasal language was typical even for descriptions of many other aspects concerning the Warsaw Treaty Organization, including activities of the Political Consultative Committee. Its sessions were emphasized in the meantime as a proof of the socialist countries' unity. However, such claims usually lacked any concrete content. Speech by the PUWP CC First Secretary E. Gierek at Polish embassy in Moscow in the presence of the CPSU CC General Secretary, L. Brezhnev, and the prime minister of the USSR, A. Kosygin, December 17, 1974 PZPR KC, XIV/592, AAN, Warsaw.

[182] *Projekt wystąpienia I Sekretarza KS PZPR tow. E. Giereka w drugim punkcie DKP*, undated 1974, XIA/586, ibid.

[183] *Zpráva o průběhu a výsledcích zasedání Politického poradního výboru států Varšavské smlouvy*, 24. 4. 1974, 1261/0/6, sv. 115, a. j. 117/1, NA, Prague.

[184] *Informace o konzultacích mezi FMZV a MID SSSR v roce 1974*, 10. 12. 1974, TO(t) 1970-1974, box 3, i. č. 89, sign. 020/112, č. j. 018.337/74-1, AMZV, Prague.

Bogumił Rychłowski and Soviet deputy foreign minister Nikolay Rodionov in November 1974. The Soviet representative traditionally declared full support for reform of the alliance's political structures. However, he now claimed that due to the existing international situation, it was impossible to open such a delicate question.[185] Regardless, Polish diplomats, in their talks with their Soviet counterparts, relentlessly suggested announcing the Warsaw Treaty Organization reform as part of its 20th anniversary celebrations.[186] In connection with the forthcoming Helsinki conference, Moscow intended to present the Warsaw Pact as a peaceful and mainly political organization. Thus, the Kremlin finally approved the changes.[187]

Reform of the alliance's political structures was discussed on January 29-30, 1975 at the meeting of deputy foreign ministers in the Soviet capital. The General Secretary of the Warsaw Treaty Organization, Nikolay Firyubin, took part in the talks on that level for the first time. The initial proposal for establishing a foreign policy commission,[188] besides the Political Consultative Committee, was refused by Romania. The Committee of Foreign Affairs Ministers was created instead.[189] The

---

[185] Pilna notatka z rozmowy z wiceministrem Spraw Zagranicznych ZSRR, tow. N.N. Rodionowem w sprawie przygotowań do jubileuszowej sesji DKP UW (Moskwa, dnia 28.XI.1974), 30. 11. 1974, in *PRL w politycznych strukturach* (doc. 36): 325.

[186] Jarząbek, *"Poland in the Warsaw Pact"*.

[187] *Protocol of the Meeting of the Deputy Foreign Ministers*, January 29, 1975, Accessed October 16, 2019. http://www.php.isn.ethz.ch/kms2.isn.ethz.ch/ serviceengine/Files/PHP/17368/ipublicationdocument_singledocument/6f4029fa -2bbf-4ab9-9a47-dfff127167b0/en/750129_Protocol_E.pdf.

[188] The measure reflected the resolution issued by the first Political Consultative Committee meeting in 1956, which supposed the foreign policy commission and the permanent secretariat to be established. According to the new proposal, the commission should have met *ad hoc*, but at least once a year. Its composition should have been variable - consisting of foreign affairs ministers, their deputies, or other authorized representatives. It was considered as a strictly advisory body alongside the Political Consultative Committee. The commission was supposed to mostly prepare an agenda for the Warsaw Pact's supreme body meetings and coordinate activities of individual member-states in the scope of a unified foreign policy determined by the Political Consultative Committee.

[189] A motivation beyond this Romanian move remains unclear. Bucharest was possibly leery that the creation of a foreign policy commission alongside the

existing, but not very operational, secretariat of the Warsaw Pact was supposed to undergo the essential transformation.[190] However, the effort to strictly define the competences of both bodies failed because of Romanian objections. In order to prevent the next deadlock, the rest of the member states agreed that specific guidelines for activities of new bodies could be adopted later, reflecting the praxis.[191] In fact, the situation facilitated Moscow to ensure that the establishing of institutions would bring no real changes in alliance mechanisms.

Warsaw Pact in the light of its 20ᵗʰ anniversary

Reform of the Warsaw Treaty Organization's political structures was meant to be only one part of a massive propagandistic campaign related to the 20ᵗʰ anniversary of the alliance foundation. In fact, stressing the Pact's political dimension reflected the actual opinion of the Eastern Bloc officials that during its existence, the alliance had gone through significant transformation. Initially, mostly military tools became the organization's appeals, which had a non-negligible impact on international development. The Pact was therefore presented as a very new form of alliance whose policy did not aim for the priorities of its members only, but to positively affect the whole European continent. Preparation of this campaign was dealt with at the above-mentioned meeting of deputy

---

Political Consultative Committee could improve the significance of the Warsaw Pact's supreme body.

[190] The last Political Consultative Committee session decided on personal enlargement of the existing secretariat. Moreover, its activity should not have been limited to the time of the Warsaw Pact's supreme body and recently established Committee of Ministers of Foreign Affairs meetings only. However, organization and technical preparation of those gatherings still remained the main task of the secretariat.

[191] *Zpráva o výsledcích jednání náměstků ministrů zahraničních věcí členských států Varšavské smlouvy*, 5. 2. 1975, 1261/0/6, sv.144, a. j. 149/2, NA, Prague; In his final speech, the Soviet representative Rodionov openly criticized Romanian attitudes, however, he had had that move approved by other delegations in backrooms in advance. *Report on the Meeting of the Deputy Foreign Ministers*, February 3, 1975, Accessed October 16, 2019. http://www.php.isn.ethz.ch/kms2. isn.ethz.ch/serviceengine/Files/PHP/17364/ipublicationdocument_singledocume nt/c7b53278-c977-4e11-94ba-aba66db0896a/en/750129_Report_E.pdf.

foreign ministers in Moscow. That was certain, although there was no fundamental shift. Remember that the range of the previous Warsaw Treaty Organization's 15th anniversary celebrations was determined strictly by the Kremlin.[192]

The main motto of the anniversary was "The Warsaw Treaty Organization - strong shield of socialism, world peace, and safety." Naturally, the pact's efforts to defuse tensions, maintain and strengthen world peace, or the role of the alliance in successfully finishing the Helsinki process were emphasized.[193] Besides minor additional measures,[194] the initial scenario supposed holding extravagant top gatherings. The spectacular 15th Political Consultative Committee session was considered; it would have been public for the first time. In fact, this was again just a typical propagandistic play for the present Western journalists. According to the proposal, regular secret talks would follow. The real agenda, consisting especially of the finalization of the upcoming reform of the alliance's political structures, would be discussed there. In addition, some military actions were supposed at first.[195] However, all

---

[192] *Zpráva o výsledcích jednání náměstků ministrů zahraničních věcí členských států Varšavské smlouvy*, 5. 2. 1975, 1261/0/6, sv. 144, a. j. 149/2, NA, Prague.

[193] *Návrh politickoideového pojetí a organizačního zabezpečení hlavních akcí k 20. výročí vzniku Varšavské smlouvy*, 11. 4. 1975, sv. 150, a. j. 154/25, ibid.

[194] The anniversary should have been accompanied by a series of socio-cultural actions. Making a propagandistic movie tracking the key milestones of the Warsaw Treaty Organization so far, naturally evaluated from actual political views, was also considered. A special commission led by the USSR prepared an edition of the alliance's political resolutions and public documents. Bulgaria held a historians' symposium dealing with the foundation of the Pact.

[195] A special meeting of the Military Council, the series of ceremonial gatherings at the Staff of the Unified Armed Forces as well as within the national armies, a military-historical conference, the ideological-propagandistic meetings in order to "strengthen international friendship", and military parades would have been held initially. Most of those events were supposed to be broadcasted by television or radio. The very strict division of military and political structures of the Warsaw Pact, which occurred after 1969, were illustrated by the fact that preparations for military celebrations were directed solely by the Staff of the Unified Armed Forces. The political structures did not interfere much. Despite a certain autonomy, the military officers respected decisive competences in the anniversary of a special commission, which was set up by the Political Consultative Committee in April 1974. *Zpráva o vyslání čs. delegace na schůzku náměstků*

those intentions were reconsidered. In the end, the anniversary was much more modest. In connection with the forthcoming signing of the CSCE Final Act, Moscow cancelled a massive military parade of the Unified Armed Forces in the Polish capital.[196] Neither did the Political Consultative Committee session take place, despite the very intensive efforts of Edward Gierek.[197]

According to the official argumentation of Moscow, the Warsaw Pact meeting should postpone the successful finalization of the Helsinki process.[198] However, the real purpose was different. It was most likely related to the sharp deterioration of Leonid Brezhnev's health. During 1974, the Soviet General Secretary fell seriously ill for long time. He was even forced to cancel the Crimean meeting of the Warsaw Pact states' leaders.[199] In December, his health worsened so rapidly that he did not appear in public for fifty straight days. Brezhnev finally met the alliance member-states' officials[200] on March 18, 1975, during quick extraordinary talks. He briefly reported on upcoming priorities in foreign policy, stressing the signing of the CSCE Final Act and explained he was not able to actively take part in the events because of health issues.[201] This was, in

---

ministrů zahraničních věcí Varšavské smlouvy v Moskvě 29.1.-30.1. 1975, 15. 1. 1975, 1261/0/6, sv. 143, a. j. 146/10, NA, Prague; *Zpráva o výsledcích jednání náměstků ministrů zahraničních věcí členských států Varšavské smlouvy*, 5. 2. 1975, sv. 144, a. j. 149/2, ibid; *Výsledky 11. zasedání Vojenské rady Spojených ozbrojených sil a porady vedoucích funkcionářů spojeneckých armád*, 25. 11. 1974, FMO GŠ/OS, k. 38, p. č. 133, č. j. 36036, VÚA, Prague.

[196] Jarząbek, *"Poland in the Warsaw Pact"*.

[197] Mastny, *A Cardboard Castle?*, 43.

[198] *Informacja z rozmów w KC KPZR*, undated, January-May 1975, PZPR KC, XIA/596, AAN, Warsaw.

[199] In April 1974, at the Political Consultative Committee session, Brezhnev still emphasized the significance of the Crimean meetings; he suggested they should have been held more often. *Vystoupení s. L.I. Brežněva na poradě politického poradního výboru Varšavské smlouvy 17. dubna 1974*, 1261/0/6, sv. 115, a. j. 117/1, NA, Prague.

[200] Nicolae Ceauşescu demonstratively did not take part in the meeting.

[201] Brezhnev did not confess to the Eastern leaders the depth of his health troubles. The CPSU CC General Secretary said he had been overworked and tired only. The Soviet leadership decided to remain publicly silent on the issue. Volkogonov, Dmitrij, *The Rise and Fall of the Soviet Empire. Political Leader from Lenin to Gorbachev* (London: HarperCollins Publishers, 1999), 307-308.

fact, the beginning of a stagnation period in Brezhnev's rule. Regarding his deteriorating health, the CPSU CC General Secretary was increasingly unable to carry out his duties. The Soviet policy thus became less determined in every way. Such a situation lasted until the death of Brezhnev in 1982 and was demonstrated during the Polish crisis in early 1980s. This inevitably resulted in the decline of foreign policy activities of the Eastern Bloc and therefore the Warsaw Treaty Organization itself.[202] After all, at the Helsinki conference, Leonid Brezhnev was in such bad condition that he signed the Final Act with great difficulties.[203]

The twentieth anniversary of the Warsaw Pact established the praxis of gatherings of the member-states' parliaments' representatives. Initially, a single meeting was proposed by Poland as a part of the celebrations.[204] Moscow agreed because the action corresponded with the Soviet effort to present the alliance as a political organization at the time prior to signing the Final Act.[205] The session took place in the Polish capital on May 14-15, 1975, symbolically in the same hall of Sejm where the founding charter was signed. However, only insignificant and actual

---

[202] Baev, *"The "Crimean Meetings" of the Warsaw Pact"*; Tejchman and Litera, *Moskva a socialistické země na Balkáně*, 87.

[203] During that period, Brezhnev did not show up at the politburo sessions for long weeks, even months. The General Secretary was quite athletic. However, that fact was neutralized by his heavy smoking. Many of his close colleagues, including Andrei Gromyko, in their retrospective testimonies, pointed to Brezhnev's excessive drinking as well. The healthy troubles of the General Secretary were also undoubtedly deepened by his initial opiate-sedatives addiction. Brezhnev actually suffered strong insomnia. After 1973, he became used to one or two sedative pills a day. However, he overdosed sometimes, and would fall into a comatose state followed by a long period of rest. According to testimony of V. Medvedev, security chief in the Kremlin, some of the politburo members suggested to Brezhnev to take the medicaments with zubrovka, a flavored vodka. Quite the contrary, such a mixture definitely worsened his condition. Zubok, Vladislav, *A Failed Empire: The Soviet Union in the Cold War from Stalin to Gorbachev* (Chapel Hill: The University od North Carolina Press, 2007), 241, 246; Volkogonov, *The Rise and Fall*, 325; Bacon, "Reconsidering Brezhnev", 11.

[204] *Protocol of the Meeting of the Deputy Foreign Ministers*, January 29, 1975, Accessed October 16, 2019. http://www.php.isn.ethz.ch/kms2.isn.ethz.ch/service engine/Files/PHP/17368/ipublicationdocument_singledocument / 6f4029fa-2bbf-4ab9-9a47-dfff127167b0/en/750129_Protocol_E.pdf.

[205] Jarząbek, *PRL w politycznych strukturach*, 40.

situation-conforming speeches were given again.[206] Altogether emphasized was the defensive character of the Warsaw Pact as well as promoting the peace policy according to the existing diplomatic tack of the Soviet Union. Due to understood reasons, there were also appeals for a quick and successful finalization of the Helsinki process. Following the last resolutions of the Political Consultative Committee, the events in the Middle East and Chile were criticized and the military victory of the Vietnamese communists appreciated.[207] An appeal to parliaments of every European country to actively take part in the process of easing tensions and disarmament documented the shift of the Eastern Bloc foreign policy, reflecting the peak of détente. The meeting also proclaimed alleged Eastern willingness to simultaneous military blocs' dissolution. For this purpose, NATO and the Warsaw Treaty Organization should dismantle their military structures at first.[208]

Despite its largely trivial effect, the gathering was appreciated by the Warsaw Pact member-states, with the exception of Romania.[209] For this reason, although the meetings of parliament representatives were initially supposed to be a one-off case, a decision was made to hold them

---

[206] For example, A. Indra, the CPCz CC secretary and member of the party presidium, declared that the Warsaw Treaty Organization during the twenty years of its existence proved a commitment to mutually defend an independence of all the member-states, neither threatened nor extorted anyone, and did not interfere with other countries' affairs. On the contrary, it allegedly guarded "revolutionary achievements of the workers' class and working people". *Projev s. Aloise Indry, člena předsednictva ÚV KSČ a předsedy Federálního shromáždění ČSSR na setkání představitelů parlamentů členských států Varšavské smlouvy při příležitosti jejího 20. výročí ve Varšavě dne 14. května 1975*, 1261/0/6, sv. 159, a. j. 162/13, NA, Prague.

[207] *Zpráva o setkání představitelů parlamentů členských států Varšavské smlouvy u příležitosti jejího 20. výročí*, 6. 6. 1975, ibid,

[208] *Za mír, bezpečnost, spolupráci a sblížení mezi národy Evropy (Závěrečný dokument setkání představitelů parlamentů členských států Varšavské smlouvy přijatý dne 15. května 1975)*, ibid.

[209] Romanian aversion was demonstrated by efforts of its delegation to block every mention that the meeting was held within the framework of the Warsaw Pact.

regularly in the future.[210] Thus, a new - though not very important - political forum was established within the alliance. It was most likely part of the effort to create another basis for presenting the common international appeals of the Eastern Bloc states. It was closely related to the intended role of the individual national ensembles in the process of easing tensions. At the time, the parliaments of the signatories of the Final Act were supposed to engage significantly in both bilateral and multilateral CSCE negotiations in the upcoming period.[211] Like the other talks within the Warsaw Pact, the meetings of parliament's representatives were strictly regulated. They became just another forum providing multilateral cover to the actual foreign policy tack of Moscow.

***

At the time of the signing of the Final Act, shaping of the Warsaw Treaty Organization's political structures was completed. From today's perspective, the outcomes of almost twenty years of development can be described as poor: Intended activation of the alliance's political framework at the turn of the 1960s and '70s was only temporary and strictly connected with alteration of the international situation at the time. That state reflected the general essence of the Warsaw Pact's existence. The Soviet leadership - undoubtedly a dominant and determining power in the organization - practically all the time adjusted the appearance and role of the alliance to the actual foreign policy priorities of the USSR. The Kremlin looked for proper use of the Pact in regard to the existing global situation. The bigger activity of the alliance's political structures that Moscow decided to try for a short period in 1969, was never meant to change the real substance of cooperation significantly, as some Soviet satellites misunderstood for a while. That was proven even by the Kremlin's own lukewarm attitude towards reform of the Pact. The USSR never intended to allow the establishment of any real multilateral mechanisms within the Warsaw Treaty Organization. As we explained, in the moments when the scope of cooperation started exceeding acceptable limits from the Kremlin point of

---

[210] *Informacja z rozmów w KC KPZR*, undated, January-May 1975, PZPR KC, XIA/596, AAN, Warsaw.
[211] *Zpráva o setkání představitelů parlamentů členských států Varšavské smlouvy u příležitosti jejího 20. výročí*, 6. 6. 1975, 1261/0/6, sv. 159, a. j. 162/13, NA, Prague.

view, Moscow undertook steps which decreased the Pact's activity again.

In the political terms, Brezhnev's leadership used the Warsaw Pact for similar purposes known from Khrushchev's era. It served mostly as a tool for covering Soviet foreign policy initiatives and appeals. Cooperation of the member-states was just much more systematic and better organized. The meetings under the auspices of the alliance were held more frequently. However, they never became a real forum for dealing with important issues and problems. This is proven by Moscow's utmost effort to pre-discuss their agenda and scenario as well as by Soviet unwillingness to use them as a real tool for solving the discord among the Eastern Bloc countries. On the other hand, the USSR succeeded in its main intention. The outward prestige of the Warsaw Treaty Organization undoubtedly increased. Frequently held meetings in reaction to similar NATO talks as well as presentations of significant international initiatives - especially related to the Helsinki process - made the Warsaw Pact a highly visible international organization in the early 1970s. The alliance's appeals were taken into account.

After the Warsaw Pact got a certain international reputation, the Soviet leadership tried to maintain that status through regular, but not very frequent meetings of the Political Consultative Committee, and later through the newly established Committee of Foreign Affairs Ministers and the gatherings of the alliance member-states' parliaments' representatives. However, as we explain in following chapters, those forums became increasingly strictly propagandistic in the second half of the 1970s.

Romanian question
in the first half of the 1970s

As previously outlined, a complicated relationship with Romania was one of the key inside challenges to the Warsaw Treaty Organization during the era surveyed in this work. Disputes between Romania and the Warsaw Pact were hardly new. The issue hailed back to the first half of the 1960s, when Romanian leadership in the background of the Sino-Soviet split pragmatically decided to reconsider its approach toward the USSR and the Eastern Bloc, in general. In April 1964, the Romanian Worker's Party Central Committee plenum passed a resolution in which the "*Romanian national and specific route to socialism*" was defined and described. This relatively risky attempt aimed to expand the space in which a Soviet satellite can operate more independently. The so-called April Declaration appealed to the USSR to respect international standards in relation to other communist countries. Those tendencies were augmented after March 1965, when Nicolae Ceausescu became head of the Romanian party. Romanian communists gradually created their own brand of dogmatism - a unique nationalist and Marxist doctrine that became the basis for the personal power of Ceausescu. Despite the fact that full independence from the Soviet Union was unrealistic, Romanian officials, throughout the next few years, stressed their distance from Moscow.[212]

During the 1960s, Romania made a few isolated moves in the international arena which were strongly denounced by the Kremlin. In 1967, the SRR, disregarding the other Eastern Bloc states and absence of Moscow's permission, established diplomatic ties with West Germany. After the Six-Day War, it refused to sever relations with Israel and the

---

[212] Giurescu, Dinu Constantin and Fisher-Galati, Stephen, *Romania. A Historic Perspective* (New York: Boulder, 1998), 458-464.

Romanian Communist Party[213] also ignored conclusions adopted at meetings of the International Communist and Worker's Movement. These leanings bothered Moscow on through the beginning of the next decade.[214] Romanian absence in the military suppression of the Prague Spring only underlined the specific position of the country within the Eastern Bloc. However, we are reminded that Ceausescu's regime had, at the same time, no sympathy for the process of reform in Czechoslovakia.[215]

The Romanian approach towards the Warsaw Treaty Organization was based on the aforementioned policy. As the Pact represented one of the tools which the USSR used to govern and shape its sphere of influence in Europe, the opinions on proper cooperation within the alliance's framework soon became a specific bone of contention between Moscow and Bucharest. After he came to power, Brezhnev strove for Warsaw Pact consolidation in order to make the alliance more effective in regards to Soviet diplomatic and military aims. On the contrary, Ceausescu's regime, at least on a declarative level, stressed the article of the Pact's founding charter that appealed for dissolution of political-military blocs after establishing some sort of collective security system.[216] Until 1968, these contradictions often resulted in tense and apprehensive behavior from the Romanian desk at alliance meetings. Using various obstructions,

---

[213] In July 1965, the Romanian Worker's Party was renamed the Romanian Communist Party at its 9th congress.

[214] *Zpráva o sovětsko-rumunských vztazích*, 13. 9. 1971, 1261/0/6, sv. 12, a. j. 11/info1, NA, Prague.

[215] On Romanian policy during 1968 for example Retagan, *1968. Ve stínu pražského jara*. In late 1968 and 1969, the Romanian diplomats in conversations with Czechoslovak officials stressed that Romania had neither taken part in the talks on the Prague Spring nor got involved in the military intervention. They assured that Bucharest did not intend to interfere with neither international nor internal issues which occurred in connection with the invasion of Czechoslovakia. *RSR – informace o současném stavu a perspektivách čs.-rumunských vztahů*, 3. 6. 1970, TO(t) 1970-1974, box 2, i. č. 82, sign. 017/112, č. j. 020.500/70-2, AMZV, Prague.

[216] The Article XI of the Warsaw Pact's founding charter bound the member-states to strive for establishing a collective security system in Europe. The existence of the Warsaw Treaty Organization was supposed to end on the day this all-European agreement came into effect.

Bucharest especially strove to prevent a reform of both political and military structures within the Warsaw Pact.[217]

The situation slightly changed after the invasion of Czechoslovakia. At least for some period, Bucharest considered the Soviet Union as the main threat to its security.[218] Fearing that Romania could be the next victim of Soviet aggression,[219] Ceausescu calmed down his behavior for a while. After four years of obstructions he allowed the passing of the Warsaw Pact's military statutes reform in March 1969. Thus, at the turn of the 1960s and '70s Romania was aware of two direct threats connected to membership in the alliance: The first came from the Czechoslovak experience and the potential usage of the Pact's military forces in an intervention against a defiant Ceausescu regime;[220] the second rooted from the escalation of tensions on the Sino-Soviet border. Bucharest feared that, in case of war between two socialist powers, Moscow could attempt to activate the Warsaw Treaty Organization mechanisms in order to involve its European satellites in conflicts in the Far East.[221] Considering sources available today, it must be noted that those fears seem to have been exaggerated.

---

[217] On Romanian attitude towards the reform of the Warsaw Pact in the 1960s, for example: Rinoveanu, Carmen, "Rumänien und die Militärrefrom des Warschauer Paktes," in *Der Warschauer Pakt. Von der Gründung bis zum Zusammenbruch 1955 bis 1991,* eds. Torsten Diedrich, Winfried Heinemann, and Christian Ostermann (Berlin: Christoph Links Verlag, 2009): 209-224; Ionescu, Mihai E., "Rumunsko a vojenská reforma Varšavské smlouvy," *Historie a vojenství,* no 3-4 (2003): 699-705; Bílý, "Počátky pokusu o reformu Varšavské smlouvy," 165-172.

[218] Nünlist, Christian, "*Cold War Generals: The Warsaw Pact Committee of Defense Ministers, 1969-90,* " Accessed October 16, 2019. http://www.php.isn. ethz.ch/lory1.ethz.ch/collections/coll_cmd/introductiond6c9.html?navinfo=14565.

[219] Regarding the sources which are available today, it can be said that Western considerations of forthcoming Soviet military intervention in Romania in the second half of 1968 and 1969 were based on irrelevant information. Baev, Jordan, "The Warsaw pact and Southern Tier Conflicts, 1959-1969," in *NATO and the Warsaw Pact. Intrabloc Conflicts,* eds. Mary Ann Heiss and S. Victor Papacosma (Kent: Kent State University Press, 2008): 202.

[220] The Romanian top leadership correctly admitted that the invasion of Czechoslovakia had not been officially conducted on behalf the Warsaw Treaty Organization.

[221] *K současnému vývoji vztahů mezi Rumunskou socialistickou republikou a členskými státy Varšavské smlouvy,* 16. 6. 1969, DTO 1945-1989, i. č. 34 RSR, e. č. 42, č. j. 023.383/69-2, AMZV, Prague.

However, Romania could not afford to openly distance itself from the Eastern Bloc, and consequently from the Warsaw Pact, not only because of geopolitical reasons, but also due to its ambitious economic plans.[222] Therefore, the official Romanian position towards the Warsaw Treaty Organization was defined largely by the exclusive claim that the existence of the Pact was only temporary. Its member-states should have striven for dissolution of all military alliances, in accordance with the founding charter.[223]

Romania in the political structures of the Warsaw Pact

The fear of repeating the Czechoslovak scenario was not the only factor that influenced Romania's more accommodating approach towards the Warsaw Pact after August 1968. The issue of convening an all-European security conference began to dominate the agenda of the alliance's political meetings. In March 1969, the Political Consultative Committee approved the course on this issue, a path that at first glance, did not contradict the main principles of Romanian foreign policy. Ceausescu's regime appreciated the declaration of the Warsaw Pact's supreme body calling for holding the conference without any preconditions. However, Romania stressed that this was only because the document had been worked out collectively by all Warsaw Pact member-states, and reflected Romanian priorities as well. In this regard, Romanian propaganda did not miss the opportunity to announce that the initiative opened a new route to the dissolution of military blocs in Europe.[224]

---

[222] Pech, Radek, "Rumunsko let sedmdesátých – od liberalismu k represi," *Slovanský přehled*, no. 3 (1992): 271.

[223] *Ohlas v RSR na budapešťské zasedání PPV VS*, 16. 4. 1969, DTO 1945-1989, i. č. 34, e. č. 42, č. j. 022.229/69-2, AMZV, Prague.

[224] Ibid; *Mimořádná politická zpráva k ohlasům na budapešťskou Výzvu členských států Varšavské smlouvy*, 30. 5. 1969, č. j. 023.010/69, ibid. The Romanian stance on possible dissolution of the military blocs linked to the all-European security conference did not resonate even among the NATO member-states. Mostly, the small Western countries called it a "perspective option," but treated it as largely immaterial. In addition, regarding the invasion of Czechoslovakia, there were opinions that this Soviet action had confirmed the

Indeed, the potential all-European security conference was seen by Bucharest as a forum of the equal, sovereign participant countries. This approach fully corresponded with Romania's continuous efforts to weaken Moscow's influence on its policy.[225] Romania also considered the conference a chance to strengthen relations with the West, as it had been striving to do since the mid-1960s.[226] From the Romanian point of view, the strengthening of inter-bloc cooperation was the most important aspect of a prospective security summit, whereas the USSR and some Warsaw Pact member-states, which had unresolved territorial disputes with West Germany, concentrated primarily on safety guarantees.[227]

As previously explained, the political meetings of the Warsaw Pact became more frequent in 1969 because of new international challenges. Romanian participation was not fully conflict-free. As in the past, SRR delegations sought to ensure that sessions would not exceed the scope of non-binding consultations.[228] They rejected taking clear positions on a variety of contentious points, stating the state and party leadership did not authorize them.[229] Ceausescu's regime later

---

legitimacy of NATO existence as a guarantee of the small European countries' independence.

[225] Mastny, *A Cardboard Castle?*, 40.

[226] Relations between Romania and West were improving at the turn of the 1960s and '70s. In August 1969, the U.S. president Nixon visited Bucharest. In December 1970, Ceausescu made his trip to the United States in return. One year later, Romania joined the GATT and in 1972 became an International Monetary Fund and World Bank member. Deletant, Dennis, "*Romania and the Warsaw Pact : Documents Highlighting Romania's Gradual Emancipation from the Warsaw Pact, 1956-1989*," Accessed October 16, 2019. http://www.php.isn.ethz.ch/lory1. ethz.ch/collections/coll_romania/introduction0445.html?navinfo=15342.

[227] *Note on the Meeting of the Deputy Foreign Ministers*, May 21, 1969, Accessed October 16, 2019. http://www.php.isn.ethz.ch/kms2.isn.ethz.ch/serviceengine/ Files/PHP/17253 / ipublicationdocument_singledocument/9b7b446c-693b-42d6-92a5-807295820045/en/690521_Note.pdf; Békés, "Records of the Meetings".

[228] For instance, in May 1969, at the beginning of the meeting of deputy foreign ministers in East Berlin, the Romanian representatives declared that they were authorized by the RCP leadership to non-binding discussion only. They demanded a formulation to be recorded that no real obligations would result from the talks. Ibid.

[229] *Zpráva o průběhu a výsledcích pražské porady ministrů zahraničních věcí členských států Varšavské smlouvy a návrhy na další postup v oblasti evropské bezpečnosti*, 14. 11. 1969, 1261/0/5, sv. 110, a. j. 181/info2, NA, Prague.

adopted a similar strategy towards consultations within the International Communist and Worker's Movement. Romania did not oppose said consultations, but stipulated conditions for an absence of criticism and the non-binding character of approved declarations.[230] Bucharest also made it clear that, in the process of convening, the all-European security conference would definitely not act through the Warsaw Treaty Organization exclusively. Romania considered the multilateral political meeting of the alliance to be no more than a forum where member-states inform each other about their talks with Western and neutral countries. It firmly refused to turn the consultations into shaping the unified foreign policy tack of the Eastern Bloc, which the USSR and some of its satellites were pushing for.[231] The rest of the Warsaw Pact was leery of this Romanian approach: bilateral diplomacy of its own design without taking into account the strategic priorities of the alliance. Such a situation could have potentially paved the way to superiority of the NATO countries due to their unanimity.[232]

Nevertheless, in comparison to the era preceding the military suppression of the Prague Spring, Romanian representatives behaved much more constructively at Warsaw Pact political meetings.[233] They did not construct any serious obstacles, except for vetoing some less important proposals.[234] At the time, Moscow's effort to compile the final

---

[230] *Dosavadní postoje RKS ke svolání Evropské porady komunistických a dělnických stran*, 30. 10. 1974, DTO 1945-1989, i. č. 34, e. č. 61, č. j. 017296/74, AMZV, Prague.

[231] Wenger and Mastny, "New perspectives of the origin of the CSCE," 11; Mastny, *A Cardboard Castle?*, 40.

[232] *Shrnutí poznatků z rozhovorů v Bukurešti*, 14. 7. 1969, DTO 1945-1989, i. č. 34, e. č. 42, č. j. 023.938/69-2, AMZV, Prague; *Mimořádná politická zpráva k ohlasům na budapešťskou Výzvu členských států Varšavské smlouvy*, 30. 5. 1969, č. j. 023.010/69, ibid. In fact, even NATO did not act unanimously at the time. Within its structures, French attitude towards negotiation on the basis of blocs was similarly negative to the Romanian stances within the Warsaw Pact.

[233] This shift was missed even by the Western press. *Ohlas XXIV. sjezdu KSSS v Rumunsku*, 14. 4. 1971, e. č. 52, č. j. 022.188/71-2, ibid.

[234] In May 1969, at the meeting of deputy foreign ministers, Romania blocked reaction to the Finnish proposal on the all-European security conference on behalf of the Warsaw Treaty Organization, insisting to reply on behalf of individual

documents in a benevolent manner obviously contributed to this. According to the official Soviet interpretation, Warsaw Pact declarations were supposed to reflect the priorities of all member-states. The Kremlin disregarded some contradictory Romanian statements at closed meetings, as they did not pose any significant threat.[235] Therefore, a blueprint for how to approach Bucharest in the future was set up within the Warsaw Pact in 1969. Before almost every alliance session, a few Soviet deputy foreign ministers visited all member-states, except for Romania. "*The Six*" - the Warsaw Pact countries without SRR[236] - unified its positions this way and then put them in front of the Romanian delegation at the meeting. Following non-binding voting often encouraged Romanian representatives either to join the proposal, or to accept some sort of compromise.[237]

---

member-states instead. It also refused to discuss collectively within the Pact's structures the issue of discrimination of the GDR at the upcoming 1972 Olympic games in Munich. Four months later, at the meeting of ministers of foreign affairs in Prague, Cornelieu Mănescu prevented discussion on the Polish draft of the European Security treaty. However, he was supported by the East-German representatives. *Note on the Meeting of the Deputy Foreign Ministers*, May 21, 1969, Accessed October 16, 2019. http://www.php.isn.ethz.ch/kms2.isn.ethz.ch/ serviceengine/Files/PHP/17253/ipublicationdocument_singledocument/9b7b446 c-693b-42d6-92a5-807295820045/en/690521_Note.pdf; *Zpráva o průběhu a výsledcích pražské porady ministrů zahraničních věcí členských států Varšavské smlouvy a návrhy na další postup v oblasti evropské bezpečnosti*, 14. 11. 1969, 1261/0/5, sv. 110, a. j. 181/info2, NA, Prague,

[235] For instance, in December 1969, at the meeting of the Warsaw Pact member-states' party leaders in Moscow, Ceausescu repeated his well-known – and undesirable, from the Soviet point of view - demands for dismantling all the foreign military bases in Europe, the dissolution of military blocs, and the reduction of armament levels or the nuclear weapons ban. However, the Romanian leader presented his opinions at a closed session only. *Dokumenty spotkania przywódców prtyjnych i państwowych siedmiu krajów socjalistycznych*, 10. 12. 1969, PZPR KC, V/89 (2899), AAN, Warsaw.

[236] In the mid-1970s, the Six started to be officially called 'closely cooperating member-states' of the Warsaw Pact. *Report on the Meeting of the Deputy Foreign Ministers*, February 3, 1975, Accessed October 16, 2019. http://www.php.isn. ethz.ch/kms2.isn.ethz.ch/serviceengine/Files/PHP/17364/ipublicationdocument_ singledocument / c7b53278-c977-4e11-94ba-aba66db0896a/en /750129_Report_ E.pdf.

[237] Békés, "Studená válka, détente a sovětský blok," 81.

Balancing policy

The situation changed during the early months of 1970. Bucharest realized that the Soviet opinion of the all-European security conference scheme was very different. Due to suspiciously frequent talks within the Warsaw Pact, Ceausescu's regime became afraid that Moscow's intention was to hold the conference on the basis of existing blocs.[238] In January, at the alliance deputy foreign ministers' session in Sofia, Romania unsuccessfully suggested the preliminary meeting of all potential countries participating in the security conference be held in its capital.[239] At the time, Moscow instead began to consider negotiating the conditions of the conference within a working group of three states only: Belgium for NATO, Poland for the Warsaw Pact, and Finland for neutral countries.[240] The USSR planned a following alliance meeting of ministers of foreign affairs to finalize the procedure.

Ceausescu refused this scenario and decided to dull relations with the Warsaw Pact. He noted that Moscow's intentions contradicted the Bucharest Declaration of the Political Consultative Committee, which presumed preparation of the security conference on the basis of all involved countries. In consequence, he attempted to prevent upcoming political meetings of the alliance. Romania threatened it would not participate. Up to this point, the Romanian leader had supported the Warsaw Pact's appeals on the all-European security conference for a single reason: Implied negotiations between sovereign states corresponded with Ceausescu's long-term foreign policy goals. When this assumption proved to be wrong, he started searching for support for his initiatives among

---

[238] *Záznam o návštěvě I. taj. ZÚ RSR s I. Georgescu u p. Picka dne 27. 3. 1970,* TO(t) 1970-1974, box 3, i. č. 89, sign. 020/112, č. j. 057/70, AMZV, Prague.

[239] *Informace o poradě náměstků k Evropské bezpečnosti,* 27. 1. 1970, box 1, sign. 017/111, č. j. 020.534/70-2, ibid.

[240] This concept was soon proven to be impassable and the USSR left it behind in the early months of 1970. *Oficiální návštěva ministra zahraničních věcí SSSR p. A.A. Gromyka v ČSSR – materiál pro předsednictvo ÚV KSČ,* 20. 4. 1970, box 1, sign. 020/112, č. j. 021.648/70-1, ibid; *Zpráva I. teritoriálního odboru ministerstva zahraničních věcí ČSSR o stanovisku SSSR k celoevropské konferenci o bezpečnosti a spolupráci,* 7. 4. 1970, box 8, sign. 020/311, č. j. 085/70, ibid.

neutral countries, e.g., Finland.[241] The Romanian minister of foreign affairs, Corneliu Mănescu, justified this move on the grounds that NATO, in reaction to a Warsaw Pact session, would certainly hold a similar meeting. In Romanian opinion, this situation would inevitably lead to an all-European security conference on the basis of blocs. The ulterior motive of Romanian obstructive behavior was an effort to prevent a new round of collective talks on possible reform of the Warsaw Treaty Organization's political structures being initiated by Hungary. The preliminary agenda of upcoming meetings included this item.[242]

Moscow quickly assured its satellites that planned consultations within the alliance would be held regardless of the Romanian stance.[243] However, the potential absence of Romania brought problems. The USSR intended to connect talks on the issue of an all-European security conference with the celebration of the Warsaw Pact's 15th anniversary. The Kremlin naturally directed the scope of this propagandistic play. The ostentatious distance of one alliance member was, from the Soviet point of view, very undesirable as it disrupted efforts to present outwardly the absolute unity of the Pact.[244]

At this point, Ceausescu visited the Soviet capital on May 19, 1970. In comparison to multilateral meetings, his bilateral talks with the highest officials of the USSR concerning Romanian policy within the

---

[241] Ibid; *Zasedání VNS RSR k zahraniční politice*, 3. 4. 1970, DTO 1945-1989, i. č. 34, e. č. 47, č. j. 022.007/70-2, AMZV, Prague.

[242] *RSR – zpráva o návštěvě ministra zahraničních věcí RSR C. Manesca v ČSSR ve dnech 8.-11. dubna 1970*, TO(t) 1970-1974, box 2, i. č. 82, sign. 017/112, č. j. 022.176/70-2, AMZV, Prague.

[243] *Oficiální návštěva ministra zahraničních věcí SSSR p. A.A. Gromyka v ČSSR – materiál pro předsednictvo ÚV KSČ*, 20. 4. 1970, box 1, i. č. 89, sign. 020/112, č. j. 021.648/70-1, ibid.

[244] *Zpráva o situaci na úseku evropské bezpečnosti pro ÚV KSČ – odložení projednání na duben 1970*. 18. 3. 1970, box 1, sign. 020/111, č. j. 010.295/70-AP, ibid; *Záznam o návštěvě p. M. Havláska ve IV. EO MZV SSSR*, 12. 5. 1970, box 3, sign. 020/112, č. j. 0109/70, ibid. The East-German documentation reveals that a special argumentation for purposes of the Warsaw Pact member-states' ambassadors to Bucharest was prepared. They were supposed to present it during the talks with the Romanian officials in order to assert another alliance's meeting of foreign affairs' ministers. *Politbüro des ZK Reinschriftenprotokoll nr. 9*, 24. 2. 1970, DY 30/ J IV 2/2/1269, BArch, Berlin.

Warsaw Pact were, to say the least, tumultuous. The Romanian leader did not hesitate to openly stand up against many Soviet stances. According to the testimony of the RCP Executive Committee secretary Ştefan Voicu, after some of Ceausescu's sharp responses the Communist Party of the Soviet Union (CPSU) leadership members literally turned green.[245] Angry Brezhnev then snapped at the Romanian leader about whether his country intended to stay within the Warsaw Pact. "*If you do not want to go with us, go straight to hell*," he stormed. For the first time, Brezhnev generously offered for Ceausescu to leave the alliance. Otherwise, Romania was urged to make no waves. The bid of the CPSU CC General Secretary definitely did not include the possibility of full abandonment of the Soviet sphere of influence. In fact, he prioritized the urgent approval of a new bilateral Soviet-Romanian allied treaty.[246] The Soviet leader was extremely critical. He pointed out Romanian rhetorical warnings against conducting Warsaw Pact maneuvers near the borders of the SRR. He asked whether Ceausescu considered the allied states as enemies. He also rebuked Romania for stressing the need for dissolution of military-political blocs, the absence of Romanian troops on joint military exercises, as well as their blocking

---

[245] This happened in the moment when Brezhnev criticized the Romanian demonstrative actions which were not discussed within the Warsaw Pact in advance. In his reply, Ceausescu provocatively noted that Romania was not the only country which failed to consult its foreign policy. He reminded them that the Soviet government had begun talks in Vienna with the United States on strategic arms reduction in the same way. The Romanian leader added that Bucharest had not been informed about the proceeding of the negotiation yet. *Summary No. 10 of the Executive Bureau of the CC of the RCP*, May 20, 1970, Accessed October 16, 2019. [http://www.php.isn.ethz.ch/kms2.isn.ethz.ch/serviceengine/Files/PHP/16490/ipublicationdocument_singledocument/4ffbc115-d300-43b4-a47a-b21bc4 022244/en/700520_summary.pdf. It is good to note that Romanian lack of information about SALT was unique in the scope of the Eastern Bloc. In fact, the Kremlin briefed at least its loyal satellites, the GDR for instance, on the matter of the talks. *Politbüro des ZK Reinschriftenprotokoll nr. 51*, 22. 12. 1969, DY 30/ J IV 2/2/1259, BArch, Berlin.

[246] Brezhnev initially planned to personally oversee a ceremony of signing the mutual allied treaty in Bucharest. In the end, he absented. According to the Soviet interpretation, the CPSU CC General Secretary demonstrated his disapproval of Romanian foreign policy this way. *Současný politický postoj RSR ve vztahu ke státům Varšavské smlouvy*, 11. 1. 1971, DTO 1945-1989, i. č. 34, e. č. 52, č. j. 020.206/71, AMZV, Prague.

of reform of the Pact's political structures. As before, Ceausescu firmly rejected those accusations with the claim that his country fully adhered to the text of the alliance's founding charter. After all, he liked to refer to its vague articles which allowed for various interpretations. He called the question of leaving the alliance senseless. The Romanian leader declared his interest to continue in cooperation with the socialist states. However, he warned the Soviet leadership that Romania would continuously impose a veto on all documents dealing with military, economic, and political integration of the Eastern Bloc.[247] Brezhnev retroactively complained in a talk with the Polish Unified Worker's Party (PUWP) CC secretary and the politburo member Zenon Kliszko on Ceausescu's strategy. "*He behaved like the true gypsy*," the Soviet leader raged.[248]

Romanian resistance against Moscow's effort to improve cohesion of the Eastern Bloc never grew into an open rift. However, it complicated the Kremlin's policy within its sphere of influence, as well as its approach towards many international issues.[249] Due to the absence of key Soviet documentation, the question why Moscow chose an apparently indecisive solution to the Romanian problem can be hardly answered. In general, we can state that the USSR had four options: ostracism, which would almost certainly have led to the exclusion of Romania from the Eastern Bloc; military intervention; initiating an internal coup; or maintaining a continuous long-term leash thus affecting Ceausescu and keeping his policy within the acceptable limits. An episode of rift with Albania poked holes in the first variant. In 1961, the impulsive and not so pre-calculated actions of Khrushchev's leadership caused the defection of this strategically important country from Soviet influence. Moscow considered

---

[247] *Summary No. 10 of the Executive Bureau of the CC of the RCP*, May 20, 1970, Accessed October 16, 2019. [http://www.php.isn.ethz.ch/kms2.isn.ethz.ch/serviceengine/Files/PHP/16490/ipublicationdocument_singledocument/4ffbc115-d300-43b4-a47a-b21bc4022244/en/700520_summary.pdf.

[248] According to Brezhnev's retrospective interpretation, Ceausescu took the floor for a long time at the beginning of the talks in Moscow. However, he spoke in very general terms and tried to avoid any topics of conflict, even assuring that his policy basically corresponded with the Soviet positions. *Zapis wypowiedzi tow. L.Breżniewa podczas spotkania z tow. Z. Kliszko w dniu 2 czerwca 1970 roku*, PZPR KC, XIA/88, AAN, Warsaw.

[249] Tejchman and Litera, *Moskva a socialistické země na Balkáně*, 7.

a recurrence of this scenario undesirable. In a situation where the USSR was extremely interested in easing tensions with the West, any military solution was out of the question as well. Western powers did not take any hard steps after the invasion of Czechoslovakia. However, the Kremlin recognized that another similar action would either seriously complicate, or immediately terminate, the process of détente.[250] Removal of Ceausescu and his replacement by the representatives of pro-Soviet orientation never exceeded the scope of lackadaisical debate. From this point of view, the long-term influencing of Romanian policy by diplomatic means seemed to be the most suitable solution.

An approach of Ceausescu's regime helped this strategy. Romania never took any firm action that straightforwardly led to leaving the Eastern Bloc. Perhaps, it wanted to follow neither the Yugoslavian nor the Albanian path. Moreover, the possibility that Ceausescu considered Romanian inclusion in the Soviet sphere of influence a guarantee of preservation of the Leninist-Stalinist regime in his country cannot be ruled out. The experience of the years 1956 and 1968 suggested that Moscow would not allow any significant changes to the socioeconomic system in Warsaw Pact member-states. However, the Romanian leader was well aware of how much his international activity irritated Moscow. He often defused its impact. Although Ceausescu did not change the essence of his foreign policy at the turn of the 1960s and '70s, he did try to avoid some actions which Moscow considered the most provocative. His approach towards the Eastern Bloc became more flexible. He often informed the Kremlin about his intentions in advance. According to Soviet intelligence, Yugoslavian leader Josip Broz Tito advised Ceausescu on this strategy.[251]

The typical and constant phenomenon of Romanian foreign policy in the 1970s became "balancing" – alternating between leaning towards the Eastern Bloc, China, and the West. The USSR still considered Romania as a part of its sphere of influence. Moscow tolerated this

---

[250] Madry, Jindřich, "Sovětské zájmy v pojetí obrany Československa (1965-1970)," *Historie a vojenství*, no. 5 (1992): 126-140.

[251] *Zpráva o současném vývoji na Balkáně (se zvláštním zřetelem k Rumunsku)*, 23. 10. 1972, TO(t) 1970-1974, box 8, i. č. 89, sign. 020/311, č. j. 026.083/72-1, AMZV, Prague.

development, among others, because of the stable position of the Leninist-Stalinist regime within the country and the Soviet model of socialism remaining a cornerstone of Ceausescu's policy. The RCP maintained full control over all social processes in Romania.[252] In this regard, Moscow even warned against over-strengthening Nicolae Ceausescu's "cult of personality" during the early 1970s. In praxis, the Romanian course caused the most concern to Moscow in propagandistic and ideological spheres; the nationalist rhetoric of the Romanian regime undermined the phrases about "proletarian internationalism". Romania's permanent and ostentatious proclamations of state sovereignty did not correspond with the idea of a unified foreign policy within the Warsaw Treaty Organization. However, the USSR at the beginning of the decade assured the other members of the alliance that Romania would not leave the Eastern Bloc and would henceforth participate in its organizations. Nevertheless, Moscow bore in mind that this situation would create many problems in the future. Regarding Ceausescu's policy straining the unity of the alliance, the Soviet Union even admitted that Romanian membership in the Warsaw Pact was favorable to the West.[253] In accordance with this, Brezhnev told Polish First Secretary Edward Gierek in the early 1970s: "*Nationalism twisted the mind of the great leader of great Romania, but we are patient. We believe he will finish his song and then will go with us.*"[254]

The very dynamic alteration of Romanian behavior within the Warsaw Pact was significantly influenced by economic factors.[255] Plans of Ceausescu's leadership on economic development proved to be unrealistic. Economic complications thereby forced Romania to keep close ties with the Eastern Bloc. Actually, the impact of cooperation with

---

[252] Ibid; *Celkový obraz hlavních aspektů vnitřní i zahraniční politiky RKS a RSR za rok 1972*, 19. 12. 1972, DTO 1945-1989, i. č. 34, e. č. 54, č. j. 027280/72-2, AMZV, Prague.

[253] *Zpráva o sovětsko-rumunských vztazích*, 13. 9. 1971, 1261/0/6, sv. 12, a. j. 11/info1, NA, Prague.

[254] Durman, *Útěk od praporů,* 117.

[255] Pech, "Rumunsko let sedmdesátých," 271; Deletant, *"Romania and the Warsaw Pact"*.

the West did not bring such benefit as Bucharest had expected.[256] And so, in mid-1970, Romania instrumentally revised its intention to block political meetings of the Warsaw Pact and dampened its rhetoric for a while. There was also a calming effect brought on by the visit of a Soviet governmental delegation in Bucharest in order to sign a new bilateral allied treaty.[257]

At the following two Political Consultative Committee sessions held in the second half of 1970 in Moscow and East Berlin, Romania behaved within acceptable limits. The most conflicting item came from the above-mentioned effort of Hungary and the other member-states to give the more frequent political talks within the Warsaw Pact some formal rules. Yet again, the Romanian delegation blocked any discussion on the issue. During an alliance meeting in the Soviet capital in August, Ceausescu calmly but firmly declared that the position of his country had remained constant since the mid-1960s. However, he agreed that mutual consultations on both European and global challenges, at the level of ministers of foreign affairs, were necessary. Nevertheless, he repeated the well-known Romanian stance that individual state and party leaderships should have exclusively maintained the key competences in foreign policy issues. This slightly confrontational behavior of the Romanian First Secretary was interpreted by some Eastern diplomats as a result of the isolation of his opinions within the Warsaw Pact. In fact, it was rather part of Bucharest's temporary shift towards the Eastern Bloc in terms of the previously-described policy balancing. Romania remained in the margins, but was not obstructive at that particular moment. It was ready to support some initiatives which at least partially corresponded with its foreign policy approach. In December, at the Berlin Political Consultative Committee session, Ceausescu, for the first time, put his signature on the

---

[256] Romania got into troubles mostly because of the dwindling of its oil and gas reserves. Bucharest was forced to buy those raw materials from Iran and to pay in U.S. dollars. The country therefore asked Moscow repeatedly for enhancement of Soviet deliveries of both strategic materials. A poor situation occurred also in food supplies; Romania had no choice but to ask the USSR for deliveries of grain, again.

[257] *Současný politický postoj RSR ve vztahu ke státům Varšavské smlouvy*, 11. 1. 1971, DTO 1945-1989, i. č. 34, e. č. 52, č. j. 020.206/71, AMZV, Prague.

Warsaw Pact's supreme body resolution denouncing the policy of Israel.[258] The RCP Executive Committee also appreciated both the proceedings and the outcomes of the meeting.[259]

Ceausescu made it clear that he was in favor of talks within the Warsaw Pact if they were limited to support for holding a Conference on Security and Cooperation in Europe (CSCE) based on the sovereign states as participants. He categorically refused joint actions on behalf of the Warsaw Treaty Organization, as they, in praxis, meant nothing more than support for the foreign policy tack formulated by the Kremlin.[260] In terms of this, Romania demanded the final documents of the Political Consultative Committee not to be presented on behalf of the alliance, but rather its individual member-states.[261]

---

[258] *Informace o průběhu diskuse na zasedání Politického poradního výboru Varšavské smlouvy v Moskvě dne 20. srpna 1970,* 1261/0/5, sv. 135, a. j. 213/1, NA, Prague; *Minutes of the Hungarian Party Politburo Meeting on the August 1970 PCC Meeting,* August 25, 1970, Accessed October 16, 2019. http://www.php.isn.ethz.ch/kms2.isn.ethz.ch/serviceengine/Files/PHP/18034/ipu blicationdocument_singledocument/68088b93-5008-4321-9170-0320a0989479/ hu/Minutes_hungarian_Party_1970A_11.pdf; *Současný politický postoj RSR ve vztahu ke státům Varšavské smlouvy,* 11. 1. 1971, DTO 1945-1989, i. č. 34, e. č. 52, č. j. 020.206/71, AMZV, Prague.

[259] *Circular Letter by George Macovescu,* December 8, 1970, Accessed October 16, 2019. http://www.php.isn.ethz.ch/kms2.isn.ethz.ch/serviceengine/Files/PHP/ 16355/ipublicationdocument_singledocument/082ec923-19b4-4ffa-8550-b3cc33 d1d28c/en/701208_circular_letter.pdf.

[260] The Romanian dissentious position also affected activities of the editorial commission of deputy foreign ministers who worked simultaneously with the Political Consultative Committee plenary session. Bucharest's objections against proposed documents on European security and the situation in Africa, Indochina, and the Middle East all resulted in establishing another commission on the level of ministers of foreign affairs. Both commissions then lost almost all day discussing the vast number of Romanian remarks. In fact, the behavior of Romanian representatives ominously resembled a previous obstructive strategy. The rhetoric of Bucharest did not change; it continuously stressed a conception of the countries as the international sovereigns. *Informace o činnosti redakční komise na zasedání politického poradního výboru států Varšavské smlouvy v Berlíně dne 2. prosince 1970,* 1261/0/5, sv. 146, a. j. 225/1, NA, Prague.

[261] *Circular Letter by George Macovescu,* December 8, 1970, Accessed October 16, 2019. http://www.php.isn.ethz.ch/kms2.isn.ethz.ch/serviceengine/Files/PHP/

Chinese factor and Balkan initiatives

In spring 1971, the relations between Romania and the Warsaw Pact deteriorated yet again. The shift could be seen at the 24th CPSU congress. In the backrooms, Ceausescu once more openly criticized military intervention in Czechoslovakia. Romanian press afterwards significantly reduced the transcript of Brezhnev's main speech: Specifically, its foreign policy parts and the sections venerating the importance of the Warsaw Pact's existence were not published.[262] The exact reasons can be revealed by an analysis of the top Romanian leadership documentation only. However, diplomats of the Eastern Bloc countries did not miss the fact that the more constructive Romanian approach at the previous Warsaw Pact's sessions was also recorded by the West. Considering this, they suspected that Ceausescu's current sharper rhetoric was strictly auxiliary and he had only been attempting to demonstrate his continuous specific positions.[263] Regarding this, Petre Opriş assumes that Romanian opposition within the Warsaw Pact after 1968 was also motivated by an effort to make the country more attractive to Western eyes in order to more easily get American and West-European loans along with access to modern technologies.[264]

In June 1971, the spectacular journey of the Romanian leader to China and other Asian socialist countries marked another provocative moment.[265] The Warsaw Pact member-states perceived Ceausescu's

---

16355/ipublicationdocument_singledocument/082ec923-19b4-4ffa-8550-b3cc33 d1d28c/en/701208_circular_letter.pdf.

[262] The shift in attitude of Romanian press was obvious mostly in comparison to the commentaries on the recent Warsaw Pact meeting of ministers of foreign affairs in Bucharest, where the media in the country had neglected the usual stressing of the principles of non-interference, sovereignty, and state independence. *Ohlas bukurešťské porady ministrů zahraničních věcí členských zemí VS*, 26.2.1971, DTO 1945-1989, i. č. 34, e. č. 52, č. j. 021.324/71-2, AMZV, Prague.
[263] *Ohlas XXIV. sjezdu KSSS v Rumunsku*, 14. 4. 1971, č. j. 022.188/71-2, ibid.
[264] Opriş, Petre, "Die rumänische Armee und die gemeinsemen Manöver des Warschauer Paktes," in *Der Warschauer Pakt. Von der Gründung bis zum Zusammenbruch 1955 bis 1991,* eds. Torsten Diedrich, Winfried Heinemann, and Christian Ostermann (Berlin: Christoph Links Verlag, 2009): 198.
[265] Some scholars see a direct connection between the rise of Ceausescu's megalomania and his journey to the Asian communist countries. He was actually

actions as a signal to both the West and particularly Beijing that the alliance was not unanimous in its attitude.[266] At the time, the Kremlin considered China a serious threat not only because of mutual disputes and the Sino-Western convergence, but also due to its potentially disruptive influence on the Warsaw Treaty Organization.[267] However, some parts of Ceausescu's conversation with the Chinese leadership actually concerned the Pact. The Romanian leader said his country was ready to cooperate progressively within the framework of the alliance, but in accordance with the vague founding charter only. He was determined to prevent transformation of the alliance into a supranational organization, which would deepen political, economic, and military integration of the Soviet sphere of influence, as Moscow intended. He accused China of helping to found the Pact, as Beijing had not opposed this step in 1955 and even accepted stature as an observer.[268] During his visit to Mongolia, Ceausescu again verbally assaulted the Warsaw Pact. He refused Jumdzgin Cedenbal's claim that this "peaceful" organization strove for the imposition of European security.[269] Romanian distance from the Warsaw Pact was also demonstrated following Ceausescu's absence from the

---

deeply impressed by the oriental cult of personality of communist leaders there. It gave him imagination of means and tools to affect the people and methods to maintain a *noblesse oblige* at the top of political power. Before the end of the 1970s, formation of the cults and rituals and related practices in Romania was finished. Deletant, "*Romania and the Warsaw Pact*"; Tejchman and Litera, *Moskva a socialistické země na Balkáně*, 116.

[266] *Informace o návštěvě stranické a vládní delegace RSR v asijských socialistických zemích*, 6. 6. 1971 TO(t) 1970-1974, box 1, i. č. 82, sign. 017/111, č. j. 023.362/71-2, AMZV, Prague.

[267] Mastny, *A Cardboard Castle?*, 43.

[268] The Chinese observers led by Mao Zedong withdrew from the Political Consultative Committee sessions in 1961. Mastny, Vojtech, "*China, the Warsaw Pact, and Sino-Soviet Relations under Khrushchev*," Accessed October 16, 2019. http://www.php.isn.ethz.ch/lory1.ethz.ch/collections/coll_china_wapa/intro_mas tnyf409.html?navinfo=16034].

[269] *Minutes of Conversation of the Executive Committee of the Central Committee of the Romanian Communist Party*, June 25, 1971, Accessed October 16, 2019. http://www.php.isn.ethz.ch/kms2.isn.ethz.ch/serviceengine/Files/PHP/16347/ipu blicationdocument_singledocument/29ce4e90-afa7-4eab-a83d-76928bb1131b/en/710625_minutes.pdf.

informal summer talks of party leaders held by Brezhnev in Crimea.[270]

Some Soviet satellites reacted to the situation more vigorously than the USSR itself. The first attempts to streamline Romanian foreign policy occurred in 1970. A few Warsaw Pact member-states were dissatisfied not only by the threats of Romanian withdrawal from alliance political meetings, but also by Ceausescu's Balkan initiatives. In March 1970, the Romanian minister of foreign affairs Mănescu formally called for the creation of a nuclear-free zone and a significant improvement of mutual cooperation in the region.[271] The Soviet Union and primarily some of its satellites considered this undesirable. However, the Warsaw Pact member-states' appeals for action against the policy of Bucharest temporally faded-out as Romania dampened its activity in mid-1970. In the wake of worsening mutual relations the next year, Hungary was the first to step out against Romanian maneuvers. In August, Hungarian diplomats were briefed to consistently refuse all attempts at aiming to disrupt either the Warsaw Pact or Comecon unity. This directive had a strictly anti-Romanian subtext.[272] The Communist Party of Czechoslovakia CC General Secretary Gustav Husák went even further: He accused the RCP of leaving Marxist-Leninist positions, and supported Soviet opinions of Romania harming the Warsaw Pact's interests. Unlike the Soviets, he also criticized the Romanian standpoint on the need to reduce the Pact's ability to affect its member-states' policy.[273]

Taking into account geographical factors, the Eastern Bloc states considered Bulgaria - which remained fully loyal to Moscow - as a natural

---

[270] *Vzťahy NDR-RSR - informácia,* 3. 11. 1971, TO(t) 1970-1974, box 1, i. č. 68, sign. 0344/111, č. j. 025495, AMZV, Prague.

[271] Ceausescu's leadership officially intended to follow up the initiatives presented during the years 1957-1959 by then Romanian Prime Minister Chivu Stoica. However, those proposals had striven for different aims.

[272] In addition, the directive occurred in a time of significant deterioration of Romanian-Hungarian relations, which had led to the cancellation of a planned meeting of Ceausescu and Kadar in July 1971. *K výsledkům srpnového společného zasedání ÚV MSDS a vlády MLR v oblasti zahraniční politiky,* 26. 8. 1971, TO(t) 1970-1974, box 5, i. č. 58 MLR, sign. 015/311, č. j. 024.374/71-2, AMZV, Prague.

[273] *Návrh zprávy ÚV KSČ k mezinárodním otázkám,* 13. 10. 1971, 1261/0/6, sv. 16, a. j. 15/2, NA, Prague.

bulwark against Romanian tendencies in the Balkan area.[274] In this, the Bulgarian leader Todor Zhivkov acted most proactively. During confident diplomatic talks, he called for solving the issue of Romania-Eastern Bloc relations. Zhivkov suggested working out a coherent strategy in order to influence Ceausescu's policy towards the Warsaw Pact.[275] Harsh criticism was voiced also from East Berlin.[276] The German Democratic Republic (GDR) sharply denounced Romanian-Balkan initiatives. They were marked as an attempt to establish some sort of "Balkan Pact" after the expected death of Josip Broz Tito, at a moment when Ceausescu had no strong competitor in the region. Regarding his visit to China, SED leadership warned against the creation of a Beijing-Bucharest-Belgrade-Tirana axis. The East-German stance mostly corresponded with Bulgarian and Soviet positions. The GDR therefore supported Zhivkov's proposal to the Warsaw Pact bodies to start dealing with Romanian policy. East-German diplomats suggested at least holding a deputy foreign ministers' meeting. Simultaneously, the Eastern Bloc countries should have striven for developing bilateral contacts with Romania as much as possible in

---

[274] *Politický vývoj vztahů BLR s balkánskými zeměmi*, 8. 10. 1970, TO(t) 1970-1974, box 7, i. č. 12, sign. 013/311, č. j. 025537, AMZV, Prague.

[275] *Záznam o rozhovoru p. Jiřího Kučery, velvyslance zdejšího ZÚ se p. Jevgenievem Gromuškinem, velvyslaneckým radou ZÚ SSSR v Sofii*, 22.10. 1971, DTO 1945-1989, i. č. 4, e. č. 3, č. j. 025.263/71-2, AMZV, Prague

[276] In the early 1970s, the relations between East Germany and Romania were merely sporadic. This situation was unique within the Warsaw Pact. The GDR-SRR liaison was made complicated by the Romanian general approach towards the Soviet Union and the Eastern Bloc. The main reason was the Romanian establishing of diplomatic relations with West Germany on the level of ambassadors in February 1969. Also, Romanian-East-German economic cooperation remained limited to a minimum. Moreover, the countries had no valid bilateral allied treaty - it was the exception within the Eastern Bloc. The document was finalized in September 1970. However, Bucharest postponed the signature for almost two years. This was probably also caused by East-German demands that the "West-German militarism" must be mentioned at least in the preamble of the treaty. Romania, which strove for friendly relations with the West, opposed. Not only Romanian obstructions, but undoubtedly also the radical positions of Ulbricht's leadership created obstacles to concluding the agreement. *Politbüro des ZK Reinschriftenprotokoll nr. 15*, 8. 4. 1969, DY 30/ J IV 2/2/1223, BArch, Berlin.

order to create favorable conditions for general improvement of mutual relations.[277] During a visit of the Bulgarian Prime Minister Stanko Todorov in Hungary, Janos Kadar's leadership also accepted this strategy. In late 1971, Zhivkov presented those intentions in person to Brezhnev.[278]

Romanian-Balkan initiatives apparently raised concerns in Moscow, which considered them part of the attempts to disintegrate both the Warsaw Pact and Comecon. The Soviet leadership could not rule out that Bucharest, according to its proclaimed long-term effort to dissolve power blocs, was creating the background for a potential future military-political arrangement on the peninsula. The Kremlin correctly suspected that the essence of Ceausescu's Balkan policy was to reduce the superpower's influence on processes in the region. Instead of full suppression of Romanian initiatives, the USSR tried to shift them in a more favorable direction from its point of view. Moscow considered the issue an integral part to ensuring the Warsaw Pact's influence on the Balkans. It was reminded that Romania was the only member of the alliance who had relatively normal relations with all countries of the peninsula.[279] Romanian activity therefore seemed to be a potentially

---

[277] *Vzťahy NDR-RSR - informácia*, 3. 10. 1971, TO(t) 1970-1974, box 1, i. č. 68, sign. 0344/111, č. j. 025495, AMZV, Prague. Possibility that the East-German officials presented the stances at the direct order of the Kremlin cannot be ruled out. After all, it was typical behavior of the Kremlin. At the time, in diplomatic talks with its satellites, the Soviet leadership actually warned against a coalition of China, Yugoslavia, Albania, and Romania, which could have weakened the Warsaw Pact's southern flank and potentially led to establishing some sort of 'Balkan bloc' with a sealing element of Anti-Sovietism. Baev, *"The "Crimean Meetings" of the Warsaw Pact"*.

[278] *Záznam o rozhovoru p. Jiřího Kučery, velvyslaneckého rady zdejšího ZÚ se p. Nikolajem Černevem, vedoucím 2.t.o. ministerstva zahraničních věcí BLR*, 2. 12. 1971, DTO 1945-1989, i. č. 4, e. č. 3, č. j. 026.066/71-2, AMZV, Prague; *BLR-informace o setkání stranických a státních představitelů SSSR a BLR v Moskvě*, 3. 12. 1971, TO(t) 1970-1974, box 1, i. č. 12, sign. 013/111, č. j. 026.067/71-2, AMZV, Prague.

[279] In comparison to other Warsaw Pact member-states, Romanian relations with Albania were the least tense. However, they were far from being smooth. Among other things, Enver Hoxha's regime constantly criticized Bucharest for its continuing membership in the Warsaw Treaty Organization. *Informace o VI. sjezdu Albánské strany práce*, 15.12. 1971, 1261/0/6, sv. 24, a. j. 25/info3, NA, Prague.

appropriate tool for spreading Warsaw Pact policy in the area. This was obviously possible only under the condition that the Six would be able to affect Bucharest and prompt it to implement a unified course for the alliance.[280]

In the first half of the 1970s, Todor Zhivkov formally stood at the frontlines of the effort to shape Bucharest's foreign policy. Brezhnev himself, who noted that Ceausescu had crossed the line, secretly entrusted him to this role. The BCP CC first secretary was supposed to act as a "moderator" in the attempt to prompt Romanian turnabout.[281] In the 1970s, Bulgaria was the most loyal and most dependent satellite of Moscow.[282] This could clearly be seen in Bulgarian foreign policy. The country always acted under the aegis of a unified course of the Warsaw Pact; guidelines of the alliance meetings became axioms to Sofia.[283] Bulgarian approach towards Romania was not solely determined by the more independent orientation of Ceausescu's leadership. Sofia approved of neither positive Romanian relations with Tito's Yugoslavia, nor its stance on the so-called Macedonian question.[284] Regarding Romanian Balkan initiatives, Zhivkov's regime, using clichéd ideological language, warned against the penetration of nationalism, imperialism, and Maoism, and called for "intended counter-pressure" from the Warsaw Pact.[285]

In January 1972, the RCP leadership received information that the rest of the alliance members intended to use the upcoming Political

---

[280] *Rumunsko a balkánská otázka*, 19. 10. 1971, DTO 1945-1989, i. č. 34, e. č. 52, č. j. 025.188/71, AMZV, Prague; *Záznam o návštěvě delegace FMZV vedené 1. námětkem ministra p. Fr. Krajíčkem v Moskvě*, 16. 10. 1971, TO(t) 1970-1974, box 2, i. č. 89, sign. 020/112, č. j. 0441/71, AMZV, Prague.

[281] Baev, "The Warsaw pact and Southern Tier Conflicts," 200; Baev, *"The "Crimean Meetings" of the Warsaw Pact"*.

[282] Tejchman and Litera, *Moskva a socialistické země na Balkáně*, 111.

[283] *Charakteristika postojů BLR k problematice evropské bezpečnosti*, 18. 2. 1971, TO(t) 1970-1974, box 7, i. č., sign. 013/311, č. j. 021.200/71-2, AMZV, Prague; *Zahraniční politika BLR*, undated 1973, DTO 1945-1989, i. č. 4, e. č. 10, č. j. 025.229, AMZV, Prague.

[284] *Vývoj styků BLR s NDR, PLR, MLR a RSR*, 12. 1. 1970, TO(t) 1970-1974, box 7, i. č. 12, sign. 013/311, , č. j. 020276/70-2, AMZV, Prague.

[285] *Bulharská zahraniční politika v oblasti Balkánského poloostrova*, 2. 2. 1972. DTO 1945-1989, i. č. 4, e. č. 7, č. j. 022.859/72-2, AMZV, Prague.

Consultative Committee session in Prague to initiate a harsh assault on
Romanian policy. In order to defuse expected criticism, Ceausescu asked
Zhivkov for a regulatory meeting. Bulgaria refused his call. The BCP
Central Committee First Secretary insisted that the issue was too serious
and therefore had to be dealt with multilaterally at the Warsaw Pact
supreme body meeting.[286] However, before the summit in the
Czechoslovak capital, Romania sent Moscow some signals that this time
it would fully cooperate.[287] At the meeting itself, Romanian representatives
acted relatively constructively and made compromises in order to avoid
criticism. Actually, they did not even oppose a new proposal to intensify
political talks within the Warsaw Pact's framework. For the first time,
Romania roughly admitted the possibility of formalization of such
consultations.[288] The USSR subsequently appreciated that, unlike the
previous year, Ceausescu also took part in the summer Crimean meeting.
His presence was considered by Moscow to be more important than the
fact that he again stated different positions there.[289] This new Romanian
trend towards the Warsaw Pact was caused, inter alia, by the failure of
Ceausescu's Balkan policy: In early 1972, his attempts to start official
multilateral talks on closer cooperation between the countries of the
peninsula had failed.[290]

---

[286] *Záznam o rozhovoru p. Jiřího Kučery, velvyslaneckého rady zdejšího ZÚ se p.
Andrášem Šárdim, velvyslaneckým radou ZÚ MLR v Sofii ze dne 19. ledna 1972*,
e. č. 8, i. č. 020.483/72-2, ibid.

[287] For instance, the Romanian officials were unusually interested in the
preparation of the meeting's agenda. *Telegram from Romanian Deputy Foreign
Minister George Macovescu to the Romanian Ambassador in Moscow*, January
10, 1972, Accessed October 16, 2019. http://www.php.isn.ethz.ch/kms2.isn.ethz.
ch/serviceengine/Files/PHP/16345/ipublicationdocument_singledocument/3908f
529-d54d-46ad-bdf5-8bc766221c9e/en/720110_telegram.pdf.

[288] *Minutes of the Meeting of the Hungarian Socialist Workers' Party Politburo
on the January 1972 PCC Meeting*, February 1, 1972, Accessed October 16, 2019.
http://www.php.isn.ethz.ch/kms2.isn.ethz.ch/serviceengine/Files/PHP/18105/ipu
blicationdocument_singledocument / a19328f7-c508-42f2-8f85-593a871d6293/
en/Minutes_Hungarian_Party_1972_en.pdf.

[289] *Notatka z przebiegu Spotkania I-szych sekretarzy bratnich partii na Krymie
/31 lipca 1972/*, PZPR KC, XIA/612, AAN, Warsaw.

[290] *Vývoj politiky RSR vůči Balkánu v poslední době*, 12. 5. 1972, DTO 1945-
1989, i. č. 34, e. č. 54, č. j. 023.034/72-2, AMZV, Prague.

In the wake of this shift, the Warsaw Treaty Organization did not deal with the Romanian question collectively in 1972,[291] despite the general importance given to it by Moscow. At the time, the Kremlin had decided not to use official structures of the Warsaw Treaty Organization to solve disputes between its member-states. Therefore, all such activities were put on a bilateral level. Bulgaria proved its full dependency on actual Soviet course. As the Kremlin appeared satisfied with the latest shift in Romanian policy, Sofia also relented in its engagement of the issue. Hence, criticism resonated mostly from the GDR. East-German officials correctly stated that the essence of Romanian foreign policy remained unchanged. The recent mitigation of Romanian policy and improvement in mutual relations were considered as just a tactical retreat of Ceausescu's leadership.[292] In fact, Bucharest actually feared isolation within the Eastern Bloc and there were some serious warnings sent by Romanian allies. For instance, the PUWP CC First Secretary Gierek refused to visit the SRR as part of his protest against its foreign policy[293] and

---

[291] Even the speech of Zhivkov, who initiated the move, actually remained limited to unfocused warnings against NATO's efforts to penetrate the Balkan area through its intensified influencing of Yugoslavia and Albania. *Speech by the General Secretary of the Bulgarian Communist Party*, January 25, 1972, Accessed October 16, 2019. http://www.php.isn.ethz.ch/kms2.isn.ethz.ch/service engine/Files/PHP/18104/ipublicationdocument_singledocument/21225070-bcb3 -464a-86f2-edacf3a23fe3/en/Speech_Zhivkov_1972_en.pdf.

[292] In 1972, the relations between Romania and East Germany improved. During the visit of the East-German party and government delegation in Bucharest, the Romanian officials without any obstacles supported the GDR's stances on the future of West Berlin and after two years of obstructions they signed the bilateral allied treaty. *K výsledkům návštěvy stranické a vládní delegace NDR v RSR-informace*, 21.6. 1972, TO(t) 1970-1974, box 1, i. č. 68, sign. 0344/111, kr. 1, č. j. 023.817/72-4, AMZV, Prague; *Záznam ZÚ Berlín o vztazích NDR-RSR*, 5. 4. 1972, č. j. 022.277/72-4, ibid.

[293] Cancellation of Gierek-Ceausescu meeting was initiated by Poland itself, not by Moscow. Warsaw only informed the Kremlin about its intention. Afterwards, the PUWP politburo member Józef Tejchme was sent to Romania in order to explain the reasons which had led to revocation of the planned visit. In his reaction, Ceausescu accused Poland of unacceptable duress on his country; among other things, he protested against the assaults on Romanian policy in the Polish press. The incident later resulted in weakening economic cooperation between

Czechoslovak minister of foreign affairs Bohuslav Chňoupek was also very critical during his journey to the country. Romania actually reacted by toning down its rhetoric, at least for a while.[294]

Poland also did not suggest influencing Romania through open polemics. It preferred unofficial personal contacts with Romanian officials. The main aim remained not to expose disputes within the Eastern Bloc publicly.[295] However, Polish strategy proved to be less effective. For instance, in November 1972, Ceausescu assured a Polish delegation that he would coordinate his next moves in the CSCE process with the Warsaw Pact. In fact, at the following party plenum he declared a separate course of action in order to insert into the agenda of the conference issues of disarmament and withdrawal of foreign troops from territories of European countries. The Warsaw Treaty Organization, actually, strongly dismissed those principles.[296] Despite the fact that Romania invariably supported official alliance resolutions concerning the Helsinki process, its real policy differed in many aspects. The Eastern Bloc countries, therefore, believed that Bucharest intended to use CSCE to further weaken its ties to both the Warsaw Pact and the Soviet sphere of influence, in general.[297]

---

both countries. *Tezy do rozmów z towarzyszami radzieckimi*, undated 1973, PZPR KC, XIB/126, AAN, Warsaw.

[294] *Celkový obraz hlavních aspektů vnitřní i zahraniční politiky RKS a RSR za rok 1972*, 19.12. 1972, DTO 1945-1989, i. č. 34, e. č. 54, č. j. 027280/72-2, AMZV, Prague; *RSR – druhá informace o plnění závěrů kolegia ministra zahr. věcí v relaci s RSR ze 30. prosince 1971*, 3. 4. 1973, TO(t) 1970-1974, box 2, i. č. 82, sign. 017/112, č. j. 022.038/73-2, AMZV, Prague.

[295] *Informácia k vzťahom PĽR – RSR*, 18. 6. 1973, box 2, i. č. 79, sign. 016/111, č. j. 023801, ibid.

[296] *Celkový obraz hlavních aspektů vnitřní i zahraniční politiky RKS a RSR za rok 1972*, 19. 12. 1972, DTO 1945-1989, i. č. 34, e. č. 54, č. j. 027280/72-2, AMZV, Prague.

[297] Romanian stances tended not only against the Warsaw Treaty Organization, but also against general integration within the Eastern Bloc, as the new demands of the Romanian foreign affairs minister Macovecu showed. In July, during the CSCE negotiations in Helsinki, he proposed dissolution of not only military, but also economic blocs. *Postoj RSR k přípravným jednáním KEBS v Helsinkách*, 13. 4. 1973; ibid, č. j. 024.505/73, *Ke stanovisku RSR na I. fázi KEBS*, 18. 7. 1973, DTO 1945-1989, i. č. 34, e. č. 56, č. j. 022518/73, AMZV, Prague.

In 1973, the relations between Romania and the Warsaw Treaty Organization reached a new low point. Despite some failures, Ceausescu still maintained his Balkan ambitions. In the first months of the year, in connection with shifts in the CSCE process and the upcoming Vienna disarmaments talks,[298] he once more tried to mobilize the countries of the peninsula into closer cooperation, regardless of their geopolitical ties. Bulgaria unambiguously stood up against this effort. Sofia refused Romanian calls for consultation and noted that the issues were supposed to be discussed on the level of the Warsaw Pact only.[299] In fact, Zhivkov's regime intended to support solely Soviet stances. Indeed, Bulgarian activities in the Balkans mainly served to protect the interests of Moscow. One of these interests was also the elimination of the Romanian regional policy impact.[300] An important clash between both of the Balkan Warsaw Pact member-states occurred at a Crimean meeting of party leaders in July 1973. Zhivkov decided to plainly criticize Romanian-Balkan policy. The Bulgarian leader assigned Romania as pro-China, Maoist, and thus an extremely hostile axis. This, together with his other verbal assaults, almost resulted in an open rift. In his reaction, Ceausescu theatrically threatened to leave the session. The situation was calmed down by Brezhnev's personal intervention. However, he only managed to iron the issue out with serious difficulty.[301]

The Crimean incident was crucial for future development. Brezhnev himself also denounced some Romanian stances, such as their claim of Chinese contribution to détente or their appeals to the start of practical moves towards the simultaneous dissolution of military blocs.[302]

---

[298] In 1973, Vienna talks on the reduction in armed forces and armaments in Central Europe began between NATO and the Warsaw Treaty Organization. Romania once more tried to prevent negotiations on the basis of blocs.

[299] *Záznam o přijetí rady ZÚ BLR p. G. Georgieva vedoucím 2. t.o. J. Hesem dne 3. ledna 1973*, DTO 1945-1989, i. č. 4, e. č. 16, č. j. 020.038/73-2, AMZV, Prague.

[300] *Informace o problematice vztahů BLR s balkánskými zeměmi*, 3. 9. 1973, e. č. 10, č. j. 025230/73, ibid.

[301] Baev, *"The "Crimean Meetings" of the Warsaw Pact"*; Tejchman and Litera, *Moskva a socialistické země na Balkáně*, 87.

[302] *Wystąpienie końcowe Tow. Breżniewa*, undated, probably July 1973, PZPR KC, XIA/613, AAN, Warsaw.

In addition, he suspected that Ceausescu failed to inform the wider structures of the RCP about the results of the Crimean meeting of party leaders which defined, behind closed doors, the short term international priorities of the Eastern bloc.[303] Despite those facts, Brezhnev considered the aberrations of Romanian foreign policy not fundamentally problematic enough to justify risking an open clash and a new split within the Eastern Bloc. He believed that collectively dealing with the issue at the Warsaw Pact meeting was an unnecessary and dangerous move. In terms of this, the Soviet General Secretary also altered the scope of the unofficial Crimean meetings. After the 1973 row, he opted to invite the party leaders individually, but never together.[304] Remember, at the time, official multilateral political meetings under the auspices of the Warsaw Pact were held less frequently as well. This alteration was probably not caused exclusively by the Romanian factor, as it reflected a general shift in Moscow's approach towards interaction with the countries in the Soviet sphere of influence.

In fact, the Soviet leadership ignored all appeals for vigorous solutions. Thus, these were not presented solely by the agile Zhivkov.[305]

---

[303] Furthermore, unlike the others party leaders, Ceausescu in praxis totally ignored the results of the Crimean meeting. *Rumunská zahraniční politika ve světle komuniké z krymského jednání 30.-31. července 1973*, 17. 1. 1974. DTO 1945-1989, i. č. 34, e. č. 61, č. j. 010.459/74, AMZV, Prague; *Zpráva o sovětsko-rumunských vztazích*, 3. 4. 1974, TO(t) 1970-1974, box 9, i. č. 89, sign. 020/311, č. j. 012545, AMZV, Prague.

[304] Baev, *"The "Crimean Meetings" of the Warsaw Pact"*.

[305] It happened, for instance, after the visit of the Bulgarian minister of foreign affairs Mladenov in Romania when he argued with Ceausescu on implementation of the recent Political Consultative Committee resolution. At the turn of 1973 and 1974, Zhivkov constantly warned that the Romanian nationalist course had reached a level which negatively affected the situation within both the Warsaw Pact as well as the International Communist and Worker's Movement. From his point of view, some countermeasures were necessary; he suggested at least consulting the issue on the alliance's level. *Záznam z informace Nikolaje Černěva, ved. 2. t.o. MZV BLR o oficiální návštěvě ministra ZV BLR v RSR*, 22. 12. 1973, DTO 1945-1989, i. č. 4, e. č. 11, č. j. 020.516, AMZV, Prague; *Záznam o rozhovoru velvyslaneckého rady ZÚ MLR v Sofii A. Šárdiho s velvyslaneckým radou čs. ZÚ v Sofii J. Kučerou*, 4.1. 1974, e. č. 19, č. j. 010177/74, ibid; *Podkladové materiály pro návštěvu ministra zahraničních věcí BLR Petra Mladenova v ČSSR*, 31. 1. 1974, e. č. 17, č. j. 011.023/74-2, ibid.

The Polish leadership also assaulted Romanian policy harder than Moscow. During his conversation with Brezhnev, Gierek even broached the possibility of establishing closer cooperation with "proletarian internationalism" forces in Romania and deeper integration of the Romanian army into the Warsaw Pact. Although the Polish leadership realized that Ceausescu knew where the limits of his more independent policy were,[306] they carefully probed whether Moscow would not try to replace him.

Along with its strategy so far, Moscow intended to influence Ceausescu in backrooms only, even at the Political Consultative Committee session held in Warsaw, April 1974.[307] However, Edward Gierek an Erich Honecker decided to break the silence on Romania's approach. After all, a few months before, both leaders were unanimous in their criticism of Romania and their bad relations with Ceausescu's regime.[308] Thus, at the alliance's supreme body meeting, both leaders openly denounced Bucharest for helping NATO and damaging the interests of socialist countries through its moves regarding disarmament talks.[309] However, their verbal assault actually failed; Ceausescu used it to

---

[306] Romania rejected Polish criticism of its Balkan initiatives. Ceausescu's regime referred to previous Polish actions and appeals for settlement of the Central-European region. *Informácia o některých poznatkoch z návštevy viceministra p. Trepczynského v Rumunsku*, 22. 5. 1973, TO(t) 1970-1974, box 2, i. č. 79, sign. 016/111, č. j. 023134, AMZV, Prague; *Informácia k vzťahom PĽR – RSR*, 18. 6. 1973, č. j. 023801, ibid; Handouts for conversation of the PUWP CC First Secretary E. Gierek with the CPSU CC General Secretary L. Brezhnev, undated, perhaps 1973, PZPR KC, XIB/126, AAN, Warsaw.

[307] *Zpráva o sovětsko-rumunských vztazích*, 3. 4. 1974, TO(t) 1970-1974, box 9, i. č. 89, sign. 020/311, č. j. 012545, AMZV, Prague.

[308] *Návštěva stranické a vládní delegace PLR v NDR*, 3. 7. 1973, box 10, i. č. 68, sign. 0344/311, č. j. 024.098/73-4, ibid.

[309] East Germany was especially caught by surprise when they realized that the Romanian representative at the Vienna disarmament talks intended to propose international supervision of involved countries' territories through a network of control posts; these would also have supposedly clarified the delineation of the areas of no military activity. According to the East-German delegation, Romania in this manner threatened both success of the Vienna talks as well as security and sovereignty of the socialist states. Telegram to SED CC on Romanian proposals put forward at Political Consultative Committee session, April 18, 1974, DY 30/2351, BArch, Berlin.

accuse the GDR of exceeding the rules of the Warsaw Treaty
Organization. Moreover, regardless of such criticism of Romania,
Ceausescu still continued the presentation of different routes.[310]

In the wake of this fail, the Soviet strategy of long-term and
systematic influence appeared to be much more effective. At the meeting,
Romania, in fact, announced its preliminary consent to establishing the
Committee of Ministers of Foreign Affairs within the Warsaw Pact. In
spite of Bucharest agreeing with this step under the condition of
maintaining every party and government the right to shape its own foreign
policy, Romanian officials admitted that the meetings of the body could
be held even thrice a year. Ceausescu stated that, in the current situation,
stressing the importance of the alliance's political activities rather than its
military dimension was necessary. This opinion essentially corresponded
with the then claims of the Soviet leader.[311] In fact, Romania ceased
stalling reform of the Pact's political structures after almost ten years
without being pushed firmly into it.[312] The Six actually expected that the
creation of new bodies would bring increased opportunity to bulwark
divergent Romanian tendencies more effectively.[313] As a result of the

---

[310] Not only were Romanian proposals on Vienna disarmament talks unacceptable
for the rest of the Warsaw Pact member-states, but Ceausescu also acknowledged
the positive role of China in the détente process and declared support for the
Egypt-Israel treaty. Furthermore, he refused to label Pinochet's coup in Chile as a
fascist putsch. On the contrary, the Romanian leader urged the final communiqué
to appeal for simultaneous dissolution of NATO and the Warsaw Treaty
Organization once more. The Romanian representative in the editorial
commission again tried to alter the documents in preparation in terms of
Ceausescu's speech. The situation resulted in the failure of Secretariat's work; the
texts had to be compiled during a separate negotiation of the Soviet and Romanian
delegations.
[311] *Zpráva o průběhu a výsledcích zasedání Politického poradního výboru států
Varšavské smlouvy*, 24.4. 1974; *Stručná charakteristika vystoupení vedoucích
jednotlivých delegací na zasedání politického poradního výboru*, 24. 4. 1974,
1261/0/6, sv. 115, a. j. 117/1, NA, Prague.
[312] After 1970, the USSR instructed its satellites to be patient and not to put the
issue in front of Romania in any ultimate way. *Zpráva o sovětsko-rumunských
vztazích*, 13. 9. 1971, sv. 12, a. j. 11/info1, ibid.
[313] *Zpráva o výsledcích jednání náměstků ministrů zahraničních věcí členských
států Varšavské smlouvy*, 5. 2. 1975, sv. 144, a. j. 149/2, ibid.

Political Consultative Committee meeting in Warsaw, individual ministries of foreign affairs of Warsaw Pact member-states were instructed by Moscow to avoid any activity against Romania. The Soviet satellites were only supposed to analyze Romanian policy towards both the Warsaw Treaty Organization and Comecon.[314]

This conciliatory approach of Moscow to the Romanian question was apparent even towards the beginning of the disintegration of the entire Eastern Bloc. In the first half of the 1970s, the USSR had forged no methods on how to pacify easily and without serious international complications an undesirable policy of some country within its sphere of influence. The Kremlin did not even realize the severity of the problem until the beginning of the next decade when it was unable to force Polish party leadership to suppress their opposition movement.

Isolation of Romania

The relations between Romania and the Eastern Bloc remained strained. In the first half of 1974, the Six had to admit that the attempts to streamline Romanian foreign policy towards the unified course of the Warsaw Treaty Organization had failed. In fact, Romania became isolated. This was one of the reasons why Ceausescu's regime began to invite top-level party and state delegations of the Warsaw Pact countries to Romania intensively. Some of the alliance members refused such visits, leering that the Romanian leadership could have interpreted these as acts of approval and sympathy for Bucharest's foreign policy.[315] If the talks even took place in the end, Ceausescu actually tried to avoid any international topics. He was well aware that a significant difference of opinions prevailed in that field. However,

---

[314] *Zpráva o návštěvě ministra zahraničních věcí RSR p. Macovesca v ČSSR*, 2. 5. 1974, TO(t) 1970-1974, box 2, i. č. 82, sign. 017/112, kr. 2, č. j. 012.472/74-2, AMZV, Prague.

[315] *Čtvrtá informace o plnění závěrů kolegia ministra zahraničních věcí v relaci s Rumunskou socialistickou republikou ze dne 30.12. 1971*, 28. 8. 1974, PK 1953-1989, kat. č. 705, k. č. 156, č. j. 014.631/74-2, AMZV, Prague; *NOTAKA dotycząca wymiany patyjnej pomiędzy PZPR i RPK w latach 1974-1975*, undated, PZPR KC, XII 3056, 937/8, AAN, Warsaw.

he did not intend to escalate the situation anyway.[316] A dubious
relationship between Romania and the Warsaw Pact member-states was
also proven by the fact that in the summer of 1974, after the series of
explosions in the Romanian industrial complexes, rumors of the possible
participation of "allies" spread across the country.[317] In autumn of the
same year, even the USSR did not believe a restoration of the Romanian
standard position within the Eastern Bloc was possible. Assurances from
the Romanian side about its alleged intention to cooperate closely with the
Warsaw Pact member-states during finalization of the Helsinki process
were considered by Moscow to be part of Bucharest's attempts to quit an
overall isolation. The proposals of Ceausescu's regime neither
corresponded with the unified course of the Warsaw Treaty Organization
nor resonated with other CSCE participants.[318]

At the time, the USSR had already fully calculated the specific
position of Romania within its sphere of influence. For example, the
country was not involved in the recently intensified cooperation of the
Warsaw Pact member-states' embassies.[319] After all, at the beginning of
the 1970s, Bucharest had been excluded in a similar way from the
alliance's collaboration in the field of press departments; this should have

---

[316] In these intentions, Ceausescu behaved during his meeting with Zhivkov in
May 1974. However, the Bulgarian leader once again criticized mostly the
Romanian policy towards the Balkans and refused Ceausescu's appeals for closer
cooperation in the region. *Information über die Gespräche zwischen den
Genossen Todor Shiwkow und Nikolae Ceausescu in Woden, 11.-12. Mai 1974*,
19. 5. 1974, DY 30/J IV 2/2J/5294, BArch, Berlin.

[317] Information about development in Romania by Pilowitcz, the Polish
ambassador to the Soviet Union, undated, most likely 1974, PZPR KC, XIA/718,
AAN, Warsaw.

[318] *Návštěva náměstka ministra SSSR N.N. Rodionova v RSR*, 14. 11. 1974, DTO
1945-1989, i. č. 34, e. č. 61, č. j. 017.644/74, AMZV, Prague.

[319] The situation of individual Pact member-states sharing information about their
diplomatic activities uncoordinatedly - sometimes verbally, sometimes in writing
- should have been replaced by a unified system of written information exchange
between the ambassadors in the Warsaw Treaty Organization countries. The
information for the USSR was supposed to be the most detailed, of course.
*Záznam o jednání velvyslance SSSR v ČSSR V.V. Mackeviče s ministrem
zahraničních věcí ČSSR B. Chňoupkem*, 6. 11. 1974, TO(t) 1970-1974, box 6, i.
č. 89, sign. 020/117, č. j. 017.434/74-1, AMZV, Prague.

prevented a different evaluation of the events by the media in individual member-states. It was part of the measures to promote the joint foreign policy course of the Pact. The main reason of such Romanian ostracism was the different attitude of Ceausescu's regime towards the invasion of Czechoslovakia.[320]

Such isolation forced Romania to look for the alternatives. Ceausescu focused on improving relations with the West, mostly with the United States.[321] Moreover, in the early months of 1975, he played the Balkan card gain. The Romanian leader called for holding the top-level summit of all the countries of the peninsula and potentially establishing some sort of a political alliance.[322] Meanwhile, he started bouncing around the idea of Romanian admission to the Non-Aligned Movement, disregarding their membership in the Warsaw Pact. Bulgaria was supposed to take action against those tendencies. Most likely on the Soviet instructions, Zhivkov's regime changed its tactic. Henceforward, Sofia tried to affect Ceausescu through improving mutual contacts. Bulgaria pursued multiple goals: To the Warsaw Pact members, Sofia presented its policy as part of efforts to maintain Romania on the positions of the joint course of the alliance. Nevertheless, Romania was the only Bulgarian territorial link to the Eastern Bloc and an important transit country for economic cooperation with the USSR.[323] Keeping good relations

---

[320] It is interesting that the strengthening of cooperation between the press departments was affirmed by official protocols. However, for the Soviet Union the verbal agreements on close collaboration were enough. *Zhodnocení spolupráce TO s tiskovými odbory MZV pěti zemí Varšavské smlouvy*, 7. 10. 1970, PK 1953-1989, kat. č. 596, k. č. 140, AMZV, Prague.

[321] *Zahraničně-politická aktivita RSR na úseku vojenských styků v roce 1974 a v I. čtvrtletí 1975*, 14. 4. 1975, DTO 1945-1989, i. č. 34, e. č. 64, č. j. 012.989/75, AMZV, Prague. At the time, for example, Washington promised Ceausescu's regime that U.S.-Romanian mutual trade contacts would run under better conditions in comparison to other Warsaw Pact member-states. Miroslav Tejchman considers the 'most favored nation' statute as an American reward for more independent Romanian policy on Moscow. Tejchman, *Nicolae Ceausescu*, 88.

[322] *Wizyta przyjáźni towarzysza Todora Żiwkowa w Socjalistycznej Republice Rumunii, 14-17 luty 1975 rok*, PZPR KC, XIA/718, AAN, Warsaw.

[323] Sufficient capacities for the military naval shuttle connection between the USSR and Bulgaria in the Black Sea did not exist until 1978. Wanner, *Brežněv*, 73.

with Ceausescu's regime was therefore very urgent from the Bulgarian point of view.

Romania welcomed the opportunity of breaching its isolation. In May 1975, during his visit to Bulgaria, the Romanian foreign affairs minister Macovescu once more openly and vocally supported the strategy in the CSCE process which had been collectively approved at the last Political Consultative Committee session. In this way, he tried to convince Bulgaria that recent negotiations of the Romanian diplomats with the Western countries' representatives fully respected the joint course of the Warsaw Pact. However, the Bulgarian side was aware that there had been no shift in Ceausescu's diplomacy: despite the assurance of joint alliance's tactic, he still asserted own policy during the talks with the West.[324] Unlike before, Zhivkov's regime did not assault Romania at the time. The Bulgarian First Secretary himself visited Bucharest in June 1975. He talked Ceausescu out of his intentions concerning the Non-Aligned Movement. According to Zhivkov's argumentation, some NATO members could have followed Romanian step, which would have resulted in an undesirable shift in the policy of the Movement (from the socialist countries' point of view). Such an explanation was most likely just auxiliary. Zhivkov's real effort was to prevent further weakening of Romanian ties to the Eastern Bloc. In terms of his attempts to affect Bucharest's policy through improving mutual relations, the Bulgarian leader was even compliant in signing a communiqué talking about the possible simultaneous dissolution of the military-political blocs. In the past, he actually vehemently tried to convince Ceausescu on the need for an additional military strengthening of the Warsaw Pact.[325]

In praxis, Bulgarian activity brought only little success. In the end, the Romanian delegation behaved at the Helsinki conference like the

---

[324] *Záznam rozhovoru mezi vedoucím 2. t.o. MZV BLR I. Kjulevem s velvyslaneckým radou čs. ZÚ v Sofii J. Kučerou*, 27. 5. 1975. DTO 1945-1989, i. č. 4, e. č. 19, č. j. 013984, AMZV, Prague.

[325] *Vztahy BLR s balkánskými zeměmi*, 29. 4. 1975, e. č. 27, č. j. 013288, ibid; *Oficiální přátelská návštěva bulharské stranické delegace v Rumunsku*, 30. 6. 1975 č. j. 014776, ibid; *Společné prohlášení o oficiální přátelské návštěvě stranické a vládní delegace Bulharské lidové republiky v Rumunské socialistické republice*, 20. 6. 1975, příloha k č. j. 0677/75, ibid.

country was not a member of any military-political alliance. Their stances almost corresponded with the positions of the neutral states. However, by firm rejections of dislocation of the nuclear weapons in the foreign territories, Romania went even further.[326]

A cold Romanian attitude towards the Warsaw Pact in the mid-1970s was well illustrated by reluctance and strong unwillingness, which characterized participation of the country in the pompous 20th anniversary of the alliance's foundation. Romania refused to organize a manifest meeting of the social organizations on that occasion.[327] Ceausescu's regime opposed supranational celebrations and pushed for their limitation to the national level only. The Romanian representatives expressly stated that the extent of military actions must be determined by the state authorities, not by the Unified Command of the alliance. Bucharest proposed an alternative: instead of the anniversary, the ongoing CSCE process should have been celebrated by the Warsaw Treaty Organization - a scenario Romania diplomatically called "unwise and unhelpful".[328] However, in planning of the upcoming propagandistic performance, Moscow took no account of Romanian positions.[329] Ceausescu's regime therefore realized that some minimal, at least formal, participation in the event was inevitable and thus promised to hold a ceremonial gathering. Symptomatically, the event was to be totally overshadowed by the official visit of the Dutch queen Juliana, who arrived to Romania just one day before the celebration. The articles in Romanian press concerning the Warsaw Pact's anniversary stressed a temporary character of the alliance and a need for dissolution of the military blocs.[330] That was a typical tactic

---

[326] *Postoj delegace RSR na KBSE*, undated, e. č. 64, č. j. 011.868/75, ibid.

[327] The Czechoslovak normalization regime gladly took up this task.

[328] *Protocol of the Meeting of the Deputy Foreign Ministers*, January 29, 1975, Accessed October 16, 2019. http://www.php.isn.ethz.ch/kms2.isn.ethz.ch/serviceengine/Files/PHP/17368/ipublicationdocument_singledocument/6f4029fa-2bbf-4ab9-9a47-dfff127167b0/en/750129_Protocol_E.pdf.

[329] *Zpráva o výsledcích jednání náměstků ministrů zahraničních věcí členských států Varšavské smlouvy*, 5. 2. 1975, 1261/0/6, sv.144, a. j. 149/2, NA, Prague; *Záznam o návštěvě rady ZÚ BLR D. Eneva u vedoucího ZPO L. Handla*, 11. 4. 1975, DTO 1945-1989, i. č. 4, e. č. 16, č. j. 89.299/75-SPO, AMZV, Prague.

[330] *Oslava 20. výročí Varšavské smlouvy v Rumunsku*, 30. 5. 1975, i. č. 34, e. č. 64, č. j. 014045/75, ibid.

of the RCP leadership. During the official visits, in their speeches and in the propaganda, the Romanian politicians usually tried to avoid any notes on the Warsaw Pact. Eventually, they reduced those to either necessary minimum or emphasized Romanian stance on temporary existence of the alliance.

With similar reluctance, Bucharest took part in the first meeting of the Warsaw Pact's member-states' parliaments representatives which took place in the Polish capital as a part of the alliance's anniversary. Unlike other countries, Romania did not send the chairman of its national assembly, but just his deputy, Stefan Mocutu. Romanian delegates were the only ones whose speeches lacked the adoration of the Soviet Union or the importance of the Warsaw Treaty Organization and rather emphasized the benefits of the CSCE and the principles of sovereignty and independence of individual states.[331]

Contradictions in disarmament talks

In praxis, Romania also refused to follow a unified strategy of the Warsaw Pact in the disarmament talks.[332] Intensified consideration of some form of disarmament agreement regarding a shift in détente at the beginning of the 1970s therefore created another cause of disagreement between the Eastern Bloc and Bucharest. Taking into account the Soviet positions, the majority of the Warsaw Treaty Organization members were more inclined towards the division of the disarmaments' negotiations into specific aspects. The political meetings of the Pact should have been the main forum for coordination of a proper approach. On the contrary, Romania quite unrealistically proposed overall and total disarmament. The country also promoted extending collaboration in the issue to the socialist

---

[331] *Zpráva o setkání představitelů parlamentů členských států Varšavské smlouvy u příležitosti jejího 20. výročí*, 6. 6. 1975, 1261/0/6, sv. 159, a. j. 162/13, NA, Prague.

[332] *Zpráva o nejdůležitějších opatřeních sovětské zahraniční politiky v období ledna – června 1970*, 27. 5. 1970, TO(t) 1970-1974, box 8, i. č. 89, sign. 020/311, č. j. 0122/70, AMZV, Prague.

countries outside the Warsaw Pact.[333] At the beginning of the 1970s, Ceausescu's regime at least agreed with the allies that the complicated question of the disarmament and reduction of military contingents in Europe could not be linked to the Helsinki process. Bucharest was aware that negotiations on those aspects would be extremely lengthy. Thus, the prospect of holding the all-European security conference, which was one of the top Romanian priorities, would move dangerously away. Therefore, in his talks with the neutral countries, Ceausescu personally opposed putting disarmament on the agenda of the CSCE, as the U.S. pushed for until 1972.[334]

Soon after Washington and Moscow made a deal on a separation of the Helsinki process and the disarmament talks, Ceausescu totally reconsidered his position. As we already noted, after 1973 he strove for including the issue in the CSCE program. One of the reasons was probably the fact that the Vienna talks on reduction of the armed forces and armaments between NATO and the Warsaw Pact were strictly limited to Central Europe, and therefore did not concern Romania. The country was assigned to the group of states with an observer status only, so Ceausescu had no opportunity to directly affect the negotiations. Despite the initial obstruction, which will be explained in detail in the chapter dealing with the Vienna talks, Bucharest's approach in fact brought no fatal troubles for the Warsaw Treaty Organization. By 1975, Romania was *de facto* passive and objected neither proposals nor tactics of the East. Its limited criticism aimed at the framework of the negotiations which centered around a narrow group of states consisting of the USA and USSR as the main powers of both alliances, accompanied by two representatives of NATO as well as the Warsaw Pact.[335]

---

[333] *Zpráva o poradě ministrů zahraničních věcí členských států Varšavské smlouvy konané ve dnech 18.-19. 2. 1971 v Bukurešti*, 26. 2. 1971, 1261/0/5, sv. 154, a. j. 238/7, NA, Prague.

[334] *ZÚ Helsinki k návštěvě Ceaušesca ve Finsku*, 7. 7. 1971, TO(t) 1970-1974, box 1, i. č. 82, sign. 017/111, č. j. 023.735/71-2, AMZV, Prague.

[335] *Informace o průběhu IV. etapy jednání o vzájemném snížení ozbrojených sil a výzbroje ve střední Evropě*, 3. 3. 1975, PK 1953-1989, kat. č. 723, k. č. 158, č. j. 018660/74, AMZV, Prague; *Pozice RSR v otázkách odzbrojení a na vídeňských*

Contradictions between official policy of the Warsaw Treaty Organization and Romanian attitude towards the disarmament issue in the first half of the 1970s appeared not only during Vienna talks. Romania worked out its own 'disarmament programme' and tried to promote it at every international forum. The USSR and its satellites dismissed it as unrealistic. The Romanian representatives therefore acted totally independently regardless of the strategy of the Warsaw Pact, especially in the UN Committee on Disarmament in Geneva. That was helped by the fact that those talks officially took place on the basis of the sovereign countries rather than blocs. Ceausescu loudly attacked the American-Soviet chairmanship. He urged the forum not to deal with partial questions like the chemical weapons or the nuclear tests, but move straight forward to negotiations on overall nuclear disarmament. In May 1975, Romania – most likely purely instrumentally – supported the claims of some of the developing countries on a revision of the Treaty on the Non-Proliferation of Nuclear Weapons.[336] This move was seen by the Warsaw Pact as a direct violation of the alliance's founding charter.[337] Article II, among other things, committed the signatories to *collaboration* in an attempt to ban nuclear weapons.

Romania and the Warsaw Pact military structures

As already noted, Romania allowed the approval of reform of the Warsaw Pact's military structures in March 1969. However, collaboration of the Romanian armed forces with the allies was not unproblematic, similarly to the Pact's political framework. In the early months of 1970, Ceausescu provocatively stated that, according to his interpretation of the new alliance's statutes, the Romanian army is accountable to the party,

---

jednáních o snížení ozbrojených sil a výzbroje ve střední Evropě, 1. 7. 1975, DTO 1945-1989, i. č. 34, e. č. 72, AMZV, Prague.

[336] Romania voted similarly to the Non-aligned Movement countries. That position was probably linked to ongoing rapprochement between Bucharest and this organization.

[337] *Přístup RSR k otázkám odzbrojení*, 1. 8. 1975, DTO 1945-1989, i. č. 34, e. č. 72, AMZV, Prague.

government, and State Defense Council only, not to the Unified Command.[338] He also refused the suggestions of General Schtemenko, the chief of Staff of the Unified Armed Forces, about establishing the supranational, nationally mixed units and divisions within the alliance's troop formations.[339] The Romanian foreign affairs minister later confirmed military commitments of his country to the Warsaw Pact allies, but only in case of *"imperialistic aggression in Europe."*[340]

Despite those statements, behavior of the Romanian delegations at the meetings of the Warsaw Pact's military bodies missed the obstructive features. The representatives of Romania regularly took part in sessions of the Committee of Defense Ministers as well as the Military Council. In case they did not agree with some of the measures proposed there, they required their stances to be included in special protocols. In this manner, Romania actually enforced the opt-outs for itself; ensuring that some (for Ceausescu's regime) undesirable rules of military cooperation within the alliance would not concern the country. Neither the Six nor the Unified Command's representatives usually confronted Romania. The final protocols of the military bodies' meetings actually contained Romanian stances. In this case, the supreme commander of the Pact, Marshal Ivan Yakubovski, or the Soviet minister of defense, Marshal Andrei Grechko, often expressed regrets at Romanian unwillingness to participate in "joint matters." However, there was an absence of any duress. The records from the meetings suggest that they did not significantly try to alter the positions of the Romanian delegations.[341] In some cases, there was an additional commentary on Romanian-specific stances, declaring that the rest of the Warsaw Pact member-states did not

---

[338] *Porada základních kádrů ministerstva ozbrojených sil Rumunské socialistické republiky. Projev s. N.Ceausesca*, 6. 2. 1970, e. č. 47, č. j. 021.026/70, ibid.
[339] Wanner, *Brežněv*, 41.
[340] *Zasedání VNS RSR k zahraniční politice*, 3. 4. 1970, DTO 1945-1989, i. č. 34, e. č. 47, č. j. 022.007/70-2, AMZV, Prague.
[341] *Zápis o 6. zasedání Výboru ministrů obrany členských států Varšavské smlouvy*, undated, most likely February 1974, FMO GŠ/OS, k. 10, p. č. 46, č. j. 39038, VÚA, Prague; Report by the Czechoslovak defense minister, M. Dzúr, to the CPCz CC General Secretary, G. Husák, on the 7th meeting of the Committee of Defense Ministers of the Warsaw Pact member-states, 20. 1. 1975, k. 41, p. č. 138, č. j. 36041, ibid.

agree with those.[342] In this way, the Unified Command strove for isolation of Romania and neutralization of its potentially disruptive influence. Romanian positions should not have jeopardized the smooth realization of the measures within the Six.[343] Using that tactic, Romania managed to prevent deployment of the so-called representatives of supreme commander in its armed forces. As we explain later in more detail in the chapter dealing with the Warsaw Pact's military structures, this was one of the key mechanisms used by the USSR to supervise the armies of its European satellites since the beginning of the 1970s. Romania also refused to pay for the activities of these officers through financial contributions to the joint budget of the alliance. The other Warsaw Pact member-states largely stayed silent on this opt-out.[344]

At the Warsaw Pact military bodies' meetings, a unanimous consensus on their intentions was achieved much more often than during the political talks.[345] The Romanian representatives also usually took part in bilateral consultations on the issues of armaments and the joint plans of army development, which took place either in the Staff of the Unified Armed Forces or directly in the Soviet Defense Ministry.[346] Those talks were only lengthier in comparison to the other Pact member-states.[347] In 1973, Romania also agreed with the beginning of the work on the

---

[342] *8. zasedání Vojenské rady Spojených ozbrojených sil*, 24. 5. 1973, k. 37, p. č. 131, č. j. 36034, ibid.

[343] *Zpráva ministra národní obrany o výsledku jednání Výboru ministrů obrany států-účastníků Varšavské smlouvy*, February 1972, k. 40, p. č. 136, č. j. 36039, ibid.

[344] *Páté zasedání Vojenské rady Spojených ozbrojených sil*, 1. 11. 1971, k. 36, p. č. 130, č. j.36033, ibid.

[345] *3. zasedání Vojenské rady Spojených ozbrojených sil, informační zpráva pro s. prezidenta ČSSR armádního generála L. SVOBODU*, November 1970, k. 35, p. č. 129, č. j. 36032, ibid; *Výsledky 11. zasedání Vojenské rady Spojených ozbrojených sil a porady vedoucích funkcionářů spojeneckých armád*, 25. 11. 1974, k. 38, p. č. 133, č. j. 36036, ibid.

[346] *KRUG, KUB – zpráva z konsultace v SSSR*, 4. prosince 1970; tamtéž, k. 38, p. č. 133, č. j. 36036, *Změna v programu 11. zasedání Vojenské rady Spojených ozbrojených sil*, 21. 10. 1974, k. 15, p. č. 52, č. j. 39044, ibid.

[347] For example, Romania negotiated the protocol on development of its armed forces in the years 1976-1980 for the longest time of all the alliance member-states. *Informace o výsledcích 13. zasedání Vojenské rady Spojených ozbrojených sil*, 1. 11. 1975, k. 38, p. č. 133, č. j. 36036, ibid.

controversial wartime statute.[348] Despite stressing the statement that the military-political blocs had to be dissolved, Romanian leadership never put forward a proposal on unilateral cancellation of the Warsaw Treaty Organization. In the period analyzed in this work, the country also never really considered leaving the alliance's military structures.[349] Despite Ceausescu pushing for disarmament in the first half of the 1970s, at the Warsaw Pact meetings he did not seriously question the ongoing armament programs. In the spring of 1974, at the Political Consultative Committee session in Warsaw, he actually supported the Soviet claims that the socialist countries had to improve their defense capabilities until a concrete disarmament agreement was achieved.[350]

Romania and the Warsaw Pact's military drills

The most visible problem was reluctance of the Romanian leadership to send troops to joint alliance exercises. Since 1961, the country sent to the ground maneuvers of the Unified Armed Forces either small groups of generals and officers or limited numbers of military equipment, and to the demonstrations only.[351] Two years later, Bucharest stopped allowing the Warsaw Pact's drills to be held on its territory. In the following period, Romania sent just military observers to the exercises in the other alliance member-states. The Soviet propaganda, however, called their presence a "participation of the Romanian troops." In 1967, Ceausescu's regime altogether quit taking part in the alliance's exercises. In addition, it banned the Romanian officers from studying at Soviet military schools. Bucharest also announced the beginning of work on its

---

[348] *Porada vedoucích funkcionářů generálních štábů*, 18. 6. 1973, k. 15, p. č. 52, č. j. 39044, ibid.

[349] Baev, "The Warsaw pact and Southern Tier Conflicts," 202

[350] Ceausescu ostentatiously failed to mention the Warsaw Treaty Organization as an aegis for improving the defense capabilities, but talked solely about individual countries. *Stručná charakteristika vystoupení vedoucích jednotlivých delegací na zasedání politického poradního výboru*, 24. 4. 1974, 1261/0/6, sv. 115, a. j. 117/1, NA, Prague.

[351] Opriş, "Die rumänische Armee," 187

own military doctrine that would have been independent from the Warsaw Pact.[352]

Romanian apprehension over joint maneuvers of the alliance was deepened by the episode of the suppression of the Prague Spring. Ceausescu's regime was leery that a stay of the 'allies' during the maneuvers in Romania could have been the first step towards military intervention.[353] Nevertheless, in sensitive military matters, Bucharest tried not to provoke Moscow excessively. Therefore, Romania did not categorically reject the Warsaw Pact's drills on own territory. However, according to the new statute of the Unified Armed Forces, the country reserved the right to host such actions only under condition of an approval by particular constitutional bodies.[354] The USSR attempted to break Romanian positions in September 1969. The Chief of Staff of the Unified Armed Forces, General Sergei Schtemenko, called for the tactical-operational maneuvers on the Romanian territory to include participation of foreign allied troops. The Romanian military delegation therefore traveled to the Soviet capital and persuaded the general that such intention was not realizable for many reasons. The officers proposed a tactical exercise on the maps instead, which should have taken place under the command of the Romanian defense minister in the following year. Although Schtemenko admitted a possible postponing of the action, he firmly refused any change of its intended form. He argued that the drills, including direct deployment of the troops, took place in all the Pact member-states and to make an exception for Romania was therefore unacceptable.[355] As the chief of Romanian general staff Ion Gheorghe realized, the matter of the entire issue was a signal for the Warsaw Pact members as well as the West, rather than the military drill itself. It was a

---

[352] Tejchman and Litera, *Moskva a socialistické země na Balkáně,* 51.

[353] Wanner, *Brežněv,* 201.

[354] In terms of the nature of Ceausescu's regime, that in fact meant an approval of the top party leadership. *Ohlas v RSR na budapešťské zasedání PPV VS,* 16. 4. 1969, DTO 1945-1989, i. č. 34, e. č. 42, č. j. 022.229/69-2, AMZV, Prague.

[355] Remember that the drills in the Warsaw Pact's southern flank were held much less frequently than in the so-called western direction. Only nine of fifty big military maneuvers of the Unified Armed Forces held during 1955-1976 took place in the south region. Litera, "Unifikace sovětského bloku," 90.

question of prestige for the Soviet Union whether it would be able to restore the Unified Armed Forces maneuvers in Romania.[356] However, the Soviet duress brought no results. In the end, facing firm Romanian stances, the representatives of the Unified Command gave in. They must have consented to the promise of a staff exercise only; also in this case, Romania made approval of its parliament a condition for future participation of foreign military personnel.[357]

From the Soviet point of view, the situation was worsening. In spring 1970, Romania began to require the military drills on its territory to be held on the basis of particular bilateral agreements between individual participating countries; in consequence, joint maneuvers should not have been a result of the alliance itself. This unprecedented praxis that - according to Davies Deletant - aimed to prevent the Soviet leadership from justifying potential military aggression by refering to the Pact's documents.[358] However, that was a logical result of Romanian long-term efforts. Petre Opriş notes that the country had struggled with deployment of foreign troops on the basis of the Warsaw Treaty Organization in the era of Ceausescu's predecessor Gheorghe Gheorghiu-Dej already. Since 1963, Romania always tried to back up the movement of the armed forces on the foreign territories by special government conventions, unrelated to the Pact. However, Marshal Grechko, who served as the supreme commander of the alliance at the time, let Romanian requests expire without any answer.[359]

---

[356] At the time, the Romanian side signaled to NATO via diplomatic channels its constant position on the Warsaw Pact's maneuvers. For instance, in October 1969, General Popa assured the Turkish ambassador to Bucharest that no alliance military exercises would be held on the Romanian territory in that year. He also ruled out the Romanian armed forces' participation in such actions, except for sending observers. Cable from the British embassy in Bucharest, October 1969, FCO 25/606, TNA, London.

[357] *Letter from General Ion Gheorghe, Deputy Minister of Romanian Armed Forces, to Nicolae Ceauşescu*, September 9, 1969, Accessed October 16, 2019. http://www.php.isn.ethz.ch/kms2.isn.ethz.ch/serviceengine/Files/PHP/16492/ipu blicationdocument_singledocument/9424ec81-72b3-48a6-81f0-47c8676850f7/en/690909_letter.pdf.

[358] Deletant, "*Romania and the Warsaw Pact*".

[359] Opriş, "Die rumänische Armee," 191.

In the 1970s, Ceausescu's regime was able to use participation of the Romanian troops in the alliance's drills as a part of its above-explained balancing tactic. In September 1970, for example, in terms of a goodwill gesture towards the Warsaw Pact, Romania sent their armed forces to the "Brotherhood in Arms-70" maneuvers in the GDR.[360] Six months before, nothing suggested their participation. A proper ideological indoctrination of the soldiers and instillation of "joint principles of the socialist command to the troops" were significant aspects of the drills. Those terms were in praxis a guise for persuading the soldiers to fight not only in the scope of their national armies, but mainly alongside the Soviet armed forces.[361] Such a principle was strongly rejected by Romanian leadership. In decisions on participation in the maneuvers, Bucharest reflected actual needs rather than clear ultimate stances and flexibly changed its positions. In 1972, for instance, Romanian troops took part in the "Tunja-72" drills in Bulgaria but absented from the "Shield-72" maneuvers in Czechoslovakia, despite it being the main military action of the Warsaw Pact in that year,[362] with more than 100,000 soldiers participating.[363]

The irregular sending of Romanian army units to the exercises of the Unified Armed Forces also reflected Bucharest's generally negative attitude towards military drills in foreign countries. Ceausescu's regime labeled them as an unacceptable demonstration of force contributing to the persistence of international tensions.[364] Therefore, it was no surprise when

---

[360] *Současný politický postoj RSR ve vztahu ke státům Varšavské smlouvy*, 11. 1. 1971, DTO 1945-1989, i. č. 34, e. č. 52, č. j. 020.206/71, AMZV, Prague; *Vystoupení náčelníka štábu ČSLA na zasedání Vojenské rady Spojených ozbrojených sil dne 27.10. 1970 k otázce "Výsledky operační a bojové přípravy za rok 1970",* FMO GŠ/OS, k. 35, p. č. 129, č. j. 36032, VÚA, Prague.

[361] An emphasis on the ideological preparation of the troops during the drills in the GDR was much bigger than, for example, in Czechoslovakia. *Anlage Nr. 6 zum Protokoll Nr. 21/70 vom 28. 4. 1970,* DY 30/ J IV 2/2/1281, BArch, Berlin; *Informace o přípravě spojeneckého cvičení „ŠTÍT-72",* 21. 8. 1972, 1261/0/6, sv. 51, a. j. 50/10, NA, Prague.

[362] Ibid; *7. zasedání Vojenské rady Spojených ozbrojených sil,* 28. 10. 1972, FMO GŠ/OS, k. 37, p. č. 131, č. j. 36034, VÚA, Prague.

[363] Kramer, "Comentary on Soyuz-75 and Shchit-88".

[364] *Ohlas v RSR na budapešťské zasedání PPV VS,* 16. 4. 1969, DTO 1945-1989, i. č. 34, e. č. 42, č. j. 022.229/69-2, AMZV, Prague.

in 1971 the Romanian representatives at the alliance collective military bodies' meetings opposed the intentions of the Unified Command to strengthen its own role in combat preparation of individual armies, as well as the planning of joint maneuvers. At the 5th meeting of the Military Council in the Polish capital, Romanian general Marin Nicolescu objected to the rising numbers of maneuvers under the auspices of the Warsaw Pact, alleging they negatively impacted the quality of the drills. Romania stipulated conditions for the Staff of the Unified Armed Forces under which the country would join the proposed measures: Bucharest strove for weakening the powers of the supreme commander in term of planning maneuvers, mostly trying to prevent him having a right to ask the Romanian army for deployment of certain numbers of troops. Moreover, during the alliance's air defense drills, Romanian military planes should not land on the foreign territories and vice versa.[365]

Thus, Romania was reluctant to allow admission of not only allied ground forces, but also the military planes, to its territory. In this regard, the situation was more complicated for Ceausescu's regime because an integration of the anti-aircraft defense (consisting of the air forces of individual member-states) had been taking place within the Warsaw Pact since the mid-1960s already. In 1969, the unified anti-aircraft defense system was officially established. In its framework, the operational-tactical drills were held annually in the countries of the southern flank of the alliance. Besides a mutual interaction of the involved command officers, those actions also included real flights and cooperation between the military air bases in Romania, Bulgaria, Hungary, and the Soviet Union. However, the operations of foreign aircrafts over Romanian territory remained strictly limited.[366] In 1970, Romania refused to sign the new operational documents of the alliance's air defense system. The disagreement with landing allied warplanes on Romanian airports was the main reason. At the meeting of the Committee of Defense Ministers held

---

[365] *Plán realizace PROTOKOLU č. 005 jednání Vojenské rady Spojených ozbrojených sil*, undated 1971, FMO GŠ/OS, k. 8, p. č.42, č. j. 39034, VÚA, Prague; *Páté zasedání Vojenské rady Spojených ozbrojených sil*, 1.11. 1971, k. 36, p. č. 130, č. j. 36033, ibid.

[366] Opriş, "Die rumänische Armee," 195.

in February 1972 in East Berlin, the Unified Command failed to persuade Romania to revise the position.[367] At that time, the Polish general staff warned the Warsaw Pact allies that Ceausescu intended to send an unofficial military delegation to the United Kingdom in order to probe potential cooperation with the British air-industries. In that manner, Romania attempted to develop its own multipurpose fighter-bomber jet; the USSR and later also France had rejected similar Romanian proposals already.[368] At the time, it seemed that Ceausescu's regime strove for the possible ability to produce key weapons alone in the future.[369] However, the relations between Bucharest and the Warsaw Pact never reached a point where the Romanian government totally closed the airspace to the allies. Rather, Romania showed an almost demonstrative effort to get respect for its specific positions from the rest of the alliance member-states. During 1972-1974, the Soviet military aircraft usually flew over Romanian territory two to four times per month, but with permission of the country's authorities.[370] After a relatively short term, in the end, Romania joined the operational directives of the unified air defense system in 1975.[371]

In the early 1970s, due to its attitude towards the Warsaw Pact maneuvers, Romania was under constant pressure by the Soviet

---

[367] *Zpráva ministra národní obrany o výsledku jednání Výboru ministrů obrany států-účastníků Varšavské smlouvy*, February 1972, FMO GŠ/OS, k. 40, p. č. 136, č. j. 36039, VÚA, Prague.

[368] *Notatka informacyjna dotyczy palnowanej wizyty szefa Sztabu Generalnego rumuńskich sił zbrojnych w Londynie*, 20. 1. 1972, PZPR KC, XIB/173, AAN, Warsaw. Romania considered development of a combat aircraft together with Yugoslavia later. In 1973, during his visit to Romania, Grechko largely surprisingly stated that it was an internal issue of the country and the USSR would not interfere. *Informacja w sprawie opinii bukaresztańskich kół dyplomatycznych n.t. charakteru, przebiegu i konsekwencji wizyty marszałka A.Greczki w Rumunii w dniach 20-24 kietnia br.*, 15. 5. 1973, ibid.

[369] At the time, the only combat aircrafts in the Romanian air force were the Soviet fighters MiG-21. In June 1973, during his meeting with the army command, Ceausescu appealed for development of own combat equipment produced by the Romanian industry. Wanner, *Brežněv*, 207-208.

[370] Opriş, "Die rumänische Armee," 198.

[371] *Informace o výsledcích 13. zasedání Vojenské rady Spojených ozbrojených sil*, 1. 11. 1975, FMO GŠ/OS, k. 38, p. č. 133, č. j. 36036, VÚA, Prague.

leadership. In consequence, Bucharest announced its willingness to restore participation in joint drills of the alliance in March 1972.[372] At the end of the month, the Romanian military personnel took part in the staff exercise in Bulgaria.[373] However, the planning of future actions was still problematic. In autumn of the same year, the RCP leadership discussed the draft of the drill schedule for the Romanian army outlined by the Unified Command. The scenario of the "Soyuz-73" maneuvers, especially, sparked outrage within the Ceausescu leadership, causing them to oppose the proposed composition of the command group. The training for repulsing enemy aggression and the counteroffensive of Soviet, Romanian, and Bulgarian troops should have been commanded by approximately four hundred officers, but only one-quarter of them were of Romanian nationality. General Ion Ionita advised the RCP CC General Secretary to approve the drill directives for 1973. However, he should have rejected an idea that some of the actions would be led by the Warsaw Pact supreme commander, Marshal Ivan Yakubovski. According to Ionita, such a scenario could have provided the basis for direct control of the Unified Command over Romanian troops even in the case of real combat operation. Romania considered this principle unacceptable. The party as well as army leadership still insisted that the armed forces had to remain under the authority of national military bodies only. In case of either war or peace, the command of the Romanian front should have belonged solely to Ceausescu himself along with the defense ministry. Nevertheless, the Romanian leader approved Yakubovski's command role in the maneuvers. It was most likely a sop, which Ceausescu intended to exchange for the possibility to sign bilateral international conventions defining the movement of foreign troops on Romanian territory during the "Soyuz-73" exercise. However, the Unified Command still opposed this idea. During the negotiations with the alliance's military officials, Ionita therefore returned to the previously tested tactic. He proposed a change of the drill's character; only a war game on the maps should have taken place consisting of neither real operation of the Soviet troops on Romanian territory nor deployment of the Romanian army in Bulgaria. The Unified

---

[372] Rey, "The USSR and the Helsinki process," 75.
[373] Wanner, *Brežněv,* 46.

Command accepted the compromise. The movement of the troops across the borders as well as their deployment should have been determined by the one-time agreements. Unlike the initial Romanian demand, these were not convened on the governmental level, but by the military institutions only.[374] The entire issue had an unexpected solution. In the end, the drills were held solely on the maps.[375] That was probably not a result of Romanian attitude, but rather a part of overall regress of the huge military maneuvers in Europe related to deepening détente and the beginning of Vienna disarmament talks. At the same time, for instance, the scheduled extensive drills "Shield-73" were cancelled as well.[376]

The Romanian approach towards the "Soyuz-73" exercise also reveals that Ceausescu's regime did not intend to sharpen existing contradictions with Moscow, especially in sensitive military issues. Bucharest strove for compromise and easing tensions instead.[377] For example, in early 1974, at the 10th meeting of the Military Council held in Budapest, Romania approved the participation of Soviet officers in a staff-exercise on Romanian territory without any serious obstructions.[378] Since 1975, Romanian staff and troops took part in smaller ground drills of the Warsaw Pact regularly.[379] The conditions were negotiated on an *ad hoc* basis, as the country in the same year refused to join the directives for holding joint maneuvers of the Unified Armed Forces. Romania also rejected the agreement on transportation of the alliance's units across the

---

[374] At the time, during the meeting of the alliance Military Council, the Romanian general Nikolescu demanded deployment of any operational units on the foreign territories of the Warsaw Pact members to be subject to the special bilateral state conventions. *8. zasedání Vojenské rady Spojených ozbrojených sil*, 24. 5. 1973. FMO GŠ/OS, k. 37, p. č. 131, č. j. 36034, VÚA, Prague.

[375] Opriş, "Die rumänische Armee," 195-199.

[376] Mastny, Vojtech, "Imagining War in Europe. Soviet Strategic Planning," in *War Plans and Alliances in the Cold War*, eds. Vojtech Mastny, Sven Holtsmark and Andreas Wenger (London: Routledge, 2006): 33.

[377] Opriş, "Die rumänische Armee," 198.

[378] Wanner, *Brežněv*, 49.

[379] Baev, "The Warsaw pact and Southern Tier Conflicts," 201.

member-states' territories. According to Romanian officials, these documents violated territorial sovereignty.[380]

Due to similar reasons, in 1973, Romania opposed the joint organization of the Unified Armed Forces' logistic support during combat operations. Despite the document consisting of mostly general principles, it was another significant step forward to the Warsaw Pact's military integration. On the basis of the alliance, the agreement defined the rules for transfer of the Unified Armed Forces as well as the landing of warplanes at foreign airbases. At the 9th meeting of the Military Council in Prague, General Nikolescu formally disapproved the guidelines, referring to the fact that the document failed to include the amendments of his country. The supreme commander Yakubovski admitted that accusation during a following discussion. The Unified Command actually solely reflected the proposals of the Six in the material.[381] At the time, the Warsaw Pact's military headquarters most likely realized that there was no chance to alter Romanian attitude towards deployment of the troops. Therefore, the issue should have been solved within the rest of the alliance member-states. In the first half of the 1970s, Romania adopted a similar strategy towards other contentious aspects such as integration of individual navies' operations within thealliance[382] or the civil

---

[380] *Informace o výsledcích 13. zasedání Vojenské rady Spojených ozbrojených sil*, 1. 11. 1975, FMO GŠ/OS, k. 38, p. č. 133, č. j. 36036, VÚA, Praha. An extreme Romanian position on defining movement of the Warsaw Pact's troops in the foreign territories by particular conventions was illustrated by the strong requests on signing those documents even before the manifestation meeting of the soldiers at the alliance member-states' borders on the 20th anniversary. *Protocol of the Meeting of the Deputy Foreign Ministers*, January 29, 1975, Accessed October 16, 2019. http://www.php.isn.ethz.ch/kms2.isn.ethz.ch/serviceengine/Files/PHP/17368/ipublicationdocument_singledocument/6f4029fa-2bbf-4ab9-9a47-dfff127167b0/en/750129_Protocol_E.pdf.

[381] *9. zasedání Vojenské rady Spojených ozbrojených sil*, 7. 11. 1973, FMO GŠ/OS, k. 37, p. č. 131, č. j. 36034, VÚA, Prague.

[382] In march 1969, the Romanian military vessels took part in the Black Sea drills alongside the Soviet and Bulgarian navies for the last time. Already the Bulgarian commanders complained that the Romanian admiralty strongly tended towards individual actions and showed only little willingness to cooperate with the allied fleets. Since that point, Romania refused to participate in joint naval formations in the Black Sea as well as the collective analyses of the drills. For this purpose,

defense.[383] Serious problems also occurred during the talks on
construction of the protected command posts of the Unified Armed

---

the officials of the Romanian navy obstructed in various ways. In consequence,
during the first half of the 1970s, the cooperation of the Warsaw Treaty
Organization's navies in the region concerned Bulgaria and the Soviet Union.
Participation of Romanian vessels in the Black Sea drills in 1973 was a one-off
case only. Boycott of the navy exercises by Romania directly opposed the
intentions of the USSR. In the early years of the 1970s, the Soviet naval command
strove for improving coordination and unification of the command system of the
Warsaw Pact states' navies. The alliance's fleets should have been equipped with
a unified communication device. Also, the drills focusing on the anti-submarine
operations should have been intensified. At the alliance collective military
meetings, the Romanian representatives did not oppose any of these intentions,
however, they declared that their country would not participate. Baev, "The
Warsaw pact and Southern Tier Conflicts," 202; *Návrh usnesení Výboru ministrů
obrany států-účastníků Varšavské smlouvy k druhé otázce pořadu jednání - o
stavu a součinnosti válečného loďstva států-účastníků Varšavské smlouvy,*
undated, 1970, FMO GŠ/OS, k. 9, p. č. 44, č. j. 39036, VÚA, Prague; *Zpráva
ministra národní obrany o výsledku jednání výboru ministrů obrany států-
účastníků Varšavské smlouvy,* May 1970, ibid.
[383] In the early months of 1975, Romania stood up against the tendencies to
strengthen competences of the Unified Command in the field of civil defense
units. The Staff of the Unified Armed Forces was supposed to take over some
responsibilities of individual member-states in their drill. At the 7th meeting of the
Committee of Defense Ministers, Romania protested and referred to the Warsaw
Pact's military statutes, which stated that the alliance Staff had authority over the
troops within the Pact's joint formation only; the civil defense was not included.
Marshal Grechko refused such argumentation. He asserted the clause which also
empowered the Committee of Defense Ministers to deal with 'other military issues
and measures requiring joint coordination'. The Soviet defense minister
considered the civil defense one of those. *Zpráva ministra obrany ČSSR M. Dzúra
pro generálního tajemníka ÚV KSČ G. Husáka o 7. zasedání Výboru ministrů
obrany členských států Varšavské smlouvy,* 20. 1. 1975, FMO GŠ/OS, k. 41, p. č.
138, č. j. 36041, VÚA, Prague.

Forces[384] and joint communication network.[385] In some of those issues, Romania mostly opposed the common praxis of the Warsaw Pact's military structures: the Unified Command often confronted the officials of

---

[384] The issue of building the protected command posts began to be discussed within the Pact in late 1970. Two years later, at the key coordinative meeting in the Staff of the Unified Armed Forces, Romania did not oppose the intentions. However, Bucharest later made its participation conditional on either getting a fifteen-year loan or postponing the beginning of the build-up to the next decade. In autumn 1973, Romania totally reviewed its position and refused to sign the agreement. After lengthy negotiations between the supreme commander and the Romanian military officials, a compromise was achieved. The country was supposed to join the project at the planned beginning, in 1975. Nevertheless, Romania demanded assurance that it would financially contribute to the construction of the post for the south-west war field only, which should be located in Bulgaria. The Romanian officials dismissed any participation in built-up of the similar object for the western war field in Poland as well as paying for the service and operating costs of the post in Bulgaria. Bucharest also enforced the opportunity to pay a required amount after 1980, under the conditions negotiated with the alliance Staff later. The strict position of Romania is clearly contrasted with the Czechoslovak approach. Despite the project significantly complicating the economic plan of the country, the normalization regime, fully loyal to Moscow, approved funding for the protected command posts. Translation of the latter of the Czechoslovak defense minister M. Dzúr to the supreme commander of the Unified Armed Forces I. Yakubovski, July 2, 1972, FMO GŠ/OS, k. 11, p. č. 47, č. j. 39039, VÚA, Prague; *Porada k výstavbě VS pro Spojené velení*, 7. 11. 1972, p. č. 48, č. j. 39040, ibid; *Úvodní slovo s. ministra národní obrany ČSSR k otázce výstavby chráněných velitelských stanovišť Spojených ozbrojených sil*, 13. 12. 1972, ibid; *Dohoda k výstavbě velitelských stanovišť Spojených ozbrojených sil, informační zpráva pro generálního tajemníka ÚV KSČ soudruha JUDr. Gustáva HUSÁKA, CSc.*, 27. 10. 1973, ibid; *Výsledky 11. zasedání Vojenské rady Spojených ozbrojených sil a porady vedoucích funkcionářů spojeneckých armád*, 25. 11. 1974, k. 38, p. č. 133, č. j. 36036, ibid; *Porovnání protokolu o rozpočtu Spojených ozbrojených sil na rok 1975*, 19. 12. 1974, k. 18, p. č. 56, č. j. 39049, ibid.

[385] In February 1974, at the meeting of the Committee of Defense Ministers, Romania with the voice of the head of its defense resort, General Ionita, distanced itself from the build-up of the alliance's joint communication network. Bucharest did not oppose the keystone of the project, but the fact that the Staff of the Unified Armed Forces ignored the Romanian comments in the documentation. Ionita admitted that innovation of the communication network as well as construction of the protected command posts were significant and relevant tasks and that the Romanian armed forces were to participate. He only declared that his country would not join these aspects of the system which allowed the Unified Command to directly order the Romanian troops.

the member-states with a *fait accompli*. There were only relatively superficial consultations on various issues. Moreover, Romania stipulated conditions of individual government's approvals. Thus, the country opposed a model known from Czechoslovakia, for example, where the defense minister only vaguely informed the top political circles about the agreements in the Warsaw Pact's military framework.[386]

Army propaganda in Romania

Romania intended to participate in the Warsaw Treaty Organization's military structures, but solely under specific conditions. These also consisted in an ability to maintain full control over propaganda within the Romanian armed forces. At the beginning of the 1970s, "party-ideological work" in individual armies still ran largely independently from the Warsaw Pact.[387] The cooperation between the Main Political Administrations remained very limited.[388] However, Moscow sought to alter this state.[389] The consideration of more intensive collaboration occurred after 1972. The Pact was supposed to formulate a joint approach

---

[386] *Zápis o 6. zasedání Výboru ministrů obrany členských států Varšavské smlouvy*, undated, most likely February 1974, FMO GŠ/OS, k. 10, p. č. 46, č. j. 39038, VÚA, Prague.

[387] The representatives of the Main Political Administrations held collective talks in the Warsaw Pact's framework sporadically and usually during special occasions. In 1970, for example, the colloquium dealing with "global strategy of the American imperialism and the West-German imperialists and militarists at European war field" took place during the "Brotherhood in Arms-70" drills. The Soviet dominance was obvious again. The main tasks of the political indoctrination in the individual armies of the Warsaw Pact member-states were defined by the report of the Soviet general Mikhail Kalashnik. *Bericht über das Kolloquium der Politischen Hauptverwaltungen der Armeen des Warschauer Vertrages vom 31. 08. bis 02. 09. 1970 in Dresden,* DY 30/J IV 2/2J/3114, BArch, Berlin.

[388] Collaboration of the Main Political Administrations within the Warsaw Treaty Organization was uncoordinated and limited to the alliance's drills. It focused mostly on a strengthening of knowledge within the soldiers of their belonging to the Unified Armed Forces. *Směrnice ministra národní obrany a náčelníka hlavní politické správy ČSLA pro politickou a ideově výchovnou práci ve výcvikovém roce 1972,* 1. 11. 1971, 1261/0/6, sv. 17, a. j. 18/2, NA, Prague.

[389] Wanner, *Brežněv,* 46.

towards "ideological subversion" within its member-states' armed forces. Bucharest opposed such tendencies at the 4th session of the Committee of Defense Ministers. General Ionita declared that the propagandistic activities fell within the remit of individual ruling parties. On the level of the Warsaw Pact, the Political Consultative Committee should have solely dealt with the propaganda. The Romanian general threatened that his country would not take part in following the alliance's gathering of defense ministers if the items were discussed there.[390] However, that remained just words. Unlike the other member-states, Romania demonstratively failed to send the chief of the Main Political Administration to the 5th meeting of the Committee of Defense Ministers which took place in February 1973 in Warsaw. The head of the Polish defense resort, Wojciech Jaruzelski, formally proposed the regular consultations of the chiefs of the Main Political Administrations of the Warsaw Pact member-states to be established. In addition, the Staff of the Unified Armed Forces should have created its own press center and begun to issue the alliance's propagandistic bulletin. Ionita remained very passive during the discussion.[391] Once again, Bucharest did not intend to block the intentions of the rest of the member-states. Romania was content if those did not concern its armed forces.

Keep in mind that Romanian moves were not driven by any sympathy for "imperialistic propaganda." Ceausescu's regime strictly opposed the Western influences. However, it was afraid of strengthening the ideological cooperation within the Unified Armed Forces. Romanian army propaganda reflected the specific position of the country in the Soviet sphere of influence as well as the different rhetoric of the Romanian leadership. This was made apparent, for example, during the Days of Romanian Army events organized by the embassies of Romania in the Eastern Bloc countries. Usually, the Warsaw Pact member-states' military officials were not invited at all, or only invited together with the Western military attachés. In the latter case, Romanian leadership showed

---

[390] *Zpráva ministra národní obrany o výsledku jednání Výboru ministrů obrany států-účastníků Varšavské smlouvy,* February 1972, FMO GŠ/OS, k. 40, p. č. 136, č. j. 36039, VÚA, Prague.
[391] *5. zasedání Výboru ministrů obrany,* February 1973, k. 41, p. č. 137, č. j. 36040, ibid.

propagandistic movies about Romanian armed forces that totally dismissed membership in the Warsaw Treaty Organization and contained strong nationalist features. Moreover, there were warning statements that the army of Romania was ready to fight every enemy, not only "the imperialists."[392]

Mutual distrust in military issues

Skepticism of Romania by Moscow was strong, especially regarding sensitive military issues. For instance, in the spring of 1971, Soviet representatives in the UN expert committee for research on the impacts of the global armament were leery that Romania, within its campaign for massive disarmament, could have demonstratively revealed some key information about the Warsaw Pact's military programs, regardless of instructions by the Unified Command.[393] However, the USSR was mostly suspicious of potential Romanian ties to China. The Kremlin believed that the Asian power strove for strengthening its influence in the Balkan region, and in this regard Beijing was also seriously interested in the military situation there, including the role of the Romanian and Bulgarian armed forces in the Warsaw Pact's doctrine.[394] The discovery of confidential military maps in a briefcase belonging to the senior Romanian officer, who traveled to China through Moscow, supported the Soviet suspicion that Bucharest revealed the Pact's military secrets to Beijing. This assumption became a welcome pretext for

---

[392] *Vzťahy NDR-RSR - informácia*, 3. 11. 1971, TO(t) 1970-1974, box 1, i. č. 68, sign. 0344/111, č. j. 025495, AMZV, Prague; *Zpráva z filmkoktejlu na ZÚ RSR v Moskvě*, 17. 9. 1973, box 6, i. č. 89, sign. 020/117, č. j. 025440, ibid.

[393] In 1971, both NATO and the Warsaw Treaty Organization were supposed to provide data about their military programs to the UN experts from the Committee for research of impact of global armament. In the Cold War reality, naturally, this information was revealed selectively. The Warsaw Pact countries' representatives in the body coordinated their activities with the senior officials of the Unified Command. *Zpráva o výsledcích konzultace s akademikem Jemeljanovem*, 11. 5. 1971, box 4, č. j. 016 681/71-OMO, ibid; *Záznam o jednání velvyslance Mendělevičе v OMV MZV dne 15.3. 1971*, č. j. 016.427/71, ibid.

[394] *Poznámky k politice maoismu v oblasti Balkánského poloostrova*, 12. 9. 1973, box 1, i. č. 12, sign. 013/111, č. j. 025572, ibid.

exclusion of the Romanian officers from operational planning in the Unified Command.[395]

In autumn 1971, a significant stir in the Warsaw Pact member-states was caused by the rumor about dismissal of two deputies of the chief of Romanian general staff, allegedly due to their pro-Soviet stances. The diplomats of the Six shared information that some pro-Chinese generals replaced both men. Bucharest soon firmly disclaimed such reports.[396] However, purges within the Romanian armed forces actually took place during the 1970s. Since the end of the previous decade, Ceausescu's policy bothered the rest of pro-Soviet cadres in the army command. Regarding a new phase of worsening relations between Romania and the Eastern Bloc in 1971, some of the Moscow-oriented officers together with Ceausescu's opponents from the party apparatus began to plan a coup. The action, whose background has not been revealed and explained yet, failed. More than four dozen officers, led by General Ion Şerb, were arrested. According to the initial reports, Şerb was silently executed as a Soviet spy. In fact, he was sentenced to seven years in prison for giving the defensive plans of the Romanian capital to the Soviet intelligence. Actually, it is not out of the question that the entire case was just a provocation organized by the DIE, a Romanian intelligence service, which would have demonstrated to the domestic as well as foreign public how Ceausescu's more independent policy irritated Moscow.[397]

The new law of the country's defense approved in late 1972 also illustrated the lack of mutual trust between Romania and the Warsaw Pact. Its first article contained a statement that any action by a foreign state violating national sovereignty, independence, or territorial integrity of Romania, was illegal. The Eastern Bloc countries were leery that such a formulation aimed directly against further Romanian cooperation within the Warsaw Treaty Organization. According to some interpretations, various alliance rules could have been considered incompatible with the

---

[395] Mastny, *A Cardboard Castle?*, 42.
[396] *Informace o změnách v rum. armádě*, 25. 1. 1972, TO(t) 1970-1974, box 1, i. č. 82, sign. 017/111, č. j. 020.555/72-2, AMZV, Prague.
[397] Tejchman, *Nicolae Ceausescu*, 84-85.

sovereignty and independence of the member-states.[398] However, it was more likely a clear, although late, attempt to discourage Moscow from a potential recurring of the Czechoslovak scenario. At the time, regarding its appeals for renunciation of force and indefeasibility of the borders, Romania still directly referred to an episode from August 1968 in Czechoslovakia. Once again, it was proved that Bucharest did not seek to provoke Moscow without a reason. Romanian officials acknowledged the invasion during the talks with their Polish counterparts, but they avoided such an explanation in the presence of the Soviet diplomats.[399]

The mistrust between the Warsaw Treaty Organization and Bucharest was undoubtedly boosted by the strengthening of Romanian military contacts with the West. In 1974, for example, six military delegations of NATO countries visited Romania, but only three of the Warsaw Pact member-states. The Romanian army representatives traveled to the Western states four times, but made only one journey to the Soviet Union. The rest of the allied countries they did not visit at all. Also, the first foreign missions of the new chief of general staff, Ion Coman, were solely outside the Warsaw Pact. In addition, the general considered it unnecessary to establish closer personal contact with the Unified Command. On the contrary, he visited the USA in 1975. It was the first official journey of a senior military representative of the Warsaw Treaty Organization to the United States ever. However, Romania once again proved that it was not interested in any serious confrontations with Moscow over military issues. When the situation related to the visits of the military delegations appeared to be problematic, Bucharest promptly sent invitations to Romania to the commands of individual armed forces of the Warsaw Pact member-states.[400]

---

[398] *Zákon o národní obraně RSR*, 17. 1. 1973, DTO 1945-1989, i. č. 34, e. č. 56, č. j. 020.538/73, AMZV, Prague.

[399] *Informácia o některých poznatkoch z návštevy viceministra S. Trepczynského v Rumunsku*, 22. 5. 1973, TO(t) 1970-1974, box 2, i. č. 79, sign. 016/111, č. j. 023134, AMZV, Prague.

[400] *Zahraničně-politická aktivita RSR na úseku vojenských styků v roce 1974 a v I. čtvrtletí 1975*, 14. 4. 1975, DTO 1945-1989, i. č. 34, e. č. 64, č. j. 012.989/75, AMZV, Prague.

The goal of the visits of military delegations to Romania was specific, as Grechko's journey in 1973 suggests. In trying to avoid any sensitive political talks, the Romanian military officials planned a program for the Soviet marshal that consisted mostly of inspections of the troops. However, the Soviet defense minister had different intentions. Referring to alleged tiredness from traveling, he insisted on talks with military and political leaders. There he provocatively emphasized a need for unity of the Warsaw Pact's armies. He clearly opposed the claims that the conditions and time had come for a dissolution of the alliance. The Romanian defense minister acted in conversation with Grechko very moderately. He assured Grechko that the army of Romania cooperated with the partners from the Warsaw Treaty Organization, but at the same time, it did not dismiss collaboration with the other socialist countries, like China and Yugoslavia.[401] Also Ceausescu tried to avoid any controversy and strove for a maximally friendly atmosphere. He largely agreed with Grechko in the key aspects of evaluation of the military-political situation. He also personally intervened against an attempt of the Romanian military officials to evade signing the traditional final communiqué.[402] The RCP CC General Secretary did that solely in order not to bother the Soviet minister without serious reason. After all, such an approach well described the general Romanian attitude towards military cooperation within the Warsaw Treaty Organization.

---

[401] In this regard, the Romanian side informed Grechko about its serious troubles with the armament programs. Nevertheless, the Soviet defense minister was also told that Bucharest was considering development of own fighter jet in cooperation with Yugoslavia. This resulted in IAR-93 fighter-bomber which was introduced into service in the Romanian air force at the end of 1970s.

[402] *Informacja w sprawie opinii bukaresztańskich kół dyplomatycznych n.t. charakteru, przebiegu i konsekwencji wizyty marszałka A.Greczki w Rumunii w dniach 20-24 kwietnia br.*, 15. 5. 1973. PZPR KC, XIB/173, AAN, Warsaw.

The mechanisms of
the Warsaw Pact's military
structures after 1969

The 10th anniversary meeting of the Political Consultative Committee, held in Budapest on October 16-17, 1969, meant a turning point for military cooperation within the Warsaw treaty Organization. After more than five years of tumultuous internal discussions, the new statutory documents reforming the alliance military structures were finally approved.[403] These did not only provide a more precise definition of the regulations and competences in relations between the Unified Command and individual member-states during peacetime, but it also established brand-new collective bodies: the Committee of Defense Ministers, the Military Council, and the Technical Committee. The already existing Staff of the Unified Armed Forces should have gone through the significant changes.[404]

The practical realization of these measures became one of the key issues for the Warsaw Pact at the turn of the 1960s and 1970s. The approved statutes defined a scheme of the recently created bodies in general terms only. Thus, the real mechanisms were not shaped before the meetings themselves. The military structures of the alliance actually enjoyed large amounts of autonomy. The member-states' political power did not interfere with their activities significantly. There was a continual trend that had already started during the previous stage of Brezhnev's rule:

---

[403] On the attempts to reform the Warsaw Treaty Organization in the 1960s see Mastny, *A Cardboard Castle?*, 28-39; Luňák, Petr, ed., *Plánování nemyslitelného* (Praha: Ústav pro soudobé dějiny – Dokořán, 2007), 43-50; Mastny, "We are in a Bind," 230-235; Bílý, "Počátky pokusu o reformu Varšavské smlouvy," 159-171; Ionescu, "Rumunsko a vojenská reforma Varšavské smlouvy," 699-705.
[404] New Secret Statutes of the Warsaw Pact, March 17, 1969, in *A Cardboard Castle?* (doc. 62): 323.

The top state and party officials were informed about the program of upcoming military talks within the Pact, but they provided formal consent to holding those meetings only. After 1969, this approach also concerned the recently established collective bodies,[405] with the exception of rebellious Romania.

Shaping the Committee of Defense Ministers

The Committee of Defense Ministers should have become nominally the most important military institution of the organization. It consisted of the heads of the defense resorts of all the member-states, as well as the supreme commander and the chief of Staff of the Unified Armed Forces. According to official interpretation, it was an advisory body of the Political Consultative Committee. In praxis, however, the Committee of Defense Ministers worked largely independently.[406] Some scholars consider the creation of this forum as a moment when the Warsaw Treaty Organization became a real alliance or NATO counterpart.[407] The body was formally responsible for the coordination of national defense

---

[405] Letter by the CSCP CC First Secretary, G. Husak, to defense minister of Czechoslovakia, M. Dzúr, 9. 7. 1969; Letter by the Prime Minister of Czechoslovakia, O. Černík, to defense minister of Czechoslovakia, M. Dzúr, July 1969, FMO GŠ/OS, k. 9, p. č. 44, č. j. 39036, VÚA, Prague.

[406] Ślusarczyk, *Układ Warszawski,* 44. The claim by Jaczek Ślusarczyk, Polish historian and political scientist, that after establishing their collective body, the defense ministers stopped attending the Political Consultative Committee sessions is not accurate. At least in 1972, the East-German delegation contained also Heinz Hoffmann, the head of the defense resort. Also other military experts from the member-states' armies extraordinarily took part in the meeting, most likely because the issues relating to further development of the Unified Armed Forces were discussed there. However, the GDR attempted to ensure continual participation of the military officials at the Political Consultative Committee gatherings. In 1974, deputy defense minister and chief of the main staff of the East-German National People's Army (NPA), Heinz Keßler, should have attended the meeting in Warsaw. Nevertheless, the SED leadership was forced to abandon this intention. *Anlage Nr. 2 zum Protokoll Nr. 2/72 vom 18. 1. 1972,* DY 30/ J IV 2/2/1375, BArch, Berlin; *Anlage 1 zum Protokoll Nr. 12/74 vom 2. 4. 1974,* DY 30/ J IV 2/2/1497, ibid; *Politbüro des ZK Reinschriftenprotokoll nr. 14,* 15. 4. 1974, DY 30/ J IV 2/2/1499, ibid.

[407] Luňák, *Plánování nemyslitelného,* 74; Nünlist, "*Cold War Generals*".

policies as well as strengthening the defense capabilities of the entire Warsaw Pact.[408] Nevertheless, the design of its first meeting, held in December 1969 in Moscow, was little effective. The agenda contained an analysis of NATO's intentions, a development of the Unified Armed Forces during the years 1971-1975, and a report on reform of the alliance staff. However, after the main speech by Marshal Andrei Antonovich Grechko, who as the defense minister of the hosting country chaired the session, only vague additional comments by the rest of the participants and a short discussion on the next meeting's format followed. Then, a resolution was passed unanimously.[409] Moreover, it soon became clear that various key as well as sensitive and essentially controversial matters would not come within the remit of the Committee. For example, Moscow intentionally continually negotiated the very important issues of assignment of the troops with the Unified Armed Forces as well as individual national armies' development with the member-states bilaterally. The Soviet Union only had this praxis - which persisted until the Pact's dissolution - formally approved on behalf of the defense ministers.[410]

Certainly, at least the initial overestimation of the Committee of Defense Ministers' importance is also suggested by the frequency of its meeting. According to its statute, the body should have held two sessions a year, regarding the actual situation and conditions. However, such a possibility was abandoned quickly. Since 1970, the defense ministers met regularly once a year only, except for a few extraordinary events.[411] Most

---

[408] Ibid.

[409] *Zpráva ministra národní obrany pro Radu obrany státu o výsledku jednání Výboru ministrů obrany států-účastníků Varšavské smlouvy*, December 1969, FMO GŠ/OS, k. 9, p. č. 44, č. j. 39036, VÚA, Prague.

[410] Letter by M. Dzúr, defense minister of Czechoslovakia, to L. Svoboda, president of Czechoslovakia, regarding the upcoming meeting of the Warsaw Treaty Organization Committee of Defense Ministers, December 17, 1969, ibid.

[411] At the first meeting of the Committee of Defense Ministers in December 1969, the participants agreed that the following gathering would be held in May next year, in Sofia. They also defined the agenda. However, the 3rd session was outlined only very vaguely. It should have been held in the late months of 1970 in Budapest. The meeting was later postponed to 1971.

likely, there was no proper agenda for more frequent talks. The concrete task of combat preparation planning was actually discussed within another recently established collective body - the Military Council.[412] The practical issues were managed mainly by the Staff of the Unified Armed Forces as a permanent institution of the Unified Command. A statutory provision that allowed holding the Committee of Defense Ministers session twice a year was, in fact, rather a nuisance and did not reflect the real needs. Introduction of this rule was probably motivated solely by the Soviet satellites' efforts to ensure the multilateral consultative mechanisms and to prevent a similar situation from Khrushchev's era, where no official procedures for convening the meetings had existed.[413] Moreover, as historian Bohuslav Litera correctly notes, the Committee of Defense Ministers lacked any permanent secretariat.[414] Therefore, it did not extend the scope of a coordinative and consultative forum.

Nevertheless, the second meeting of the body suggested a certain shift. The military issues, mostly the condition of the air defense system and an interaction between the allied navies, were discussed in more detail.[415] The fact that the Committee of Defense Ministers really dealt with the items, not only formally and "on paper," is proven by the participants who attended the meeting. Regarding the agenda, the Czechoslovak delegation, for instance, extraordinarily contained the commanders of the anti-aircraft system and the air forces. The changes in the format of the talks were also illustrated by their approach towards the solution of many significant problems, one of which included the development of the air defense system at the turn of the 1960s and

---

[412] *První zasedání Vojenské rady Spojených ozbrojených sil*, December 1969, FMO GŠ/OS, k. 35, p. č.128, č. j. 36031, VÚA, Prague.

[413] *Zpráva ministra národní obrany pro Radu obrany státu o výsledku jednání Výboru ministrů obrany států-účastníků Varšavské smlouvy*, December1969, k. 9, p. č. 44, č. j. 39036, ibid.

[414] Litera, "Unifikace sovětského bloku," 90.

[415] According to *Protocol on establishing the Unified Command of the Treaty of Friendship, cooperation and Mutual Assistance Member-States' Armed Forces* of May 1955, all the naval units were assigned to the alliance combat formation. However, unlike the air-defense system, until 1978 there was no function of the Warsaw Pact commander-in-chief for the Navy, which virtually came under the remit of the Soviet naval forces' supreme commander, Admiral S. G. Gorschkov.

1970s.[416] In the discussion following the main report by the commander of the Warsaw Pact anti-aircraft troops, the Soviet Marshal Pavel Fyodorovich Baticki along with individual defense ministers had an opportunity to present their proposals. This was not just a formality. The content of their speeches was prepared neither by the Unified Command nor the Soviet side, but by the national military institutions.[417] During the first two Committee of Defense Ministers' sessions, the praxis was also established that the chief of general staff of each respective country could deputize themselves a new minister, when serious reasons caused a previous minister to be absent.[418]

Even the Soviet side soon realized that the initial scheme outlined by the Committee of Defense Ministers meetings was too inflexible and formal. After the 3rd session, which took place in Budapest in March 1971, Grechko therefore proposed that the next talks deal with a much wider scale of the issues. There should have been a 'free' exchange of opinions, without a necessity to approve any binding resolution.[419] In the early years of the 1970s, Moscow most likely concluded that loyalty of the satellites' military officials had reached a sufficient level. Therefore, the USSR decided to test whether a more open form of cooperation within the collective alliance military bodies could have bought more benefits. Thus, ever since the 4th session of the Committee of Defense Ministers, held in

---

[416] In this regard, the Warsaw Treaty Organization air-defense system commander-in-chief, the Soviet Marshal Baticki, stressed the need to develop anti-aircraft defenses to eliminate any NATO threat posed by having the capability to deliver the air strikes in extremely low-fly altitude. The Pact member-states were also supposed to strengthen the air-defense missile protection of their capitals and huge industrial centers.

[417] *Jednání Výboru ministrů, informační zpráva pro náčelníka generálního štábu*, 7. 4. 1970, FMO GŠ/OS, k. 9, p. č. 44, č. j. 39036, VÚA, Prague.

[418] Hungarian defense minister Lajoz Czinege did not participate in the Moscow meeting due to health problems. In Sofia, his Romanian counterpart, Ion Ionita, absented because of similar reasons. Both were deputized by the chiefs of Hungarian and Romanian general staffs, Karoi Chemi and Ion George. *Zpráva ministra národní obrany pro Radu obrany státu o výsledku jednání Výboru ministrů obrany států-účastníků Varšavské smlouvy*, December 1969; *Zpráva ministra národní obrany o výsledku jednání výboru ministrů obrany států-účastníků Varšavské smlouvy*, May 1970, ibid.

[419] *3. zasedání Výboru ministrů obrany*, March 1971, k. 10, p. č. 45, č. j. 39037, ibid.

January 1972 in Berlin, the main reports would be sent to the participants of the meeting in advance, in order to allow the national military institutions to prepare a detailed minister's speech and commentary on the discussed issues.[420] The shift in a conception of the Committee of Defense Ministers' meetings was also documented by the fact that since 1971, the media in the Eastern Bloc countries began informing on the body sessions, however briefly. Symptomatically, this was initiated by the Soviet Union and directly ordered by the chief of Staff of the Unified Armed Forces, General Sergei Matveyevich Schtemenko.[421] The participants of the 3rd meeting in Budapest also met the HSWP CC First Secretary, Janos Kadar.[422] Henceforward, after the sessions of the Warsaw Treaty Organization military bodies, the party leaders of the hosting country met their members regularly. These changes were most likely related to the alliance structures' reaction to the progress of détente. As we explain later, 1971 could be seen as a moment when easing international tensions between the East and West also began to practically affect the processes within the Warsaw Pact's military framework.

The Unified Command bodies

According to the Statute of March 1969, the activities within the alliance military structures were supervised and ruled by three collective bodies of the Unified Command: the Military Council, Staff of the Unified Armed Forces, and the Technical Committee.[423] The initial meeting of the Military Council held in Moscow, October 9-10, had mostly organizational features. Besides outlining plans of future activities by Marshal Ivan Ignatevich Iakubovsky, the alliance supreme commander, only the Statutes of Staff of the Armed Forces as well as the Technical

---

[420] *4. zasedání Výboru ministrů obrany*, January 1972, k. 40, p. č. 136, č. j. 36039, ibid.

[421] Letter by M. Dzúr, defense minister of Czechoslovakia, to L.Svoboda, the president of Czechoslovakia, regarding the upcoming meeting of the Warsaw Treaty Organization Committee of Defense Ministers, December 14, 1970, k. 10, p. č. 45, č. j. 39037, ibid.

[422] *3. zasedání Výboru ministrů obrany*, March 1971, ibid.

[423] New Secret Statutes of the Warsaw Pact, March, 17, 1969, in *A Cardboard Castle?* (doc. 62): 323.

Committee were approved. However, such outcomes were again agreed upon in advance. Despite the fact that participants of the session[424] were given a chance to comment on all the items on the agenda, the essence of the results did not change at all.[425] In comparison to the Committee of Defense Ministers, the scheme of the Military Council gatherings settled down a little bit earlier, in November 1970 at the latest, when the 3rd meeting of the body in Varna was accompanied by the allied troops' combat presentations and demonstrations of the discussed military tactics for the first time. During the following years, this became routine.[426] The place of the Military Council sessions rotated, similarly to the Committee of Defense Ministers. In this case, rotation of the hosting countries was even more formal and unimportant, as all the body's activities were fully controlled by the Soviet military officials. Neither was there a chairmanship rotation; the supreme commander always directed the meeting. Moreover, in the first half of the 1970s, organization of gatherings in individual member-states proceeded under close supervision of the chief of the alliance staff, General Schtemenko, and other Soviet officers from the Unified Command. Finally, the supreme commander himself must have approved the definite plan of meetings.[427]

---

[424] According to *Statute of the Unified Armed Forces*, the Military Council's permanent members contained the supreme commander, the chief of the alliance Staff, the chief of the Technical Committee, the chiefs of all the member-states' general staffs, the chief of the unified air-defense system, and the supreme commander's deputies for the air forces and navy. In reality, many more military officials and experts took part in the body meetings, especially related to the issues to be discussed. The documentation on preparation of such gatherings reveals that those functions of the Military Council Secretary also existed. However, his activities remained limited to the talks and were strictly technical and supportive. Collection of materials on preparation of the 15th Military Council meeting in Prague, April 1977, FMO GŠ/OS, k. 39, p. č. 134, č. j. 36037, VÚA, Prague.

[425] *První zasedání Vojenské rady Spojených ozbrojených sil,* December 1969, k. 35, p. č.128, č. j. 36031, ibid.

[426] *3. zasedání Vojenské rady Spojených ozbrojených sil,* 1. 11. 1970, p. č. 129, č. j. 36032, ibid.

[427] *Příprava 9. zasedání Vojenské rady a porady vedoucích funkcionářů armád států Varšavské smlouvy,* 6. 9. 1973, k. 37, p. č. 131, č. j. 36034, ibid.

The reformed Staff of the Unified Armed Forces - in fact, an already existing Moscow-based body, now joined by non-Soviet army officers - began working on November 25, 1969.[428] A percentage contribution to the alliance's budget was set as a representative proportionality key. After the reform, 135 generals of all the Pact member-states worked in the institution. Technical personnel, consisting of 200 people, were exclusively of Soviet nationality.[429] The Technical Committee also started its activities[430] with 68 generals supplemented by 20 assisting personnel there. These numbers of staff in both institutions were just temporary: after getting necessary experience in the new form of collective cooperation, they should have been increased to 200 and 80 officers respectively.[431] This is more evidence that before the Warsaw Treaty Organization military structures' reform, there was no real interaction between the individual member-states' representatives in the Unified Command. The situation also suggests that the Statutes approved in March 1969 reflected mostly the political interests. New numbers of the officers did not come from real military experience and needs. This was probably just a Soviet sop to the satellites which formally gained bigger representation. That interim period most likely lasted until

---

[428] The Staff of the Unified Armed Forces existed since establishing the Pact in 1955. Its members, however, almost exclusively contained Soviet officers. The reform of the Warsaw Treaty Organization military framework in March 1969, therefore, only changed the structure of the institution, rather than creating a new body. Nevertheless, the documents contain the term "establishing the coalition staff." It must be understood in a way that the new format of the body would have significantly differed from the period before 1969.

[429] Those positions in the Staff of the Unified Armed Forces were actually held by the representatives of the Soviet satellites' armies before 1969.

[430] A report by W. Jaruzelski, Polish defense minister, reveals that in 1970, the Technical Committee still did not work properly and, in fact, remained in the stage of preparations. Report by W. Jaruzelski, defense minister of Poland, to E. Gierek, the PUWP CC First Secretary, on the Warsaw Treaty Organization military bodies' activities in 1970, January 18, 1971, PZPR KC, XIB/173, AAN, Warsaw.

[431] *Zpráva ministra národní obrany pro Radu obrany státu o výsledku jednání Výboru ministrů obrany států-účastníků Varšavské smlouvy*, December 1969, FMO GŠ/OS, k. 9, p. č. 44, č. j. 39036, VÚA, Prague.

the second half of 1971.[432]

As the chief of the Technical Committee, and therefore also the deputy supreme commander for the weaponry, the Soviet general I. V. Schtepanyuk was appointed. The alliance Staff remained under command of the aforementioned General Schtemenko, who had already held this post before the reform of the body. The duties of other officers in the Staff of the Unified Armed Forces as well as the Technical Committee were defined by a new special statute. Although the Unified Command initially worked out the document, it was later discussed by the Military Council. Also, individual defense ministers had an opportunity to include their notes in the text. It is good to add that such a procedure - of approving the regulations of military cooperation within the alliance - became very usual in the upcoming years. According to this document, the personnel performing the duties in the Unified Command were transferred under direct competences of the supreme commander, although those officers still remained the members of their national armies. The length of service in the Unified Command bodies was defined vaguely - between a 4- to 6- year term. An alteration of the personnel was subject to agreement of the alliance Staff and the member-states' general staffs. The efforts to give all the Pact members equal right - however just on paper - resulted in the fact that the languages of every country could have been used for official communication within the Unified Command institutions. Russian was the only "working" language.[433] No need to point out that every military functionary working in the alliance headquarters' bodies was also a member of the communist or workers' party.[434]

---

[432] It is an estimation based on the Czechoslovak documents. *Výkon vojenské služby generálů a důstojníků ve štábu Spojených ozbrojených sil*, October 1971, k. 8, p. č. 42, č. j. 39034, ibid.

[433] *Návrh Statutu o průběhu služby ve štábu Spojených ozbrojených sil*, 10. 10. 1969, k. 42, p. č. 142, č. j. 36047, ibid.

[434] *Доклад о выполнении рещения государств-участников Варшавского Договора, принятого на совечании Политического Консултативного Комитета 17 марта 1969 года*, 12. 12. 1969, PZPR KC, XIA/106, AAN, Warsaw. In the CSPA, about 75-80 percent of commanders were the CPCz members or candidates. Bílek, Jiří and Koller, Martin, "Československá armáda v rámci sovětského bloku. 1945-1989,"

The representatives of the Soviet satellites' armies held nominally important ranks in the Staff of the Armed Forces, including deputy chiefs.[435] However, some concessions during appointment to various positions most likely had no significant impact on the Soviet control of every really vital post within the Unified Command. The personnel from allied countries still served as the liaison or deputy officers.[436] This assumption is also suggested by the composition of the alliance Staff delegation which visited Czechoslovakia in the early months of 1974. The key-posts, especially the chiefs of particular departments and their deputies, were mostly held by Soviet army members.[437] Moreover, according to the former East-German admiral Theodore Hoffmann's testimony, even after reform, the Staff of the Unified Armed Forces did not take its own initiatives on the important military issues. The East-German military officials, therefore, allegedly focused their attention on the General Staff of the Soviet army, where the majority of the significant matters were still decided.[438] Thus, the USSR's domination within the Warsaw Pact's military structures continued after 1969. The level of respect of such a state by the individual member-states' representatives varied considerably. Reportedly, the Polish and Romanian officers were the less conciliatory and their East-German colleagues the most.[439]

---

in *Od Velké Moravy k NATO. Český stát a střední Evropa*, ed. Josef Tomeš (Praha: Evropský literární klub 2002): 198; *PODKLAD pro vystoupení ministra národní obrany ČSSR u příležitosti jeho návštěvy v SSSR*, undated, 1983, FMO GŠ/OS, k. 24, p. č. 79, č. j. 41102, VÚA, Prague.

[435] *Opatření k vyslání důstojníků do štábu Spojených ozbrojených sil*, 17. 10. 1969, k. 42, p. č. 142, č. j. 36047, ibid.

[436] Wanner, *Brežněv*, 50.

[437] *Konzultace funkcionářů štábu Spojených ozbrojených sil v ČSSR*, 7. 1. 1974, FMO GŠ/OS, k. 3, p. č. 31, č. j. 39023, VÚA, Prague.

[438] On the other hand, the SED leadership at least initially paid significant attention to selection of the East-German representatives in the Unified Command's reformed institutions. General Heinz Keßler, the SED CC member, deputy defense minister and deputy chief of NPA Main Staff, was appointed the highest GDR representative within the alliance command structures. *Politbüro des ZK Reinschriftenprotokoll nr.19*, 13. 5. 1969, DY 30/ J IV 2/2/1227, BArch, Berlin.

[439] During 1969-1989, GDR regularly sent two generals and twenty officers into the Staff of the Unified Armed Forces. Schaefer, Bernd, "*The GDR in the Warsaw*

Considering the unavailability of the Unified Command documentation, it can be hardly assumed how real the daily cooperation within the body was, or what was the result of the aforementioned tensions.

In September 1970, a legal definition of the position of the Unified Command personnel, especially in the recently established bodies, was put forward by the supreme commander Yakubovski. After its approval by the member-states in April 1973, the Staff of the Unified Armed Forces became a legal person. Its members maintained the status of international functionaries. This provided them with *de facto* diplomatic privileges.[440] Clarification of the legal position of the Unified Command officials should have facilitated their activities on all the member-states' territories.[441] In addition, as a legal person, the Staff of the Unified Armed Forces had an ability to conclude the agreements or perform other legal acts, as well as the right to property and the disposal of it. All the personnel and documentation of the alliance headquarters were excluded from the state organs and judiciary competences regardless of their actual location. The potential wavier of the immunity of any Warsaw Pact military headquarters' member was decided by the chief of Staff of the Unified Armed Forces with consent of their respective defense minister. This allowed for the start of criminal penalties against the person.[442] The agreement mostly aimed to create appropriate conditions for the collective military bodies' activities.[443] Despite the measures being basically formal and technical, they illustrated well the matter of changes which the Warsaw Treaty Organization was undergoing at the turn of the 1960s and

---

Pact," Accessed October 16, 2019. http://www.php.isn.ethz.ch/lory1.ethz.ch/collections/coll_gdr/intro2644.html?navinfo=44755.

[440] Additionally, since 1979, families of the Unified Command functionaries were given ID papers which provided them with immunity. *Ustanovení o průběhu služby důstojníků, generálů a admirálů armád členských států Varšavské smlouvy*, 23. 3. 1979, FMO GŠ/OS, k. 44, p. č. 156, č. j. 36508, VÚA, Prague.

[441] *Podpis úmluvy o právní způsobilosti, výsadách a imunitách štábu a dalších orgán velení spojených ozbrojených sil členských států Varšavské smlouvy*, 26. 7. 1972, 1261/0/6, sv. 49, a. j. 49/4, NA, Prague.

[442] Śluszarczyk, *Układ Warszawski,* 38-39.

[443] *Orientační plán sjednávání mezinárodních smluv na r. 1972*, 8. 2. 1972, PK 1953-1989, kat. č. 632, k. č. 145, č. j. 016.130/72, AMZV, Prague.

'70s. Largely chaotic and often legally unfounded methods of control over the Eastern Bloc, so typical of Khrushchev's period, ended. The Warsaw Pact was supposed to become a real military alliance with official regulations, which contrasted sharply with the situation in the 1950s and 1960s.

Regarding the 1969 Warsaw Pact reform, the existing literature notes that Moscow used it to alter control of its European satellites' armed forces from direct to indirect. Vojtech Mastny points out that within the new military bodies, especially the Military Council and Committee of Defense Ministers, the Soviet supremacy was guaranteed by the supreme commander and the chief of Staff of Unified Armed Forces, who were the members.[444] However, such an outlined control mechanism is incomplete, at least. The statutory documents approved in 1969 also committed the general staffs of individual member-states to detailed periodical reports to the Staff of the Unified Armed Forces and to the Technical Committee on the condition of the troops assigned to the alliance combat formation.[445] According to Article III of the new Statute, national bodies were solely responsible for the situation within the individual member-states' armies.[446] In reality, the armed forces also remained subjects to control and inspections by the Unified Command.[447]

Moreover, the influence of so-called representatives of the supreme commander within individual national armies is totally neglected by the scholars. That function was introduced by the reformed statute of the Unified Command with an amendment stating that deployment of these officers in the member-state's armed forces had to be approved by the country's government. Their activities would have mostly reflected the

---

[444] Mastny, Vojtech, "The Warsaw Pact. An Alliance in Search of a Purpose," in *NATO and the Warsaw Pact. Intrabloc Conflicts*, eds. Mary Ann Heiss and S. Victor Papacosma (Kent: Kent State University Press, 2008): 149-150.

[445] *Předkládání periodických informačních hlášení*, 16. 11. 1972, FMO GŠ/OS, k.13, p. č. 50, č. j. 39043, VÚA, Prague.

[446] New Secret Statutes of the Warsaw Pact, March 17, 1969, in *A Cardboard Castle?* (doc. 62): 323.

[447] *Výsledky splnění úkolů v ČSLA za rok 1971 a hlavní úkoly ČSLA pro rok 1972*, 1. 11. 1971, 1261/0/6, sv. 17, a. j. 18/2, NA, Prague.

guidelines of the particular defense minister.[448] In March 1969, the Political Consultative Committee declared that the duty of the supreme commander's representatives would be regulated by a 'special statute'. However, not even outlines of such a document existed at the time. The state and party officials therefore could have only speculated about its final form. Except for Romanian politicians, they most likely did not pay any serious attention to the issue. The Unified Command worked out the document one year later.[449] The initial, very vague thesis was significantly extended and various shifts in meaning appeared. First of all, the representatives of the supreme commander became accountable to the Unified Command's top officers only. In praxis, they were supposed to act as a direct communication link between the supreme commander and the defense ministers, or the Staff of the Unified Armed Forces and the national general staffs. Although the representatives of the supreme commander had no decision-making power, they could have significantly affected the events within individual armies. Through them, the Soviet military command also closely monitored the situation within the Warsaw Treaty Organization member-states' armed forces. Actually, those authorities had access to all relevant information about the condition of the troops assigned to the Unified Armed Forces formation. They were obligated to take part in the drills and inspections which aimed at "strengthening mobilization or combat readiness of the armies,"[450] and were given all the details about the armament programs. They were also authorized to put forward their own proposal on solving the problems to the chiefs of the national general staffs. These Soviet military officials informed the supreme commander and the chief of Staff of the Unified Armed Forces in detail about their findings. In addition, they had a right

---

[448] New Secret Statutes of the Warsaw Pact, March 17, 1969, in *A Cardboard Castle?* (doc. 62): 323.

[449] The common scenario recurred again - the Unified Command sent the draft of the document to individual member-states' defense ministries and they were supposed to either approve it or present their objections and comments to be incorporated.

[450] In reality, such a flexible formulation could have covered almost all the military preparations.

to appoint their own deputies and "assistants" in individual sorts of troops and units.[451]

An extremely important role of the representatives of the supreme commander in the national armies is proven even by the fact that they, by virtue of their function, attended most of the Military Council sessions in the 1970s. However, the initial statute of the body did not envisage that at all. For instance, the Czechoslovak delegations - the representatives of the member-state at the alliance body meeting – also regularly contained the Soviet General K. G. Kozhanov and other Soviet officers appointed by him. This ingenious system, in praxis, not only allowed for supervision of the Warsaw Pact member-states' troops unification, but it also meant a sort-of return to a network of the Soviet military advisers well known from Stalin's era. Despite this, the reform of the military statutes in March 1969 should have provided more autonomy for the alliance member-states' armies. However, an additional approval of the statute of the supreme commander's representatives in spring 1970 actually reinforced Soviet control.

It is important to note, that the party and state leadership did not discuss this document in detail; an approval by the defense ministers sufficed.[452] At the alliance level, only the Military Council dealt with the issue, during its 2nd meeting in Budapest, April 1970.[453] It is not surprising that the Romanian delegation, for understandable reasons, refused to take any position towards the proposed material. However, the rest of the Pact members approved it, regardless of Bucharest's stance.[454] During the following years, Romania neither subsequently signed the Statute of the supreme commander's representatives in the national armies nor

---

[451] *Statut o představitelích Hlavního velitele v armádách VS*, January 1970, FMO GŠ/OS, k. 35, p. č. 128, č. j. 36031, VÚA, Prague.

[452] The Czechoslovak military documentation lacks any mention that the statute defining the activities of the supreme commander's representatives was discussed or approved by the political institutions of the country. *Plán realizace PROTOKOLU č. 002 jednání Vojenské rady Spojených ozbrojených sil*, undated 1971, k. 8, p. č. 42, č. j. 39034, ibid.

[453] *Zasedání vojenské rady Spojených ozbrojených sil*, April 1970, k. 35, p. č.128, č. j. 36031, ibid.

[454] *2. zasedání Vojenské rady Spojených ozbrojených sil*, April 1970, ibid.

financially contributed to their service. In May 1971, the Military Council passed a Statute of general funding of the Unified Command during its 4th session. Romania put forward a request that the document did not concern the activities of the representatives of the supreme commander within the national armies, because the SRR was not the signatory of the agreement on their services. Another delegation opposed, but the Romanian stance was included in a final protocol of the meeting.[455] In this manner, in fact, Bucharest ensured an opt-out. At the following gathering of the Military Council, the Ceausescu regime definitively refused to pay for the representatives of the supreme commander.[456]

The Technical Committee has stood outside of the scholars' interest so far. This is partially because of the relatively unavailable heuristic base, but also by the less attractive activities of the body. However, it is at least good to at basically outline how one of two permanent institutions of the alliance headquarters worked. The statute of the Unified Command of March 1969 limited itself to only a few laconic notes regarding the Technical Committee. The real procedures of the body were defined by a specific guideline, which authorized it to supervise a maximum possible unification of the armament and military equipment within the Pact member-states' armies as well as an elimination of the parallel development of similar weapons systems. Therefore, the Technical Committee closely cooperated mostly with the Permanent Commission for Defense Industry of Comecon.[457] Besides, it ensured that only the weapons corresponding to the actual strategic conception of the Unified Command would be developed. The Technical Committee

---

[455] *4. zasedání Vojenské rady Spojených ozbrojených sil*, 17. 5. 1971, k. 36, p. č. 130, č. j. 36033, ibid.

[456] *Páté zasedání Vojenské rady Spojených ozbrojených sil,* 1. 11. 1971, ibid.

[457] Due to understandable reasons, the Soviet Union attempted to spread military hardware production among its satellites. It used for this purpose the Commission for Defense Industry of Comecon, whose activities were linked to the decisions made within the Warsaw Treaty Organization's military structures. In the late months of 1970, the Soviet defense ministry tried to secure an expensive and difficult production of the modern anti-aircraft missile systems KRUG, KUB and STRELA-1 in this manner. *KRUG, KUB – zpráva z konsultace v SSSR*, 4. 12. 1970, k. 15, p. č. 52, č. j. 39044, ibid.

received the assessment proposals on the project from individual member-states. The body took none of its own initiatives. Moreover, its decisions had to be always confirmed by the supreme commander. The working groups assigned to the approved project consisted of the representatives of the interested armies, as well as the Technical Committee itself. However, costs of particular weapons development were not funded from a joint budget of the alliance. The participating states alone financed individual projects. The Unified Armed Forces' budget covered only activity of the institution itself.[458] It is also good to add that bilateral cooperation in the field of military equipment production was preserved. A joint commission for the armament projects, for instance, had been previously established in 1955 by Czechoslovakia and Poland.[459]

Funding for the Unified Command

The activities of the alliance military bodies were financed by all the Warsaw Treaty Organization member-states through annual contribution to the Unified Armed Forces' budget.[460] The ratio of their levies came out of the Political Consultative Committee's decision of 1969. Considering the fact that this ratio was also determining funding for other projects, it was one of the most important aspects of the alliance

---

[458] *Zásady pro realizaci opatření Technického výboru Spojených ozbrojených sil ke koordinaci vědecko-výzkumných a vývojových prací armád členských států Varšavské smlouvy*, 4. 6. 1974, FMO GŠ/OS, k. 13, p. č. 50, č. j. 39043, ibid.

[459] Both Warsaw Treaty Organization member-states cooperated in the manufacturing of "Skot" military transport vehicles, for example. In the 1970s, they also considered collaboration in a licensed production of the Soviet T-72 tank or Su-25 combat jet aircraft. In reality, the joint production of "Skot" was just a formality, as G. Husák complained about during his talks with E. Gierek in 1971. The situation once again illustrates that Czechoslovakia still considered better coordination of the Warsaw Pact member-states' efforts as a suitable solution for the excessive demands on the defense industry in the country. *Protokoł ze spotkania I sekretarza KC PZPR tow.Edwarda Gierka i prezesa Rady Ministrów PRL tow. Piotra Jaroszewicza z I sekretarzem KC KPCz tow.Gustavem Husakiem i przewodniczącym rządu CSRS Lubomirem Sztrougalem*, 16. 1. 1971, PZPR KC, XIA/38, AAN, Warsaw; *Notoatka w sprawie współpracy polsko-czechosłowackej w sfere produkcji obronnej*, undated, most likely 1976, XIA/595, ibid.

[460] Since 1983, the documentation starts to operate with the term "Unified Command's budget." The reason remains unknown.

military structures' mechanism. The percentage rate between the USSR and the rest of the member-states was 44.5 to 55.5. The differences between contributions of individual Soviet satellites apparently did not reflect the real numbers of their inhabitants, which were considered one of the possible criteria for dividing up the commitments within the Warsaw Pact in the 1960s and 1970s.[461] From this point of view, the set-up coefficient most obviously handicapped Czechoslovakia. The country paid 66.6 percent more than fit for the population number. To the contrary, East Germany most benefited from the system. Regarding this, its contributions were undersized more than 60 percent (see *Table 1* bellow). However, the huge gap between these numbers could be misleading. The funding system parameters actually reflected the different economic capabilities of individual Warsaw Treaty Organization member-states. Levies to the Unified Armed Forces basically corresponded with an income ratio of the trade within Comecon. From this point of view, individual Soviet satellites did not pay for each other. Nevertheless, the Soviet Union still subsidized roughly half of the military contributions of East Germany and Hungary.[462] Composition of the alliance budget, therefore, largely reflected the individual member-states' economic development level.

The Statute of financial management of the Unified Command was discussed in May 1971, at the 4th meeting of the Military Council in Berlin. Traditionally, the document was worked out by the Staff of the Unified Armed Forces. Nevertheless, it again included prompts by individual member-states' military and economic bodies.[463] The praxis,

---

[461] For example, in 1972, the funding mechanism for the Unified Armed Forces budget was internally discussed by the Czechoslovak military institutions. There was certain, although very empty and indirect, criticism of the fact that individual member-states' contributions did not reflect the population numbers. It is good to add that the Czechoslovak side dealt with the problem only because the same coefficient should have been used for levies to the expensive construction of the Unified Command protected posts on the Polish and Bulgarian territories. *Podíl jednotlivých států kromě SSSR na rozpočtu Spojených ozbrojených sil*, 5. 1. 1972, FMO GŠ/OS, k. 11, p. č. 47, č. j. 39039, VÚA, Prague.

[462] Ibid.

[463] *4. zasedání Vojenské rady Spojených ozbrojených sil*, 17. 5. 1971, k. 36, p. č. 130, č. j. 36033, ibid.

that a budget was firstly consulted within the Military Council and later put forward to the Committee of Defense Ministers for approval, was established.[464] Respect for the financial plan of the Unified Command bodies' activities was under ongoing supervision. The national military institutions, in cooperation with the alliance Staff, evaluated the situation twice a year.[465] In case of a surplus from the previous year, the levies of individual member-states were decreased by the proper amount. The budget mechanism, naturally, worked on the basis of allocating funds. Czechoslovakia, for instance, contributed 618 thousand rubles to the Unified Armed Forces' budget, but the country received back 61 thousand rubles more in non-trade payables for maintenance of the special communication lines and taking various joint measures on its territory.[466]

---

[464] *Páté zasedání Vojenské rady Spojených ozbrojených sil,* 1. 11. 1971, ibid.

[465] *Plán realizace PROTOKOLU č. 004 jednání Vojenské rady Spojených ozbrojených sil,* undated 1971, k. 8, p. č. 42, č. j. 39034, ibid.

[466] *Rozpočet SOS na r. 1973 a vyhodnocení za r. 1971,* 31. 10. 1972, k. 18, p. č. 56, č. j. 39049, ibid.

	CSSR	PRB	HPR	GDR	PPR	SRR	note
Number of inhabitants (million)	14.3	8.4	10.3	17.2	32	19.5	101.7 million inhabitants total
Portion of total population numbers of the Soviet satellites, the Warsaw Treaty Organization member-states (percent)	14.1	8.2	10.1	16.9	31.5	19.2	
Hypothetical contributions to total budget of the Unified Armed Forces reflecting population numbers (percent)	7.8	4.5	5.5	9.6	17.5	10.6	
Real contribution to the Unified Armed Forces' budget (percent)	13	7	6	6	13.5	10	
Gap between hypothetical and actual contribution to overall budget of the Unified Armed Forces (percent)	+5.2	+2.5	+0.5	-3.6	-4.0	-0.6	
The Warsaw Treaty Organization member-states' incomes from the Comecon's budget (percent)	13.25	7	10	13.25	13.25	10	
Gap between real contribution and levies reflecting population numbers (percent)	+66.6	+55.5	+9	-60	-29.6	-6	
Gap between levies to the Unified Armed Forces' budget and incomes from Comecon's budget (percent)	+0.25	0	+4	+7.25	-0.25	0	

Table I[467]

[467] The table edited according to *Podíl jednotlivých států kromě SSSR na rozpočtu Spojených ozbrojených sil*, 5. 1. 1972, k. 11, p. č. 47, č. j. 39039, ibid.

Until 1975, the budget contained 5 million transferable rubles annually. The unambiguously highest amount, 2.7 million, was used for securing the Unified Command bodies' work.[468] In this year, the alliance budged increased for the first time, by 1.391 million transferable rubles. It reflected the costs of construction of the protected command posts of the Unified Command in Poland and Bulgaria.[469] None of the Warsaw Treaty Organization member-states ever openly objected to such an established funding system. During the annual approving procedure of the budget, only Romania always opposed allocating a sum for the activities of the representatives of the supreme commander within the national armies. The rest of the alliance members tolerated that praxis; Romanian objections were only included in particular protocols and the country paid a lesser amount.[470] This is another example of the Soviet leadership's approach in the 1970s, which strove for isolation of Romania rather than causing open clashes.[471] As the previous chapter showed, this strategy was realized very consistently within the Warsaw Pact military bodies.

All of this led to the conclusion that reform of the Warsaw Treaty Organization in 1969 undoubtedly consisted of serious deficiencies. They mostly grew out of often only formal features of various changes. Nevertheless, we can agree with the claim that the Pact's military framework was improved in the early 1970s.[472] The USSR abandoned a pure dictate to its satellites. Alteration of the alliance mechanisms should have provided them with a bigger sense of partnership and therefore made

---

[468] *Zpráva o výsledku jednání představitelů armád států-účastníků Varšavské smlouvy k návrhu rozpočtu Spojených ozbrojených sil na rok 1973 a vyhodnocení rozpočtu Spojených ozbrojených sil na rok 1971*, undated 1972, k. 18, p. č. 56, č. j. 39049, ibid.

[469] *Výsledky 11. zasedání Vojenské rady Spojených ozbrojených sil a porady vedoucích funkcionářů spojeneckých armád*, 25. 11. 1974, k. 38, p. č. 133, č. j. 36036, ibid.

[470] *Zpráva o výsledku jednání představitelů armád států-účastníků Varšavské smlouvy k návrhu rozpočtu Spojených ozbrojených sil na rok 1973 a vyhodnocení rozpočtu Spojených ozbrojených sil na rok 1971*, undated 1972, k. 18, p. č. 56, č. j. 39049, ibid.

[471] Tejchman and Litera, *Moskva a socialistické země na Balkáně*, 25.

[472] Luňák, *Plánování nemyslitelného*, 74.

the Warsaw Pact a more effective instrument from Moscow's point of view.[473] First of all, the reform terminated the situation when the Pact's military structures worked ambiguously, on an *ad hoc* basis, and very vague statutes created a wide space for the Soviet gratuitousness. Previous explanations prove that the Soviet Union remained in a highly privileged position within the Warsaw Pact's military framework. However, such a state largely reflected its dominant military power and capacities which significantly outweighed the rest of the member-states. Nevertheless, after 1969, two important shifts occurred: Military cooperation was better organized because of the more active Staff of the Unified Armed Forces[474] as well as regular meetings of the Committee of Defense Ministers and the Military Council. Moreover, Moscow soon realized that the loyalty level of the Warsaw Treaty Organization member-states' military officials allowed them to design the talks on a more open, mutually beneficial basis. The Unified Command's work was given a quite strictly restricted legal framework by the series of the statutory documents. Henceforward, the military bodies' activities basically received more clearly defined regulations.[475] Approval of those statutes were probably facilitated by the fact that their content was discussed by the military officials alone. Except for Romania, the top political bodies of the member-states only consented

---

[473] Mastny, "The Warsaw Pact," 149-150.

[474] In 1971, work of the Staff of the Unified Armed Forces was praised by the head of Czechoslovak defense resort, Martin Dzúr, in his speech at the Committee of Defense Ministers. Referring to the allegedly positive experience, he suggested making that institution a center of military cooperation within the Warsaw Pact. Despite the fact that we must adopt a critical approach towards the speech of the Czechoslovak defense minister, as it could have been formulated in order to please present Unified Command functionaries, it suggests that in comparison to the era before 1969, there was a real shift in the way the alliance Staff worked. *Vystoupení ministra národní obrany k otázce "Stav spojení a prostředků velení vojskům na válčišti"*, December 1970, FMO GŠ/OS, k. 10, p. č. 45, č. j. 39037, VÚA, Prague.

[475] During the years 1969-1973, at least 14 statutory documents providing a groundwork for the alliance military cooperation were signed. It was clearly a significant shift in comparison to the situation before 1969, when all the regulations were formally defined by two very brief documents only. *Porada vedoucích funkcionářů generálních štábů*, 18. 6. 1973, k. 15, p. č. 52, č. j. 39044, ibid.

to the documents at best. Therefore, we can accept Christian Nünlist's claim that the Warsaw Pact's military officials were more disciplined and less quarrelsome than their political counterparts.[476] This is also proved by the retrospective testimonies of the Polish representatives in the Unified Command institutions.[477] At the same time, improved cooperation within the alliance's military structures created an appropriate base for training the East-European officers whose identity was situated on the edge between national states and the supranational dimension of the entire Eastern Bloc.[478] The senior officers, usually educated at the Soviet military academies,[479] maintained a significant loyalty level towards the joint interests of the alliance. But in the lower army ranks, national prejudices still flourished. The most striking was the relation between the Polish and East-German soldiers.[480]

The Warsaw Treaty Organization military structures' mechanism after 1969 had three different dimensions: According to the official documents, all the member-states were equal. Formally, the Soviet

---

[476] Nünlist, "*Cold War Generals*". In the first half of the 1970s, the Warsaw Treaty Organization military bodies meetings showed much more consensus between the member-sates' delegations than in case of the political talks. We are reminded that especially at the turn of the 1960s and 1970s, the tensions occurred not only between Romania and the rest of the alliance, but also between the USSR, GDR, and PPR, regarding the tactics in détente process and mostly the conditions for holding the all-European security conference.

[477] Nowak, *Od hegemonii do agonii,* 62.

[478] Litera, "Unifikace sovětského bloku," 91.

[479] W. Jaruzelski considered improving the education system of the command personnel and coordination of individual armies' preparation for conducting the joint combat operations to be one of the main positive effects of the 1969 reform of the Warsaw Pact. Report by W. Jaruzelski, defense minister of Poland, to E. Gierek, the PUWP CC First Secretary, on the Warsaw Treaty Organization military bodies' activities in 1970, January 18, 1971, PZPR KC, XIB/173, AAN, Warsaw.

[480] Mastny, *A Cardboard Castle?*, 42. For example, in October 1969, the Polish military institutions complained that some of the NPA officers showed their cold and disdainful attitudes towards Poles during the "Oder-Neisse-69" drills. The PPA commanders also disliked their East-German counterparts' effort to call the GDR the main ally of the USSR. Polish Army Report on East German Misbehavior during the "Odre-Neisse-69" Exercise, October 22, 1969, in: *A Cardboard Castle?* (doc. 65): 339.

generality acted in the position of *primus inter pares*. In reality, Soviet dominance prevailed. In comparison to the previous era, the level of its brashness, however, probably decreased.

Positive effects of reform of the Warsaw Pact military bodies of 1969 was confirmed by the Political Consultative Committee session, held in Prague three years later. It stated with satisfaction that military cooperation between the Pact members generally expanded and also that the command system was improved. Those were neither pure propagandistic nor obligatory statements. If we look beyond the fact that the Soviet generality remained in privileged positions within the alliance military structures, we can essentially agree with that claim. In praxis, the outcome of such evaluation was that the Political Consultative Committee *de facto* rendered some sort of autonomy to the Unified Command. Henceforward, the alliance supreme body neither intended to alter the military structures' mechanisms nor affected them significantly. Potential changes in military cooperation should have been realized by an agreement between the supreme commander and the defense ministers.[481] Therefore, the Warsaw Pact's only political institution's surveillance of the alliance military structures' practical work was extremely formal. This further strengthened the Kremlin's influence on their activities. In praxis, despite its efforts for higher autonomy, the Soviet military command, which maintained a key role within the Warsaw Pact, still followed a policy line defined by the CPSU's top echelons. Thus, Moscow kept strict control over the alliance military structures through this indirect manner.

A claim can be found in existing literature that NATO recognized the Warsaw Treaty Organization as a real counterpart primarily because of its reform of the communist alliance military framework. Regarding this, some scholars point out that communiqué of the North Atlantic Council, held in Brussels in December 1970, mentioned the Warsaw Pact as a potential partner in disarmament talks for the first time.[482] However, this fact undoubtedly primarily came out from generally increased

---

[481] *Překlad usnesení států-účastníků Varšavské smlouvy, přijatého na zasedání Politického poradního výboru 26. ledna 1972*, FMO GŠ/OS, k.8, p. č.41, č. j. 39033, VÚA, Prague.

[482] Luňák, *Plánování nemyslitelného*, 74; Nünlist, *"Cold War Generals"*.

activities of the alliance's political bodies, especially their more frequent meetings, rather than reform of the military structures itself. After all, the disarmament proposals were not presented by the military, but by the political platforms of the Pact.

Troops assignment to the Unified Armed Forces

During the 1960s, the Warsaw treaty Organization member-states' armies had already stopped being built as independent entities. To large extent, they simply became a part of the Unified Armed Forces.[483] The 1969 reform of the military framework formally boosted the member-states' rights significantly in the process of assigning their own troops to the alliance combat formation. This was actually a cornerstone which determined the general size of the national armies. After a long period, when the numbers were directly set by a single document of 1955, Article II of the new Statute granted the prerogative to every member-state government to decide the numbers of assigned troops during peacetime as well as a wartime. The national institutions could have taken into account advice by the supreme commander and the Political Consultative Committee, but were also allowed to act fully in accordance with the country's economic and financial sources available.[484] The Warsaw Treaty Organization's military structures, therefore, officially committed to respect the real capacities of individual member-states' economies.

The build-up of the Warsaw Pact armies, which after 1970 proceeded on the basis of the five-years plans - similar to general economic planning in the Eastern Bloc - was defined by two key documents. The first was the overall guidelines for development of the Unified Armed Forces over a particular five-year period. The material was worked out by the Unified Command and later formally approved by the Committee of Defense Ministers. At the 1st and 3rd meetings of the body in 1969 and 1971, the issue of "military five years" (1971-1975) was discussed. This scheme was innovative. In the past, the Unified Armed

---

[483] *Výsledky splnění úkolů v ČSLA za rok 1971 a hlavní úkoly ČSLA pro rok 1972*, 1. 11. 1971, 1261/0/6, sv. 17, a. j. 18/2, NA, Prague.
[484] New Secret Statutes of the Warsaw Pact, March 17, 1969, in *A Cardboard Castle?* (doc. 62): 323.

Forces' development was arranged for a shorter time period. A previous plan covered the years 1967-1970.[485] However, the Committee of Defense Ministers only gave its blessing to generally outlined trends. It was not involved in the negotiations on the concrete protocols which defined the national armies' commitments to the Warsaw Pact in detail. Those talks were held bilaterally and at the level of the national military institutions and the Unified Command, specifically between the special groups of the national general staffs and the Staff of the Unified Armed Forces. If some contentious questions remained, these were put forward to the chiefs of both institutions for mutual consideration.[486] The final protocols on assignment of the troops to the alliance combat formation were signed by individual defense ministers and the supreme commander.[487]

It was mentioned that the Warsaw Treaty Organization military statutes from March 1969 granted the member-state's right to assign the troops to the Unified Armed Forces in accordance with the real countries' capacities. In reality, this regulation was not fully complied with. For instance, in May 1970, the Czechoslovak People's Army (CSPA) representatives officially appreciated that during consultations on the 1971-1975 Protocol, the chief of Staff of the Unified Armed Forces, General Schtemenko, had respected the proposals and capabilities of the country.[488] However, such an evaluation was a lie. It was nothing more than part of the Czechoslovak normalization regime's general policy which tried to avoid any of the slightest hints of unwillingness to cooperate within the Warsaw Treaty Organization after August 1968. Prague did not provoke Moscow in any way, especially in the sensitive military matters

---

[485] *Zpráva ministra národní obrany pro Radu obrany státu o výsledku jednání Výboru ministrů obrany států-účastníků Varšavské smlouvy*, December 1969, FMO GŠ/OS, k. 9, p. č. 44, č. j. 39036, VÚA, Prague.

[486] *Jednání k otázkám Protokolu na rok 1971-1975*, April 1970, k. 15, p. č. 52, č. j. 39044, ibid.

[487] *Zpráva ministra národní obrany pro Radu obrany státu o výsledcích jednání výboru ministrů obrany států-účastníků Varšavské smlouvy*, March 1971, k. 10, p. č. 45, č. j. 39037, ibid.

[488] *Jednání k otázkám dvoustranného Protokolu*, 15. 5. 1970, k. 15, p. č. 52, č. j. 39044, ibid.

on which they acted with maximal loyalty.[489] In reality, this meant resigning its own initiatives and the servile realization of all the Unified Command's proposals.[490] The situation was reflected by the fact that alongside East Germany, Czechoslovakia was the first to sign the protocol on assignment of the troops to the alliance formation in the years 1971-1975.[491] A retrospective analysis worked out two years after the agreement had been concluded actually pointed to the remaining contradictions between the requirements of the Unified Command and the real Czechoslovak capacities. The material had to admit that the country had failed to fulfill the requested numbers of troops. Therefore, it suggested a

---

[489] It is also interesting to note that the Czechoslovak normalization purges affected also the Warsaw Pact command. Deputy chief of Staff of the Unified Armed Forces, General Eduard Kosmel, was withdrawn due to his "political mistakes." In the beginning of 1972, the politically competent Martin Korbela replaced him. It remains a question just how much was Kosmel's withdrawal only a demonstrative move towards Moscow. Actually, the general was neither expelled from the party nor downgraded in the armed forces, but only transferred to the post of deputy commander-in-chief of the 4th combined army of the Western military district. *Změna ve vedoucí funkci ve Štábu spojených ozbrojených sil států – účastníků Varšavské smlouvy*, 4. 1. 1972, 1261/0/6, sv. 26, a. j. 27/19, NA, Prague. Jan Wanner notes that in the first half of the 1970s, the Soviet side continuously pointed out the fact that a majority of the Czechoslovak officers had supported Alexander Dubček, either directly or indirectly. Therefore, more than half of the command corps was either dismissed or forced to leave the army. Since 1975, replacement with the new cadres, those loyal to Husák's leadership, began. However, the professional skills of this new group were significantly lower. Henceforward, completing just two years of high school was enough for holding senior posts in the Czechoslovak armed forces. Such incompetence from those officers bothered the Soviet side as well. Wanner, *Brežněv*, 72.

[490] *Vystoupení ministra národní obrany k otázce "výsledky plánování rozvoje ozbrojených sil států-účastníků Varšavské smlouvy na léta 1971-75"*, December 1970, FMO GŠ/OS, k. 10, p. č. 45, č. j. 39037, VÚA, Prague; *Vystoupení náčelníka štábu ČSLA na zasedání Vojenské rady Spojených ozbrojených sil dne 27.10. 1970 k otázce "Výsledky operační a bojové přípravy za rok 1970"*, k. 35, p. č. 129, č. j. 36032, ibid.

[491] The level of difficulties during the talks, at the same time reflected well the degree of individual member-states' loyalty: Thus, Romania was the last to approve the document and after the deadline. *3. zasedání Vojenské rady Spojených ozbrojených sil*, 1.11. 1970, ibid; *3. zasedání Výboru ministrů obrany*, March 1971, k. 10, p. č. 45, č. j. 39037, ibid.

revision in the upcoming military five-year plan.[492] Despite the negative trends in economic development, it was also out of the question to postpone the expensive rearmaments programs, especially in the air forces and air defense system.[493] Thus, in spite of the official documents talking about respecting the economic capabilities of the Warsaw Treaty Organization member-states, in the first years of the 1970s, Moscow's duress on building large armies as well as other financially exhaustive military projects largely continued. This illustrates well the matter of reform of the Warsaw Pact military structures: Disregarding the articles of the statutory documents of March 1969, the Soviet Union still kept various side-methods - enough to promote its interests within the Eastern Bloc.

The sub-changes of the above-described mechanism occurred after 1973. At the time, the supreme commander put forward *Draft of Joint Perspective of Development of the Unified Armed Forces of the Warsaw Treaty Organization member-states* based on the Political Consultative Committee's previous decision, which basically became a long-term guideline for the Eastern Bloc countries' military build-up.[494] The negotiations over the next five-year period (1976-1980) was also affected by the factor of climaxing détente. However, the main features of the system were preserved.

---

[492] According to Protocol for the years 1971-1975, Czechoslovakia was supposed to maintain the armed forces consisting 212-222,000 soldiers during peacetime. In reality, the country failed to fulfill that number. In May 1973, only 200,000 men served in the army. The CSPA command noted that Czechoslovakia had human and economic capacities to provide no more than 192,000 soldiers in peacetime and 540,000 during wartime. However, the Unified Command requests at the time expected mobilization of 700,000 troops in case of armed conflict. *Zhodnocení lidských a ekonomických možností státu ve vztahu k zámyslu výstavby ČSLA a souhrnné údaje o celkových počtech osob a nákladech na výstavbu ČSLA v letech 1976-1980*, undated 1973, k. 3, p. č. 31, č. j. 39023, ibid.

[493] *Vyjádření GŠ/SÚP k čj. 18034/ZD-OS-1970*, undated, 1970, k. 11, p. č. 47, č. j. 39039, ibid.

[494] *Zámysl výstavby ČSLA na léta 1976-1980*, 9. 7. 1973, k. 3, p. č. 31, č. j. 39023, ibid.

The Warsaw Pact military strategy in the first half of the 1970s

The changes in the Warsaw Treaty Organization military framework in March 1969 also had a certain, although very limited, impact on the shaping of the strategic conception of the alliance. Similar to many other cases, significant differences between provisions of the official statutory documents and following praxis can be seen. Those were especially demonstrated through the real competences of individual member-states on the one side, and the Unified Command and top Soviet army institutions on the other. In terms of operational and strategic plans, the national military as well as political bodies were granted strong rights by the new Statute of the Unified Armed Forces. Henceforward, according to Article V, the operational planning concerning deployment of any country's troops fell within the remit of individual defense ministries and the national general staffs. These institutions, however, were supposed to take into account recommendations by the supreme commander and even the General Staff of the Soviet army. Such a postscript, which provided the top-level of the Soviet military with some official authority over the Warsaw Pact for the first time, allowed the Soviet senior officers to keep the strategic and operational planning under their control. The final plans were signed by individual defense ministers and the supreme commander. The consent of the member-state government was also necessary.[495] But in reality, the national institutions loyally gave their political blessing only to the materials prepared by the Soviet side.[496]

---

[495] New Secret Statutes of the Warsaw Pact, March 17, 1969, in *A Cardboard Castle?* (doc. 62): 323.

[496] For example, in 1970, the Czechoslovak Council of State Defense clearly declared that the CSPA was built as a part of the Warsaw Pact and therefore strictly in accordance with the model of the Soviet army. Any duplicity in military doctrine or strategy was absolutely undesirable. It is likely that this was valid for all the Pact member-states' armed forces, with the specific exception of Romania.

Petr Luňák's claim can be accepted, that reform of the Warsaw Treaty Organization at the end of the 1960s at least partially improved the Eastern-European allies' knowledge of the Soviet military plans. At the meetings of the newly established collective military bodies, the issues of strategy were also discussed. However, the Soviet satellites still could not have been absolutely certain about its entire aspects.[497] The complete war strategy of the Warsaw Pact remained known solely by the Unified Command's highest-ranking officers and eventually by the Soviet senior military elites.[498] Research in the particular archives of former Pact member-states therefore faces the problem that Moscow provided only selective information for its allies.[499] From this point, it must be stressed that regarding the remaining unavailability of the alliance military headquarters' documentation, the overall strategy of the Warsaw Treaty Organization cannot be fully explained even today.

In the 1970s, the armed forces of individual Warsaw Pact member-states were already trained and built strictly for coalition warfare. Considering the particular armies' tasks in the Pact's strategic doctrine, their buildup differed and was highly specialized. The key factors were the objectives pursued in the war theatre, the geographic location of the country, the estimation of the strength of the anticipated enemy forces, and the potential human and economic sources of the Warsaw Pact member-

---

Zápis z jednání ministra národní obrany na operačním sále dne 4. března 1970, FMO GŠ/OS, k. 2, p. č. 29, č. j. 39021, VÚA, Prague.

[497] Luňák, *Plánování nemyslitelného,* 59.

[498] Former Czechoslovak General Mojmír Zachariáš testified that praxis, when individual fronts knew only those parts of the strategic plans which directly concerned them, also involved the Soviet units dislocated in Czechoslovakia after August 21, 1968. As the Central Group of the Soviet troops was, according to the Warsaw Pact's conception, formally subordinated to the command of the Czechoslovak Front, the overall operational documentation was held by the Czechoslovak side. The Central Group command had only the part defining its actions in the scope of the second echelon. Interviews with the Czechoslovak Generals, 2002-2003, in *Plánování nemyslitelného* (ad. 2): 415.

[499] Heuser, "Warsaw Pact Military Doctrine," 437.

states for waging a war.[500] Therefore, the East-German NPA was mostly an additional element to the main Soviet military group dispatched in the GDR. To a large extent, the armed forces of Hungary, which bordered on neutral Austria, also remained secondary in rank. This is contrary to the important roles played in the Czechoslovak and especially the Polish army, which still maintained a relatively high level of prestige among the nation. At the bottom of notional importance scale for combat in the scope of the Warsaw Treaty Organization lay the Romanian armed forces. This was determined not only by the geographic position of the country, but also by the Ceausescu regime's effort to build the army as a "national force" and an instrument of ensuring territorial security against Soviet political duress.[501] Taking into account such differentiations, the Unified Command acquainted in detail individual Unified Armed Forces' echelons with their specific tasks only. The testimonies of former East-German defense ministry officers suggest that the Warsaw Pact member-states lacked most of the information about the overall Soviet nuclear strategy. They would have just estimated it on the basis of the drill scenarios.[502]

The late 1960s and first half of the following decade brought thoughts about the chance of avoiding an armed conflict between the East and West, sourced from the temporary success of the détente policy. At the same time, changes in the approach towards the overall features of a potential war appeared as well. However, this second shift did not solely reflect improving the international situation, but mostly the general development of military technologies. As we explain later, the alteration of military thinking came very slowly within the Warsaw Treaty Organization. Moreover, it was not primarily instigated by the alliance's internal structures.

---

[500] *Vystoupení ministra národní obrany ČSSR k otázce BOJOVÁ PŘÍPRAVA A BOJOVÁ POHOTOVOST ARMÁD VARŠAVSKÉ SMLOUVY,* undated 1972, FMO GŠ/OS, k. 40, p. č. 136, č. j. 36039, VÚA, Prague.
[501] Wanner, *Brežněv,* 67.
[502] Kramer, "Warsaw Pact Military Planning," 16.

Interpretation of offensive strategy

For a better understanding of the above-mentioned shifts in thinking of the Soviet military strategists who also shaped the Warsaw Pact's doctrine, the main features of the alliance's general strategy must be outlined first. In the first half of the 1970s, the main aspect of the Pact's war plans remained a fast offensive aimed at seizing Western Europe after the outbreak of a potential conflict. Such an offensive conception, which had been established in the early 1960s during the second Berlin crisis, persisted in various modifications until 1987.[503] Central Europe remained the center of combat operations in the plans of NATO as well as the Warsaw Pact. The result of an entire war should have been mostly decided there.[504] After an initial NATO aggression, the basic strategic line of the Warsaw Treaty Organization in this war theatre supposed a massive counterattack by five echelons consisting of the groups of Soviet troops dispatched on the East-German, Czechoslovak, and Polish territories, in Belarus and Ukraine. Also, the NPA, CSPA, and PPA would have taken part in the offensive. The combat operations would have been joined by the Baltic fleets of the USSR, PPR, and GRD, as well as the air forces of other alliance members.[505] The Warsaw Pact plans were offensive, but not necessarily aggressive. Nevertheless, the Unified Armed Forces were prepared for a systemic and ideological war.[506] The drills therefore also contained the instigation of hatred towards the enemy. On the other hand, this was the same in the armies of NATO.[507]

---

[503] Mastny, "The Warsaw Pact," 145.

[504] *Vystoupení ministra národní obrany k otázce STAV NATO A NEBEZPEČÍ AGRESE V EVROPĚ*, undated 1972, FMO GŠ/OS, k. 40, p. č. 136, č. j. 36039, VÚA, Prague.

[505] Kramer, "Warsaw Pact Military Planning," 14. The question of using the air forces of some Warsaw Pact member-states remained problematic. At the end of the 1960s, many of the Eastern Bloc countries' armies still did not have their own frontal aviation. *Zpráva ministra národní obrany pro Radu obrany státu o výsledku jednání Výboru ministrů obrany států-účastníků Varšavské smlouvy*, December 1969, FMO GŠ/OS, k. 9, p. č. 44, č. j. 39036, VÚA, Prague.

[506] Nowak, *Od hegemonii do agonii*, 46.

[507] *Vystoupení ministra národní obrany k otázce STAV NATO A NEBEZPEČÍ AGRESE V EVROPĚ*, undated 1972, FMO GŠ/OS, k. 40, p. č. 136, č. j. 36039, VÚA, Prague.

Regarding the outlined offensive strategy, the existing literature often stresses the aggressive matter of the Warsaw Treaty Organization and consequently even the Soviet Union as its leading power. From the operational plan of the Czechoslovak Front of 1974, Petr Luňák concludes that even after almost a decade of attempts at easing the Cold War, the East did not abandon the opportunity to seize Western Europe and in case of armed conflict "to accomplish the historical victory of socialism" this way.[508] At the beginning of the 1990s, in the era following right after the Warsaw Pact dissolution, when the serious historical research of the alliance had just started and there was only a limited heuristic base available, Mark Kramer even presented a claim that the Warsaw Treaty Organization had counted on attacking Western Europe ever since the outburst of the conflict and therefore intended to start an offensive war. Consideration of repulsion of the potential aggression allegedly occurred no sooner than in the 1980s, after the shifts in strategic doctrine. On the basis of former East-German documents, Kramer concluded that at least the NPA had not practiced a defense against a NATO attack, except for a few moments in Gorbachev's era, because such a possibility had seemed to be highly unlikely. This fact - that the drill scenarios used to be based on the presumption that an initial aggression was conducted by the West - Kramer labeled as an ideological construct which aimed to generally excuse the offensive intentions and to dampen potential criticism by its own generality. The Warsaw Pact command allegedly never took the eventuality of a NATO attack seriously. Kramer concluded this mainly from the lack of significant practice focused on repulsion of the initial aggression in the drill scenarios. Usually, only troop mobilization and launching the Unified Armed Forces' counteroffensive were exercised. Actually, the drills of defense against enemy attack were incomparably less frequent in the Warsaw Treaty Organization armies than in comparison to NATO, where the defensive elements formed the backbone of the maneuvers.[509] However, the Western military planning came from the assumption that NATO troops would face a much stronger enemy.

---

[508] Luňák, *Plánování nemyslitelného*, 60.
[509] Kramer, "Warsaw Pact Military Planning," 14-15.

Therefore, a defensive and passive position seemed to be the more proper tactic.[510]

Kramer's claim has to be understood as a one-sided interpretation of a specific source,[511] moreover stripped of all Cold War context. Similar conclusions of the initial papers on the Warsaw Treaty Organization's strategic doctrine was actually based mostly on the drill plans as well as the scenarios of individual military maneuvers. However, the combat drills of the Warsaw Pact in the 1970s saw the typical gradual regress of massive maneuvers including huge numbers of troops in favor of operational exercises on the maps.[512] Considering the remaining unavailability of the entire strategic plans of the Unified Command, it is also good to take into account the talks at the alliance's collective military bodies' meetings, which also dealt with the issue of strategy.

At the 3rd meeting of the Military Council – the secretive high-level talks of the Warsaw Pact military elites – was held in Varna in autumn 1970, the main task was marked as preparation of the allied armies to "repulse the imperialistic aggression and totally defeat the enemy

---

[510] Durman, *Útěk od praporů*, 124.

[511] Kramer correctly noted that after German unification widespread and systematic shredding of the East-German NPA took place. Nonetheless, approximately 25,000 GDR military documents were given to the administration of the West-German defense ministry. Most importantly, materials which provided insight to the strategic and operational planning of the Warsaw Pact were missing. Kramer concluded it was an attempt to blind the trail of aggressive intentions and an offensive towards the West in preparation. However, the shredding of the East-German military documents must be rather understood as the Soviet Union's attempt to keep its military secrets during the turbulent period following the Eastern Bloc's collapse. Moreover, the Warsaw Treaty Organization still existed at the time. Kramer, "Warsaw Pact Military Planning," 13.

[512] Mastny, "Imagining War in Europe," 31; Jensen, Frade P., "The Warsaw Pact's Special Target. Planning the Seizure of Denmark," in *War Plans and Alliances in the Cold War*, eds. Vojtech Mastny, Sven Holtsmark and Andreas Wenger (London: Routledge, 2006): 95. At the beginning of the 1970s, some military drills remained massive. For instance, 40,000 and 20,000 troops, respectively, took part in "Taran-70" and "Zenit-70" maneuvers, similar to the "Neutron-70" command-staff exercise. *Vystoupení náčelníka štábu ČSLA na zasedání Vojenské rady Spojených ozbrojených sil dne 27.10. 1970 k otázce "Výsledky operační a bojové přípravy za rok 1970"* FMO GŠ/OS, k. 35, p. č. 129, č. j. 36032, VÚA, Prague.

afterwards." Regarding the CSPA and the Unified Armed Forces' first echelon in general, the supreme commander Yakubovski also stressed the capability of securing the western borders of the Warsaw Pact countries at any time.[513] It is very unlikely that the marshal would have hesitated to speak openly and "camouflage" the real intentions at a closed military meeting. Along the same lines, the discussion also took place at the 4th gathering of the Committee of Defense Ministers in Berlin, February 1972. Czechoslovak defense minister Dzúr emphasized the CSPA ability to stop an initial enemy offensive at the borderlines between Czechoslovakia and West Germany. He considered the realization of this task to be the key factor following the fast launch of an alliance troops' counterattack.[514] Preventing ground-combat operations passing into the territories of the Warsaw Pact member-states remained the primary objective. Dzúr also assured present defense ministers and the Unified Command representatives that the Czechoslovak border divisions practiced the territorial defense in the mountainous terrain of the country's western frontiers with priority.[515] During the aforementioned meeting in Varna, Karel Rusov, the chief of the CSPA general staff, also confirmed to the alliance military officials that alongside an offensive operation, the drills consisted of repulsion of the initial enemy aggression and the securing of

---

[513] Ibid; *3. zasedání Vojenské rady Spojených ozbrojených sil*, 1. 11. 1970, ibid.

[514] After dispatching the Central Group of Soviet troops in 1968, long-term numbers of the CSPA were reduced by 20,000 soldiers. However, according to former Czechoslovak General Jan Franko's testimonys testimony, the Soviet military command still did not change the conception of the independent Czechoslovak first echelon. The Central Group units were supposed to act like the second echelon, which had consisted of the Belarusian and Transcarpathian military districts in the past. There was just no need to transport them to the Czechoslovak territory. The Soviet troops therefore potentially gained almost one extra week of combat activities. Interviews with the Czechoslovak Generals, 2002-2003, in *Plánování nemyslitelného* (ad. 2): 413.

[515] *Vystoupení ministra národní obrany ČSSR k otázce BOJOVÁ PŘÍPRAVA A BOJOVÁ POHOTOVOST ARMÁD VARŠAVSKÉ SMLOUVY*, undated 1972; *Vystoupení ministra národní obrany k otázce STAV NATO A NEBEZPEČÍ AGRESE V EVROPĚ*, undated 1972, FMO GŠ/OS, k. 40, p. č. 136, č. j. 36039, VÚA, Prague.

state borders as well.[516] Therefore, it is unlikely that the Warsaw Treaty Organization armies did not practice the defense stage at all or totally neglect the option of a NATO attack. On the contrary, the collective military bodies' meetings as well as retrospective testimonies by the Eastern generality[517] prove that an initial NATO strike on the Warsaw Pact was a precondition for the counterattack and seizure of Western Europe in the 1970s already, not just after the doctrinal changes in the following decade.

The conclusion appearing mostly in analyses from the early 1990s, that the practicing of offensive operations by the East-European armed forces proves the generally aggressive intentions of the Soviet leadership (which in reality was in charge within the Warsaw Pact), is also dismissed by P. F. Jensen. According to his interpretation, the alliance's military drills cannot be considered a reflection of real war objectives. He rather takes them as a practice of the achievable limits during potential combat operations. It is good to point out that some contradictions exist between the maneuver scenarios and the operational plans available, especially regarding the offensive of the Soviet, East-German, and Polish troops against Denmark.[518] The military drills therefore did not precisely correspond with the operational plans. They rather reflected actual military thinking and provided at least a limited opportunity to test it in praxis. The strategic planning and lessons learned during the drills most likely affected

---

[516] *Vystoupení náčelníka štábu ČSLA na zasedání Vojenské rady Spojených ozbrojených sil dne 27.10. 1970 k otázce "Výsledky operační a bojové přípravy za rok 1970"*, k. 35, p. č. 129, č. j. 36032, ibid.

[517] At the conference in Stockholm, April 2006, Alexander Liakhovski commented on the issue. The former Soviet General, who in the late 1970s served in the Main Operational Department of the Soviet general staff, confirmed the above-mentioned assumptions. The military command of the USSR allegedly really saw NATO as a potential aggressor. The Warsaw Pact's plans, therefore, were strictly based on the premise of the initial enemy attack. Hoffenaar, Jan and Findlay, Christopher, eds., *Military planning for European Theatre Conflict During the Cold War. An Oral History Roundtable, Stockholm, 24-25 April 2006* (Zürich: Center for Security Studies, 2007), 61.

[518] During the peak of détente, the NPA was tasked with containing the attacking Danish and West-German units as long as possible. The following counteroffensive should have been left to the advancing Polish front.

each other. The Warsaw Pact's offensive plans to seize almost the whole of Western Europe must be considered a mostly hypothetical scenario for the worst case ever - an outbreak of a total war between the East and West, rather than a manual on how to solve potential international crises or a limited local conflict. Such a danger actually did not *fully* vanish even at the peak of détente. Also, general conclusions about the Kremlin's intentions cannot be made from considering only the partially available documentation like the Czechoslovak or Polish war plans.[519] Beatrice Heuser's work, which is based on similar sources to Kramer's, confirms that materials in German military archives provide no evidence of strictly aggressive intentions of the Warsaw Treaty Organization. They just reflect an active-defense tactic against a potential NATO attack. In addition, we are reminded that a huge gap existed between the real pursued objectives of the Soviet foreign policy and the military planning in case of war, if proven to be inevitable. However, it is still not clear whether the outlined plans would have been put into praxis.[520]

It remains the reality that even the official propagandistic claim marked the Warsaw Pact's main task to repulse "the imperialistic aggressor and to defeat him ultimately".[521] The political guidelines for drills of the Czechoslovak armed forces during the years 1972-1973 also stated that the Warsaw Pact's actual objective was to strengthen combat cooperation of the Unified Armed Forces in order to repulse an aggression by the enemy followed by his "destruction in the short period."[522] Those theses, in fact, were meant as a euphemism for the blitz-offensive towards the West. Annihilation of the enemy solely through defense of its own

---

[519] Jensen, "The Warsaw Pact's Special Target," 95-104.

[520] Heuser, "Warsaw Pact Military Doctrine," 438, 451.

[521] *Vystoupení ministra národní obrany ČSSR k otázce BOJOVÁ PŘÍPRAVA A BOJOVÁ POHOTOVOST ARMÁD VARŠAVSKÉ SMLOUVY*, undated 1972, FMO GŠ/OS, k. 40, p. č. 136, č. j. 36039, VÚA, Prague.

[522] This scenario was reflected also in concept by the "Soyuz-73" maneuvers which practiced combat operations at the southwestern war theater. The initial strike on Bulgaria by the ground as well as the air forces of NATO's southern flank remained the main presumption. The Soviet, Romanian, and Bulgarian troops should have repulsed the aggression and then launched the counteroffensive. Opriş, "Die rumänische Armee," 197.

territory was hardly achievable. Therefore, the Warsaw Treaty Organization mostly emphasized the practicing of the attack operation.[523] Moreover, the Pact's military officials were quite sure of the Eastern superiority in terms of conventional weapons. They expected no significant problems repulsing a NATO attack. In the first half of the 1970s, the Warsaw Pact also presumed, that during an initial stage of war, NATO would rely mostly on airstrikes.[524] The defense phase should have taken a disproportionately shorter period than the following counterattack. That is probably the reason why the drills paid much less attention to it. Moreover, such a conception was affected significantly by the positive experience of the Red Army's massive offensive operations during World War II. At the time, its preserving influence on Soviet military thinking, especially in terms of preparation and planning of the operations (more than three decades after the defeat of Nazi Germany), was proven by the chief of the Staff of the Armed Forces, General Schtemenko.[525] In addition, from the Soviet point of view, the Warsaw Pact's military doctrine was defensive. As the march to the west should have come after the initial NATO aggression, it was officially considered simply as passing the combat into the enemy territory.[526]

Alteration in thinking of conflict's features

The main outcomes of the Warsaw Treaty Organization strategic doctrine did not undergo any substantial changes in the first half of the 1970s. Nevertheless, development of military technologies and their consistently growing potential at the end of the previous decade had already resulted in significant reconsideration of a potential conflict

---

[523] *Zpráva pro PÚV KSČ o splnění úkolů ČSLA ve výcvikovém roce 1972-1973 a hlavních úkolech pro výcvikový rok 1973 – 1974*, 28. 9. 1973, 1261/0/6, sv. 93, a. j. 90/1, NA, Prague.

[524] *Návrh vystoupení ministra národní obrany Československé socialistické republiky k otázce plnění úkolu "Upevnění a rozvoj jednotného systému PVO členských států Varšavské smlouvy" v podmínkách ČSLA*, 28. 11. 1973, FMO GŠ/OS, k. 10, p. č. 46, č. j. 39038, VÚA, Prague.

[525] *Porada vedoucích funkcionářů generálních /hlavního/ štábů armád Varšavské smlouvy*, 24. 6. 1975, k. 19, p. č. 58, č. j. 39051, ibid.

[526] Heuser, "Warsaw Pact Military Doctrine," 447.

scenario so far. The rigid thoughts that a clash between the superpowers would inevitably be a total nuclear war receded. The initiator of this shift was obviously the West, especially the USA. The Warsaw Pact only followed the development. Moreover, its 1969 military framework reform only had a marginal impact on these changes.

In the second half of the 1960s, the constantly improving capabilities of Soviet strategic missiles to hit targets in Northern America accurately, made the Pentagon look for an alternative to its existing "massive retaliation" doctrine. This resulted in the new strategy of "flexible response," adopted by NATO in 1967. From Washington's point of view, it established a much more favorable option of a limited nuclear war in Europe.[527] Up to this point, the U.S. president faced an unenviable dilemma: His only way to use nuclear weapons was as a total strike on the enemy territory, including also strictly civilian targets. Simultaneously, such a decision would have meant a death sentence for millions of U.S. citizens, as the same Soviet counteraction would have almost certainly followed.[528] Therefore, the "flexible response" doctrine introduced that the main deterrence against the enemy aggression was the initial threat of war escalation before the first massive nuclear strike. During this stage of the conflict, the Warsaw Treaty Organization should have come to conclusion that the risks taken were not worthy of the pursued aims. The primary task was only to face an initial Eastern attack. But NATO should have remained ready to escalate the fights at any moment, even with limited use of nuclear weapons, if needed. In order to maintain an opportunity to negotiate and therefore to keep war escalation under control, after 1967, the Western

---

[527] Jensen, "The Warsaw Pact's Special Target," 105.

[528] Burr, William, "'Is this the best they can do?' Henry Kissinger and the US quest for limited nuclear operation, 1969-75," in *War Plans and Alliances in the Cold War*, eds. Vojtech Mastny, Sven Holtsmark and Andreas Wenger (London: Routledge, 2006): 122. According to much of the analytics, U.S. willingness to prevent a potential military seizure of Western Europe at the cost of risk of annihilation of American cities in a total nuclear war could not have been credible even to the Kremlin. In September 1979, former American Secretary of State, Henry Kissinger, publicly confirmed these doubts retrospectively. He stated that it had been absurd to base the Western strategy on believability of mutual suicide. Holloway, David, "Jaderné zbraně a studená válka v Evropě," *Soudobé dějiny*, no 1-2 (2011): 44.

strikes should have avoided the main political and military centers of the Warsaw Pact.[529] At the end of the 1960s, taking into account such doctrinal changes in NATO, the Soviet military elites began to abandon the dogmatic thought that a future conflict had to inevitably start with widespread nuclear Armageddon. Therefore, the Warsaw Pact's strategy was slightly adapted and began to calculate with an initial, most likely conventional, phase of war. Nevertheless, an assumption still prevailed that nuclear weapons would be used, in the end.[530]

     Outlined reaction of the Warsaw Treaty Organization to the NATO "flexible response" doctrine was actually very slow. The presumption, widely accepted by the Eastern military strategists after 1964, that a war would begin with a Western surprise nuclear assault, remained valid for at least another five years, which followed the previously-established, aforementioned NATO strategy. Afterwards, since 1971, it began to be replaced by the thesis of the potential use of nuclear weapons only after a significant rise of international tensions.[531] It seems that date was not random. Obviously, it was a direct response to the substantial shift in détente. Some scholars point out a persisting gap between the political reality and the Warsaw Pact's military planning. They conclude this from the slow strategic changes at the turn of the 1960s and '70s.[532] However, a closer insight reveals it was quite logical to wait with reconsidering the military conceptions until détente proved to be a relatively viable process.[533] In 1970, the basic direction of the Warsaw Pact armies' development still reflected an assumption that the military-political situation in Europe would not alter significantly during the

---

[529] Heuser, "Warsaw Pact Military Doctrine," 445.

[530] Holloway, "Jaderné zbraně," 42.

[531] *PROTOKOL o předání a převzetí funkce náčelníka plánování obrany a ochrany teritoria OO/OS*, 12. 2. 1971, FMO GŠ/OS, k.1, p. č. 16, č. j. 53006, VÚA, Prague.

[532] For example, Luňák, *Plánování nemyslitelného*, 60-61

[533] We are reminded that in 1971 the intensive talks on holding the CSCE had already proceeded. Also, the Soviet-U.S. negotiations SALT were in the advanced stage. In addition, the USSR and Poland signed their agreements with West Germany.

following five years at least.[534] However, new international conditions, characterized by the intensive talks between the Eastern and Western political elites, could not have been ignored even by the Warsaw Pact top military officials, who remained leery of ongoing world events. Nevertheless, they had to admit that a surprise NATO nuclear attack, which had not been totally out of the question in the dramatic era of the first half of the 1960s, was hardly imaginable at the moment.

During the Unified Armed Forces drill "West-69," the Soviet defense minister Grechko had to acknowledge that regarding the actual situation in Europe, almost no imminent threat of attacking the Warsaw Pact member-states existed. For the first time, he mentioned in front of the allies a possibility of waging a war without atomic ammunition. Formally, the Soviet defense minister also supported the CPSU "no use first policy."[535] However, he still reserved the right, in certain situations such as the enemy revealing its aggressive intentions, to preemptively strike in

---

[534] *Základní úvahy pro výstavbu ČSLA na léta 1971-1975 s výhledem do roku 1980*, undated, 1970, FMO GŠ/OS, k. 2, p. č. 29, č. j. 39021, VÚA, Prague.

[535] In July 1982, the Soviet Union officially unilaterally pledged not to use its nuclear weapons first. However, during the years 1970-1973, Moscow had already secretly negotiated with Washington on a similar mutual agreement. But the Western strategists were afraid that the USSR was only trying to strengthen the importance of its conventional forces in this manner, in the field where the Warsaw Pact maintained superiority over NATO. Thus, the talks were not successful. After this failure, Moscow started a public campaign to support such a principle which climaxed in the aforementioned move in 1982. Despite this, the West was very skeptical of the Soviet pledge; the CPSU CC, in fact, already issued a secret directive demanding the strategic planning to be based on a no first use policy in 1974. According to the course-books of the Soviet general staff academy, such a principle especially concerned the Warsaw Pact's operations. Moreover, since the end of the 1960s, the Soviet general staff was forced to consider usage of nuclear weapons as the "worst case scenario" - a move which should have been preceded by a strike on the Warsaw Pact countries and the use of atomic ammunition by the enemy. Since the mid-1970s, the Warsaw Pact military doctrine therefore increasingly oriented to the possibility of waging the large conventional war. More Garthoff, Raymond L., "New Thinking and Soviet Military Doctrine," in *Soviet Military Doctrine from Lenin to Gorbachev 1915–1991*, eds. Willard C. Frank and Philip S. Gillette (Westport: Preager, 1992): 197; Jensen, "The Warsaw Pact's Special Target," 106-107; Burr, "Is this the best they can do?" 130.

order to eliminate that threat.[536] The principles of a preemptive, undoubtedly nuclear attack and no first use policy obviously sharply contrasted. This is more proof that the Soviet military command was abandoning deeply established thoughts from the most strained Cold War period only with a strong reluctance.

Marshal Grechko was most likely one of the main opponents of any alteration in military thinking. In 1969, at the 1st Committee of Defense Ministers' session, he still warned the East-European allies against overestimating the "flexible response" doctrine. Its basis, in his opinion, remained rooted in the previous "massive retaliation" strategy. Grechko suspected that NATO, despite the claims of a limited war possibility, still relied mostly on the extensive use of nuclear weapons.[537] He emphasized that the Western alliance was preparing itself for an eventuality to start a preventative nuclear strike within six days at the latest following an outbreak of conflict.[538] The marshal's mistrust in the Western willingness to conduct conventional combat operations was undoubtedly also caused by the estimated ratio of strength between both pacts. During the above-mentioned "West-69" drills, Grechko himself stressed the numerical advantage of the Warsaw Pact's ground divisions in up to 1.6:1 rate. He also stated the significant superiority of the navies of the USSR, PPR, and GDR over the enemy in the Baltic area.[539] The marshal also

---

[536] Speech by Marshal Grechko at the "Zapad" Exercise, October 16, 1969, in *A Cardboard Castle?* (doc. 66): 342.

[537] In this issue, Grechko was vindicated also by the Warsaw Pact countries' diplomats. Regarding the North Atlantic Council meeting in December 1969, they emphasized the resolution talking about a need to prefer nuclear conflict at the European war field. In addition, NATO General Secretary, Manlio Brosio, warned against the further reduction of NATO forces as it would have resulted in significant conventional preponderance of the Warsaw Treaty Organization. *Průběh a výsledky jednání vedoucích orgánů NATO – PZ č. 3*, 8. 1. 1970, TO(t) 1970-1974, box. 8, i. č. 89, sign. 020/311, č. j. 020.090/70-1, AMZV, Prague.

[538] Speech by Grechko at the First Meeting of the Warsaw Pact Committee of Ministers of Defense, December 22, 1969, in *A Cardboard Castle?* (doc. 70): 356.

[539] According to presumption of the "West-69" exercise, due to its preponderance, the Warsaw Pact's Northern Front was supposed to advance 200-800 km into the enemy territory without being reinforced by the second echelon units. The heavier fighting was expected for the Central and Southern Fronts. Their regrouping was supposed to happen already after advancing to 50-200 km. Actually, the first

expected that following military development would further move the correlation of forces in favor to the East.[540] According to Grechko's belief, NATO was unable to make a successful conventional strike and therefore remained reliant on its nuclear arsenals.[541] In this regard, during the analysis of the maneuvers, he criticized the commanders of troops representing the Western attacking units by stating that in their operation on the East-German and Czechoslovak territories they "unrealistically" reached the depth of 600 km without using atomic munitions.[542] The official scenario of NATO's "flexible response" doctrine assumed that the Western conventional forces would be able to withstand the Warsaw Pact attack up to 90 days. In reality, however, almost no officials of both hostile blocs believed it.[543]

Despite the fact that the Warsaw Pact command dealt with the possibility of the initial stage of war without using weapons of mass destruction at the turn of the 1960s and '70s, a surprise nuclear attack by NATO remained the main presumption. In response, a similar strike should have come, followed by the beginning of the march to Western Europe. In terms of a general strategic concept, only the speed of the Unified Armed Forces' advance changed. According to the "Taran-70" drills scenario, the Czechoslovak Front was supposed to reach the borders of France in eleven or seventeen days, depending on the use of conventional ammunition only or nuclear weapons. The exercises were still based on the assumption that conflict sooner or later grew into the nuclear stage.[544] This was obviously also an outcome of rigid Leninist-Stalinist thinking from a certain part of the senior Soviet military officials. If the next world war should have been the last decisive "class conflict"

---

echelons of these Fronts had to create a sufficient-enough shield for the regrouping of other Warsaw Pact troops in the East-German and Czechoslovak territories.

[540] Speech by Marshal Grechko at the "Zapad" Exercise, October 16, 1969, in *A Cardboard Castle?* (doc. 66): 342.

[541] Mastny, "Imagining War in Europe," 31.

[542] Luňák, Petr, "War Plans from Stalin to Brezhnev. The Czechoslovak Pivot," in *War Plans and Alliances in the Cold War*, eds. Vojtech Mastny, Sven Holtsmark and Andreas Wenger (London: Routledge, 2006): 89.

[543] Durman, *Útěk od praporů*, 124.

[544] Luňák, *Plánování nemyslitelného*, 59-60.

between socialism and capitalism, it logically must have resulted in a total nuclear inferno at some moment.[545] Such a deep-rooted assumption persisted in the Soviet military thinking through the mid-1970s.[546]

Thus, at the beginning of the 1970s, the Warsaw Treaty Organization troops simultaneously practiced for conventional as well as nuclear war. However, the center of the drills remained in various nuclear operations,[547] mainly in the option to strike first a massive blow which would have *de facto* devastated the enemy and eliminated their similar effective counteraction at the same time.[548] During the Warsaw Pact's maneuvers, great emphasis was especially placed on tactical nuclear strikes[549] facilitating the breakthrough in the enemy defense and allowing

---

[545] Fully in accordance with this belief, in September 1971, the Soviet General M. I. Cherednichenko stated in the party military journal *Kommunist vooruzhenych sil* that the next war would begin with a surprise nuclear strike by the "imperialistic aggressor." Scott, William F., *Selected Soviet Military Writings 1970-1975* (Washington: U.S. Government Print Office, 1976), 6. The analysis of the Soviet military journals undoubtedly requires a very cautious approach and critical evaluation. It is largely a tendentious source, moreover written with an extreme emphasis on the ideological basis. In addition, deeper insight could be provided only by a wide analysis of the Soviet military journals, which significantly transcends the scope of this work. However, publication of the aforementioned opinions proves that at least part of the Soviet top military still kept them.

[546] Heuser, "Warsaw Pact Military Doctrine," 438.

[547] The Warsaw Pact drills simulated the option of a first-strike surprise nuclear blow or an answer to a similar Western attack. The scenarios also contained an eventuality that nuclear weapons would be used during the fighting in order to finalize devastation of any remaining enemy troops and reserves. Kramer, "Warsaw Pact Military Planning," 15.

[548] *Vystoupení náčelníka štábu ČSLA na zasedání Vojenské rady Spojených ozbrojených sil dne 27.10. 1970 k otázce "Výsledky operační a bojové přípravy za rok 1970"*, FMO GŠ/OS, k. 35, p. č. 129, č. j. 36032, VÚA, Prague; *Vystoupení ministra národní obrany ČSSR k otázce BOJOVÁ PŘÍPRAVA A BOJOVÁ POHOTOVOST ARMÁD VARŠAVSKÉ SMLOUVY*, undated 1972, k. 40, p. č. 136, č. j. 36039, ibid; *Návrh rozkazu ministra národní obrany pro přípravu vojsk ČSLA v roce 1972*, 1. 11. 1971, 1261/0/6, sv. 17, a. j. 18/2, NA, Prague.

[549] In military terminology, a tactical nuclear strike generally means the use of atomic weapons at a war field in a combat situation. On the contrary, the term "strategic nuclear attack" refers to a nuclear strike on an enemy's cities, logistic networks of centers behind a frontline.

for the fast advance of attacking troops.[550] The strikes were supposed to aim at mostly military targets. The objective was to eliminate NATO backup divisions and to force some of the Western alliance member-states either to surrender or to declare neutrality.[551] In addition, the drill plans considered an option that after seizure of the Western cities, the loyal governments consisting of local communists and leftist social democrats would be established.[552] The same sources suggest that an occupation of the extensive Western territories remained an important pursued aim even at the end of the 1980s.[553]

The superpowers' SALT negotiations, in reality, had no influence on the Warsaw Pact's nuclear strategy, as the overall global Soviet nuclear doctrine did not directly concern the alliance. Moscow considered the Unified Armed Forces as an instrument for combat operations at the European war fields.[554] The intercontinental ballistic missiles capable of hitting targets in the U.S. and Canada remained under exclusive control of the Soviet strategic forces. According to testimony of the former Soviet General, Adrian Danilevich,[555] in the first half of the 1970s, a potential war period was sorted into four phases. The first assumed conventional combat. During the second, limited use of mostly tactical nuclear weapons was expected. In the third, the conflict escalated into a total nuclear war, including use of all the strategic arsenals. The final result should have been decided in the fourth, undefined stage. During the 1970s, the first phase significantly expanded in strategic thinking, both timely and territorially.[556] The Warsaw Pact accepted this periodization of war in 1972 at the latest, when the 4th Committee of Defense Ministers session in Berlin discussed the issue. The combat exercises of the Unified Armed

---

[550] Kramer, "Warsaw Pact Military Planning," 15.

[551] Heuser, "Warsaw Pact Military Doctrine," 441.

[552] The Surrender of Hannover according to the Polish Army's "Bison" Exercise, April 21-28, 1971, in *A Cardboard Castle?* (doc. 73): 380; Nowak, *Od hegemonii do agonii,* 43.

[553] Heuser, "Warsaw Pact Military Doctrine," 441-442.

[554] Mastny, "Imagining War in Europe," 31; Mastny, "The Warsaw Pact," 142.

[555] In the 1970s, Danilevich held the post of chief of main operational planning in the Soviet army. Later, during the years 1984-1990, he also served as deputy chief of the Soviet general staff for doctrine and strategy.

[556] Holloway, "Jaderné zbraně," 42-43.

Forces should have reflected the military intelligence assumption that NATO considered three possible scenarios: to start a conflict with conventional weapons, nuclear weapons, or prospectively use atomic ammunition later during combat operations.[557]

The military documents available do not answer a key question: Which political decision would have turned the war into a nuclear stage? At the beginnings of the 1990s, Mark Kramer suggested that such competence fell within the remit of the CPSU General Secretary. Based on his analysis of the East-German documents, Kramer also noted that from the moment when conflict reached the nuclear phase, offensive use of atomic charges could have been authorized by the commanders of individual army fronts. Moreover, even the division command allegedly had a right to order a defensive tactical nuclear assault.[558] However, this was probably only a specific case reflecting the generally weak position of the East-German NPA, which was never supposed to fight alone, and in comparison to the rest of the member-states' armies, kept substantially less autonomy. The Soviet military commanders even considered the East-German officer corps to be their direct subordinates.[559] Within the Warsaw Treaty Organization, the Soviet side exclusively held the so-called nuclear weapons key. In fact, it is extremely unlikely that the division commanders

---

[557] *Vystoupení ministra národní obrany ČSSR k otázce BOJOVÁ PŘÍPRAVA A BOJOVÁ POHOTOVOST ARMÁD VARŠAVSKÉ SMLOUVY*, undated 1972, FMO GŠ/OS, k. 40, p. č. 136, č. j. 36039, VÚA, Prague.

[558] Kramer, "Warsaw Pact Military Planning," 15-16.

[559] In general, the NPA was mostly a bridgehead for the Northern Group of the Soviet troops in Germany. Cooperation of both armies was definitely highest within the entire Warsaw Pact; since its establishing, the East-German armed forces strictly represented the alliance army, not the national force. Moreover, the Soviet as well as East-German military officials feared direct confrontation between the troops of East and West Germany. The NPA units, therefore, were supposed to fight mainly the U.S. and Benelux countries' forces - which were paradoxically better equipped - and to capture West Berlin. Only one tank division, air force, and navy should have clashed with the FRG troops. Mostly, the Soviet army was supposed to eliminate Bundeswehr. Moreover, in case of war, the East-German armed forces would have been fully placed under the Soviet command. Schaefer, *"The GDR in the Warsaw Pact"*.

of other member-states had any real expertise on this issue.[560] Thus, unlike the situation in NATO, Moscow maintained the nuclear monopoly in possession of nuclear weapons within the Warsaw Pact.[561]

In the first half of the 1970s, the Warsaw Treaty Organization's military strategists expected the initial conventional phase of war to be very limited in terms of time. Transition of troops from conventional to nuclear combat was practiced during the "Neutron-73" drills, for example. According to the maneuver scenario, the first stage - including preparations for the operation - took no more than four days.[562] These considerations were certainly also fueled by the intelligence. The East-German military analytics, for instance, concluded from NATO's "Wintex-73" drills that the West would most likely use its nuclear arsenal at the last moment, when the Warsaw Pact troops reached a depth of 100 km in the territory of the FRG.[563] However, some Unified Command senior officials no longer categorically excluded the possibility of a strictly conventional operation. In mid-1973, at the gathering of the chief of the Pact member-states' general staffs, in his report on actual strategic shifts, General Schtemenko labeled it as a relevant alternative to nuclear war. Nevertheless, in this case, regarding the overall strategic plan of the Warsaw Pact - to conduct a counteroffensive into Western Europe - only the speed of advance changed. The attacking fronts were supposed to penetrate at least 700-800 km into enemy territories. An absence of nuclear assaults would have accelerated their march. In that situation, advances up to 100 km per day were expected, but 40-60 km less, in the case of use of

---

[560] The Soviet nuclear weapons were subjected to a very special régime, as shown by the extraordinary measures concerning the nuclear depots constructed on the Czechoslovak territory during the "Javor" operation. The Council of State Defense dealt with the procedures on March 4, 1970. Afterwards, the material should have been presented to Gustáv Husák by defense minister Martin Dzúr in person. No other top political or military representatives were supposed to take part in the meeting, except for the chief of general staff and the chief of operational planning group. *Zápis z jednání ministra národní obrany na operačním sále dne 4. března 1970*, FMO GŠ/OS, k. 2, p. č. 29, č. j. 39021, VÚA, Prague.

[561] Wanner, *Brežněv*, 66.

[562] *Informace o přípravě velitelsko-štábního cvičení „NEUTRON-73"*, 20. 4. 1973, 1261/0/6, sv. 77, a. j. 74/1, NA, Prague.

[563] Kramer, "Warsaw Pact Military Planning," 17.

atomic ammunition.[564] These unrealistic calculations[565] from today's perspective prove well that the Eastern military command was clearly aware of its conventional superiority to NATO troops. The claims about an existing mutual balance of military power between two pacts, presented by the Warsaw Treaty Organization representatives during the Vienna talks on a reduction of troops and armaments in Central Europe at the time, were therefore just intentional lies.

Despite Schtemenko's considerations suggesting that a conventional operation could have been favorable for the Warsaw Pact, nuclear weapons remained a key principle in the first half of the 1970s. Nor did the alliance military strategists expect any significant shifts in this issue on the mid-term horizon. According to the conclusion of 1972, the nuclear arsenal, respectively its carriers, was supposed to define the main perspective of the Warsaw Treaty Organization armies' development even in the years 1985-1990.[566] In line with that, the supreme commander Yakubovski later, at the 9th meeting of the Military Council in Prague, urged strengthening the missile units - especially the fronts - to be equipped with accurate tactical rockets like the R-17, capable of carrying a nuclear warhead.[567] The situation is documented by the CSPA war plan of 1974. In comparison to the previous version of 1964, there were no significant changes. It was still based on the assumption that NATO would attack the Warsaw Pact first, using nuclear weapons. Afterwards, the Unified Armed Forces' first echelon was supposed to start its advance into Western Europe at the extreme speed about 60-90 km per day.[568] Conventional conflict was

---

[564] *Porada vedoucích funkcionářů generálních štábů*, 18. 6. 1973, FMO GŠ/OS, k. 15, p. č. 52, č. j. 39044, VÚA, Prague.

[565] In this regard, former Czechoslovak general Anton Slimák testified that estimation of the offensive speed did not undergo a significant realistic revision until the 1980s. Interviews with the Czechoslovak Generals, 2002-2003, in *Plánování nemyslitelného* (ad. 2): 419.

[566] *Vojensko-politické úvahy a tendence vojensko-technického rozvoje ozbrojených sil protivníka a záměry ve výstavbě ČSLA*, undated, 1972, FMO GŠ/OS, k. 2, p. č. 29, č. j. 39021, VÚA, Prague.

[567] *9. zasedání Vojenské rady Spojených ozbrojených sil*, 7. 11. 1973, k. 37, p. č. 131, č. j. 36034, ibid.

[568] Plán týlového zabezpečení operace ČSF, 1974, in *Plánování nemyslitelného* (doc. 10): 241.

considered to be likely, however, just as a short prologue to a decisive nuclear clash.[569]

Obviously, at least some part of the senior Eastern officers were aware of the exorbitance and over-confidence of these plans. In the first half of the 1970s, they voiced their opinions at the Warsaw Pact collective bodies' meetings, however, evidently to only a limited scale. In October 1970, at the 3rd Military Council session in Varna, General Rusov correctly stated that the troops' activity in radioactively contaminated locations as well as their protection against weapons of mass destruction "had not been practiced enough yet." In fact, this was just a euphemism for their infeasibility.[570] At the 4th Committee of Defense Ministers gathering in Berlin, sober predictions were presented that especially the enemy air and nuclear strikes would cause significant damage and chaos, not only in NATO countries', but also in the Warsaw Pact's territories.[571]

Despite the aforementioned partial shifts in the Eastern military strategists' thinking, in the first half of the 1970s, the direction of strategic changes was still indicated mostly by the United States. At the time when Soviet war planners were still slowly reacting to the "flexible response" doctrine, the Nixon administration had already introduced other strategic innovations. In March 1971, the defense secretary Melvin Laird announced a new "realistic deterrence" doctrine.[572] Among other things, it was probably just another Western impulse to accelerate changes in war considerations of the Soviet top military. At the turn of the 1960s and '70s, American analytics actually correctly doubted the USSR's willingness to accept the concept of limited nuclear war. Nixon's national security adviser, Henry Kissinger, kept trying to weaken the destructive impact of a potential use of atomic arsenals. He intended to extend planning of limited nuclear war so far, especially to improve the possibility of

---

[569] Luňák, "War Plans from Stalin to Brezhnev," 89-90.

[570] *Vystoupení náčelníka štábu ČSLA na zasedání Vojenské rady Spojených ozbrojených sil dne 27.10. 1970 k otázce "Výsledky operační a bojové přípravy za rok 1970"*, FMO GŠ/OS, k. 35, p. č. 129, č. j. 36032, VÚA, Prague.

[571] *Vystoupení ministra národní obrany ČSSR k otázce BOJOVÁ PŘÍPRAVA A BOJOVÁ POHOTOVOST ARMÁD VARŠAVSKÉ SMLOUVY*, undated 1972, k. 40, p. č. 136, č. j. 36039, ibid.

[572] "Strategy: Realistic Deterrence," *The Palm Beach Post*, March 10, 1971.

controlling its escalation. Kissinger believed that even nuclear war could remain under control. In June 1972, he warned the administration that existing plans were not flexible enough and in the case of an outbreak of conflict, did not adequately provide a wide-enough scale of actions for Washington. The situation resulted in a new strategy of nuclear weapons usage, formally approved and announced by the new defense secretary James Schlesinger in January 1974.[573] It contained in total 11 different options of striking the enemy. At the same time, it included the pledge not to attack targets or nearby cities with more than 100,000 inhabitants.[574] The capitals and targets in heavily populated zones should not have been automatically assaulted. Henceforward, the main plan of a NATO strike against the Warsaw Pact forces utilized new ways to keep the conflict under control and escalate it slowly if necessary. The Pentagon thought that its new approach towards nuclear war would also be more acceptable for the U.S. allies in Europe, among others, because it estimated mostly strikes against the mighty conventional forces of the Warsaw Pact.[575]

Reaction of the Warsaw Treaty Organization, and the Soviet senior military respectively, to the aforementioned NATO strategic innovations was negative. At its 5th session held in Warsaw, February 1973, the Committee of Defense Ministers marked the new "realistic deterrence" doctrine as much more dangerous than the previous "flexible response" strategy, as it assumed a substantial strengthening threat of force during the solving of crises as well as bigger involvement of NATO members in potential combat operations.[576] Actually, at the time, the Vienna disarmament talks between both pacts were beginning and

---

[573] The key initiative in establishing the Schlesinger doctrine was taken by Henry Kissinger.

[574] These objects were defined such that a strike on them would have hit more than 10 percent of a nearby populated area. It remained clear that in the case of massive assault, numbers of civilian casualties would have been enormous. During a preemptive strike on the Soviet nuclear forces, every target - missile silos, bombers, and submarine bases - would be hit by more than one warhead. This obviously also meant annihilation of nearby towns and villages. Burr, "Is this the best they can do?" 133.

[575] Ibid, 119-130.

[576] 5. zasedání Výboru ministrů obrany, February 1973, FMO GŠ/OS, k. 41, p. č. 137, č. j. 36040, VÚA, Prague.

provided at least a hypothetical chance for the reduction of U.S. military presence on the Old Continent. Therefore, any improvement in the role of NATO European member-states was undesirable from the Eastern point of view. The Soviet strategy, at least outwardly, refused to accept the concept of controlled nuclear conflict. Formally, it still adhered to the claim of a suicidal balance of power; there would have been no winners in nuclear war, but losers only. In his extremely propagandistic essay published in Kommunist newspaper soon after the Schlesinger doctrine announcement, Grechko warningly emphasized the 24th CPSU congress conclusion that any nuclear strike on the USSR would provoke a totally devastating counterattack by similar means.[577] During his negotiations with the West, Brezhnev also intended to abide by the clear premise of mutual destruction: "*Who uses nuclear weapons first, dies second. You can turn the enemy into dust, but you still will not win the war,*" stated the CPSU CC General Secretary. This rhetorical assumption remained an integral part of his foreign policy.[578] The Soviet leadership therefore suspected that Washington used NATO doctrinal changes to force them to abandon the devastating global war theory.[579]

A twin-track approach can be seen in Soviet military thinking. First of all, there was a difference between the publicly declared concept of mutual assured destruction, including also refusing the claims of limited conflict, and the real military preparations and build-up. In 1971, evaluating the "realistic deterrence" doctrine, the Soviet General M. I. Cherednichenko stated in a military journal that the Americans obviously continued to rely on massive nuclear strikes. The West was reminded of the danger of mutual destruction.[580] In fact, a couple months earlier, the alteration of the Soviet consideration of potential conflict had been put into place: A decision had been made to adjust the obsolete plans of 1965 on managing territories of some Warsaw Pact member-states in wartime.

---

[577] Grechko, Andrei A., "On Guard over Peace and Socialism," *Kommunist*, no. 7, May 1973, quoted in *Selected Soviet Military Writtings*: 11-12.
[578] Volkogonov, *The Rise and Fall*, 313.
[579] Burr, "Is this the best they can do?" 130.
[580] Cherednichenko, M. I., "Modern War and Economics," in *Kommunist vooruženych sil*, no. 18, September 1971, quoted in *Selected Soviet Military Writtings*: 47.

Actually, these were closely linked to the overall estimation of combat features.[581] Despite its public denunciation of the limited nuclear war concept, the Soviet military command itself was preparing the plans for such a conflict; in 1973 the Pentagon also realized this, due to documents obtained through espionage.[582]

The second difference appeared in the Soviet consideration of global nuclear war as well as limited nuclear or conventional conflict in Europe. The Soviet military planning probably relied on the assumption that the results of war would be decided in a global nuclear missile clash between the superpowers. However, the Soviet generality also came to conclusion that during a total war, it could have benefited from lessons previously learned in limited nuclear conflict - some initial phase of the clash.[583] Military importance of the Warsaw Pact for the entire Soviet strategic plans therefore grew proportionate to strengthening thoughts of conventional or limited nuclear conflict in Europe. It appears not to be just incidental that the Soviet duress on absolute unification of the armies and maximal possible integration of the Unified Armed Forces command system coincided with an outlined alteration of the imaginary character of potential war.

After 1973, the Warsaw Pact armies increasingly trained for conventional or just limited nuclear offensive war. Some of the Unified Command top officials began to consider these scenarios more favorable for the march to the west. Therefore, preparation for the upcoming military five-year plan (1976-1980) emphasized strengthening offensive capacities. The CSPA as a part of the first echelon, for example, was ordered to maintain airborne troops, rearm its fighter-bomber air division, and most importantly, to reinforce the tactical missile units. Defense was represented mostly by the anti-aircraft system.[584] The rise of thoughts of conventional war was also reflected by the emphasis on introducing

---

[581] *PROTOKOL o předání a převzetí funkce náčelníka plánování obrany a ochrany teritoria OO/OS*, 12. 2. 1971, FMO GŠ/OS, k. 1, p. č. 16, č. j. 53006, VÚA, Prague.

[582] Burr, "Is this the best they can do?" 126.

[583] Ibid, 130.

[584] *Konzultace funkcionářů štábu Spojených ozbrojených sil v ČSSR*, 7. 1. 1974, FMO GŠ/OS, k. 3, p. č. 31, č. j. 39023, VÚA, Prague.

expensive combat helicopters into service, as a necessary instrument for ground forces support.[585] In fact, their effectiveness in strictly nuclear war was almost negligible. Such increasing offensive capabilities of the Warsaw Treaty Organization prompted NATO to develop "deep fire attacks" against the Unified Armed Forces' second echelon dispatched mostly in western parts of the USSR.[586]

Despite the undeniable fact that the initial limited phase of war was acquiring importance in strategic considerations during the 1970s, the claims of victory in nuclear missile war persevered, at least partially. It seems that there was a lack of compliance within the top Soviet military strategists in the middle of the decade. For example, at the 7th Committee of Defense Ministers' meeting held in Moscow on January 6-8, 1975, the Soviet general A. T. Altutin stressed the need for a proper preparation of back-territories to support troops fighting for victory in total nuclear war. Therefore, cooperation of the civil defense units[587] should have been intensified within the Warsaw Treaty Organization alongside the establishment of the system of mutual notification on the enemy nuclear strikes.[588] Soon before the Helsinki conference, and at the peak of détente, the Soviet military press still allowed room for opinions that the Americans, in reality, continuously relied on a surprise nuclear attack, which was supposed to maximize damages to the primary targets on the

---

[585] *8. zasedání Výboru ministrů obrany*, 24. 11. 1975, k. 41, p. č. 139, č. j. 36042, ibid; *Rozkaz ministra národní obrany pro přípravu vojsk ČSLA ve výcvikovém roce 1976-1977*, 15. 10. 1976, 1261/0/7, sv. 17, a. j. 19/3, NA, Prague.

[586] In the mid-1970s, NATO still planned to stop the Warsaw Pact second echelon's advance by nuclear strikes. Since the 1980s, modern weapons systems would have allowed the elimination of a formation like this by using solely conventional ammunition. Holloway, "Jaderné zbraně," 43.

[587] The civil defense was supposed to protect the population from the effects of weapons of mass destruction strikes mostly due to an instrumental reason: Civilians were expected to be engaged in elimination of the enemy's attacks in order to speed up restoration of production needed for supplying the troops.

[588] Report by M. Dúzr, defense minister of Czechoslovakia, to G. Husák, the CPCz General Secretary, on the 7th meeting of the Warsaw Treaty Organization Committee of Defense Ministers, January 20, 1975, FMO GŠ/OS, k. 41, p. č. 138, č. j. 36041, VÚA, Prague.

USSR's territory, especially to the atomic ammunition depots, missile units, and military command centers. However, the concepts of conventional and limited nuclear war were increasingly distinguished. Part of the generality considered the latter eventuality to be more dangerous. Not because of its destructiveness in comparison to conventional conflict, but because it created psychologically easier conditions for the turn into a total nuclear phase.[589] Contradictions between some text published in military journals and Altutin's speech at the Warsaw Pact's closed meeting prove not only the existence of difference between frequented public declarations and real military strategy, but perhaps also certain disagreement of the Soviet senior military in terms of willingness to alter their thinking to the new international and military situation.

Impact of Yom Kippur war on the Warsaw Pact military strategy

Non-negligible indicators of how much the Warsaw Treaty Organization began to take the possibility of conventional war seriously in the first half of the 1970s are the emphasis and precision of the alliance senior military's analysis of the course of the Yom Kippur war, and a conflict between Israel and the Arab countries' coalition in October 1973. At the time, the Warsaw Pact attached great importance to the lessons learned from local wars in general.[590] In late 1969, at the 1st Committee of Defense Ministers meeting, Grechko already warned against the opportunity that the enemy - the West - did not need to attack the Warsaw Treaty Organization directly, given the existing international situation.

---

[589] Seymenko, L. S., "New Forms, but the Same Content," in *Krasnaya Zvezda*, April 8, 1975, quoted in *Selected Soviet Military Writtings*: 58.

[590] At the 6th meeting of the Committee of Defense ministers in Budapest, for instance, the Warsaw Pact accepted the assumption based on observing the local wars, that during the initial conventional stage of conflict, NATO intended to rely mainly on airstrikes. *Návrh vystoupení ministra národní obrany Československé socialistické republiky k otázce plnění úkolu "Upevnění a rozvoj jednotného systému PVO členských států Varšavské smlouvy" v podmínkách ČSLA*, 28. 11. 1973, FMO GŠ/OS, k. 10, p. č. 46, č. j. 39038, VÚA, Prague.

Some local hotbed of tensions, especially Indochina or actually the Middle East, could have been used for this purpose.[591]

From a military point of view, the Yom Kippur war was a snapshot of what the presumable character of an initial phase of East-West conflict on the European war field might look like: it consisted of the use of all the conventional sort of troops and both sides waged the war with modern military equipment, all according to existing tactical and operational principles. Therefore, the Arab-Israeli clash allowed for identifying the strong and weak aspects of both Cold War blocks' military doctrines and plans. The conflict mostly affected the limited war theory.[592]

The military analysis of the Arab-Israeli war was worked out by the Warsaw Pact supreme commander Yakubovski himself in cooperation with the alliance staff. Then he sent out the material to individual defense ministers. According to Marshal's instructions, the commands of individual military districts should have learned the conclusions in detail. Based on the course of the fights in the Middle East, Yakubovski deduced that the Warsaw Pact troops' preparation fully corresponded to the features of contemporary warfare.[593] Military failures of the Arab coalition were ascribed solely to its bad cooperation, lack of troop training, and poor organization of the command, rather than the widely-used Soviet military equipment. On the contrary, the supreme commander marked modern weaponry as a key factor which prevented the total destruction of Egyptian and Syrian armed forces. The Warsaw Pact Unified Command linked the spectacular defeats of the Arabic tank corps to the inappropriate use by the Egyptian and Syrian generality.[594] After all, Brezhnev spoke in the same vein at the Political Consultative Committee gathering in 1974. He declared confidently that the Soviet military equipment had crushed the myth of invincibility of the

---

[591] *Zpráva ministra národní obrany pro Radu obrany státu o výsledku jednání Výboru ministrů obrany státu-účastníků Varšavské smlouvy*, December 1969, k. 9, p. č. 44, č. j. 39036, ibid.

[592] *Rozbor konfliktu na Středním východě*, undated 1974, k. 19, p. č. 58, č. j. 39051, ibid.

[593] *Vyhodnotenie bojovej činnosti na Strednom východe*, 30. 4. 1974, ibid.

[594] *Rozbor bojové činnosti na Středním východě v říjnu 1973*, undated 1974, ibid.

Israeli army.[595] One of the pursued aims was undoubtedly to calm the Pact member-states' military officials down. Massive penetration of the tank divisions through the enemy defense was an important part of the Warsaw Pact military doctrine at the time.[596] Thus, the failure of the Soviet armament in a direct confrontation with their American and British counterparts had a significant negative psychological impact on the Eastern Bloc states' armed forces. In turn, the Unified Command initially emphasized a very fast advance of the Arab coalition troops. This resulted only in further strengthening the thoughts of the veracity of a "permanent uninterrupted offensive" doctrine.[597]

The Yom Kippur war therefore instigated no changes in the Warsaw Treaty Organization doctrine. The Soviet military command considered the victory of Israel only as warning against its own deficiencies in the field of the modern communication technologies. In February 1974, at the 6th Committee of Defense Ministers meeting in Bucharest, Grechko stressed the need to focus on building-up technologically a very sophisticated alliance communication system, including combination of the cables, radio, tropospheric, and space transmit devices. He pointed to the fact that NATO spent almost 30 percent of its entire military budget on similar systems.[598] Thus, the Middle East conflict provided, among other things, an opportunity for the Soviet military circles to assert increasing spending on armaments programs.

Collective evaluation of the enemy

In the aforementioned work by Mark Kramer, an assumption appeared in the Warsaw Treaty Organization countries that there existed an organized effort by the top military circles to trick both ordinary soldiers and public. The aggressive meanings of NATO and its intention to attack the Warsaw Pact were stressed in the Eastern Bloc. The Western

---

[595] *Vystoupení s. L.I. Brežněva na poradě politického poradního výboru Varšavské smlouvy 17. dubna 1974* 1261/0/6, sv. 115, a. j. 117/1, NA, Prague.
[596] *Základní úvahy pro výstavbu ČSLA na léta 1971-1975 s výhledem do roku 1980*, undated, 1970, FMO GŠ/OS, k. 2, p. č. 29, č. j. 39021, VÚA, Prague.
[597] *Rozbor konfliktu na Středním východě*, undated 1974, k. 19, p. č. 58, č. j. 39051, ibid.
[598] Luňák, "War Plans from Stalin to Brezhnev," 90.

troops' capabilities were constantly and intentionally exaggerated and the defensive character of the NATO doctrine was twisted. Allegedly, only the highest-ranking officers of the Warsaw Pact command knew about a massive defense system which NATO built along the Czechoslovak and East-German borders. Supposedly, its existence was concealed even from the drills' participants. Information about it was one of the most top-secret materials in the East-German NPA.[599] Enough importance cannot be given to this fact: this was sensitive information about a presumable enemy which is ordinarily kept secret. The Warsaw Pact military bodies actually discussed the issue collectively. After all, in April 1970, at the 2nd Military Council meeting in Budapest, the supreme commander Yakubovski already emphasized the need to practice overcoming the nuclear mine roadblocks by the alliance troops.[600] Two years later, at the aforementioned 4th session of the Committee of Defense Ministers, Martin Dzúr also warned against the system in place to destroy the roads by both conventional and 0.5-100 kiloton nuclear charges in the West-German border areas.[601]

Kramer noted that information about the real NATO strength was intentionally hidden as they allegedly revealed the defensive nature of the Western alliance. Therefore, among the Warsaw Pact countries' broader army command, there could have potentially appeared doubts about the necessity and legitimacy of the offensive military doctrine.[602] However, it is obvious that estimated NATO strength as well as its military strategy were discussed not only during the Warsaw Pact military institutions' meeting. At least the narrower circle of individual armies' commanders was most likely also informed. The Unified Armed Forces bodies dealt with NATO combat preparation since 1973, at the latest. The drills in Western Europe were actually carefully monitored not only by the Soviet military intelligence, but also by their colleagues from the rest of the Pact

---

[599] Kramer, "Warsaw Pact Military Planning," 16-17.

[600] 2. zasedání Vojenské rady Spojených ozbrojených sil, April 1970, FMO GŠ/OS, k. 35, p. č. 128, č. j. 36031, VÚA, Prague.

[601] Vystoupení ministra národní obrany k otázce STAV NATO A NEBEZPEČÍ AGRESE V EVROPĚ, undated 1972, k. 40, p. č. 136, č. j. 36039, ibid.

[602] Kramer, "Warsaw Pact Military Planning," 18.

member-states.[603] For the following year, the alliance military bodies discussed the results of the surveillance of NATO's condition multilaterally, although relatively good cooperation between the Warsaw Pact member-states' intelligence services had existed already.[604] This was a significant aspect, because according to the official guidelines, the Unified Armed Forces drills were based on the real estimation of NATO strength and its operational plans obtained by the intelligence.[605]

Protected posts of the Unified Command

The Warsaw Treaty Organization began to deal with the opportunity to build the Unified Command protected posts at the end of 1970. The intention was to protect the top commanders of the alliance troops from the effects of weapons of mass destruction and to secure the capabilities of their further communication in case of war. Up to this point, the object used as the command post, in fact, provided almost no protection against such weapons.[606] This is another proof of the general growth of the importance of alliance military structures after 1969.

The project postulated building two protected command posts in Poland and Bulgaria. In a potential conflict, the Unified Armed Forces' combat operations in western and southwestern war theatre, respectively, should have been commanded from there. According to the original plans, both objects should have been finished by 1975.[607] However, problems

---

[603] In May 1973, at the 8th Military Council session in Sofia, the U.S. forces drill "Reforger-4" was analyzed collectively. *8. zasedání Vojenské rady Spojených ozbrojených sil*, 24. 5. 1973, FMO GŠ/OS, k. 37, p. č. 131, č. j. 36034, VÚA, Prague.

[604] *Plán realizace Protokolu čís. 010 z 10. zasedání Vojenské rady Spojených ozbrojených sil*, undated 1974, k. 8, p. č. 42, č. j. 39034, ibid.

[605] Jensen, "The Warsaw Pact's Special Target," 100.

[606] *Vystoupení ministra národní obrany k otázce "Stav spojení a prostředků velení vojskům na válčišti"*, December 1970, FMO GŠ/OS, k. 10, p. č. 45, č. j. 39037, VÚA, Prague; *Vystoupení náčelníka generálního štábu ČSLA k otázkám systému velení v Československé lidové armádě*, October 1970, k. 35, p. č. 129, č. j. 36032, ibid.

[607] *Zpráva ministra národní obrany pro Radu obrany státu o výsledcích jednání výboru ministrů obrany států-účastníků Varšavské smlouvy*, March 1971, k.10, p. č. 45, č. j. 39037, ibid.

occurred soon. In late 1970, the supreme commander Yakubovski asked the Polish defense ministry for consent to build a command post in the northwestern part of the country. It was supposed to protect about a hundred alliance officers from a nuclear blast of pressure up to 20 kg/cm^2. The construction costs, including a nearby radio station, should have been about 15-16 million transferable rubles. Poland, Czechoslovakia, and the Soviet Union were supposed to pay the amount together.[608] The Polish side did not welcome the Soviet intentions warmly. Warsaw considered them as a continuation of Moscow's efforts to shape Polish territory into the center of combat operations and to create an additional opportunity to directly command the PPA units in praxis. Since the second half of the 1950s, there had been almost continual duress on constructing the Warsaw Pact's fortified communication hubs under exclusive control of the Staff of the Unified Armed Forces in Poland. However, Gomulka's leadership was able to oppose. In the early 1970s, Polish officials tried to prevent the construction of the protected command post again. Nevertheless, they formulated their objections very cautiously.[609] The political fall of Wladyslaw Gomulka and the arrival of Edward Gierek, who was more compliant in considering cooperation within the Warsaw Pact, probably contributed to the fact that Poland approved the object on its territory in the end.

The financial aspect of the project proved to be the more serious complication. Various member-states lodged the objection that they had not reserved particular monies in their economic plans for funding the expensive construction of the objects.[610] Troubles can be illustrated with the Czechoslovak case. The supreme commander's request to share 50 million valuta crowns was delivered after the finalization of the five-year

---

[608] Letter by Marshal I. Yakubovski, the Warsaw Pact Unified Armed Forces supreme commander, to General W. Jaruzelski, defense minister of Poland, October 7, 1970, PZPR KC, XIA/106, AAN, Warsaw.

[609] Poland asked the alliance Staff for consultation and especially planned to point out the problems related to the project's funding. *Notatka w sprawie budowy umocnionego stanowiska dowodzenia Zjednoczonego Dowództwa Układu Warszawskiego*, 7. 12. 1970, ibid.

[610] *Výstavba chráněných velitelských stanovišť SOS*, 4. 7. 1973, FMO GŠ/OS, k. 11, p. č. 48, č. j. 39040, VÚA, Prague.

plan in 1975. Husák's comments suggest that Czechoslovakia had serious problems with releasing the requested money. However, the CPCz CC First Secretary, fully loyal to Moscow, did not dare to protest against the Unified Command's intentions.[611] In the early days of 1972, the supreme commander asked the Warsaw Pact member-states' institutions for the means of construction of the protected command posts, which in many cases went beyond the approved military budgets.[612] The only collective coordinative meeting took place in November at the alliance Staff in Moscow.[613] The entire project turned out to be extremely costly. In September, Yakubovski informed the defense ministers that in order to create a compact system of command posts, constructing another object nearby the Soviet capital was necessary. The overall costs were supposed to reach 50 million transferable rubles. Reimbursement of this amount should have corresponded to percentage levies of individual countries to the Unified Armed Forces' budget.[614]

In the end, releasing the monies from the already approved five-year plans of the Warsaw Pact member-states showed itself to be too problematic. Therefore, in mid 1973, the supreme commander announced

---

[611] The amount should have been allocated by the defense ministry from its budget. This move, however, threatened the ongoing rearmament programs. Therefore, the situation should have been discussed by Dzúr and Marshal Yakubovski during the upcoming Committee of Defense Ministers gathering. The Czechoslovak side intended to ask the supreme commander to postpone its payment. *Výstavba předsunutého velitelského stanoviště na území PLR*, 18. 2. 1971; ibid, *Výstavba předsunutého velitelského stanoviště na území PLR*, February 1971, p. č. 47, č. j. 39039, ibid.

[612] *Výstavba chráněných míst velení na západním a jihozápadním směru*, 3. 1. 1972, ibid.

[613] Problematic aspects of the project should have been discussed solely with the Unified Command. A special commission including the Staff of the Unified Armed Forces representatives and the general staffs of individual member-states was established. *Výstavba předsunutého velitelského stanoviště a vysílacího radiového střediska pro Spojené velení na území Polské lidové republiky*, June 1971, ibid; *Porada k výstavbě VS pro Spojené velení*, 7. 11. 1972, p. č. 48, č. j. 39040, ibid.

[614] *Výstavba velitelských stanovišť pro Spojené velení, informační zpráva pro s. generálního tajemníka ÚV KSČ*, 1. 10. 1972, ibid; *Stanovisko odboru VIII-FMF ke zvláštnímu úkolu čj. 18034/ZD-OS-1970: Záznam ze dne 4. 12. 1972*, p. č. 47, č. j. 39039, ibid; Translation of a letter by M. Dzúr, Defense Minister of Czechoslovakia, to I. Yakubovski, the Unified Armed Forces' Supreme Commander, July 2, 1972, ibid.

postponing the beginning of construction to 1976, to the following five-year plan.[615] This move was probably also a reaction to a shift in détente. The immediate risk of an outburst of nuclear war had decreased. The project therefore lost its urgency. However, the entire situation proved a weakness of the strictly centrally directed economies of the Eastern Bloc countries, which were not able to react flexibly to actual needs.

The approval of the statute of the protected command posts' construction proceeded strictly on the line of the Unified Command and the defense ministries.[616] During the internal discussion, the Czechoslovak military representatives noted the fact that intended funding of operational costs by all the Pact members contradicted the Unified Armed Forces Statute. The document of 1969 had actually strictly stated that every alliance command's activities be paid from a joint budget. However, the Czechoslovak side did not object to approval of the agreement; it only initiated the multilateral consultation with the chief of Staff of the Unified Armed Forces, where this contradiction would have been formally solved.[617] This only demonstrated the then effort to make the Warsaw Pact work on the background of existing law and not to break it in any way. The agreement on operating, servicing, and protection of the Unified Armed Forces' command posts initially assumed that a definite funding

---

[615] Considering Czechoslovakia, the move caused serious troubles to the ongoing five-year economic plan, despite the fact that the country was supposed to pay only a negligible amount of 6.5 million rubles. After negotiation with Moscow, the first tranche was postponed to 1973. *Výstavba chráněných velitelských stanovišť SOS*, 4. 7. 1973, ibid.

[616] In the Czechoslovak case, the Council of State Defense gave its blessing to the document on December 17, 1973. Unlike many other negotiations within the Pact military structures, Gustáv Husák was now receiving relatively detailed reports about progress on the issue. The likely reason was that in comparison to other military questions, which were discussed exclusively by the Unified Command and the defense ministries, construction of the protected command posts also affected the financial planning of the country. *Výstavba chráněných velitelských stanovišť SOS*, 3. 4. 1973; *Výstavba chráněných velitelských stanovišť Spojených ozbrojených sil*, 29. 9. 1973, ibid.

[617] *Úvodní slovo soudruha ministra národní obrany k bodu 3/ Dohoda o provozu, údržbě a ochraně velitelských stanovišť Spojených ozbrojených sil*, undated, 1972, p. č. 48, č. j. 39040, ibid.

mechanism would be discussed at the meeting of the chiefs of the general staffs with General Schtemenko. However, soon after approving the document in the early months of 1974, the chief of Staff of the Unified Armed Forces declared that the issue was considered finished. Prospective ambiguities should have been solved during the bilateral talks.[618] The Unified Command typically did not want to open the sensitive and potentially contentious questions at the multilateral meetings. It preferred negotiations with individual member-states, which probably better allowed it to isolate any display of discontent.

The post in Bulgaria was successfully finished in 1975 with the estimated costs of 22.9 million rubles. However, after receiving documentation from the Unified Command, the Polish side calculated the construction costs on its territory at 89.7 million, although the price of only 27.1 million had been estimated initially. The question should have been assessed by the Staff of the Unified Armed Forces once more. It concluded that the post in Poland could possibly be built for 50 million rubles. The gap was supposed to be paid by the member-states.[619] Therefore, construction of the protected command posts during the years 1978 and 1982 caused the Unified Armed Forces' budget to constantly increase as

---

[618] The army of the country, where the command post was located, was responsible for servicing the objects. However, it used it sporadically. The aim was to "save" the command posts for war. As the objects belonged to the Unified Armed Forces, the units tasked with its maintenance were transferred under direct competences of the Unified Command. *Výstavba chráněných velitelských stanovišť Spojených ozbrojených sil*, 10. 9. 1973, ibid; *Informace ministra národní obrany ČSSR o návrhu "Dohody o provozu, údržbě a ochraně velitelských stanovišť Spojených ozbrojených sil"*, undated 1973, ibid; *Výstavba chráněných velitelských stanovišť Spojených ozbrojených sil*, 23. 1. 1974, ibid.

[619] *Výstavba chráněných velitelských stanovišť Spojených ozbrojených sil*, 24. 8. 1977, k. 19, p. č. 59, č. j. 39052, ibid. In the Czechoslovak case, this meant to increase the amount to almost 3 million transferrable rubles, which was equal to more than 44 million valuta crowns. However, it was also unacceptable to really discuss rising the originally estimated amount. Gustáv Husák refused to deal with the question at the regular Council of State Defense meeting, where inconsistent opinions would have been presented. Therefore, he rather insisted the issue to be discussed and approved in advance. *28. schůze Rady obrany státu – podklady*, 10. 9. 1977, ibid.

the project was actually funded by these resources.[620] It was happening at a time when most of the Warsaw Pact countries began to experience strong economic troubles as a result of gradually rising demands on armaments. As the alliance budget amounted to 5 million rubles in 1975, seven years later it had doubled and also included the same amount in non-trade payables.[621] This amount was not lowered until 1983, when the costs of the protected command posts were finally paid.[622] The high cost of the project was increased by the fact that Romania refused to fund both the construction and operating of the command post for the western war theatre. As already mentioned, the Six did not confront Bucharest on this issue.

---

[620] The post in Poland was finished in late 1979. However, the costs of maintenance of the object for the southwestern war theater in Bulgaria were raised afterwards, as it was necessary to cover increasing the Bulgarian People's Army personnel's salaries.

[621] *16. zasedání Vojenské rady Spojených ozbrojených sil*, 27. 9. 1977, FMO GŠ/OS, k. 40, p. č. 135, č. j. 36038, VÚA, Prague; Information by M. Dzúr, Defense Minister of Czechoslovakia, to G. Husák, the CPCs CC General Secretary, September 26, 1978, k. 34, p. č. 125, č. j. 36025, ibid; *Převedení zůstatku finančních prostředků z roku 1978 na rok 1979*, 10. 4. 1979, k. 44, p. č. 156, č. j. 36508, ibid; *Zpráva k návrhu rozpočtu Spojených ozbrojených sil na rok 1980 a vyhodnocení rozpočtu Spojených ozbrojených sil na rok 1978*, 8. 10. 1979, k. 45, p. č. 153, č. j. 36511, ibid; *ZPRÁVA k návrhu rozpočtu spojených ozbrojených sil na rok 1981 a vyhodnocení rozpočtu Spojených ozbrojených sil za rok 1979*, 27. 11. 1980, kr. 48, p. č. 178, č. j. 37023, ibid; Information by M. Dzúr, Defense Minister of Czechoslovakia, to G. Husák, the CPCs CC General Secretary, September 14, 1981, k. 50, p. č. 186, č. j. 37512, ibid; *Návrh rozpočtu Spojených ozbrojených sil na rok 1983 a vyhodnocení rozpočtu Spojených ozbrojených sil za rok 1981*, 7. 10. 1982, kr. 53, p. č. 204, e. č. 38013, ibid.

[622] The budget increased by 200 thousand rubles in non-trade payables, but decreased by 3.5 million transferrable rubles.

The armament programs and disarmament initiatives at the peak of détente

The relaxation of international tensions at the turn of the 1960s and 1970s was reflected by the Warsaw Treaty Organization's military structures significantly slower and to a lesser extent than in the case of its political framework. The military elites took a reserved position towards détente. In fact, skepticism about the possibility of a permanent peaceful coexistence or even cooperation between the East and West was typical of the Warsaw Pact's top military officials in the entire 1970s. Their thinking underwent partial changes only - especially the Soviet generality, supported by a radically-oriented part of the CPSU leadership[623] and military-industrial complex representatives,[624] who asserted new rounds of massive armaments even at the peak of détente.

Considering the real mechanisms of the Warsaw Treaty Organization, it is understood that the level of armaments was decided in Moscow. The alliance itself only provided the formal auspices for particular programs. Thus, the following chapter has to inevitably deal

---

[623] Especially, the CPSU CC International Department, led by Boris Ponomarev, took a reserved and even hostile attitude towards the idea of cooperation with the West. He also showed strong personal antipathy to Brezhnev. The International Department top officials, including Ponomarev and the CPSU CC secretary, Mikhail Suslov, soon began to present their mistrust as well as open disagreement with the CSCE process. In terms of an ideological world view, even the sphere of international relations should have remained under the principle of class struggle. From this point of view, easing relations with the Western "imperialists" was considered a weakening of Soviet positions. Rey, "The USSR and the Helsinki process," 72-74

[624] The interests of this group, relatively problematic to be defined, were advocated in political terms by the commission on arms control established in 1968 within the Soviet politburo. Its most important members were Andrei Grechko, the defense minister, and his future successor, the politburo candidate and the CPSU CC secretary overseeing the armed forces and weapons industry, Dmitri Ustinov. Zubok, *A Failed Empire,* 204.

with the processes within the Soviet leadership at the time. As we do not have a comprehensive collection of the archival documents, it is based mostly on knowledge of the secondary literature. Therefore, some outlined conclusions cannot be taken in any ultimate way. It is likely they will be particularly revisited in future.

Intensified armament at the turn of the 1960s and '70s.

In the long term, the Warsaw Pact's conventional forces were stronger than the NATO units dispatched in Europe, at least considering their quantity. Moreover, at the end of the 1960s, the Soviet Union caught up with the existing U.S. advantage in nuclear weapons. One of the significant aspects of Leonid Brezhnev's appointment as the CPSU leader in 1964 was also the strengthening role of the army and the military-industrial complex representatives in a decision-making process within the Soviet leadership. This resulted in substantial modernization and extension of the armed forces as the Kremlin began to adjust its policy more to the Soviet generality's strategic demands. Unlike in Khrushchev's era, both the army and weapons industry could have successfully asserted their claims in the second half of the 1960s. Thus, at the beginning of the following decade, the Nixon administration had to admit that through a quick and complex buildup of troops, the Soviet Union had achieved a global strategic parity. An antimilitaristic public atmosphere that engulfed the U.S. as a result of the dragging war in Vietnam, however, provided no chance for regaining former American superiority in the short-term horizon.[625] It would be misleading to claim that the CPSU leadership decided to improve Soviet military capacities solely because of the army circles' duress. Undoubtedly, a non-negligible part of the top politicians, including Brezhnev himself, actively supported achieving the strategic parity.[626]

---

[625] Volkogonov, *The Rise and Fall*, 266; Durman, *Útěk od praporů*, 19, 123.
[626] After all, Brezhnev was in charge of the supervisor for the military-industrial complex already during Khrushchev's rule. In this position, he established close cooperation with one of the top advocates of the arms race, Dmitri Ustinov. As a head of the military-industrial commission, the future minister of defense already

The strategic parity also meant that possession of nuclear weapons was increasingly difficult to relate to the superpowers' political purposes. During the years 1969-1972, the U.S. as well as Soviet activities were primarily motivated by the effort to maintain the existing balance of forces in preserving an atmosphere of mistrust. In this stage, détente was mostly a side-product of general stabilization of the global situation rather than a genuine attempt to ease international tensions.[627] According to Brezhnev's idea, significant military strengthening of the Warsaw Pact should have preceded relaxation of tensions between the blocs. The Soviet leader also operated with additional pressure against the West through support of national liberation fights in Third World countries. In the beginning, Brezhnev's strategy of parallel armament and international dialogue seemed to be more successful than Khrushchev's pompous, but not serious, demilitarization campaigns. Instead of unrealistically proposing mutual dissolution of two military blocs, Moscow now sought to legitimize political and military *status quo* in Europe.[628] Both superpowers of course still considered détente as a mutual battle, a continuing of the Cold War by less dangerous means. Therefore, each side sought to take any possible advantage over its rival.[629]

The Soviet generality's approach in this stage of détente was described well by the speech of Marshal Grechko at the 1st Committee of Defense Ministers of the Warsaw Pact in December 1969. The head of the Soviet defense resort had to admit that there was no imminent risk of assault on the alliance member-states at the time. Nevertheless, he warned against continual preparation of war by the "aggressive circles".[630] The marshal's report, to a large extent, predetermined following the development of armament projects in the Eastern Bloc. Despite persistent tensions on the Soviet-Chinese borders, which had resulted in armed

---

opposed any reduction of armed forces in the era following the Cuban missile crisis. Volkogonov, *The Rise and Fall,* 277; Zubok, *A Failed Empire,* 151.

[627] Mastny, "Superpower Détente", 19.

[628] Nünlist, "*Cold War Generals*".

[629] Zubok, *A Failed Empire,* 230.

[630] *Zpráva ministra národní obrany pro Radu obrany státu o výsledku jednání Výboru ministrů obrany států-účastníků Varšavské smlouvy*, December 1969, FMO GŠ/OS, k. 9, p. č. 44, č. j. 39036, VÚA, Prague.

clashes on the Ussuri river nine months earlier, the Soviet defense minister assured his Warsaw Pact counterparts that NATO remained the most likely enemy. After all, even the preamble to the founding charter called the Western alliance the biggest threat to the "peaceful countries."[631] By the example of the U.S. strategy in Indochina, Grechko traditionally condemned the Western "imperialism" and pointed to the strengthening of NATO troops. In the following years, the references to improving military capacities of the West actually provided a justification for massive development of the Unified Armed Forces, as the Warsaw Pact's military buildup, especially after 1972, did not correspond with the Eastern propaganda which increasingly called for limitation of the arms race and the relaxation of international tensions.

At the end of the 1960s, the Soviet generality got the Committee of Defense Ministers' formal blessing to continue the expensive armaments programs in the Eastern Bloc countries and the expected enlargement of drills - everything under the auspice of the universal (and in the following years, frequently-used) terminology of "development and strengthening of the Warsaw Pact's defense capabilities." For the Soviet satellites, Grechko's laconic statement that the USSR was no longer able to sufficiently supply all the Pact member-sates' armies with military equipment was very important. Therefore, individual defense ministers were tasked with appeals to the proper political institutions in their countries for consent to increase weapons industry capacities, especially in the technologically unsophisticated sectors. The key for the Warsaw Pact's further development should not only have been increasing the troop numbers, which was in praxis no longer bearable, but a wider modernization of arms and reduction of mobilization terms.[632] The Soviet

---

[631] Gizicki, Wojciech, *Od Układu do Paktu: (r)ewolucyjna zmiana w polityce bezpieczeństwa Polski* (Lublin: Institut Sądecko-Lubelski, 2011), 12.

[632] VÚA, f. FMO GŠ/OS, k. 9, p. č. 44, č. j. 39036, *Zpráva ministra národní obrany pro Radu obrany státu o výsledku jednání Výboru ministrů obrany států-účastníků Varšavské smlouvy*, December 1969. In 1970, at the 3rd Military Council session, the slow reduction of time norms needed to reach combat and mobilization readiness along with the lack of the drills focused on these aspects were criticized by Schtemenko, the chief of Staff of the Unified Armed Forces. Later, the emphasis on extremely short mobilization terms were especially given

Union, however, intended to keep its exclusive position in the development of key military devices - especially aircrafts, rockets, navigations systems, and of course, nuclear weapons.[633] In this regard, Czechoslovakia was an exception. At the beginning of the 1970s, production in the country was reorienting from heavy weaponry to lighter, but more sophisticated weapon systems.[634]

The Warsaw Pact collective military bodies began to work at the time, when direct threat of armed conflict between the West and East was receding fastest since the outbreak of the Cold War. Nevertheless, at the initial alliance meetings, Grechko presented the opinion that a risk of new war was not imminent only because of existing Soviet military superiority. Such a claim contained the clear message that maintaining it was crucial for preserving peace in the future. This gave him the right to apply for other massive investments in the armed forces. In order to support his thesis, the marshal, for example, persuaded the Warsaw Pact allies that Washington had agreed with SALT negotiations solely due to the actual Soviet nuclear advantage.[635] Until his death in 1976, Grechko emphasized on every occasion a need to "*keep, even in the future, high vigilance and to strengthen the economic as well as defensive power*" of the Soviet Union and the entire Warsaw Treaty Organization, by extension.[636]

---

in the countries directly neighboring to NATO. Their geographical location potentially allowed the Western alliance to launch an attack on the Warsaw Pact troops dispatched there without a long stage of preparations. *3. zasedání Vojenské rady Spojených ozbrojených sil*, 1.11. 1970, k. 35, p. č. 129, č. j. 36032, ibid; *Vystoupení ministra národní obrany ČSSR k otázce BOJOVÁ PŘÍPRAVA A BOJOVÁ POHOTOVOST ARMÁD VARŠAVSKÉ SMLOUVY*, undated 1972, k. 40, p. č. 136, č. j. 36039, ibid.

[633] Wanner, *Brežněv*, 42.

[634] In the CSSR, this process was accompanied with the transfer of production from the Czech lands to Slovakia. Despite the fact that dependency on the primary Soviet military production was generally increasing, Czechoslovakia remained unique in the world in terms of its weapons industry. The country with 15 million inhabitants developed, produced and serviced the impressive spectrum of ground as well as air armaments in huge series. Bílek and Koller, "Československá armada," 207.

[635] Mastny, *A Cardboard Castle?*, 41; Mastny, "Imagining War in Europe," 31.

[636] Wanner, *Brežněv*, 20.

Stances of Grechko, who had no official authority towards the Warsaw Treaty Organization at the time,[637] were formally presented within the alliance structures by the supreme commander, Marshal Yakubovski. He identified himself with the Soviet minister's claim that despite some improvement, the international political-military situation remained complicated and strained, and therefore strengthening the Warsaw Pact's defense capabilities was necessary in the future. This was supposed to continue in the ongoing armament projects. In March 1970, at the 2nd Military Council meeting in Budapest, the supreme commander presented concrete demands on significant acceleration of rearmament of the units assigned to the Unified Armed Forces.[638] The plans were designed up to 1975, in order to coincide with the *Protocols on build-up of the armies* of individual alliance member-states.[639] However, the marshal's appeals appeared at a time when many of the Warsaw Pact countries was struggling with decline of their economic growth. To the contrary, such a situation forced them to slow down the tempo of the troops' modernization.[640] As explained before, the armament programs were not funded by the Pact's budget. Individual countries bore the financial burden.[641] It is good to add that within the Warsaw Treaty Organization, defense spending remained strictly a matter of the national states. Their military budgets were never subjects of official information, comparisons, exchange of data, or even joint alliance overviews. Only a narrow group of people within the Unified Command had stats about

---

[637] During the years 1960-1967, Grechko held the post of the Warsaw Pact supreme commander.

[638] *2. zasedání Vojenské rady Spojených ozbrojených sil*, April 1970. FMO GŠ/OS, k. 35, p. č.128, č. j. 36031, VÚA, Prague; *3. zasedání Vojenské rady Spojených ozbrojených sil*, 1. 11. 1970, p. č. 129, č. j. 36032, ibid.

[639] *Schválení Plánu realizace protokolů z jednání Výboru ministrů obrany*, 15. 11. 1971, k. 8, p. č. 43, č. j. 39035, ibid.

[640] For example, the Czechoslovak military officials solved the problem through accelerating rearmament of the key units of the first echelon only. *Základní úvahy pro výstavbu ČSLA na léta 1971-1975 s výhledem do roku 1980*, undated, 1970, k. 2, p. č. 29, č. j. 39021, ibid.

[641] *3. zasedání Vojenské rady Spojených ozbrojených sil*, 1. 11. 1970, k. 35, p. č. 129, č. j. 36032, ibid.

individual armies' buildup. Until Mikhail Gorbachev came to power in the USSR, the Pact members had not shared information about their military spending at all.[642]

Moscow never ordered its satellites to increase military spending directly. However, it demanded realization of measures which inevitably resulted in such a step. In this regard, it needs to be noted that the warnings against the permanent strengthening of the enemy voiced at the alliance military bodies were in fact, at least at the beginning of the 1970s, only a false pretext for continual armaments. On the contrary, the Warsaw Pact's own military analytics stated that NATO was lying on its historic bottom. The intelligence stressed its weak flanks in comparison to the central formation of troops, insufficient strategic depth, vulnerability of supply lines, inadequate command and control systems, pressure on reduction of military budgets in the member-states, or chronic disagreement between the U.S. and their European allies. In the light of this, some room for reduction, or at least a slowdown of the ambitious armament programs opened for the Warsaw Treaty Organization,[643] especially taking into account an adverse economic situation of some member-states. However, Yakubovski's and Grechko's behavior at the alliance sessions strove for the exact opposite.

Propagandistic peace and disarmament initiatives

In this stage of the Cold War, Moscow tried to solve the aforementioned contradiction between peaceful or disarmament propaganda preached by the Eastern Bloc and the increasing armaments of the Warsaw Pact through a presentation of unilaterally favorable international initiatives. It was mostly obvious in advance that they were unlikely to succeed. Especially until 1973, these initiative were simply a propagandistic show for the world as well as the home public. Despite the ghost of a chance to conclude such agreements, Moscow left no avenue unexplored. In any case, it issued only such proposals whose implementation

---

[642] Fučík, Josef, "Rozpočtové výdaje na obranu v zemích Varšavské smlouvy v letech 1955–1989," in: *Vojenské výdaje v letech studené války a po jejím skončení*, ed. Miroslav Krč (Praha: Ústav mezinárodních vztahů, 2000): 70.
[643] Mastny, *A Cardboard Castle?*, 41; Mastny, "Imagining War in Europe," 31.

would have strengthened the Eastern positions in praxis. For this purpose, the Warsaw Treaty Organization often provided the suitable auspices for the Kremlin. In June 1970, for instance, the USSR had its appeal for reduction of foreign troops dislocated on the European countries' territories formally approved on behalf of the entire alliance at the meeting of ministers of foreign affairs. This proposal should have been an alternative to previous NATO appeals for reduction of total armed forces' numbers of both alliances.[644] Considering the Warsaw Pact's conventional superiority, this was unfavorable to the Eastern Bloc and therefore of course unacceptable. On the contrary, the withdrawal of some foreign troops from European countries consisted also in the retreat of the U.S. units beyond the Atlantic. In fact, such a move would have resulted in another substantial strengthening of the Warsaw Pact's military position.[645] Removal of foreign troops from the Old Continent would have naturally weakened NATO much more. In case of escalation of tensions or outbreak of war, transit of divisions from the western borders of the USSR to the territories of its allies would have been incomparably easier than the transfer of the American units across the ocean. It is necessary to emphasize that a similar strategy - to weaken the presumed enemy - was also pursued by Western initiatives.

Moscow also often issued disarmament proposals which seemed to be neutral at first glance. However, their concrete conditions made them worthless. By way of illustration, in the case of the appeals for underground nuclear explosion tests ban, the Kremlin refused any supranational inspections of their compliance with the agreement. As the measure lacked any effective control mechanism, the USA correctly

---

[644] The Warsaw Treaty Organization acted on previous direct instructions by Moscow. *Zpráva o průběhu a výsledcích budapešťské porady ministrů zahraničních věcí členských států Varšavské smlouvy*, 14. 7. 1970, 1261/0/5, sv. 134, a. j. 211/info4, NA, Prague; *Zpráva o nejdůležitějších opatřeních sovětské zahraniční politiky v období ledna – června 1970*, 27. 5. 1970, TO(t) 1970-1974, box 8, i. č. 89, sign. 020/311, č. j. 0122/70, AMZV, Prague.

[645] This kind of objection soon resonated in various NATO member-states, but the Soviet proposal was surprisingly appreciated by Great Britain. *Ohlas v západních zemích na poradu ministrů zahraničních věcí členských států Varšavské smlouvy v Budapešti*, undated, 1261/0/5, sv. 134, a. j. 211/info4, NA, Prague.

rejected the treaty as absolutely otiose.[646]

Contradictions between the publicly declared political line and the conclusions presented at the closed meetings of the Warsaw Pact's military structures began to appear more frequently after 1971. At the turn of March and April, the 24th CPSU congress took place. In reaction to the alleged NATO weakness, as well as the real shift in political talks between both blocs, the Soviet political leadership presented a new set of peace initiatives. The Kremlin, also aware of the Warsaw Pact's military capabilities, acted very confidently in international affairs. Gromyko enthusiastically announced that there was no question in the world, which could have been decided without the Soviet Union or in opposition to it.[647] The congress' conclusions contained the obligatory warning that despite continuing process of détente, "*the aggressive NATO bloc - led by the U.S. - which is every year getting more armaments, getting stronger, and raising the risk of a new world war*" remained the biggest threat to global peace.[648] Nevertheless, in terms of the so-called "peaceful program," Brezhnev for the first time generously admitted a possibility to reduce the armed forces and weaponry numbers in Europe. The move was supposed to concern not only selective armies of individual countries, but also the

---

[646] *Návrh rámcové směrnice pro postup čs. delegace ve Výboru pro odzbrojení v roce 1972*, 8. 2. 1972, PK 1953-1989, kat. č. 632, k. č. 145, č. j. 016149/72, AMZV, Prague.

[647] Mastny, *A Cardboard Castle?*, 41.

[648] This was a certain propagandistic contradiction. According to the actual needs, the U.S. or FRG were alternately called the most aggressive NATO member. In 1972, at the Committee of Defense Ministers gathering, the head of the East-German defense resort, Heinz Hoffmann, stressed an aggressive character not only of NATO, but mostly of the West-German state. *Vystoupení ministra národní obrany ČSSR k otázce BOJOVÁ PŘÍPRAVA A BOJOVÁ POHOTOVOST ARMÁD VARŠAVSKÉ SMLOUVY*, undated 1972, FMO GŠ/OS, k. 40, p. č. 136, č. j. 36039, VÚA, Prague; Nünlist, "*Cold War Generals*". It is good to add that rise of West-German military spending was really massive at the time. During the 1970s, it increased from 6.2 to 21.4 million USD. Within NATO European countries, the FRG achieved the top in this indicator. Krč, Miroslav, "Vojenské výdaje v letech studené války a jejich determinanty s důrazem na NATO a Varšavskou smlouvu," in: *Vojenské výdaje v letech studené války a po jejím skončení*, ed. Miroslav Krč (Praha: Ústav mezinárodních vztahů, 2000): 43.

alliance formations.[649] The Soviet leader talked about the alteration of the international situation that was allowing room for reduction of nuclear as well as conventional arsenals, in his opinion. On the contrary, only one month earlier, Grechko warned the Warsaw Pact's military officials that international tensions were increasing.[650] The entire situation most likely reflected deepening animosity between the supporters of détente within the CPSU, including the General Secretary, and the Soviet generality backed by the ideologically dogmatic core of the party elites. This was the result of the obvious alteration of the USSR's foreign policy strategy.[651]

At the time, the Soviet top military sought to temper optimism of some Eastern political elites. According to their opinion, the risk of assault on the USSR and other socialist countries had not been fully eliminated yet. The generality used an ideological construct about the constant "aggressive matter of imperialism."[652] Let us add that such thinking was not typical only of the military. For example, just ahead of his political fall, Water Ulbricht also presented similar claims during a meeting with the Warsaw Pact Generals after the 4th Military Council session in Berlin.[653] The Soviet military command, respectively its orthodox-thinking part, was allowed to voice their opinions publicly, despite them often differing from the official political line.[654] For instance, in his essay published in September 1971 in the party military press, General Cherednichenko urged a necessity to design the economy fully in line with

---

[649] Wanner, *Brežněv*, 42-43.

[650] Mastny, "Imagining War in Europe," 32.

[651] Zubok, *A Failed Empire,* 214-215. Vladislav Zubok notes that at the 24th CPSU congress, Brezhnev finally established his leadership in foreign affairs. His "Peace Program" and opening approach towards the FRG received unanimous support. This allowed Brezhnev to silence his political critics. In addition, Gromyko, Brezhnev's ally, harshly spoke against anonymous figures inside the party who interpreted any agreement with the Western capitalists as some kind of conspiracy.

[652] Cherednichenko, "Modern War and Economics," 47.

[653] *4. zasedání Vojenské rady Spojených ozbrojených sil,* 17. 5. 1971, FMO GŠ/OS, k.36, p. č.130, č. j.36033, VÚA, Prague.

[654] It is good to add that tolerance of public presentation of Stalinist opinions, which largely did not correspond with the officially declared as well as implemented policy of the Soviet leadership, was typical of Brezhnev's era.

the army's strategic intentions. The military and civil economic sectors should have been closely linked.[655] The criticism also aimed at the allegedly insufficient military budget, despite it reaching an impressive 17.9 billion rubles in 1971, according to the official stats.[656]

However, unlike the radicals, the Soviet political leadership was aware of a need to not focus on the weapons industry only. The economy should have been developed in complexity. Moreover, the USSR desperately needed a source of hard currency as well as Western technologies. The leading forces within the CPSU actually saw a proper solution in deepening détente.[657] Also, Brezhnev himself placed high hopes on the development of economic cooperation with the U.S. which

---

[655] The economy should have been oriented to maintain the armed forces on the highest level. Cherednichenko stressed development of such an economy, which would be able to produce mostly weapons and military equipment, allowing it to keep a permanent high-mobilization readiness and provided the army with enough supplies during war. According to the General, allegedly more than 42 percent of the Soviet weapons industry did not produce strictly military material at the time. Scott, *Selected Soviet Military Writings*, 6.

[656] Estimates of the real military spending in the Warsaw Treaty Organization countries are very problematic. The Soviet political officials as well as the alliance command considered all the stats of defense character in the states' budgets as a top secret strategic indicator. Therefore, it was not desirable to subject military spending to any economic analyses or comparisons with other macroeconomic data, as it could have raised critical stances within the state apparatus. Thus, all the information about the real situation - especially in the USSR - were available to a narrow circle only and kept carefully secret. During the years 1975-1984, officially admitted military spending by the Soviet Union oscillated around 17 and 17.4 billion rubles. However, at the end of the 1980s, the authors of the significant economic analyses estimated that the real monies were multiples higher. The retrospective estimation for 1975 was somewhere between 43 and 73.8 billion. Fučík, "Rozpočtové výdaje na obranu," 64, 82.

[657] For example, in May 1972, at a meeting of the Soviet Defense Council, Brezhnev attacked - in an unusually hard way – Grechko, who had tried to slowdown SALT negotiations through his military co-workers. The General Secretary asked the defense minister whether he could clearly guarantee that in the case of a continual arm race, the USSR would get clear superiority over the United States. The marshal's negative answer provided a suitable pretext for Brezhnev to denounce the existing approach, when the military spending literally wrung the Soviet economy. Zubok, *A Failed Empire,* 220-223.

he expected to come after improvement of mutual relations.[658] However, after 1972, the initial idea that the arms race control under the conditions set by the Soviet Union would result in quick economic benefits, was proven to be wishful thinking. Even one year after the conclusion of SALT, there was still no breakthrough on trade and economic cooperation between the USA and USSR, which Brezhnev was striving for through détente.[659] It is symptomatic that at that moment, the voice of the Soviet generality imminently raised itself again. They once more prompted the utmost cautious approach and warned against the undermining of defense capabilities of the USSR and the entire Warsaw Treaty Organization.[660]

Relations between Brezhnev and the Soviet top military

In the light of the aforementioned, it is necessary to at least generally explain the quite specific relationship between Brezhnev and the Soviet generality. The CPSU CC General Secretary himself undoubtedly supported negotiations with the West from the position of strength.[661] He paid extreme attention to defense capabilities. But he totally dismissed the claim by the chief of general staff, Marshal Viktor Kulikov,[662] which was repeatedly stated by other top military officials, that the general staff is an operative brain of the USSR.[663] Brezhnev adhered to the highest principle

---

[658] Stenogram narady aktywu partyjnego poświęconej spotkaniu Krymskiemu przywódców krajów socjalistycznych, 3. 7. 1973 PZPR KC, XIA/404, AAN, Warsaw.

[659] Zubok, A Failed Empire, 230.

[660] Wanner, Brežněv, 51.

[661] Zubok, A Failed Empire, 243. In late 1969, during his talk with the Polish delegation on the top level, Brezhnev, in accordance with this principle, complained about the enormous military costs; but, he also called them necessary. They not only should have guaranteed safety, but also provided an essential strength which the West had to take into account during negotiations. Protokół z wizyty polskiej delegacji partyjno-rządowej w Moskwie, w dniach 1-3 października 1969, PZPR KC, XIA/87, AAN, Warsaw.

[662] During the years 1971-1977, Kulikov served as the chief of General Staff of the Soviet army. After the death of the Warsaw Pact supreme commander Yakubovski, he was moved to this post which he held till 1989.

[663] Durman, Útěk od praporů, 42.

that the political strategy must be decided by the party, not by the army. He outlined contradictions between his conception of détente, leading to the significant alteration of Soviet foreign policy, and therefore the radical, largely Stalinist approach of the generality, led by Defense Minister Grechko, inevitably had to come to light, sooner or later. In this regard, it is necessary to emphasize that the discord did not proceed between two equal partners. The army never attained such a power in the Soviet Union that allowed it to set foreign policy parameters.

In December 1969, at the CC plenum, the Soviet political leadership gave to the top military a sop: They repeatedly supported increasing monies for the armament programs. In return, they expected that the generality would not object to SALT negotiations. However, this assumption proved to be wrong. The military soon began to voice their disagreement with the actual political line.[664] Defense Minister Grechko, Brezhnev's former superior at the time of World War II, raised the biggest concerns between the Soviet supporters of the relaxation of international tensions. The marshal, who felt nothing but contempt for the West,[665] did not hide his opinion that the general staff saw no significant benefit in SALT.[666] On the contrary, at the beginning of the 1970s, the Soviet military circles, similar to their American counterparts, demanded a completely free hand in continuing the arms race. For long years, Brezhnev's friends in the Soviet military command and military-industrial complexes considered the U.S. as the archenemy. A new perspective of armaments control and negotiating a compromise with Washington did not correspond with their orthodox anti-American thinking. In addition, the still valid military doctrine focused on a victory in nuclear war, whose basis went back to Khrushchev's era, which made the situation even worse. Thus, Grechko as well as his successor Ustinov were worthy counterparts of American hawks, who took a similarly negative attitude towards détente.[667] Considering this, it is good to add, that significant contradictions between the Western declarations and real measures aiming

---

[664] Mastny, "Imagining War in Europe," 31.
[665] Zubok, *A Failed Empire,* 205.
[666] Durman, *Útěk od praporů,* 126.
[667] Zubok, *A Failed Empire,* 205, 215-216.

at strengthening NATO, were also perceived by the general staffs of the Warsaw Pact countries.[668] The entire situation contributed to a deepening of mutual distrust.

Since the beginning of the 1970s, the Soviet top military watched with a big mistrust the group of Soviet diplomats, who rejected the claims of possible victory in nuclear war and aimed to make some sort of coexistence deal based on mutual trust of the superpowers. Fully along these lines, Grechko complained at the CPSU CC politburo meeting that the main Soviet negotiator in SALT, deputy foreign minister Vladimir Semyonov, was surrendering to the Americans. Nor was Brezhnev particularly supportive of the diplomats at first. In terms of preserving defense capabilities of the country, he told them sternly not to reveal military secrets.[669] During SALT negotiations, a clearly existing borderline between competences of Soviet diplomats and top military was proven significantly. For example, one participant - the future chief of the Soviet general staff, General Nikolai Ogarkov - explained to the American side in the backrooms that in the USSR, the army solely dealt with military issues, and information for the ministry of foreign affairs' civilians was limited to what the military considered strictly necessary.[670] According to Brezhnev, such a state was largely natural. He respected the special position of the military elites for a long time. He reconsidered this approach around 1972, in the moment when danger of a possible alliance of marshals and the ideologically dogmatic part of the party apparatus based on their opposition towards détente emerged. This, in fact, not only threatened Brezhnev's line of the relaxation of tensions, but mostly his power statute. The disputes between the Soviet leader and the generality, climaxing in the first half of the 1970s, clearly showed the limits of his understanding of the army. He supposed the huge armament program,

---

[668] *Komunikat wywiadowczy za okres od 1 do 31.08. 1972 r.*, PZPR KC, XIB/173, AAN, Warsaw.

[669] In October 1969, before the SALT negotiations began in Helsinki, the CPSU CC General Secretary had warned the Soviet delegation that the KGB would be closely monitoring their activities. Zubok, *A Failed Empire,* 215-216.

[670] Taking into account the following course of the negotiations on reduction of troops and armament between the Warsaw Treaty Organization and NATO, we can say that Ogarkov's words were largely true.

which in reality had been running since he came to power, to be a sop that should have secured a strict loyalty of the military circles to both party and his person. The right to streamline some political decision concerning détente claimed by the generality, Brezhnev considered as a breaking of this unwritten agreement.[671] However, at the politburo meetings, he never openly confronted Grechko as well as his successor Ustinov on the issue of further armaments.[672] At least in the case of the former, it was undoubtedly some sort of mutual silent deal related to a power struggle within the Soviet top ruling circle.[673] It is necessary to point out that outlined disagreements never grew into an open rift or even to a crisis of Brezhnev's leadership.

As already noted, the CPSU CC General Secretary was a supporter of negotiation from the position of strength. His attitude, similar to many others members of the then-ruling Soviet nomenklatura, had been shaped by the lesson of World War II, which had caught the USSR absolutely unprepared in terms of military measures. Thus, Brezhnev's leadership asserted a policy of maintaining military power and readiness for any new war. The head of the party never objected to such a principle. But he was realistically aware of the disastrous impact of a potential global nuclear conflict. Therefore, he sought to ensure peace between the superpowers. Nevertheless, he considered the strengthening of military capabilities as a necessary prologue to any international agreements. He was not able to understand why the West saw in the Soviet strategic forces buildup in the 1970s a threat to its own security. However, Soviet military circles perceived NATO countermeasures in the same way. The closed circle was

---

[671] Durman, *Útěk od praporů,* 42, 153.

[672] Zubok, *A Failed Empire,* 243.

[673] In his work based mostly on Sovietological literature, Jan Wanner put forward an assumption that the dispute over the approach towards détente was usually just a pretext and result of the power intrigues within the Soviet leadership. In this regard, he notes that one of the top Soviet political officials, the CPSU CC secretary and the politburo member, Mikhail Suslov, spoke against Grechko's claim about the link between the increase of Soviet military power and strengthening the peace. But Suslov, in fact, did not agree with Brezhnev's line emphasizing détente and disarmament. He attacked the defense minister as Grechko, despite certain mutual discords, belonged to the General Secretary's backers in the politburo. Wanner, "Sovětská hierarchie a politika SSSR," 168.

established. The Soviet leader's thought that strength and peace were not mutually contradictory caused many complications later. This was especially the case in the moment when the American neoconservatives and the Pentagon's experts began to unanimously claim, referring to the continual expansion of Soviet strategic forces, that Moscow was striving for absolute dominance. In the second half of the decade, this campaign became one of the key factors of détente failure. Nonetheless, the Soviet leader was not so naive to think that détente would bring an "eternal peace." In October 1971, he lectured his speechwriters that negotiations between the blocs could have postponed war by 25 years, probably even by a century.[674] Almost two years later, in front of the Warsaw Pact member-states' leaders at the Crimean meeting, he pronounced preventing the nuclear apocalypse one of the main "historical" tasks of the communist movement.[675] Despite that these statements were probably not meant literally, they reflected the real alteration of both Brezhnev's attitude and Soviet foreign policy in general, which was occurring at the time.

The Soviet generality's reaction to the reduction of nuclear arsenals was confused. There was probably no clear consensus on the issue. The military elites were not permitted to harshly criticize their political superiors, of course. Therefore, the concept of coexistence of countries with different social systems and solving problems by political means was formally appreciated by the Soviet military circles. But in reality, they urged further strengthening of the Soviet nuclear capabilities, referring to the continual U.S. armament.[676] These demands once again directly contradicted the Soviet foreign policy appeals for nuclear disarmament.[677] However, during the first half of the 1970s, Brezhnev was

---

[674] Zubok, *A Failed Empire*, 202-203; 215; 243.

[675] *Stenogram narady aktywu partyjnego poświęconej spotkaniu Krymskiemu przywódców krajów socjalistycznych*, 3. 7. 1973, PZPR KC, XIA/404, AAN, Warsaw.

[676] Kulish, Vasili Mikhailovich, "'Conclusion' from Military Power and International Relations," in: Mezhdunarodnyye otnosheniya, 1972, quoted in *Selected Soviet Military Writtings*: 30-35. Colonel Vasili Kulish was considered to be one of the most reputable Soviet military-political analysts at the time.

[677] In June 1971, the USSR presented its proposal on holding the conference of five nuclear powers which would have dealt with nuclear disarmament. In

constantly improving his position. Since about 1974, he became so privileged within the Soviet leadership that his stances were absolutely crucial in shaping Moscow's attitude towards détente.[678]

The Warsaw Pact's auspice to the armament programs

The aforementioned SALT agreement, which affected the Warsaw Treaty Organization only indirectly because of the Soviet nuclear monopoly, brought some new element of stability into relations between the superpowers.[679] On the other hand, as the talks were proceeding, along with its earlier explained strategy, Moscow permanently asserted strengthening the military cooperation within the alliance, especially with quick modernization of member-states' armies as well as further unification and standardization of the weaponry.[680] In January 1972, the contrast between political declarations and real military measures, clearly showed up at the Political Consultative Committee session in Prague. Brezhnev outlined his visions on the future of the alliance: It should have been a political rather than military organization. The CPSU CC General Secretary also anticipated the Warsaw Pact's gradual rapprochement with

---

October of the same year, in the UN General Assembly, Moscow initiated convening a worldwide disarmament conference. Ślusarczyk, *Układ Warszawski*, 83.

[678] In March 1973, Brezhnev already managed to push P. E. Shelest and G. L. Voronov off the politburo. They were the long-term critics of his foreign as well as domestic policy. One year later, the General Secretary was able to repulse an attempt in this body to substantially revisit the Soviet attitude towards the West. Wanner, *Brežněv*, 18-20.

[679] After the Cold War, the Soviet backers of détente retrospectively declared that bigger reduction of strategic arsenals would have been rejected by the hawks. This applies also to the American side. For example, initially, SALT should have significantly reduced, even totally banning, an anti-missile defense. Such a step would have made the ongoing installation of independent warheads on Minuteman III rockets pointless as they had been designed mostly to overcome the Soviet anti-missile system. The main U.S. negotiator, Gerard C. Smith, commented on the situation that in case of a more ambitious deal, it would have been problematic for the USA to find proper targets for the 7-9 thousand MIRVs soon available. Ibid, p. 158; Burr, "Is this the best they can do?," 123.

[680] Jarząbek, *PRL w politycznych strukturach*, 77.

NATO, once an agreement on the key issues of the Helsinki process would be reached, especially inviolability of borders, nonuse of force or the threat of force and noninterference in each other's internal affairs.[681] At the same meeting, however, also the supreme commander Yakubovski presented his report on filling the plans of the Unified Armed Forces' development in the years 1971-1975. Despite Brezhnev's previous claims, the Political Consultative Committee unanimously agreed with continuance of the massive armament programs focused on the modernization of military hardware and equipment.[682] Everything should have proceeded in the scope of a new "socialist economic integration", which had been approved shortly before at the Comecon gathering.[683]

Thus, the Warsaw Treaty Organization's supreme body reflected the shift in détente only verbally. There were no modifications in course of armaments. It gave its formal political blessing to the further strengthening of alliance troops, without any complications or asking the unpleasant questions.[684] Generally, it can be said that the Political Consultative Committee sessions dealt with the military issues at a universal, almost even vague level. During preliminary talks, Moscow even excluded any profound discussion on the supreme commander's report.[685] Mostly, the alliance military bodies focused on concrete questions.[686] However, it is hard to imagine that the initiative presented by

---

[681] Mastny, *A Cardboard Castle?*, 42.

[682] *Zasedání Politického poradního výboru*, 13. 1. 1972, FMO GŠ/OS, k. 8, p. č. 41, č. j. 39033, VÚA, Prague.

[683] In the new Comecon's plans, great attention was paid to development of the weapons industry. Wanner, *Brežněv*, 51.

[684] *Překlad usnesení států-účastníků Varšavské smlouvy, přijatého na zasedání Politického poradního výboru 26. ledna 1972*, FMO GŠ/OS, k. 8, p. č. 41, č. j. 39033, VÚA, Prague.

[685] *Note Regarding the Presentation of the Soviet Draft Document Relating to European Security to be Adopted at the Prague Conference of the PCC of the Warsaw Treaty*, January 15, 1972, Accessed October 16, 2019. http://www.php.isn.ethz.ch/kms2.isn.ethz.ch/serviceengine/Files/PHP/18096/ipu blicationdocument_singledocument / 1da64635-b971-458f-b43e-532c02f671d3 / en/720115_note.pdf.

[686] *Spolupráce s armádami Varšavské smlouvy – usnesení Politického poradního výboru z 26.1. 1972*, 27. 10. 1973, FMO GŠ/OS, k. 8, p. č. 41, č. j. 39033, VÚA, Prague.

Yakubovski was not approved by the Soviet political leadership in advance. In fact, it fully reflected Brezhnev's tactic of negotiation from the position of strength. By the decision of the Political Consultative Committee, the Warsaw Treaty Organization adopted the formal auspices over the ongoing armament programs in the Eastern Bloc countries. Like in many similar cases, the practical impact of this step was more or less non-existent.[687] It is also well illustrated by a lax approach from Yakubovski towards the application of a particular directive of the alliance's highest body. The marshal, who was very agile in other cases, probably considered the order to analyze further opportunities for development of military cooperation between the member-states pointless. In fact, the situation that developed in the Warsaw Pact's military structures after 1969 was fully satisfying for the supreme commander. To work out a brief document took Yakubovski almost 17 months. Moreover, as a result, he only repeated previously-set objectives complemented by a vague note that these measures should have been the backbone of the armies' buildup in the following five-year military plan (1976-1980).[688]

Brezhnev's vision on the mostly political character of the Warsaw Treaty Organization and its possible cooperation with NATO did not resonate in the alliance military structures at all. In February 1972, the 4th Committee of Defense Ministers in Berlin presented totally different views: It publicly stated that despite certain improvements of the military-political situation in Europe, which was naturally ascribed mostly to the effort by the USSR and the Warsaw Pact, the "imperialistic" feature of NATO and especially the U.S. policy had not changed.[689] Such phrases

---

[687] For instance, the Operational Department of the General Staff of the CSPA declared that the resolution of the Prague session of the Political Consultative Committee, in praxis, required no additional significant measures. *Plán realizace jednání Výboru ministrů a Politického porad.výboru*, 20. 4. 1972, p. č. 43, č. j. 39035, ibid.

[688] *Usnesení Politického poradního výboru k upevnění družby mezi armádami Varšavské smlouvy, informační zpráva pro soudruha generálního tajemníka ÚV KSČ*, 2. 11. 1973, p. č. 41, č. j. 39033, ibid.

[689] In this regard, in his speech, Martin Dzúr, the Czechoslovak defense minister, marked the conclusion of the agreement between Poland and West Germany and the powers' deal on West Berlin as the key factors for easing tensions in Europe. *Zpráva ministra národní obrany o výsledku jednání Výboru ministrů obrany*

were not limited to the propagandistic resolution. The ideas of a deepening struggle between socialism and imperialism could have also been heard during internal discussions. These revealed even more concrete intentions: The Warsaw Pact military officials clearly admitted they considered détente mostly as an opportunity for pushing the West into being defensive on the international stage. Therefore, in terms of the process of easing tensions, they suggested especially striving for a realization of such measures that would have additionally moved the correlation of the forces in favor of the East. They also contemplated the support for the Third World countries' activities playing straight into the hands of the Soviet Union's global political aims.[690]

The military elites were well aware of the fact that the main military ties, as well as the countries' foreign policy orientation, would be preserved, even if - in an extreme scenario - the eventual success of the East-West negotiation paved a way to the dissolution of NATO and the Warsaw Treaty Organization.[691] Nevertheless, détente remained an untrustworthy and potentially dangerous process for the Soviet army command and the heads of the Warsaw Pact member-states' defense resorts. They feared that through the relaxation of tensions, in fact, the West was striving for weakening the socialist countries' unity and attempting to unleash internal instability there. In addition, they considered a side-result of détente, U.S.-Chinese rapprochement, to pose

---

státú-účastníků Varšavské smlouvy, únor 1972; *Vystoupení ministra národní obrany k otázce STAV NATO A NEBEZPEČÍ AGRESE V EVROPĚ*, undated 1972, k. 40, p. č. 136, č. j. 36039, ibid.

[690] Ibid. The Warsaw Pact's military support to the Third World countries significantly increased during the first half of the 1970s. In 1970, the USSR supplied them with arms and equipment to the value of 995 million USD, while the Eastern Bloc states received 75 million only. Five years later, value of the Soviet supplies to the Third World rose to 6.6 billion USD, and the rest of the Warsaw Treaty Organization member-states sent weapons worth about 525 million. Litera, "Unifikace sovětského bloku," 89.

[691] *Vojensko-politické úvahy a tendence vojensko-technického rozvoje ozbrojených sil protivníka a záměry ve výstavbě ČSLA*, undated, 1972, FMO GŠ/OS, k. 2, p. č. 29, č. j. 39021, VÚA, Prague.

an even more significant threat.[692] Therefore, in 1972, the Committee of Defense Ministers, with typical notes to increasing NATO capabilities, not only justified another arms race, but also decided on an intensification of preparation of the Warsaw Pact countries' territories for war. The possibility of using the civil infrastructure for military purposes should have been improved. In praxis, the Staff of the Unified Armed Forces was supposed to supervise this process again. It only subsequently informed the Committee of Defense Ministers about its activities in this area.[693] The situation once more proves the solely coordinative and consultative role of later institutions.

The peak of détente and shift in the military elites' thinking

At least until 1973, the defense ministers as well as many others of the Warsaw Pact's top military representatives were very skeptical about détente. Up to this point, they even considered the propagandistic and apparently unrealistic disarmament proposals issued by the Eastern Bloc countries risky. They posed a threat, however just *hypothetical*, of weakening the military capabilities of the alliance.[694] These were very irrational fears and a typical example of military thinking. In reality, the alliances' military analytics clearly showed that the West would accept any disarmament agreement only if the existing correlation of forces was preserved or moved against the East.[695] As previously explained, the Eastern Bloc countries actually presented unilaterally favorable initiatives

---

[692] *Vystoupení ministra národní obrany k otázce STAV NATO A NEBEZPEČÍ AGRESE V EVROPĚ*, undated 1972, k. 40, p. č. 136, č. j. 36039, ibid.

[693] *Zpráva ministra národní obrany o výsledku jednání Výboru ministrů obrany států-účastníků Varšavské smlouvy*, únor 1972, ibid.

[694] Nünlist, *"Cold War Generals"*.

[695] The Warsaw Pact's military analytics correctly noticed that the U.S. would have agreed with reduction of their military presence in Europe solely in the case of compensation in form of either weakening the Warsaw Pact's capabilities or strengthening the armies of the European members of NATO. *Vojensko-politické úvahy a tendence vojensko-technického rozvoje ozbrojených sil protivníka a záměry ve výstavbě ČSLA*, undated, 1972, FMO GŠ/OS, k. 2, p. č. 29, č. j. 39021, VÚA, Prague.

only. In addition, for the longest time, the Warsaw Pact's top military considered disarmament a purely declarative, but not really pursued, aim. Despite the shifts in the U.S.-Soviet talks, the internal analysis and estimation of further development emphasized that during the planning of the armies' buildup, a potential alteration of the international environment was unnecessary to take into account.[696] The question is whether these officially stated opinions reflected only an effort to take the ideologically orthodox positions at all costs. In fact, the prognosis for future military-political development was also justified by "unchallengeable" Marxist-Leninist thoughts from V. I. Lenin's work.[697]

However, progress in the Soviet-U.S. dialogue was obvious. At the turn of 1972 and 1973, neither Grechko nor other hardliners could continually totally ignore the international development as it was in praxis affecting even the Warsaw Pact's activities. In Autumn 1972, at the 7th Military Council session, the chief of Staff of the Unified Armed Forces announced the cancellation of the scheduled "Shield-73" drills.[698] Moreover, an upcoming meeting of the body was postponed in order to better react to dynamical international development at the time.[699] Given the absence of Soviet documents, the question of whether the military was ordered to take these steps by the CPSU political leadership cannot be answered; nevertheless, this scenario is very likely. Given this situation, the Soviet defense minister had to admit in 1973 that a certain easing of tensions in Europe and a shift in détente had really occurred. But he triumphantly announced that the Soviet army managed to improve its defense capabilities at the same time. The Warsaw Treaty Organization member-states' armies were supposed to follow it. In Grechko's opinion, further relaxation of international tension was posing another threat: Competition between both blocs was increasingly moving to the sphere of

---

[696] *Prognóza výstavby čs.lidové armády do roku 1990*, undated 1973, p. č. 30, č. j. 39022, ibid.

[697] *Vojensko-politické úvahy a tendence vojensko-technického rozvoje ozbrojených sil protivníka a záměry ve výstavbě ČSLA*, undated, 1972, p. č. 29, č. j. 39021, ibid.

[698] Mastny, "Imagining War in Europe," 32-33.

[699] *7. zasedání Vojenské rady Spojených ozbrojených sil*, 28. 10. 1972, FMO GŠ/OS, k. 37, p. č. 131, č. j. 36034, VÚA, Prague.

political struggle and its result was undoubtedly much harder to influence from the military point of view.[700] According to this thinking scheme, NATO allegedly recognized that the Warsaw Pact was too strong a military rival. Therefore, it was expected to try to move the center of the contest between the two blocs to an ideological dimension.[701] The military analysis of potential tendencies of international development actually began to admit that the effort to avoid direct East-West armed conflict would intensify during the upcoming years. The military clashes were supposed to increasingly transfer into the world's peripheries. In the future, the main forces, "imperialism and socialism" were expected to struggle mostly by political means.[702]

---

[700] In this regard, the Warsaw Pact's top military feared that intensified "hostile propaganda" could have weakened the Eastern soldiers' resistance in potential war. In fact, the proposed measures against the Western propaganda within the armed forces of the Warsaw Treaty Organization states were unbelievably vague; they included only closer cooperation of individual Main Political Administrations and an increase of political propaganda within the armies under the motto "upbringing the soldiers to a class hatred towards imperialism." *Vystoupení ministra národní obrany na 5. zasedání Výboru ministrů obrany států-účastníků Varšavské smlouvy k 1. bodu programu na téma "O opatřeních v oblasti boje proti imperialistické propagandě vedené proti spojeneckým armádám.",* draft, January 17, 1973, k. 41, p. č. 137, č. j. 36040, ibid; *Plán realizace protokolu č. 005 jednání Výboru ministrů obrany států-účastníků Varšavské smlouvy,* 2. 7. 1973, k. 8, p. č. 43, č. j. 39035, ibid.

[701] Nünlist, *"Cold War Generals".* These statements resonated not only during the meetings of the Warsaw Pact military bodies. In August 1973, Colonel I. Sidelnikov published a strongly ideological essay in the "Red Star" military journal. He expected that as a result of reconciliation between the East and West as well as the Helsinki process, the "class struggle" between socialism and capitalism would inevitably move into an economic and political-ideological dimension. An absolute peaceful coexistence between the socialist and capitalist camps he called unthinkable. He also blamed the Western top military for efforts to escalate a nuclear arms race and accused the Maoists of undermining détente. The ideological character of this text was underlined by the claim that nuclear weapons had no chance to change the "political nature" of war. Scott, *Selected Soviet Military Writings,* 5.

[702] *Vojensko-politické úvahy a tendence vojensko-technického rozvoje ozbrojených sil protivníka a záměry ve výstavbě ČSLA,* undated, 1972, FMO GŠ/OS, k. 2, p. č. 29, č. j. 39021, VÚA, Prague.

In 1973, the Warsaw Pact's military officials began to fear that détente could have threatened the ongoing armament projects, resulting in a reduction of troop numbers in the future. In this regard, at the 5th Committee of Defense Ministers' meeting, held in Warsaw in February 1973, Martin Dzúr noted that the West increasingly pointed out the contradictions between the real easing of international tensions and the continual strengthening of the Eastern Bloc's armies.[703] At the 8th Military Council gathering in Sofia, the supreme commander Yakubovski warned against such development. However, he confirmed that the issue would be finally decided by the central committees of the ruling parties, hence by the political power.[704] Thus, no nominally high military official of the Warsaw Treaty Organization was in position to question the political institutions' competences in the key matters.

At this point, the most powerful man of the Eastern Bloc, the CPSU CC General Secretary, openly admitted the possibility of the Warsaw Pact forces' reduction. In summer 1973, during his meeting with the Soviet satellites' leaders on Crimea, Brezhnev stated that the aim of foreign policy of the USSR was to force the West to "accept our game rules" in détente. He told the party leaders that under the current situation, it was possible to think about the reduction of both nuclear and conventional forces of the Warsaw Treaty Organization in Europe by 5-10 percent. However, in accordance with his previously explained thinking, Brezhnev emphasized that defense capabilities must not be forgotten and therefore their strengthening could be necessary once again in the future.[705] His optimism at this point was undoubtedly fueled mainly by concluding the U.S.-Soviet agreement on the Prevention of Nuclear War. The General Secretary and the President of the United States, Richard Nixon, signed it in the same year, on June 22, symbolically on the

---

[703] *Vystoupení ministra národní obrany na 5. zasedání Výboru ministrů obrany států-účastníků Varšavské smlouvy k 1. bodu programu na téma "O opatřeních v oblasti boje proti imperialistické propagandě vedené proti spojeneckým armádám. "*, draft, 17. 1. 1973, k. 41, p. č. 137, č. j. 36040, ibid.
[704] *8. zasedání Vojenské rady Spojených ozbrojených sil*, 24. 5. 1973, k. 37, p. č. 131, č. j. 36034, ibid.
[705] Mastny, "Imagining War in Europe," 33.

anniversary of the Nazi invasion of the USSR.[706] The chief of the Kremlin fully appreciated the deal. At the Crimean meeting with the Warsaw Pact member-states' leaders, he actually triumphantly labeled it as the American waiver of nuclear war against the socialist countries.[707]

Détente and change of the armament plans for the years 1976-1980

In this situation, the military five-year plans of 1976-1980 were completing. The buildup of the Warsaw Pact's armed forces, of course, primarily reflected the military-political analysis of trends within NATO, which at the time focused mostly on the improvement and better efficiency of nuclear missiles, according to Eastern observers and informants.[708] The key elements of the Unified Armed Forces' development in the upcoming five-year period was outlined by the supreme commander Yakubovski at the aforementioned 8th Military Council session in Sofia, May 1973. In

---

[706] The agreement formally followed the treaty on Basic Principles of Relations between both superpowers which had been concluded one year earlier in Moscow. However, the USA and USSR were just vaguely obligated to refrain from any actions potentially leading to nuclear conflict. The document, therefore, did not regulate the first use of atomic arsenals in potential conflict. *Agreement between the United States of America and the Union of Soviet Socialist Republics on the Prevention of Nuclear War*, Accessed October 16, 2019. http://www.fas.org/ nuke/control/prevent/text/prevent1.htm. There was a significant gap between Washington and Moscow in the understanding of this treaty. The Americans initially considered the Soviet initiative as a dangerous move aiming to elicit the U.S. renouncement of nuclear weapons usage. However, the entire defensive doctrine of NATO was based on it. Kissinger also feared that the document would provide a pretext for Moscow to preemptively strike China. The American side, therefore, considered the agreement largely symbolic even at the time of its signing. Zubok, *A Failed Empire*, 230.

[707] *Stenogram narady aktywu partyjnego poświęconej spotkaniu Krymskiemu przywódców krajów socjalistycznych*, 3. 7. 1973, PZPR KC, XIA/404, AAN, Warsaw. Brezhnev's words can be treated with caution. It is good to add that according to testimony of former Polish diplomat, J. Nowak, the Crimean meetings were affected by massive alcohol consumption. Thus, some ideas presented there cannot be taken literally, although these obviously reflected an alteration in the Soviet foreign policy. Nowak, *Od hegemonii do agonii*, 50.

[708] *Zámysl výstavby ČSLA na léta 1976-1980*, 9. 7. 1973, FMO GŠ/OS, k. 3, p. č. 31, č. j. 39023, VÚA, Prague.

keeping with the strategy of recent years, he stressed the modernization of weapons systems rather than extensive military buildup.[709] Regarding the changes in thinking about the features of potential conflict explained in the previous chapter, the analysis of the Yom-Kippur war became a basis for specifying already existing armament plans. The center of weaponry programs moved to air defense, radio-electronic warfare devices,[710] anti-tank weapons, artillery, missile units, as well as motor rifle and tank divisions.[711] Significant resources should have been invested in the buildup of a complex communication system.[712] It is understood that in the Warsaw Pact member-states' armies, primary attention was paid to rearmament of the units assigned to the Unified Armed Forces.[713] However, these represented their vast majority.

Modernization and general improvement of the air-defense system were especially set as the main objectives. Vast monies should

---

[709] *8. zasedání Vojenské rady Spojených ozbrojených sil*, 24. 5. 1973, k. 37, p. č. 131, č. j. 36034, ibid.

[710] The chief of Staff of the Unified Armed Forces, General Schtemenko, became one of the biggest advocates of development of radio-electronic warfare devices within the Warsaw Pact. At the military bodies meetings during the years 1973-1975, he warned that this combat technique would become a key factor in NATO military doctrine. He also pointed to the Yom Kippur war experience. When the Israeli jammers absented, the Arab coalition needed 1.5 missiles on average to shoot down one hostile aircraft. Otherwise, the number increased to 6.5 and more. In 1975, in order to strengthen radio-electronic equipment, Schtemenko put forward a draft of "Guidelines for Organization and Waging REW During Joint Activities of the Unified Armed Forces of the Warsaw Pact member-states." Ibid; *12. zasedání Vojenské rady Spojených ozbrojených sil*, 23. 5. 1975, k. 38, p. č. 133, č. j. 36036, ibid.

[711] *Zámysl výstavby ČSLA na léta 1976-1980*, 9. 7. 1973; ibid, *Úvodní vystoupení náčelníka generálního štábu - 1. ZMNO k výstavbě ČSLA v letech 1976-1980 na zúženém kolegiu ministra národní obrany dne - prosince 1973*, k. 3, p. č. 31, č. j. 39023, ibid.

[712] Based on the experience of the Yom Kippur war, Grechko stressed the importance of a complex communication network including a broad spectrum of devices - from cables to space satellites. *Zápis o 6. zasedání Výboru ministrů obrany členských států Varšavské smlouvy*, undated, perhaps February 1974, k. 10, p. č. 46, č. j. 39038, ibid.

[713] *Informace členů Kolegia ministra národní obrany o zasedání Politického poradního výboru členských států Varšavské smlouvy*, undated 1974, k. 8, p. č. 41, č. j. 39033, ibid.

have been put into this area.[714] In the mid-1970s, the backbone of the anti-aircraft system in most member-states' armies was represented by obsolete gun cannons, rather than modern missiles. Their significantly-less effectiveness in praxis was demonstrated clearly during the Arab-Israeli conflict.[715] The Warsaw Pact's top military was aware of own lagging in this field. They expected neither scheduled modernization would be able to protect the Pact members' airspace adequately. The overall capabilities of the alliance air-defense system to cause losses to the first massive airstrike of the enemy was supposed to increase from 20 to 52 percent, respectively from a pathetic 16 to at least 33.4 percent in the first day of conflict. Previous development of these weapons had failed to meet the plans. In the early months of 1974, Yakubovski had to admit that despite significant investments, strengthening of air-defense as well as air forces had proceeded very slowly during the current five-year plan. Thus, some of the Warsaw Pact member-states' armies were still able to track and effectively navigate no more that 20 percent of their fighter jets altogether.[716] From this point, air defense would become the core of

---

[714] Integration of air-defense within the Warsaw Treaty Organization had already taken place in the 1960s. The alliance unified air-defense system was established in 1969. However, the Pact's commanders were still worried about its insufficient ability to eliminate strikes of the enemy air forces. Moreover, cooperation between individual member-states' air-defense units lagged. The massive Warsaw Pact drills, held regularly on the Soviet territory after 1970, brought only small improvement. Thus, during the 1970s, integration of the air-defense units was the fastest in the entire Unified Armed Forces. In 1972, an agreement on their unification was concluded. *3. zasedání Vojenské rady Spojených ozbrojených sil*, 1. 11. 1970; ibid, k. 10, p. č. 45, č. j. 39037, *Vystoupení ministra národní obrany k otázce "výsledky plánování rozvoje ozbrojených sil států-účastníků Varšavské smlouvy na léta 1971-75"*, December 1970, FMO GŠ/OS, k. 35, p. č. 129, č. j. 36032, VÚA, Prague; *Informace o zpracování oper.dokumentace Jednotného systému PVO států VS*, 3. 2. 1972, k. 40, p. č. 136, č. j. 36039, ibid; *Zámysl výstavby ČSLA na léta 1976-1980*, 9. 7. 1973; *Perspektiva výstavby systému PVO ČSSR*, undated 1973, k. 3, p. č. 31, č. j. 39023, ibid.
[715] *10. zasedání Vojenské rady Spojených ozbrojených sil*, 2. 4. 1974, k. 37, p. č. 132, č. j. 36035, ibid.
[716] *Zápis o 6. zasedání Výboru ministrů obrany členských států Varšavské smlouvy*, undated, perhaps February 1974, k. 10, p. č. 46, č. j. 39038, ibid; *Informace členů Kolegia ministra národní obrany o zasedání Politického*

Warsaw Pact's armament programs. As it represented mostly defensive means, this was a significant shift in comparison to the past, when the main resources had flown into offensive weaponry. However, rather than international development, the measure probably reflected actual deficiencies in the Warsaw Pact's armies. We are also reminded that without the ability to proceed against enemy air forces, a possibility of victory in any military conflict could not have been taken seriously.

In 1974, the Warsaw Pact's analytics stated that the NATO budget had been permanently growing during the last three years.[717] Combat capacities of the estimated enemy increased by 18-19 divisions – more than 12 percent.[718] This provided Grechko with an opportunity to constantly roar about "reactionary forces of imperialism" continuing in preparation for war. The Soviet minister once more urged improving the defense capabilities which, in praxis, resulted in other massive investments into the development of the Unified Armed Forces.[719] Despite the fact that until 1976, the military spending in NATO countries outweighed those in the Warsaw Treaty Organization; the finance flowing into the armies per capita was constantly rising in the Eastern Bloc during the entire 1970s. There was an opposite process in the West.[720] After 1973, the political statements often justified further

---

poradního výboru členských států Varšavské smlouvy, undated, perhaps April 1974, k. 8, p. č. 41, č. j. 39033, ibid.

[717] In the 1970s, NATO countries increased their military spending from 113.1 to 190.4 billion USD. This growth mostly concerned the European members. Annually, defense investments rose by 3 percent. However, their proportion in GDP fell from 5.7 to 4.2 percent. The extent of military investments from government spending decreased in similar way - from 28.1 to 15.1 percent. Krč, "Vojenské výdaje v letech studené války," 42-43.

[718] 10. zasedání Vojenské rady Spojených ozbrojených sil, 2. 4. 1974, FMO GŠ/OS, k. 37, p. č. 132, č. j. 36035, VÚA, Prague.

[719] Zápis o 6. zasedání Výboru ministrů obrany členských států Varšavské smlouvy, undated, perhaps February 1974, k. 10, p. č. 46, č. j. 39038, ibid.

[720] At the beginning of the decade, 286 USD per capita were spent on military purposes in the scope of the entire Warsaw Treaty Organization. In 1979, it was 360 USD. Krč, "Vojenské výdaje v letech studené války," 42-43. A direct comparison of military spending between NATO and the Warsaw Pact is very complicated. Understanding of the state's military strength in NATO countries as well as methods for quantification of defense spending significantly differed from

"strengthening of the Warsaw Pact's defense capabilities" by the remaining absence of disarmament agreements.[721] Thus, in the period of climaxing détente, soon before finalization of the Helsinki process, preparations of the Unified Armed Forces for war, however primarily defensive, not only continued, but were even accelerated and intensified. At the meetings of the Warsaw Treaty Organization military institutions, the USSR put pressure on the member-states. Moscow still *de facto* demanded increasing military spending and improving combat readiness of the armies. The submissive satellites, with some exception by Romania, could not oppose.[722]

Czechoslovakia provides a good example of the entire situation. In late 1973, after a consultation with the Staff of the Unified Armed Forces representatives, Defense Minister Dzúr informed the CPCz CC General Secretary Husák that the Unified Command officials, using the side-mechanisms, were trying to persuade the Czechoslovak side about the need to reinforce the selected CSPA units beyond the scope of previous supreme commanders' official recommendations.[723] But realization of these suggestions would have exceeded the "economic and human capacities of the country". That was a euphemism already used back in the 1960s in connection with Moscow's urging to a militarization of the Czechoslovak economy and maintaining unbearable troop numbers. In the early months of 1974, the Soviet duress resulted

---

the Eastern Bloc. However, there are analytic tools which allow revealing basic development tendencies in both alliances. Ochrana, František, "Pojetí vojenské síly států Varšavské smlouvy," in: *Vojenské výdaje v letech studené války a po jejím skončení*, ed. Miroslav Krč (Praha: Ústav mezinárodních vztahů, 2000): 54, 58.

[721] *Stručná charakteristika vystoupení vedoucích jednotlivých delegací na zasedání politického poradního výboru*, 24. 4. 1974, 1261/0/6, sv. 115, a. j. 117/1, NA, Prague; *Deklarace o upevňování přátelství a prohlubování bratrské spolupráce mezi Komunistickou stranou Československa a Jednotnou socialistickou stranou Německa a mezi ČSSR a NDR*, 17. 10. 1974, TO(t) 1970-1974, box 5, i. č. 68, sign. 0344/112, č. j. 017236/74-4, AMZV, Prague.

[722] Mastny, "The Warsaw Pact," 150.

[723] Regarding the five-year plan of 1976-1980, the supreme commander recommended not only keeping troop numbers on the existing level, but also creating new units, which would require additional investments. *Návrh společné perspektivy rozvoje SOS - využití*, 10. 5. 1973, FMO GŠ/OS, k. 2, p. č. 30a, č. j. 39023, VÚA, Prague.

in an alteration of attitude of the Normalization regime, which had been completely loyal up to this point. Similar to the previous decade, Czechoslovakia began to alert Moscow to its economic problems caused by the massive military spending. Symptomatically, the Czechoslovak military officials did not intend to open the issue at the alliance collective bodies' meetings. They tried to deal with it through parallel consultations with the Soviet general staff. Dzúr asked his Soviet counterpart for these, pointing out that this institution was the most qualified for the issue.[724] This suggests certain CPSA officials' disillusion with the practical work of the reformed Warsaw Treaty Organization's military framework. They initially had hoped it would also result in better distribution of the armament burden among the Pact member-states.[725] True, there was some real improvements in cooperation of the Eastern Bloc countries' weapons industries after 1969.[726] However, requirement for new and more sophisticated weapons inevitably brought a need to increase military spending significantly. Related economic troubles of the Soviet satellites should have been solved by enlargement of the licensed production, deeper specialization, and collaboration of the Warsaw Treaty Organization's

---

[724] *Otázky k projednání s ministrem obrany SSSR*, 14. 12. 1973; ibid, *Informace o "Zámyslu výstavby ČSLA na léta 1976-1980" - informační zpráva pro generálního tajemníka ÚV KSČ soudruha JUDr. Gustáva HUSÁKA, CSc*, 14. 1. 1974, k. 3, p. č. 31, č. j. 39023, ibid.

[725] Up to this point, Czechoslovakia had striven to solve the armament issue fairly within the alliance structures. In 1973, for example, the Czechoslovak general staff and its Soviet counterpart were the first to send to the Unified Command a summary of military equipment to be introduced in service during the years 1976-1980. In this regard, the supreme commander Yakubovski commended the CPSA as an example to other armies. *8. zasedání Vojenské rady Spojených ozbrojených sil*, 24. 5. 1973, k. 37, p. č. 131, č. j. 36034, ibid.

[726] In 1973, General Schtepanyuk, the chief of the Technical Committee, talked about successfully dealing with the previously-existing duplicity in military production by individual Warsaw Pact member-states as well as its increasing effectiveness. His word probably corresponded to the truth, as deficiencies in the armed forces were pointed out relatively openly during the closed meetings of the alliance military bodies. *5. zasedání Výboru ministrů obrany*, February 1973, k. 41, p. č. 137, č. j. 36040, ibid.

individual members. Restriction of expenditures was never discussed as an option.[727]

Such a state was shown well in the GDR. Despite a constantly declining population, the country managed to increase the numbers of its troops during the years 1967-1978. The officially-admitted direct military spending rose from 3.9 to 5.1 percent of East-German GDP. This trend also continued into the second half of the 1970s. And yet, in 1975, these stats, however rough, stopped being published. Also, Poland increased numbers of its armed forces by 26 percent, to 401,000 men. These were mostly defensive units, not the attack divisions belonging to the Warsaw Pact's offensive echelon. Officially shown military spending decreased from 4.4 to 3.5 percent of GDP, but this was largely caused by the inflation and impressive economic growth which Poland experienced under Gierek's leadership at the time.[728]

Naturally, not only the satellites, but also the Soviet Union itself took part in the arms race. At the time, the Eastern superpower had developed its own version of MIRV warheads,[729] which meant a significant leap in evolution of the atomic arsenal. Modern, in terms of technical parameters, the then unmatchable fighter jets MiG-25 and supersonic intercontinental nuclear bombers Tu-22M reinforced the Soviet air forces. Works began on development of an "Akula" class, a heavy atomic submarine missile carrier, in order to improve the poor condition of the navy. During the decade after 1972, the USSR produced 4,125 intercontinental ballistic missiles, while the U.S. made 929 only. The American military strategists were above all worried of the increasing number of warheads on a single Soviet carrier.[730] Since the beginning of the decade, numbers of the Soviet ground forces increased by one-third, to 590,000 men in 1977. In addition, the units were equipped with a more-recent type of tanks, the T-72, in addition to infantry fighting vehicles,

---

[727] *12. zasedání Vojenské rady Spojených ozbrojených sil*, 23. 5. 1975, k. 38, p. č. 133, č. j. 36036, ibid.
[728] Wanner, *Brežněv*, 71.
[729] MIRV (Multiple Independently Targetable Reentry Vehicle) is a ballistic missile which carries a number of warheads capable of hitting different targets.
[730] Zubok, *A Failed Empire*, 242.

guns, rocket launchers, and mobile anti-aircraft systems.[731] During the 1970s, the Soviet annual military spending rose from 108 to 145 billion USD.[732]

Despite certain modifications, the Warsaw Pact's top military did not abandon its rhetoric even in the period of final preparation for signing the Helsinki Final Act. The only exception was Romania. Unlike their alliance colleagues' statements, the military officials of the country emphasized the continuing of détente in the mid-1970s. Moreover, they did not call the "imperialistic forces" the main obstacle in relaxation, but rather the existence of the blocs or dislocation of troops and holding drills on foreign territories. However, in June 1975, these opinions of George Locafetu, the chief of the Romanian general staff, were vigorously denounced by General Schtemenko.[733] Much more typical was a speech by Fritz Streletz, the chief of the East-German Main Staff, at the 7th Committee of Defense Ministers' session in Moscow, January 1975. He once again pointed to constantly increasing NATO spending as well as the fact that despite certain improvements of the international situation, the alliance still kept the awaiting troops in Western Europe strong, modern-equipped, and in high combat readiness.[734] Heinz Hoffmann, defense minister of the GDR, added that NATO was intensifying military preparations against the Warsaw Pact and remained the main instrument of the "imperialistic policy" on the Old Continent. Such a state should have proven that in spite of the relaxation of international tensions, a possibility of sudden escalation of the military-political situation and unexpected aggression had not vanished. These claims, in fact, noticeably resembled Marshal Grechko's speech five years earlier, analyzed in the beginning of the chapter. They document how little the real opinions and rhetoric of the Warsaw Pact's top military changed in the first half of the 1970s.[735] This

---

[731] Wanner, *Brežněv*, 70.

[732] Krč, "Vojenské výdaje v letech studené války," 43.

[733] *Porada vedoucích funkcionářů generálních /hlavního/ štábů armád Varšavské smlouvy*, 24. 6. 1975, FMO GŠ/OS, k. 19, p. č. 58, č. j. 39051, VÚA, Prague.

[734] Report by M. Dzúr, the Czechoslovak defense minister, to G. Husák, the CPCz CC General Secretary, on the 7th Committee of Defense Ministers meeting, January 20, 1975, k. 41, p. č. 138, č. j. 36041, ibid.

[735] Nünlist, "*Cold War Generals*".

fact undoubtedly contributed to the continuation of the massive arms race, which became one of the key factors for the fall of détente at the end of the decade.

Détente and the first consideration of disarmament

As suggested earlier, at a certain point, gradual progress in détente also started bringing a shift in considerations of disarmament. Originally the totally expedient, mostly propagandistic proposals of the Soviet Union and the entire Warsaw Treaty Organization respectively presented at the turn of the 1960s and '70s, slowly turned to the belief that the relaxation of international tensions might really result in some form of a deal on the reduction of troops and armament in the future. Given the size of the Unified Armed Forces, such an aspect obviously concerned the Warsaw Pact much more than SALT negotiations on nuclear weapons, which took place strictly on the superpowers' basis.

At the beginning of these thoughts lay two moments. First of all, regarding progress in the Soviet-U.S. talks, Brezhnev probably found a strong connection between the political side of détente and possible disarmament. In the Eastern Bloc, it soon became known under the "military relaxation" euphemism.[736] As already explained, military superiority over the West, or at least obvious parity in this area was an axiom to the Soviet leader. However, the CPSU CC General Secretary also knew that escalation of the arms race caused many economic troubles. The second and decisive reason why the Eastern Bloc began to take disarmament more seriously in the first half of the 1970s, was found in the pragmatic Soviet effort for the success of the CSCE. The center of the U.S. visions of détente lay in conventional disarmament – in the elimination of the existing advantages of the Warsaw Pact. Washington stipulated beginning some sort of talks about the reduction of troops and armaments as a condition for convening the all-European security

---

[736] Mastny, "Imagining War in Europe," 33; *Superpowers Détente*, p. 23. In 1971, Brezhnev still did not take such an option very seriously. After his meeting with W. Brandt, for example, he vaguely appealed for a reduction of conventional forces, but took no real steps.

conference. The successful holding of the conference was one of the main Soviet foreign policy priorities at the time.[737] Some concession, therefore, seemed to be inevitable. At the beginning of the 1970s, U.S. diplomats were stressing that they considered the Warsaw Pact's effort to hold the all-European security conference as nothing more than a propagandistic play. Washington intended not to allow the summit to take place until at least concrete talks on the burning international issues began. These included the status of West Berlin and the actual reduction of conventional troops of both alliances, among other issues.[738]

A question of potential disarmament related to détente began to be very cautiously discussed at the Warsaw Pact's forums in the early months of 1971. At the meeting of ministers of foreign affairs in Bucharest, all the member-states - except for Romania - supported the opinion that efforts to split the disarmament talks into individual and concrete aspects were necessary.[739] Like in many other cases, the Soviet satellites' approach should have been coordinated at the meetings under the auspices of the

---

[737] There were significant contradictions between the U.S. and its West-European allies. A majority of the European members of NATO wanted to engage in the CSCE process which they considered as an opportunity to improve their relations with the Eastern Bloc. Some fears appeared in Western Europe that détente combined with encroachments onto existing military parity could have been destabilizing and might decrease security. This opinion was presented mostly by France which asserted political relaxation without disarmament. NATO European members also thought that during disarmament talks with the Warsaw Pact, the United States, as the dominant alliance power, would have pursued mainly its own aims. Wenger and Mastny, "New perspectives of the origin of the CSCE," 14; Morgan, Michael Fotry, "North America, Atlanticism, and the making of the Helsinki Final Act," in *Origins of the European security system: the Helsinky process revisited, 1965–75*, eds. Andreas Wenger, Vojtech Mastny and Christian Nuenlist (Abingdon: Routledge, 2008): 28.

[738] *Zpráva I. teritoriálního odboru ministerstva zahraničních věcí ČSSR o stanovisku SSSR k celoevropské konferenci o bezpečnosti a spolupráci*, 7. 4. 1970, TO(t) 1970-1974, box 8, i. č. 89, sign. 020/311, č. j. 085/70, AMZV, Prague.

[739] A proposal occurred during the discussion that the Warsaw Pact countries participating in the Disarmament Committee in Geneva should have tried to split the issue of chemical and bacteriological weapons into two separate items during the following session of the body.

Warsaw Treaty Organization.[740] At this point, Moscow intended to keep disarmament initiative on the Eastern Bloc's side mostly due to propagandistic reasons. It stressed the reduction of atomic arsenals and verbally supported an idea of establishing regional nuclear-free zones.[741] Not only did such rhetoric echo strongly in the world public, but also the NATO military doctrine relied heavily on nuclear forces. As mentioned before, the main Soviet aim in terms of the reduction of conventional arms and troops remained the withdrawal of the U.S. divisions from Europe in the largest possible numbers. Thus, in January 1971, Grechko admitted that some negotiations on reduction of foreign military units dislocated on the European countries' territories could be launched during the CSCE.[742] For these reasons, Moscow considered it desirable that a potential disarmament agreement would concern an extensive territory: the entire continent, in the best case. Unlike NATO, the USSR initially rejected holding disarmament talks on the basis of the pacts.[743] As France remained outside NATO military structures, approved measures would not have affected its strong-armed forces.[744] In addition, according to Moscow's opinion, an official negotiation in the scope of the blocs could have resulted in negative publicity.[745]

---

[740] In contrast, Romania emphasized the importance of general and total disarmament and urged broadening cooperation with the socialist states outside the Warsaw Pact on this issue. *Zpráva o poradě ministrů zahraničních věcí členských států Varšavské smlouvy konané ve dnech 18.-19. 2. 1971 v Bukurešti*, 26. 2. 1971, 1261/0/5, sv. 154, a. j. 238/7, NA, Prague.

[741] *Politická zpráva ZÚ Moskva č. 10 o výsledcích XXIV. sjezdu KSSS*, 22. 7. 1971, TO(t) 1970-1974, box 8, i. č. 89, sign. 020/311, č. j. 023.940/71-1, AMZV, Prague.

[742] *Vermerk über die Konsultation des Ministers für Auswärtige Angelegenheiten der DDR, Genossen Otto Winzer, mit dem Minister für Auswärtige Angelegenheiten der UdSSR, Genossen A.A. Gromyko, am 11. 1. 1971 in Moskau*, 12. 1. 1971, DY 30/J IV 2/2J/3289, BArch, Berlin.

[743] *Zestawienie stanowisk Układu Warszawskiego i NATO w sprawach związanych z Europejską Konferencją Bezpieczeństwa i Współpracy*, undated, after 11. 6. 1971, PZPR KC, XIB 171, AAN, Warsaw.

[744] *Politická zpráva ZÚ Moskva č. 14/71 – Postup SSSR v oblasti odzbrojení po XXIV. sjezdu KSSS*, 23. 6. 1971, TO(t) 1970-1974, box 8, i. č. 89, sign. 020/311, č. j. 023.400/71-1, AMZV, Prague.

[745] *Informace o vývoji situace v otázkách evropské bezpečnosti a spolupráce*, 6. 10. 1971, f. PK 1953-1989, kat. č. 621, k. č. 144, AMZV, Prague.

The aforementioned 24th CPSU congress in the spring of 1971 brought a significant break for disarmament progress. It called the European Soviet satellites to intensify their diplomatic activities in favor of détente. Alongside the CSCE, another dimension was officially set - an attempt to reduce troops and armament in Europe. Since the beginning, the Kremlin intended to independently deal with the issue of the Helsinki process. Brezhnev's interest was to hold the all-European security conference as soon as possible. At this point, he even considered that it could have taken place before the end of 1972.[746] Therefore, a special separate body was supposed to discuss the complicated question of disarmament in Europe at the intended summit, or another parallel forum should have been established. Moscow also showed some willingness to start preliminary, but absolutely non-binding, negotiations within the CSCE on the simultaneous dissolution of NATO and the Warsaw Pact, respectively, and the abolition of their military structures in the first stage.[747] It is important to add that this was nothing more than a purely symbolic gesture that was expected to have no real immediate consequences. This was proven not only by the ongoing intensive development of military cooperation within the Warsaw Treaty Organization, but also by the fact that Moscow continually criticized Romania due to its appeals for termination of the military-political block's existence.[748]

---

[746] Mastny, *A Cardboard Castle?*, 43.

[747] In May 1971, the Soviet government summed up its statements in an aide-mémoire on the actual development in Europe addressed to the continental countries. Symbolically, it was presented just ahead of the North Atlantic Council meeting in Lisbon. NATO preliminarily accepted the proposal, but stipulated the condition of solving the issue of West Berlin first. *Informace o vývoji situace v otázkách evropské bezpečnosti a spolupráce*, 6. 10. 1971, PK 1953-1989, kat. č. 621, k. č. 144, AMZV, Prague.

[748] In 1973, during the Crimean meeting, Brezhnev still considered it necessary to confront Ceausescu due to his effort to start practical moves towards the simultaneous dissolution of NATO and the Warsaw Treaty Organization. On the contrary, the Soviet leader defined the strengthening of the alliance as the right course. *Wystąpienie końcowe Tow. Breżniewa*, undated, perhaps July 1973, PZPR KC, XIA/613, AAN, Warsaw.

The first negotiation between NATO and the Warsaw Pact

In October 1971, the first direct talks between the hostile pacts began. NATO decided to send its representatives to the Warsaw Pact countries in order to discuss actual disarmament initiatives of both blocs. The Warsaw Treaty Organization planned to establish similar expert commissions to negotiate in NATO states and vice versa. It was soon agreed that the disarmament talks would be limited to the region of Central Europe. The reduction of troops and armament would have concerned West Germany and Benelux countries on one side, and Czechoslovakia, Poland, and East Germany on the other. The reduction of both U.S. and Soviet forces dislocated on the territories of these states was expected as well.[749] At the turn of 1971 and 1972, the Warsaw Treaty Organization was under gradual nonpublic duress: Through diplomatic channels, NATO was sending a clear message that it would not allow any shift in preparation of the CSCE unless parallel negotiations on the reduction of armed forces was launched.[750] Thus, in April 1972, Henry Kissinger managed to make a secret deal in Moscow. In exchange for an American promise to end the obstructions in preparation for the all-European security conference, the USSR agreed to begin talks on the mutual reduction of NATO and Warsaw Pact troops beginning in 1973. The superpowers approved this as a way of separation between the disarmament issue and the Helsinki process.[751] Therefore, the resulting establishment of the Vienna talks[752] on mutual reduction of troops and armament cannot be seen as a goodwill gesture by the Eastern Bloc. It was

---

[749] *Informace o vývoji situace v otázkách evropské bezpečnosti a spolupráce,* 6. 10. 1971, PK 1953-1989, kat. č. 621, k. č. 144, AZV, Prague.

[750] *Zpráva o stavu přípravy a svolání celoevropské konference o bezpečnosti a spolupráci a o úkolech FMZV,* 18. 4. 1972, kat. č. 639, k. č. 146, č. j. 010.441/72, ibid.

[751] Wenger and Mastny, "New perspectives of the origin of the CSCE," 15.

[752] Perhaps, the Austrian capital was chosen as a hosting place because it had been previously considered as an alternative city for holding the CSCE. *Zpráva o stavu přípravy a svolání celoevropské konference o bezpečnosti a spolupráci a o úkolech FMZV,* 18. 4. 1972, PK 1953-1989, kat. č. 639, k. č. 146, č. j. 010.441/72, AMZV, Prague.

the rather direct outcome of U.S. pressure and effort to link the question of disarmament with the CSCE program. In accordance with policy defined by Moscow, the Warsaw Pact member-states, except for Romania,[753] strictly opposed such an eventuality.[754]

The Eastern Bloc countries were obviously not unanimous in their opinion whether the move was right. Poland was one of the Soviet satellites which in the early 1970s strongly supported disarmament or at least a reduction of military spending. Soon after Edward Gierek became a head of the PUWP, Warsaw - striving for its own economic rise - declaratively took up an old Khrushchev claim that the arms race escalation was an intentional Western tactic aiming to slowdown the socialist countries' economic development.[755] At the beginning of 1971, during his visit to the GDR, the Polish minister of foreign affairs, Stefan Jędrychowski, advocated the idea of armed forces' reduction within the CSCE process. Taking into account the sensitivity of the issue, he did not forget to point out that a similar consideration had been previously outlined by Moscow. But the East-German officials vigorously refused these Polish thoughts and marked them as a return to the Rapacki plan. Berlin also criticized the potential weakening of the defense capabilities of East Germany, Poland, and Czechoslovakia.[756] Based on this experience and many others, by the end of 1971, Gierek's leadership still believed that unlike in NATO, there would be a lack of willingness within the Warsaw Treaty Organization to even start negotiations on

---

[753] Romania, similarly to the FRG in the West, wanted to connect the disarmament talks with the CSCE. France opposed the Vienna talks in general, mostly because it ran on the basis of blocs. Wenger and Mastny, "New perspectives of the origin of the CSCE," 15.

[754] Jarząbek, *PRL w politycznych strukturach*, 77.

[755] On the other hand, the Eastern Bloc countries' diplomats noticed that Poland was aware of the global reality. It did not envisage a short-term possibility of significantly reducing the arms race in *both* NATO and the Warsaw Pact. *Informácia o výhľadovom zameraní poľskej zahraničnej politiky*, 11. 11. 1971, TO(t) 1970-1974, box 1, i. č. 79, sign. 016/111, č. j. 025.103/71-2, AMZV, Prague.

[756] *Informace o jednání ministra zahraničních věcí PLR s. Jendrychowského v NDR ve dnech 6.-8. 1. 1971*, 13. 1. 1971, box 1, i. č. 68, sign. 0344/111, č. j. 020.190/71-4, ibid.

disarmament in Europe.[757] At this point, however, Brezhnev himself became interested in that step. In the summer of 1972, at the Crimean meeting, he complained to the Warsaw Pact member-states' leaders that no deal with the West had yet been made which would have allowed him to reduce the arms race costs. He claimed these costs were unbearably high. The Soviet general secretary correctly noted that further military modernization was increasing spending significantly, as the new weapons systems based on the most-recent technologies were much more expensive.[758]

In comparison to the Helsinki process, the activity of the USSR and the entire Warsaw Treaty Organization regarding parallel disarmament talks, substantially lagged behind the West. In 1972, NATO began to put forward concrete proposals and Moscow was initially unable to react. In April, for instance, the Czechoslovak diplomats had to admit that there was no opportunity to get involved in the issue, because of the necessity to wait for an official joint statement from the Warsaw Pact. Its absence clearly reflected the overall unpreparedness of the Kremlin. Perhaps also due to this reason, Moscow refused a request by NATO General Secretary, Manlio Brosio, for a trip to the Warsaw Treaty Organization countries, as the concrete details of the scheduled disarmament talks should have been discussed.[759] In this stage, given the absence of a coherent strategy, the Soviet European satellites were only supposed to secure that the Helsinki process and Vienna negotiations would not be linked after all. If the West had resorted to obstructions in the CSCE preparation, the same Eastern reaction would have followed in the disarmament issue.[760] Finally, a more comprehensive disarmament

---

[757] Comparison of Warsaw Treaty and NATO Positions concerning the European Security Conference, December 1, 1971, in *A Cardboard Castle?*, (doc. 75), 390.
[758] *Notatka z przebiegu Spotkania I-szych sekretarzy bratnich partii na Krymie /31 lipca 1972/*, PZPR KC, XIA/612, AAN, Warsaw.
[759] *Zpráva o stavu přípravy a svolání celoevropské konference o bezpečnosti a spolupráci a o úkolech FMZV*, 18. 4. 1972, PK 1953-1989, kat. č. 639, k. č. 146, č. j. 010.441/72, AMZV, Prague.
[760] *Zpráva o vyslání čs. delegace na multilaterální jednání k přípravě konference o evropské bezpečnosti a spolupráci a směrnice pro její činnost*, 31. 10. 1972, 1261/0/6, sv. 57, a. j. 58/4, NA, Prague.

tactic began to arise within the Warsaw Pact in the early months of 1973. Moscow intended to limit the talks with the West to strictly procedural questions. It suspected that NATO was planning to exploit the negotiation and put forward an unacceptable proposal in Vienna such as a disproportional reduction of troops, thus harming the Warsaw Treaty Organization and therefore blocking the talks.[761] Afterwards, the West would have presented this failure as the Eastern unwillingness to reduce the armed forces, and in this regard, could have started a slowdown of the CSCE process.

Since the moment the Kremlin agreed with the Vienna talks, it planned for every proposal put forward and asserted by the Eastern Bloc countries to be pre-discussed and unified within the Warsaw Treaty Organization. In this regard, some of the member-states, for example, Czechoslovakia and Poland, naively expected they would have much more autonomy in Vienna than they were given in reality.[762] Moscow's strict supervision, which was much tighter than in the CSCE process, was only one limiting factor.[763] The key surveillance over individual states'

---

[761] The warning against the Western effort to force the Warsaw Treaty Organization into a much bigger reduction of its troops was already voiced in the summer of 1970 at the alliance meeting of deputy foreign ministers in Budapest. *Draft of the Speech by Hungarian Deputy Foreign Minister Frigyes Puja for the Meeting of the Deputy Foreign Ministers*, June 19, 1970, Accesed October 16, 2019. http://www.php.isn.ethz.ch/kms2.isn.ethz.ch/serviceengine/Files/PHP/173 00/ipublicationdocument_singledocument/da490bf2-98aa-46fc-ab20-123b86e70 032/en/700619_Draft_Speech_ENG.pdf.

[762] This was undoubtedly also fueled by the Soviet approach. During unofficial talks, for instance, the Soviet representatives offered to the Czechoslovak delegation a chance to work out its own proposal on troops and armament reduction. Despite this step fading away, it is still possible that the Kremlin wanted to take advantage in certain stages of negotiation and make the West face a draft which was not a Soviet document. *Účast ČSSR v současných jednáních o odzbrojení*, 26. 1. 1973, PK 1953-1989, kat. č. 657, k. č. 149, AMZV, Prague; *Předporada k jednání o snížení ozbrojených sil a výzbroje ve střední Evropě*, 27. 10. 1973, FMO GŠ/OS, k. 7, p. č. 39, č. j. 39031, VÚA, Prague.

[763] In the 1970s, the so-called unified line of the Warsaw Pact countries was asserted even more rigorously at the Vienna talks. Nowak, *Od hegemonii do agonii*, 66

diplomats was performed by the military.[764] After all, the U.S. representatives in Vienna were given confident warning that in his decisions on the approach towards the disarmament talks, Brezhnev was to some extent limited by the sops to the generality which demanded its right to "guard the country's security."[765]

Talks on reduction of troops and armaments in Central Europe

On the neutral soil of Austria, the negotiations on reduction of troops and armaments in Central Europe proceeded very slowly and with huge complications. This scenario was suggested already during January and June 1973, at the preliminary meetings.[766] At the beginning, NATO and the Warsaw Treaty Organization made some sub-deals. First of all, a round of the direct participants was narrowed to eleven states. Since October, the USSR, CSSR, PPR, and GDR should have negotiated the concrete disarmament measures on behalf of the Warsaw Pact. NATO was represented by the United States, Canada, Great Britain, West Germany, Belgium, Netherlands, and Luxembourg. The rest of the member-states of both alliances acted as observers. Also, the procedural questions were defined: the talks should have been held within the special working body and in a secret regime. The results were supposed to neither harm nor jeopardize the security of both pacts. At these points, however, the ability, or even willingness, to reach an agreement ended. Significant troubles occurred right away when defining the geographical scope of a potential

---

[764] During preparation of concrete proposals for the Vienna talks, both the foreign affairs and defense ministries of the individual member-states were supposed to cooperate. Formally, the military were the subordinates to the diplomats and they should have followed their instructions. The head of the Czechoslovak delegation, for example, was a diplomat of the ministry of foreign affairs named Radoslav Klein, who worked at the embassy in Moscow. *Informace o průběhu a výsledcích přípravných konzultací k jednání o snížení ozbrojených sil a výzbroje ve střední Evropě*, 18. 8. 1973, 1261/0/6, sv. 90, a. j. 84/info2, NA, Prague; *Zápis o pohovoru s pplk. ing. Radovanem PRAŽÁKEM*, 28. 10. 1973, FMO GŠ/OS, k. 7, p. č. 39, č. j. 39031, VÚA, Prague.
[765] Durman, *Útěk od praporů*, 152.
[766] Ślusarczyk, *Układ Warszawski*, 81.

treaty. The West asserted reduction concerning all the Warsaw Pact countries where the Soviet troops were dispatched. But on the NATO side, the measures should have involved West Germany and Benelux despite the fact that the U.S. military units were also dislocated in other West-European states. Soviet fears that NATO would promote asymmetric disarmament came true as well. The West rejected the Warsaw Pact's claim about an alleged balance of forces and wanted the agreement to reflect the real troop numbers. The Eastern negotiators soon realized that the proposed disproportional measures aimed to eliminate the Unified Armed Forces' conventional advantage. They naturally refused such a scenario. Indeed, from that perspective, the very vague, final communiqué could be seen as a success of the East.[767] Nevertheless, complications during preliminary talks confirmed Moscow's previous assumption that the Vienna negotiations would be the tough and long-term issues and thus the West must be prevented from linking it directly with the Helsinki process.[768]

In the summer of 1973, at the Crimean meeting, Brezhnev stipulated conditions for his willingness to make a deal in Vienna: For the Warsaw Treaty Organization, the existing favorable balance of forces should be preserved at any cost. Thus, in accordance with Brezhnev's thinking, only reduction of the destructive potential of both blocs was eligible.[769] A question to be answered is whether the Soviet leader in fact

---

[767] The document was worked out by the Soviet, Czechoslovak, American, and British delegations. The Warsaw Pact countries appreciated that the text basically corresponded to a resolution of the alliance meeting of Ministers of Foreign Affairs held in January.

[768] At the time, Brezhnev strictly opposed connections between disarmament and the CSCE during his visits to the U.S. and France. He operated with claims about difficulties of the disarmament process which could have unnecessarily slowed down the political relaxation. *Informace o průběhu a výsledcích přípravných konzultací k jednání o snížení ozbrojených sil a výzbroje ve střední Evropě*, 18. 8. 1973, 1261/0/6, sv. 90, a. j. 84/info2, NA, Prague; *Informace o návštěvě gen.taj. ÚV KSSS L.Brežněva v USA a Francii, přednesená čl. PB ÚV KSSS a min.ZV SSSR A.A. Gromykem*, 26. 7. 1973, TO(t) 1970-1974, box 1, i. č. 89, sign. 020/111, č. j. 024006, AMZV, Prague.

[769] *Stenogram narady aktywu partyjnego poświęconej spotkaniu Krymskiemu przywódców krajów socjalistycznych*, 3. 7. 1973, PZPR KC, XIA/404, AAN, Warsaw.

thought that such a form of disarmament was really possible. He had to be aware of strong mistrust, mutual fear, and both blocs' efforts to prevent the other side from getting any advantage through the conclusion of the agreement in the Austrian capital. Perhaps that assumption was one of the reasons why he did not order reviewing the ongoing armament programs of the Warsaw Pact.

Regarding the Vienna talks, the Soviet Union acted with absolute dominance within the Warsaw Treaty Organization. As the Kremlin urged a joint approach of the Eastern Bloc countries to foreign policy in the 1970s, this was applied even more rigorously in terms of sensitive military matters.[770] On the other hand, the Soviet hegemony in the disarmament talks resulted from the simple fact that the country maintained a nuclear weapons monopoly within the Warsaw Pact.[771] There was no possibility for the rest of the alliance to take initiative or perform more independently. Therefore, the alliance coordinative meeting was held in Moscow soon before the beginning of the first regular round of Vienna talks. The representatives of the defense and foreign affairs' ministries participated.[772] Oleg Khlestov, the head of the Soviet delegation in Vienna and a member of the Soviet diplomatic resort collegium, urged a totally unified approach from the Warsaw Pact countries. He argued using a standard threat of the time, which was that the West would undoubtedly strive for smashing their coherence. The delegations were supposed to mutually consult and coordinate their moves with the Soviet representatives, who played an absolutely key role.[773] In reality, the rest of the alliance should have solely backed their positions. The Soviet

---

[770] Nowak, *Od hegemonii do agonii,* 66.

[771] Jarząbek, *PRL w politycznych strukturach,* 77.

[772] Considering the estimated scheme of the Vienna talks, the meeting in Moscow from October 23-25 was split. On the first day, only the direct participants - the USSR, CSSR, PPR, and GDR - negotiated. The rest of the Warsaw Pact - "the countries with a special status" - joined them following day.

[773] *Memorandum on the Meeting of the Representatives of the WP States' Ministries of Foreign Affairs on 24 October 1973 in Moscow,* Accessed October 16, 2019. http://www.php.isn.ethz.ch/kms2.isn.ethz.ch/serviceengine/Files/PHP/17350/ipublicationdocument_singledocument/b176c403-aa4d-4700-a883-886554c0e816/en/730521_Memorandum_E.pdf.

satellites' diplomats were explicitly forbidden from taking any personal initiatives.[774] The USSR therefore only informed the rest of the Pact about its stances and received their promises of unified approach. In light of this, it was symptomatic that no top military officials, but only the second-rank authorized garniture, were sent to the preliminary meeting in the Soviet capital.[775] The Soviet attitude to the Vienna talks largely reflected the conclusions presented earlier by Brezhnev during the Crimean summer meeting, which assumed aggregative reductions of all troops in Central Europe – except the Navy - by 15 percent. Moscow mostly used a misleading claim that an absolute balance of forces between NATO and the Warsaw Treaty Organization had developed in the region. The real Soviet intentions were well illustrated by the fact that the Eastern representatives were instructed to try to limit the talks to the most general and vague level possible. It was explicitly forbidden to deal with the issues which exceeded commonly-known information about the Warsaw Pact's armed forces.[776]

In the light of this, it seems almost paradoxical that in 1973, the Warsaw Pact's military structures considered making some, however limited, deals at the Vienna disarmament talks possible. They even adapted the plans of the armed forces' buildup in the upcoming five years (1976-1980) to reach such an eventuality. Thus, the CSPA received an order from Moscow to prepare for potential reduction of its tank and motor rifle divisions or artillery - the conventional and in some sense obsolete weapons. In the substantial area of air-defense, reduction should have involved only gun cannons, inefficient for a long time. On the contrary,

---

[774] *SOSV - Vídeň* (telegram), 23. 11. 1973; *Zápis o pohovoru s pplk. ing. Radovanem PRAŽÁKEM*, 28. 10. 1973, FMO GŠ/OS, k. 7, p. č. 39, č. j. 39031, VÚA, Prague.

[775] Ibid. Acting of individual delegations was probably strictly defined by the national political institutions, based on Soviet instructions. In the case of Czechoslovakia, it was a directive by the CPCz CC Presidium and the Czechoslovak government from November 22, 1973. *Informace o posledním průběhu a dosavadních výsledcich jednání o snížení ozbrojených sil a výzbroje ve střední Evropě*, 21. 5. 1976, 1261/0/7, sv. 6, a. j. 4/info, NA, Prague.

[776] *Předporada k jednání o snížení ozbrojených sil a výzbroje ve střední Evropě*, 27. 10. 1973; ibid, *Zápis o pohovoru s pplk. ing. Radovanem PRAŽÁKEM*, 28. 10. 1973, FMO GŠ/OS, k. 7., p. č. 39, č. j. 39031, VÚA, Prague.

the Vienna negotiations were not expected to affect the key units of the first echelon or the border patrol.[777] Therefore, some willingness to reduce the army numbers has to be understood as a political and cosmetic move pliant to the international atmosphere of ongoing détente. The internal military documentation reveals that a potential agreement in Vienna would not have jeopardized the smooth continuing of the Warsaw Pact's armament projects in any way. The Eastern top military intended to assure that reduction of troop numbers and the decommissioning of obsolete weaponry was just a tactical move guising a further strengthening of the Unified Armed Forces through introducing new, technologically sophisticated arms.[778] However, a similar strategy had been already followed by NATO at the time.[779] Let us add that this form of disarmament was welcomed by some Warsaw Pact member-states as it, to some extent, solved the issues of the lack of inductees and servicing the obsolete hardware.[780]

The Soviet plans operated with the reduction of the armed forces in stages. In 1975, a symbolic cut of 20,000 troops and corresponding weapons in the armies of the direct participants in the Vienna talks should have taken place. In the following two years, the troop numbers were supposed to decrease by another 5 and 10 percent, respectively.[781] Such a conception had no chance to succeed of course, because NATO military officials pursued similar aims as their Eastern counterparts. They unambiguously intended to weaken the enemy through disarmament. Therefore, the Western alliance wanted to push a deal which would have included the ground forces only, the area where the Warsaw Pact had an

---

[777] *Úvodní vystoupení náčelníka generálního štábu – 1. ZMNO v výstavbě ČSLA v letech 1976-1980 na zúženém kolegiu ministra národní obrany dne prosince 1973,* k. 3, p. č. 31, č. j. 39023, ibid.

[778] *Podkladové materiály pro možnou realizaci "smlouvy" o snížení ozbrojených sil a výzbroje ve střední Evropě,* undated 1973, k. 7, p. č. 39, č. j. 39031, ibid.

[779] One year earlier, at the Committee of Defense Ministers gathering in Berlin, Heinz Hoffmann pointed to NATO strategy to reduce its armed forces' numbers, but to strengthen their capabilities through modernization at the same time. Nünlist, *"Cold War Generals".*

[780] *Smlouva o opatřeních ke snížení ozbrojených sil a výzbroje ve střední Evropě* (Soviet draft), 1973, FMO GŠ/OS, k. 7, p. č. 39, č. j. 39031, VÚA, Prague.

[781] Ibid.

advantage. The air forces and missile units, the fields where the West dominated due to its technological superiority, would have remained intact. In the end, the Vienna talks inevitably froze. Instead of dealing with factual aspects, mostly quarrels took place there. The dispute between the British and Soviet delegations on whether the Warsaw Pact was of an offensive or defensive nature, provides a good example.[782] Taking into account this development, in the early months of 1974, the USSR, with at least the certain although rudimentary assistance of Poland, Czechoslovakia, and East Germany, worked out a draft of a concrete disarmament agreement which was, however, only a conclusion of the Eastern stances thus far. In terms of its traditional tactics, Moscow naturally presented it in Vienna on behalf of the entire Warsaw Treaty Organization. Afterwards, a widespread media campaign for its support should have been launched, possibly including also the Western press.[783]

The biggest complication resulting from the Vienna talks for the Warsaw Treaty Organization was therefore the deepening discord between Romania and the Six. The Ceausescu regime soon stepped up against the negotiation format. It was not satisfied with the fact that the potential agreement was not supposed to include Romanian armed forces. Moreover, having an observer status only, Bucharest had no chance to substantially affect its parameters.[784] As a result, Romania ignored the tactic of the Warsaw Pact countries' unified approach during the following talks in Vienna. It also objected to negotiations on the basis of the blocs.[785] Romanian obstructions required holding many separate meetings of the

---

[782] *SOSV - Vídeň* (cable), 23. 11. 1973, ibid.

[783] Cable by R. Klein from the Vienna talks on reduction of troops and armaments in Central Europe, March 4, 1974, ibid.

[784] *Informace o průběhu a výsledcích přípravných konzultací k jednání o snížení ozbrojených sil a výzbroje ve střední Evropě*, 18. 8. 1973, 1261/0/6, sv. 90, a. j. 84/info2, NA, Prague.

[785] Romania even opposed even principle that the organizational issues of the Vienna talks were negotiated by two countries only, each representing one alliance. *Předporada k jednání o snížení ozbrojených sil a výzbroje ve střední Evropě*, 27. 10. 1973, FMO GŠ/OS, k. 7, p. č. 39, č. j. 39031, VÚA, Prague; *Zpráva o návštěvě ministra zahraničních věcí RSR s. Macovesca v ČSSR*, 2. 5. 1974, TO(t) 1970-1974, box 2, i. č. 82, sign. 017/112, č. j. 012.472/74-2, AMZV, Prague.

Six, as Bucharest put forward various amendments to the Eastern proposals. Ceausescu tried to ensure that a deal was aimed mostly at the withdrawal of foreign military contingents from the European countries' territories. Romania sought to prevent disarmament solely in the region of Central Europe by asserting the clause that committed the signatories of the potential treaty to start similar negotiations on the rest of the continent immediately.[786]

At the beginning of 1974, the U.S. delegation in Vienna managed to prompt the East to unofficial, off-record talks.[787] These allowed at least a little bit more open discussion. The Warsaw Pact's tactic remained the same. It was based on criticizing *de facto* all the Western proposals, and a willingness to only approve a draft of the agreement if it was previously worked out by the Soviet side.[788] But NATO also took rigid positions. It insisted on a two-stage approach. In the first, reduction should have solely concerned the U.S. and Soviet troops in Central Europe. This was unacceptable to Moscow for two reasons: First of all, the USSR opposed the deal not applying to British and French armed forces, as both would have played an important role on the Central-European battlefield in case of war. Moreover, they had their own atomic ammunition. The USSR also feared that reduction of its military presence in the Warsaw Pact countries neighboring the NATO states would create a suitable space for the intensified Western ideological influence on society there. In the 1970s, dislocation of the Soviet divisions in Central Europe was also designed as a sort of "ideological police." During advanced détente, it would have

---

[786] *Zpráva náměstka ministra zahraničních věcí pro náčelníka generálního štábu ČSLA Karla Rusova*, 5. 2. 1974, FMO GŠ/OS, k. 7, p. č. 39, č. j. 39031, VÚA, Prague.

[787] The unofficial negotiations, proceeding alongside the main Vienna talks, were held by six countries - the U.S., West Germany, and Netherlands for NATO; and the USSR, Poland, and East Germany for the Warsaw Pact. Romania rejected such an approach, stating that reduction of involved countries would result in strengthening the bloc-to-bloc conception. Cable by R. Klein from the Vienna talks on reduction of troops and armaments in Central Europe, March 4, 1974, ibid.

[788] Cable by R. Klein from the Vienna talks on reduction of troops and armaments in Central Europe, February 8, 1974, ibid.

made the impression that August 1968 could happen again.[789] Therefore, the Kremlin still insisted on unblocking the negotiations through the "first step," a purely symbolic reduction of all the direct participants' troops by 20,000 soldiers. From the military point of view, this was a totally minor measure. However, it could have been used well by the propaganda. Such a step would have provided a solid argument for the East against the Western opponents of détente who claimed that the disarmament talks were a dead end.[790] On the other hand, NATO also played a propagandistic game. The North Atlantic Council, for instance, publicly refused a common opinion at the time that the Warsaw Pact's conventional advantage was balanced by better Western nuclear arsenals. This aimed to convince the world public that the Soviet demands on the symmetric reduction of armed forces based on the assumption of alleged military parity was not well founded.[791]

In terms of continual political relaxation, especially shifts in the Helsinki process, at the Political Consultative Committee session in April of 1974, Brezhnev suggested a shift in the approach towards disarmament. He stated that given the actual situation, a propagandistic benefit from the presentation of proper initiatives was no longer essential. Henceforward, he stressed the real moves aiming to eliminate a risk of war and reduction of the arms race. These efforts should have become the main aspects of the Soviet policy towards the United States. It would be a fundamental change in Khrushchev's position to issue unrealistic and mostly propagandistic peace initiatives. Nevertheless, there was a parallel with the massive reduction of ground forces during Khrushchev's rule in the

---

[789] Mathew Ouimet claims that the danger of the Soviet military intervention hung over the Eastern Bloc countries like the "sword of Damocles" during the entire 1970s. However, such a conclusion is extremely simplified and does not reflect the real development. Ouimet, Matthew J, *The Rise and Fall of the Brezhnev Doctrine in Soviet Foreign Policy* (Chapel Hill: The University od North Carolina Press, 2003), 65.

[790] *Informace o zahájení čtvrté etapy vídeňských jednání o snížení ozbrojených sil a výzbroje ve střední Evropě*, 1. 10. 1974, PK 1953-1989, kat. č. 710, k. č. 157, č. j. 016.261/74, AMZV, Prague.

[791] Cable from the North Atlantic Council session in Paris, February 15, 1974, FMO GŠ/OS, k. 7, p. č. 39, č. j. 39031, VÚA, Prague.

second half of the 1950s: The money made available was also in this case supposed to help the Soviet economy.[792]

Brezhnev's visions, however, had no real impacts on the Vienna talks. In the beginning of 1975, a stalemate in the Austrian capital was evident. At this point, Hungarian political officials did not hide their opinion from the Warsaw Pact allies that the negotiations would not bring any concrete results.[793] The talks froze due to a relatively banal question: Both pacts' representatives were not able to agree on the defining criteria for the terms "ground troops" and "military air forces."[794] The following development suggests that despite the fine words of the CPSU CC General Secretary, this situation, to some extent, suited Moscow at the time. It seems that the reduction of troops in Central Europe remained no priority to the Kremlin. The Warsaw Treaty Organization maintained clear conventional superiority. The superpowers' negotiation on the reduction of strategic nuclear weapons (SALT), which was Brezhnev's greatest interest, continued in parallel. In addition, the Kremlin achieved its main goals in détente through signing the Helsinki Final Act. Thus, a breakthrough in Vienna lost its original political value.[795] Nevertheless, the East was encouraged by the CSCE progress. On the international stage, it started to operate with a claim that the process of political relaxation in Europe had to be supplemented by some military deals. Therefore, in Vienna, the Warsaw Treaty Organization more vehemently than any time before pushed forward mutually symmetric reduction of armed forces, now by 15 percent during a three-year period. Nevertheless, its representatives had to know that given the existing situation, there was no

---

[792] *Vystoupení s. L.I. Brežněva na poradě politického poradního výboru Varšavské smlouvy 17. 4. 1974,* 1261/0/6, sv. 115, a. j. 117/1, NA, Prague.

[793] *Informace o návštěvě s. Živkova v MLR,* 10. 2. 1975, DTO 1945-1989, i. č. 4, e. č. 27, č. j. 011229, AMZV, Prague.

[794] Even the definition of the term "military" proved to be problematic. In the armed forces of the Warsaw Treaty Organization countries, the military personnel did many works which were carried out by the civilians within NATO. Also, paramilitary organizations like border and territorial guards played an important role in the entire composition of the Warsaw Pact forces. *Informace o posledním průběhu a dosavadních výsledcích jednání o snížení ozbrojených sil a výzbroje ve střední Evropě,* 21. 5. 1976, 261/0/7, sv. 6, a. j. 4/info, NA, Prague.

[795] Wanner, *Brežněv,* 54.

chance to make such a deal. NATO still was not able to reach the agreement with its counterpart on the procedural questions,[796] and even less in an objective evaluation of the correlation of forces on the European battlefield. Both alliances continually put forward the proposals which weakened their rival in praxis.[797]

      The Vienna disarmament talks remained in a trap even at the end of the 1970s.[798] However, their indirect outcome was the temporary, but obvious, weakening of total military activity in Central Europe. The Warsaw Treaty Organization proposed the scheduled drills of significant extent to be announced in advance. As a goodwill gesture, since 1973, only small and local military maneuvers were held in Poland, Czechoslovakia, and Hungary. The Warsaw Pact's conference on joint drills and unified exercise norms, held at the turn of October and November the same year in Prague, closed a book on massive maneuvers.[799] These were restored at the beginning of the 1980s in accordance with the fall of détente.

---

[796] For example, the definition of the subject matter of the negotiations; stages and methods of the reduction; the issue of the weapons industry; or the preliminary freezing of troop numbers in the region.

[797] *Informace o posledním průběhu a dosavadních výsledcích jednání o snížení ozbrojených sil a výzbroje ve střední Evropě,* 21. 5. 1976, 1261/0/7, sv. 6, a. j. 4/info, NA, Prague. *Informace o průběhu jednání o snížení ozbrojených sil a výzbroje ve střední Evropě v 2. pololetí 1976,* 11. 2. 1977, sv. 30, a. j. 33/info1, ibid.

[798] Ślusarczyk, *Układ Warszawski,* 81.

[799] Wanner, *Brežněv,* 47-49.

The Warsaw Pact's political structures on the path to the fall of détente

The signing of the CSCE Final Act on August 1, 1975, can be considered as the biggest political achievement of the Warsaw Treaty Organization during the first two decades of its existence. The roots of the Helsinki process went back to 1966, to the Bucharest declaration of the Political Consultative Committee. Hence, convening the all-European security conference became the highest priority of the Pact's foreign policy. Along these lines, the Final Act was evaluated by Gromyko, the Soviet minister of foreign affairs, with two months' distance. However, the USSR was aware that the agreement also contained various - from the Eastern Bloc's point of view -undesirable aspects. This mostly included the so-called third basket which consisted of the issue of respect for human rights, among other things. Some fears were raised also by the announcement of military drills in advance and participation of the other side there. Moscow intended to deal with these problems operatively during the time. It had no comprehensive strategy and only envisaged that the Warsaw Pact member-states would mutually consult the situation.[800]

The Kremlin did not take the Helsinki accords without any fears. Some in the Soviet leadership, for example, Nikolai Podgorny, Yuri Andropov, Alexei Kosygin, or Mikhail Suslov enunciated doubts whether the third basket was not an unacceptable concession which opened door for the Western "imperialistic interference" with the Soviet sphere of interest. The Soviet diplomacy led by Gromyko opposed these arguments. On the contrary, the Minister of Foreign Affairs emphasized that the Yalta settlement in Europe, in fact, had been guaranteed. He also believed that

---

[800] *Zpráva o návštěvě člena politického byra ÚV KSSS a ministra zahraničních věcí Sovětského svazu A.A. Gromyka v ČSSR*, 1. 10. 1975, 1261/0/6, sv. 169, a. j. 172/12, NA, Prague.

deepening cooperation with the West could have helped the Soviet Union to overcome the beginning economic difficulties. Gromyko expected that Moscow would be strong enough to prevent Western interference with events in the Eastern Bloc.[801] The prevailing fact was that the USSR and many Warsaw Pact countries, especially Poland, Czechoslovakia, and East Germany, achieved their main foreign political aims through the Final Act. In reality, a confirmation of *status quo* and inalterability of the borderlines in Europe were among the main motives why Brezhnev's leadership supported détente.[802]

In fact, the Final Act was full of contradictions.[803] These reflected not only a necessary compromise between the East and West, but also between individual NATO members.[804] The document was limited to the political and economic field, in praxis. It brought only marginal measures in the military area. The elements of confidence building, including an announcement of military drills and dislocation of troops nearby state borders in advance, did not correspond to Moscow's idea of "military relaxation".[805] Some persons in the Soviet leadership, especially Arvid Pelshe, Mihail Suslov, and Boris Ponomarev, pointed this out. They argued that the change of the international atmosphere caused by détente could not be considered irreversible until some concrete military agreements were concluded.[806] That was not just a Soviet specificity. Immediately after the Helsinki summit, the claims that political détente had to be supported by military relaxation also strongly resonated in some parts of the American political spectrum.[807]

---

[801] Durman, *Útěk od praporů*, 169.

[802] Bowker, Mike, "Brezhnev and Superpowers Relations," in *Brezhnev Reconsidered*, eds. Erwin Bacon and Mark Sandle (New York: Palgrave, 2002): 93; Tejchman and Litera, *Moskva a socialistické země na Balkáně*, 98.

[803] Both Cold War blocs were fully agreed on just the issue of trade and industrial cooperation. Wanner, *Brežněv*, 22.

[804] Suri, Jeremy, "Henry Kissinger and the reconceptuailization of European security, 1969-75," in *Origins of the European security system: the Helsinky process revisited, 1965–75*, eds. Andreas Wenger, Vojtech Mastny and Christian Nuenlist (Abingdon: Routledge, 2008): 60.

[805] Mastny, "Imagining War in Europe," p. 34.

[806] Wanner, *Brežněv*, 23.

[807] Volkogonov, *The Rise and Fall*, 277.

The Czechoslovak leadership's report suggests that the Warsaw Treaty Organization countries considered implementation of the Helsinki conference outcomes to be a long-term process. It was expected that individual alliance member-states would actively engage and negotiate bilaterally with the Western countries. The priority was to deepen détente and to develop positive relations between the countries with different social systems. In terms of the above-mentioned, the East gave great emphasis to the military area. It was meant to be an overall transition to the aforementioned military relaxation. The efforts to reduce and later to stop the arms race, disarmament in general, and easing military tensions in Europe were all pronounced as official policy. The USSR and countries of its sphere of interest also considered it desirable to achieve at least partial results at the Vienna talks, to start negotiations on holding the world disarmament conference, or to make the work of the Geneva committee for armament control more effective. Nevertheless, realization of these steps was not supposed to harm the socialist countries' interests at any costs. Even due to this reason, a unified foreign policy was assumed to be continually coordinated within the Warsaw Pact.[808]

As the previous chapters explained, Moscow already began to restrict activities of the Warsaw Pact's political structures significantly in the period before the Helsinki conference. In the light of this, it is not surprising that despite the proclamations about the need for further consultations, no top multilateral meetings were held. The talks of deputy foreign ministers were an exception. Since the mid-1970s, they took place even more frequently and their significance was rising.[809] In 1975, the establishing of the Committee of Ministers of Foreign Affairs was approved. However, the body still remained on paper, largely because of the Romanian stances. In the mid-1970s, in terms of its balancing policy,

---

[808] Evaluation of the Helsinki Final Act by the Czechoslovak Party Presidium, April 28, 1976, in *A Cardboard Castle?* (doc. 78): 397-401.

[809] Besides preparation of the Political Consultative Committee and Committee of Ministers of Foreign Affairs' sessions, another meeting of the deputies took place annually which usually had no direct connections with the agenda of both bodies. Unlike in the past, not only the first deputy of the minister of foreign affairs took part in these meetings, but rather a deputy in charge of the issues to be discussed. Békés, "*Records of the Meetings*".

Romania found itself in a centrifugal phase towards the Eastern Bloc.[810] In December 1975, Moscow tried to convene a meeting of the new institution. In the changing international situation, the USSR sought to unify its satellites' positions this way. Also, some Warsaw Pact members called for alliance consultations regarding the Final Act.[811] However, Romania opposed holding a meeting under the alliance auspices. The Soviet Union surrendered as it considered the consultations necessary. The

---

[810] Romania began to call itself a "developing socialist country." In February 1976, it joined G77, a group of the developing states. At the time, the Soviet Union considered Romanian foreign policy less useful than Cuban or Vietnamese. Romania also intensively strove for joining the Non-aligned Movement, but this was impossible as the country remained a Warsaw Pact member. Therefore, Bucharest tried to trivialize its membership. During the talks with non-aligned states' officials, Romania put NATO and the Warsaw Treaty Organization on the same level of a threat to world peace and demanded their immediate dissolution. This was a temporary and instrumental shift in the Romanian policy motivated mostly by the perspective of beneficial economic cooperation with the developing countries. However, Ceausescu's effort to get an observer statute for his country at the upcoming non-aligned summit in Colombo was blocked by both Cuba and Vietnam, probably on the direct instruction of Moscow. On the contrary, Yugoslavia pointed out the fact that despite the membership in the Warsaw Pact, Bucharest took positions mostly corresponding to the non-aligned countries. Nevertheless, Beograd at the same time assured the Eastern Bloc states that in its support for Romania, it did not plan to cross the line, which could have been considered as appeals for leaving the Warsaw Treaty Organization. After the long negotiations, Bucharest were given a "permanent host" statute which allowed it to participate in the Non-aligned Movement and at the same time did not complicate its membership in the Warsaw Pact. *Zahraničně-politický vývoj RSR*, 1. 7. 1975, DTO 1945-1989, i. č. 34, e. č. 72, AMZV, Prague; *Zpráva o sovětsko-rumunských vztazích*, 5. 6. 1976; Ibid, č. j. 016.572/76, *Přístup RSR k nezúčastněným zemím*, 4. 10. 1976, e. č. 73, č. j. 014.834/76, ibid.

[811] The Bulgarian ministry of foreign affairs, for example, consulted the Helsinki accords' impact with the Warsaw Pact members except for Romania in the second half of 1975. Afterwards, Sofia suggested holding at least the alliance meeting of deputy foreign ministers. In the beginning of the following year, also Czechoslovakia proposed to coordinate a further approach towards the third basket of the CSCE within the alliance. *Záznam o rozhovoru náměstka ministra zahraničních věcí ČSSR Bergera s náměstky ministrů zahraničních věcí BLR a RSR*, 16. 1. 1976, i. č. 4, e. č. 16, č. j. 76.504/76 ibid; *Záznam o rozhovoru rady ZÚ BLR D. Josifova s zástupcem vedoucího OKS Žabokrtským*, 13. 1. 1976, č. j. 76.285/76-OKS, ibid.

Soviet side agreed that the final communiqué would not be issued on behalf of the Pact. Nevertheless, the Kremlin was afraid that this would set a dangerous precedent.[812] It probably still saw the alliance as an important foreign policy instrument. Along these lines, at the 25th CPSU congress in February 1976, Brezhnev actually appealed for further integration of the socialist countries within the Warsaw Treaty Organization and the Comecon.[813]

The first significant alliance meeting after the Helsinki conference did not take place until the end of 1976. At this time, the USSR began to feel the first negative impacts of the Final Act. It planned to face them through holding the Political Consultative Committee session, among other things. Outwardly, the gathering should have fully reflected the triumph of détente. The agenda was set once again by the CPSU CC and the Soviet Ministry of Foreign Affairs.[814] Then, the scenario was only slightly adjusted collectively by the deputy foreign ministers, just ahead of the summit itself. The importance of their meeting was mostly the fact they managed to harmonize positions with the Romanian side which in turn did not act confrontationally at the highly-publicized summit. No key instigations of the Soviet satellites were put into the agenda of the Political Consultative Committee. On the contrary, some of the most loyal member-states kowtowed to the Kremlin. Czechoslovakia, for instance, obsequiously proposed the special amendment to the prepared resolution adoring Brezhnev's and the CPSU leadership's contribution to a peaceful policy. The value of this proposal is illustrated by the reaction of Nikolai Rodionov, the Soviet diplomat, who rejected it as counterproductive and exaggerated.[815]

---

[812] The Soviet information for the Polish leadership about holding the meeting of deputy foreign ministers of the Warsaw Pact member-states, undated, perhaps 1975, PZPR KC, XIA/600, AAN, Warsaw.

[813] Wanner, *Brežněv*, 23.

[814] *Oficiální přátelská návštěva A. Gromyka v BLR*, 29. 11. 1976, DTO 1945-1989, i. č. 4, e. č. 37, č. j. 018009, AMZV, Prague.

[815] *Reports (3) by Hungarian Deputy Foreign Minister István Roska on the Meeting of the Deputy Foreign Ministers*, November 23, 1976, Accessed October 16, 2019. http://www.php.isn.ethz.ch/kms2.isn.ethz.ch/serviceengine/Files/PHP/17372/ipublicationdocument_singledocument/e70f7928-5e79-4505-97fd-666fcebaa8a0/en/761123_Reports_E.pdf.

Before the Political Consultative Committee session in Bucharest, Leonid Brezhnev stated that the meeting would be a "milestone" for the further approach of the Warsaw Treaty Organization towards détente and even for strengthening socialism worldwide. These were not just empty phrases. In fact, the CPSU CC General Secretary probably had high hopes for the resolution. In this regard, he traveled to the Romanian capital in advance. He wanted to personally ensure Nicolae Ceausescu's consent with the prepared declaration.[816] The Soviet leader actually met his aim and Romania really acted in accordance with the Soviet intentions at the gathering.[817]

The Political Consultative Committee was held in the Romanian capital on November 25-26, 1976,[818] an abysmal two and half years after its last meeting. A perspective of continuation of détente and the new peace initiatives were the main items on the program. The CPSU CC General Secretary's speech was in the limelight again. Like in 1974, the other leaders confined themselves to unqualified consent with his claims or traditionally declared their support for further strengthening the

---

[816] *Sprawozdanie z wizyty delegacji partyjno-państwowej PRL w ZSRR*, 17. 11. 1976, PZPR KC, XIA/595, AAN, Warsaw.

[817] After a long time, none of the typical disputes about the text of the final documents took place. Despite the fact that Romania put forward some adjustments again, this time the Six considered these constructive rather than obstructive. Some differences were preserved in the form of Romanian unwillingness to harshly condemn the events in West Germany. But in his speech, Ceausescu appreciated the role of the USSR and stated his firm belief about the possibility of additional improvement in Soviet-Romanian cooperation. The Warsaw Pact member-states also noted that the Romanian statement contained a claim about the socialist countries' unity for the first time. The Romanian approach, including providing perfect conditions for the meeting, was publicly praised by Erich Honecker, on behalf of the other delegations. *Zpráva o zasedání politického poradního výboru států Varšavské smlouvy konaném v Bukurešti 25. – 26. 11. 1976*, 1. 12. 1976, 1261/0/7, sv. 23, a. j. 26/1, NA, Prague.

[818] The Political Consultative Committee session took place in the Romanian capital symbolically ten years after issuing the Bucharest declaration which had called for holding the all-European security conference. This fact was massively stressed by the Eastern propaganda, of course. *Informace o ohlasech v BLR na zasedání PPV VS v Bukurešti*, 23. 2. 1977, DTO 1945-1989, i. č. 4, e. č. 43, č. j. 82.386, AMZV, Prague.

alliance.[819] Brezhnev appreciated the international development at the time. He reiterated that the West was finding itself in a complicated position on the international stage, especially regarding the developing world. This fully corresponded with the Soviet strategy in this stage of the Cold War. In terms of increasing Soviet involvement in Africa, Brezhnev spoke against "closing in a narrow circle of the Warsaw Pact." The Eastern Bloc was supposed to develop relations with all the socialist countries.[820] After all, this strategy had been confirmed by the previous participation of Mongolian, Laotian, and Vietnamese delegates at the Deputy Foreign Ministers' meeting in July, which should have harmonized the Warsaw Pact's tactic in the UN. A direct engagement of the alliance in Asia and the Third World remained unreal. Nevertheless, the USSR tried to also coordinate foreign policy with its allies over there.[821] The Political Consultative Committee resolution stated the Warsaw Pact's willingness to support the national liberation movements and to cooperate with them. Thus, as the German question had been solved, the Soviet leadership was thinking about an extension of the geopolitical range of the alliance. It clearly still considered the Pact as the significant tool of influencing the international development. On this purpose, it was also decided to continue the gatherings of the parliaments' representatives under the auspices of the Warsaw Treaty Organization. The cooperation within its framework should have acquired a more civil and systematic dimension. In addition, closer collaboration in the ideological area was supposed to send a demonstrative signal to the West that the alliance was determined to face any interference with the internal affairs of its members. In this regard, Brezhnev's speech appealed for a common fight against the pressure of the Western propaganda in respect for human rights, to which the Eastern Bloc countries calculatedly committed to in the Helsinki Act.

---

[819] *Zpráva o zasedání politického poradního výboru států Varšavské smlouvy konaném v Bukurešti 25. – 26. 11. 1976*, 1. 12. 1976, 1261/0/7, sv. 23, a. j. 26/1, NA, Prague. This scheme became already typical. But in the second half of the 1970s, speeches by the Soviet satellites' leaders were increasingly only flattering features. Nowak, *Od hegemonii do agonii*, 51.

[820] *Projev vedoucího sovětské delegace na zasedání PPV Varšavské smlouvy*, 25. 11. 1976, 1261/0/7, sv. 23, a. j. 26/1, NA, Prague.

[821] Wanner, *Brežněv*, 55, 57.

In fact, the Political Consultative Committee resolution also formally advocated the respect for human rights, however in passing only. But the document mostly exalted the contribution of previous appeals by the Warsaw Pact supreme body to the process of easing international tensions.[822]

Along with the planned peace initiatives, Brezhnev supported demands for the simultaneous dissolution of NATO and the Warsaw Treaty Organization.[823] But it should have just been a political phrase without any impact on the further strengthening of the Eastern alliance. It is not surprising then that Ceausescu especially appreciated the shift to claims about dissolution of the blocs in his speech.[824] As usual, Romania went farthest within the Warsaw Pact; it also repeated a demand of foreign troop- and bases-withdrawal and called for deepening economic cooperation with the West.[825] It is not out of the question that the more accommodating Romanian policy towards the alliance was actually influenced by the Soviet emphasis on the military relaxation. Bucharest saw it as a welcomed opportunity to reduce its own military spending. Ceausescu also appreciated his last bilateral meeting with Brezhnev in Crimea,[826] where the Soviet leader had given his blessing to a more active

---

[822] *Za nové cíle v uvolnění mezinárodního napětí, za upevnění bezpečnosti a rozvoj spolupráce v Evropě. Deklarace členských států Varšavské smlouvy*, 26. 11. 1976, 1261/0/7, sv. 23, a. j. 26/1, NA, Prague.

[823] By this decision, The Political Consultative Committee confirmed the previous conclusions of the conference of the European communist and workers' parties in Berlin which three months earlier supported the Soviet declarative proposal on simultaneous dissolution of the military-political alliances and withdrawal of nuclear weapons from territories of the foreign countries all around the world. Wanner, *Brežněv*, 55.

[824] *Stručná charakteristika vystoupení jednotlivých delegací na zasedání PPV Varšavské smlouvy*, 25. 11. 1976, 1261/0/7, sv. 23, a. j. 26/1, NA, Prague.

[825] The Six considered the Romanian behavior as a good-will gesture and a manifestation of Bucharest's willingness to further development of mutual cooperation in international issues. *SPRAWOZDANIE z narady Doraczego Komitetu Politycznego paśtw-stron Układu Warszawskiego w Bukareszcie*, 27. 11. 1976, PZPR KC, XIA/589, AAN, Warsaw.

[826] Unlike the other Warsaw Pact's party leaders, Ceausescu did not spend his holiday in Crimea. He regularly traveled there to meet Brezhnev, but after the talks the Romanian leader immediately flew back to his country.

Romanian involvement in the issues of both disarmament as well as simultaneous dissolution of NATO and the Warsaw Treaty Organization.[827]

A relatively positive attitude of the U.S. towards a potential establishing of the Organization for Security and Cooperation in Europe perhaps led Brezhnev to the preliminary thoughts that some sort of collective security system could have been created to replace NATO and the Warsaw Pact.[828] In fact, dissolution of the blocs did not exceed the scope of the useful and contributory détente phrase. It was also suggested by the military measures approved at the Political Consultative Committee meeting. Instead of the seriously ill supreme commander, the report on improvement of the Warsaw Pact's military framework so far and outlines of further development of the Unified Armed Forces was presented by the new chief of the alliance Staff, General Anatoly Ivanovych Gribkov. This material was also worked out by the top CPSU leadership. This praxis was preserved even into the future.[829] The presented conclusions were approved without any discussion.[830] The Political Consultative Committee only issued a

---

ИНФОРМАЦИЯ *о крымской встрече и беседе Л.И. Брежнева с Н. Чаушеску I августа 1979 года,* August 20, 1979, Accessed October 16, 2019. http://digitalarchive.wilsoncenter.org/document/114160.

[827] *Informacja o wizycie I Sekretarza KC PZRP tow. Edwarda Gierka w Socjalistycznej Republice Rumunii,* 12. 10. 1976, PZPR KC, XIA/702, AAN, Warsaw.

[828] Zubok, *A Failed Empire,* 229.

[829] In his memoirs, Gribkov states that since the second half of the 1970s at the latest, when he became the Chief of Staff of the Unified Armed Forces, all the drafts of documents for Political Consultative Committee sessions came from the CPSU CC Politburo. This also included the supreme commander's reports. The military institutions only sent some thesis which served as grounds for the formulation of these documents. The Unified Command top officials were present when these materials were being discussed in the Kremlin. However, Gribkov himself stayed in an anteroom and the defense minister, Dmitri Ustinov, conveyed his thoughts to the politburo. Грибков, Анатолий Иванович, *Судьба Варшавского Договора. Воспоминания, документы, факты,* (Москва: Русская книга 1998), 24-25.

[830] *Zpráva o zasedání politického poradního výboru států Varšavské smlouvy konaném v Bukurešti 25. – 26. 11. 1976,* 1. 12. 1976, 1261/0/7, sv. 23, a. j. 26/1, NA, Prague.

nonpublic declaration which set a wide rearmament of the troops as a key task in the upcoming period. The armies should have first created the elite - by recent weapons-equipped units - so-called *yacheykas*. They were supposed to instruct the rest of the alliance formation how to use the sophisticated arms and equipment later. The key investments flew into modern tanks, an air-defense system, and aircraft – all technologically sophisticated as well as very expensive weapons. Also, future improvement of the alliance command structures was vaguely announced.[831]

The first meeting of the Warsaw Treaty Organization supreme body after the Helsinki conference showed that the Pact did not intend to give up on its military role anyway. The decision was traditionally justified with reference to a presumed strengthening of NATO. The goal set by the Warsaw Pact - to stop the arms race and following disarmament - was understood by the Soviet leadership as a long-term process without any immediate impact. On the contrary, the Eastern Bloc states were supposed to get involved mostly in pushing the Soviet proposal, thus obligating the signatories of the Final Act not to use nuclear weapons first. Typically, Moscow issued it on behalf of the entire alliance. It was formally presented as the beginning of an effort that would result in a destruction of nuclear arsenals in the end. A conciliatory tone of the 15th Political Consultative Committee session resolution clearly reflected the atmosphere of climaxing détente as well as Brezhnev's successes in foreign policy. Rather than the previous obligatory and rhetorical assaults against the West, the document consisted, for example, of a neutral announcement of scheduled fights of the socialist countries' astronauts in the Soviet cosmic ships within the "Interkosmos" program.[832]

---

[831] Information by M. Dzúr, the Czechoslovak defense minister, for G. Husák, the CPCz CC First Secretary, January 18, 1977, FMO GŠ/OS, k. 8, p. č. 41, č. j. 39033, VÚA, Prague.

[832] *Za nové cíle v uvolnění mezinárodního napětí, za upevnění bezpečnosti a rozvoj spolupráce v Evropě. Deklarace členských států Varšavské smlouvy*, 26. 11. 1976, 1261/0/7, sv. 23, a. j. 26/1, NA, Prague.

Reform of the alliance political structures in praxis

The decision about establishing the Joint Secretariat and the Committee of Ministers of Foreign Affairs was made already in January 1975. In context of the Warsaw Pact's history, it is not surprising that another 22 months elapsed before the Political Consultative Committee was held and blessed the intention. Moreover, during the meeting in Bucharest, the issue was discussed at length once again.[833] Nevertheless, both bodies were established successfully in the end,[834] yet the member-states failed to define their competences and mechanisms. In this regard, the USSR instrumentally used the hesitating position of Romania. Moscow terminated the discussion on the topic, referring to a possible renewed deadlock in the question.[835] Thus, it did not put forward the Polish proposals, which defined activities of both institutions in detail. Romanian unwillingness to strengthen the competences of multilateral bodies once again played into the hands of Moscow, which itself had no interest in other changes of the Pact's mechanism.[836]

---

[833] *Stručná charakteristika vystoupení jednotlivých delegací na zasedání PPV Varšavské smlouvy*, 25. 11. 1976, ibid.

[834] The Committee of Ministers of Foreign Affairs should have been held according to need, and at least once a year. Gierek's proposal to define the date of these meetings was not approved. All the alliance member-states were supposed to alternately host the sessions in alphabetic order. The hosting country also paid all the costs. The body had no decisive powers. Its task was to consult the international issues, to prepare proper materials for the Political Consultative Committee gatherings and to help implementation of its decisions afterwards. The Committee could have convened the Deputy Foreign Ministers' meeting, if needed. The Joint Secretariat was supposed to provide technical support for the Political Consultative Committee and Committee of Ministers of Foreign Affairs' sessions. It was continually chaired by an elected official and every member-state was represented by one functionary. *Rozhodnutí o vytvoření výboru ministrů zahraničních věcí a spojeného sekretariátu /Přijaté na poradě politického poradního výboru členských států Varšavské smlouvy v Bukurešti, dne 26. listopadu 1976/*, ibid.

[835] *Zpráva o zasedání politického poradního výboru států Varšavské smlouvy konaném v Bukurešti 25. – 26. 11. 1976*, 1. 12. 1976, ibid.

[836] Jarząbek, *"Poland in the Warsaw Pact"*.

As previously explained, Poland was the country eminently interested in reform of the Warsaw Pact's political structures. Edward Gierek personally had high hopes of the new collective political bodies' activities. He believed they could become a new, really effective instrument of cooperation in the field of foreign policy.[837] But the final form of the "reformed" Joint Secretariat did not satisfy Poland. After some short episodes[838] and disputes about its location[839], the body officially started to work in 1977. Warsaw naively hoped that in accordance with the previous Polish proposals of the 1960s and early 1970s, the Secretariat could have played an important role in the alliance mechanism. However, it was limited to dealing with purely technical and organizational questions.[840] The institution lacked any political influence. Its core represented a group of Soviet officials, security, and interpreters who traveled to the meetings of the Political Consultative Committee, Committee of Defense Ministers, and Committee of Ministers of Foreign Affairs. In addition, according to the testimony of the former Polish diplomat Jerzy Nowak, Poland had no own representatives in the Secretariat; even the interpreters from the satellite countries still traveled to the meetings with individual delegations. The Warsaw Treaty Organization General Secretary, Nikolai Firyubin, remained an official

---

[837] *Tezy na rozmowy plenarne w Moskwie*, undated 1976, PZPR KC, XIA/719, AAN, Warsaw.

[838] The Joint Secretariat was established by the Political Consultative Committee decision of April 1974, in fact. Its first action was preparation of the 20th anniversary of the Warsaw Pact foundation and a reform of its political structures. *Zpráva o vyslání čs. delegace na schůzku náměstků ministrů zahraničních věcí Varšavské smlouvy v Moskvě 29.1.-30.1. 1975*, 15. 1. 1975, 1261/0/6, sv. 143, a. j. 146/10, NA, Prague.

[839] Poland proposed the Joint Secretariat to be located in Warsaw or at least meet for the first time there. This was rejected by the Soviet representatives. They told the Polish diplomat they considered it very important that the institution work in Moscow. The Polish side was at least assured that its initiatives in the body would be taken into account. Pilna notatka z rozmowy z wiceministrem Spraw Zagranicznych ZSRR, tow. N.N. Rodionowem w sprawie przygotowań do jubileuszowej sesji DKP UW (Moskwa, dnia 28.XI.1974), November 30, 1974, in *PRL w politycznych strukturach* (doc. 36): 325.

[840] Jarząbek, "*Poland in the Warsaw Pact*".

without any political competences and had no right to act on behalf of the alliance.[841]

Cold showers - not only for Polish expectations - were especially practical features of the first Committee of Ministers of Foreign Affairs' session held on May 25-26, 1977, in the Soviet capital. The delegation of Poland as well as its East German and Czechoslovak counterparts wanted to present their specific demands concretely, to start a serious debate on the possibility of disarmament in conditions of détente or the Warsaw Pact's policy towards developing countries. However, Moscow showed an absolute disinterest. The discussion was limited to a very general level. Individual member-states just informed each other about progress in their effort to implement the Helsinki Act measures. No key decisions were made. The unanimously approved resolution, in fact, reflected exclusively an actual foreign political line of Moscow.[842] In backrooms, the Soviet diplomats used the standard excuses referring to potential Romanian obstructions, but for example, the Polish delegation did not consider them convincing any more. After he came back from the meeting, the new Polish Minister of Foreign Affairs, Emil Wojtaszek, voiced fears that despite establishing the collective body, the USSR intended to return to the praxis so far, when the issues were pre-negotiated bilaterally in a dubious way, instead of being discussed in multilateral institutions.[843] Thus, setting up the Committee of Ministers of Foreign Affairs did not cause any radical change of the Warsaw Pact's mechanism. On the contrary, an existence of the body provided Moscow with a suitable tool against potential complaints by the satellites about their limited influence on the alliance activity.[844] Moreover, the mechanism as well as the real role of this forum remained unclear. The USSR dealt with this problem

---

[841] Nowak, *Od hegemonii do agonii*, 48.
[842] *Notatka informacyjna z pierwszego posiedzenia Komitetu Ministrów Spraw Zagranicznych państw - stron UW w Moskwie w dniach 25-26 maja br.*, May 27, 1977, Accessed October 16, 2019. http://www.php.isn.ethz.ch/kms2.isn.ethz.ch/serviceengine/Files/PHP/111538/ipublicationdocument_singledocument/eebc17 17-1814-4f01-bcad-b8f17dbbaaf0/pl/1977_Meeting_CMFA.pdf.
[843] Jarząbek, *PRL w politycznych strukturach*, 41.
[844] Tejchman and Litera, *Moskva a socialistické země na Balkáně*, 88.

with the vague claim that everything would be solved spontaneously by establishing praxis during the following meetings.[845]

In fact, the importance of the Committee of Ministers of Foreign Affairs did not exceed previously unattached multilateral gatherings of the heads of diplomatic resorts.[846] In reality, the purpose of its first session was solely to unify the Warsaw Treaty Organization member-states' positions ahead of the CSCE summit in Beograd. A joint line was practically defined by the opening speech of Andrei Gromyko. The aim was to assert the conclusions of the previous Political Consultative Committee's resolution, especially a treaty on renunciation of the first nuclear strike or to stop expansion of existing military blocs as a first step towards their future dissolution.[847] In the center of the Warsaw Pact's attention remained the military relaxation. However, Moscow alone wanted to define the parameters of this process. It strictly limited any activities of its satellites going beyond the scope of propagation of the alliance proposals. At the time, almost all the Pact members showed their

---

[845] Czechoslovakia unsuccessfully proposed a regular term for the Committee sessions, for example, during the spring every year. *Notatka informacyjna z pierwszego posiedzenia Komitetu Ministrów Spraw Zagranicznych państw - stron UW w Moskwie w dniach 25-26 maja br.*, May 27, 1977, Accessed October 16, 2019. http://www.php.isn.ethz.ch/kms2.isn.ethz.ch/serviceengine/Files/PHP/111 538/ipublicationdocument_singledocument/eebc1717-1814-4f01-bcad-b8f17dbbaaf0/pl/1977_Meeting_CMFA.pdf.

[846] Anna Locher is wrong in her claiming that the anchoring of the Ministers of Foreign Affairs' meetings was the Soviet reaction to their success in the détente process. In fact, it was rather an implementation of the efforts which were rooted back to the mid-1960s. We are reminded that until 1974, Romania blocked any attempts to realize this intention. Furthermore, Locher speculates that the establishing of the Committee of Ministers of Foreign Affairs was related to the Soviet fears of too much independence of the Warsaw Pact member-states in the international environment. In reality, the Soviet satellites had been already bounded enough by the existing alliance mechanisms in the first half of the 1970s. From the Soviet point of view, establishing of the body thus strengthened an outward impression that the Warsaw Pact was a real alliance and in this manner had to be better understood in the next stages of détente. Moscow actually still preferred negotiations on the basis of the blocs which were much easier to control. Locher, "*Shaping the Policies of the Alliance*".

[847] *Informace o zasedání Výboru ministrů zahraničních věcí Varšavské smlouvy*, 7. 6. 1977, 1261/0/7, sv. 41, a. j. 46/info7, NA, Prague.

interest to engage in disarmament talks.[848] Traditionally, the Romanian proposals went farthest. As implementation of the Final Act's measures had so far disappointed the Ceausescu regime,[849] its initiatives increasingly contradicted the Warsaw Pact's unified policy. On the contrary, these began moving closer to the Non-aligned Nations. The only exception was the issue of respect for human rights. In this area, given its nature, the Romanian regime of course acted in line with the other Soviet satellites.[850] Considering the rise of domestic opposition movements, the human rights phenomenon was more and more acute for the Eastern Bloc countries. Thus, at the Committee of Ministers of Foreign Affairs' meeting, for example, Bohuslav Chňoupek, the head of the Czechoslovak diplomatic resort, informed the allies about the activities of Charta 77. He blamed NATO for its plotting and urged the Warsaw Pact to fight against the Western *"pseudo-arguments on the human rights violation."*[851] Moscow also considered the emphasis given on human rights by the new U.S. president, James Carter, as a direct attack on the legitimacy of its sphere of influence.[852]

---

[848] For example, at the Committee of Ministers of Foreign Affairs, the Czechoslovak normalization regime, which in fact remained passive on the international stage in the first half of the 1970s, also appreciated some of the Western proposals on confidence-building measures in the military area. *Notatka informacyjna z pierwszego posiedzenia Komitetu Ministrów Spraw Zagranicznych państw - stron UW w Moskwie w dniach 25-26 maja br.*, May 27, 1977, Accessed October 16, 2019. http://www.php.isn.ethz.ch/kms2.isn.ethz.ch/serviceengine/Files/PHP/111538/ipublicationdocument_singledocument/eebc17 17-1814-4f01-bcad-b8f17dbbaaf0/pl/1977_Meeting_CMFA.pdf.

[849] At the meeting, George Macovescu, the Romanian Minister of Foreign Affairs, objected the alleged overestimation of the Final Act's implementation. Romania wanted to stand up for significant shifts in this area at the upcoming CSCE summit in Beograd.

[850] *Informace o zasedání Výboru ministrů zahraničních věcí Varšavské smlouvy*, 7. 6. 1977, 1261/0/7, sv. 41, a. j. 46/info7, NA, Prague.

[851] *Vystúpenie B. Chňoupka na zasedání ministrov zahraničných vecí Varšavskej zmluvy dňa 25. mája 1977,* ibid.

[852] Munteanu, Mircea, "The Beginning of the End for Détente: The Warsaw Pact Political Consultative Committee," Cold War International History Project e-Dossier, no. 24 (2011).

An effort to move détente towards a desirable direction, especially on the issues of renunciation of first nuclear strike and symmetrical disarmament, should have also been supported by the second gathering of the representatives of the Warsaw Treaty Organization member-states' parliaments.[853] It took place on July 5-6, 1977, in Leningrad, purely instrumentally just ahead of the CSCE summit in Beograd. In accordance with the Soviet strategy to widen the Warsaw Pact's range, the observers of Mongolia, North Korea, Vietnam, and Cuba were invited to the meeting. But only the delegation of "the liberty island" attended. In fact, a majority of the speeches just repeated claims which could have been already heard during the previous Warsaw Pact sessions.[854] Nevertheless, an idea appeared that the gatherings of the parliaments should have become a sort of unofficial advisory body of the Political Consultative Committee. Bulgaria especially, loudly tried to improve such a cooperation. The country also proposed widening the scope of consultation under the auspices of the Warsaw Treaty Organization to the level of the parliamentary committees, depending upon the issues to be discussed. But this initiative remained without any response. Nevertheless, some regular, however not exactly specified meetings, were expected. This shows that the USSR intended to use the Warsaw Pact as a cover for almost all the diplomatic activities of its satellites towards the West. The contacts between individual Eastern Bloc countries' legislature presidiums and their Western counterparts were obviously increasing at the time.[855] Moscow probably planned to use a traditional method: The parliament

---

[853] *Materiał informacyjny z polecienia tow. E. Babiucha*, 3. 6. 1977, PZPR KC, XIA 560, AAN, Warsaw.

[854] Especially emphasizing the need for continuation of the process of easing international tensions, an approval of concrete measures at the ongoing disarmament talks, and a criticism of the campaign pointing to violation of human rights in the Eastern Bloc were voiced. Romanian behavior differed marginally. Unlike the first meeting two years earlier, Bucharest even sent the chairman of its National Assembly, Nicolae Giosan. Despite that, Romania initially strictly rejected admitting that the meeting had any connections to the Warsaw Treaty Organization; it surrendered after the Soviet diplomatic intervention.

[855] *Zpráva o průběhu a výsledcích konzultativního setkání představitelů parlamentů členských států Varšavské smlouvy v Leningradě ve dnech 5. a 6. července 1977*, 1261/0/7, sv. 46, a. j. 51/7, NA, Prague.

representatives' meetings under the alliance auspices should have defined a joint policy - in fact shaped by the Kremlin - for the Eastern legislatures in their negotiations with the West.[856] Thus, this forum in reality limited again the Warsaw Pact member-states' opportunities to act more independently in détente process. Therefore, it is not surprising that the Leningrad gathering's resolution addressed to the parliaments of the signatory countries of the Helsinki Act, only repeated actual claims and initiatives of the Soviet foreign policy.[857]

Slow fall of détente

The Warsaw Treaty Organization countries expected they would use the aforementioned CSCE summit in Beograd to officially put forward the drafts of the treaty on renunciation of first nuclear strike as well as suspending validity of both NATO and the Warsaw Pact's articles which talked about the possibility to enlarge the alliances by new members. These were in praxis the recent initiatives of the Political Consultative Committee. However, the Eastern and Western conceptions of détente began to differ significantly in the second half of 1977. The effort to push forward a no first use policy was appreciated, especially by the neutral countries and the Western peace movement. NATO, still relying on its slight nuclear superiority, refused this for the understood reasons.[858] Rather than pompous but questionable peace initiatives, the Western alliance sought to focus on concrete confidence-building measures in the military area. The parameters for the announcement of drills in advance should have been tightened alongside improving the conditions for

---

[856] In this regard, it is good to remember that the parliaments in the Leninist-Stalinist regimes in the countries of the Soviet sphere of influence largely played a puppet role. All the key decisions were made in the party apparatus and legislatures only formally approved the proposed laws.

[857] *Provolání k parlamentům a členům parlamentů zemí, které podepsaly Závěrečný akt konference o evropské bezpečnosti a spolupráci, přijaté dne 6. července 1977 v Leningradě na závěr konzultativního setkání představitelů parlamentů členských států Varšavské smlouvy*, 1261/0/7, sv. 46, a. j. 51/7, NA, Prague.

[858] *Tezy na rozmowy plenarne*, undated 1977, PZPR KC, s. XIA/788, AAN, Prague.

participation of the other side's observers. NATO countries also proposed the announcement of movement of huge military contingents and publishing the military budgets. These did not correspond with the Soviet aims in this stage. Therefore, the Warsaw Pact members should have rejected NATO proposals, declaring that these substantially exceeded the level of existing trust between the blocs.[859] In reality, the East was driven into a defensive position. Its effort to exclude some contentious items from the agenda, especially the issue of human rights, and "to focus on more important themes" failed. The Soviet delegate, Yuri Voroncov, even had to threaten many times that he would leave the meeting in protest.[860]

The CSCE summit in Beograd took place in the moment when the apparently developing Soviet-U.S. relations began to cool down rapidly. This process climaxed after the Soviet invasion of Afghanistan, resulting in American sanctions against the USSR as well as NATO military countermeasures.[861] Since 1975, détente started being a popular target of criticism from both political camps in the United States.[862] The real deterioration of the U.S.-Soviet relations began in the late months of the following year. It is obvious that despite some doubts in the Soviet leadership, the United States was first to begin reviewing the attitude towards the process of relaxation in the second half of the 1970s.[863]

Interpretations of this move differ. The first theory claims that the USSR failed to adjust its policy to the spirit of détente. The country refused to accept the concept of mutual assured destruction and remained oriented

---

[859] *Výsledky přípravných jednání a postup čs. delegace na bělehradské schůzce představitelů účastnických států Konference o bezpečnosti a spolupráci v Evropě*, 2. 9. 1977, 1261/0/7, sv. 49, a. j. 53/8, NA, Prague.

[860] Wanner, *Brežněv*, 58.

[861] Westad, Odd Arne, "The Fall of Détente and the Turning Tides of History," in: The Fall of Détente. Soviet-American Relations during Carter Years, ed. Odd Arne Westad (Boston: Scandinavian University Press, 1997), 4.

[862] Zubok, *A Failed Empire*, 228.

[863] The key discussion about relations between both blocs took place in the Kremlin during the years 1966-1971. In that period, Brezhnev was just building his position within the Soviet politburo. Bowker, "Brezhnev and Superpowers Relations," 92.

to a victory in nuclear war.[864] Moscow tried to spread communism by force through its activities in the Third World and used weakening of the United States in the international environment to intensify the "class-struggle." According to this concept, the USA was not able to continue in détente and it was the right decision to abandon it after the invasion of Afghanistan. The second, more rational view, ascribes the fall of détente to the absolutely different aims pursued by both superpowers in this process. Moscow strove for confirmation of its position as the global power equal to the United States. Washington, on the contrary, tried to face the Soviet ambitions. The incompatibility of these positions logically resulted in new growing tensions. Actually, the undoubted Soviet successes instigated the beginning of discussions on the U.S. political stage about the correctness of foreign policy up to this point. Détente clearly did not have the effect that Nixon and Kissinger expected at its start. Domestic troubles made the USA unable to catch up with the initiative of the USSR in this area in the first half of the 1970s. On the other hand, the Soviet side in somewhat simplified terms considered the fall of détente mostly as the result of the rise of the so-called New Right and the American military-industrial lobby.[865]

In reality, the Soviet-U.S. relations deterioration after 1977 was caused by a broad complex of factors: It undoubtedly consisted of the Soviet actions in Africa, slow and deadlocked talks on armament control and increasing anti-Soviet feelings in the American political spectrum.[866] The election of James Carter as the U.S. president also meant an important moment. His declarative emphasis on respect for human rights combined with the changing balance of military strength between NATO and the Warsaw Pact fueled the West-East tensions. Continually strained Soviet-Chinese relations also increased the growth of global instability.[867]

---

[864] We are reminded that this concerned the Soviet army command only. On the contrary, the key officials in the CPSU leadership, including Brezhnev, officially promoted the claims about impossible victory in a nuclear war, as explained in the previous chapter.

[865] Bowker, "Brezhnev and Superpowers Relations," 103-104.

[866] Zubok, *A Failed Empire,* 257.

[867] Munteanu, "The Beginning of the End for Détente".

The concrete military build-up by the Warsaw Treaty Organization, which obviously also contributed to the fall of détente, is analyzed in a special chapter. Considering the aforementioned effort to expand the geopolitical range of the alliance to the Third World areas, it is necessary at least briefly to explain the Soviet military engagement in the developing countries at the time. During the 1970s, the USSR was clearly increasing its influence outside Europe. The beginnings of this process date back to Khrushchev's era. But no sooner than a decade after the fall of Stalin's successor, the Soviet Union was able to act as a really important player in the Third World.[868] Since the mid-1970s, it got involved in Angola, Mozambique, Ethiopia, Afghanistan, South Yemen, Nicaragua, and Granada. The aims pursued by Moscow are disputed. The definite explanation would require analyzing the primary Soviet documents. Some analytics claim that the CPSU leadership never gave up the thesis about the class struggle. It began to take more initiative, because this was allowed by the growth of Soviet military strength. According to that concept, détente only opened room for the Soviet expansion. The second theory suggests that the Soviet involvement in the Third World did not reflect an expansionism. The higher activity of the USSR in those areas is seen as a result of international development, which could not have been influenced by Moscow directly. The Soviet leadership only tried to support and preserve the leftist regimes and groups, which achieved power largely alone. In this regard, a bitter lesson from the Pinochet coup in Chile played an important role.[869] Moreover, in accordance with their world-view, some part of the Soviet politburo thought that the USSR had a historical obligation to support anti-capitalist regimes, like Cuba or Vietnam. One of

---

[868] Webber, Mark, "'Out of Area' Operations: The Third World," in *Brezhnev Reconsidered*, eds. Erwin Bacon and Mark Sandle (New York: Palgrave, 2002): 112.

[869] The American support for the coup in Chile, where General Augusto Pinochet toppled the leftist government of Salvador Allende in September 1973, outraged the Soviet leadership. The event allowed some radically oriented Soviet party-members to criticize again the claims about a peaceful road to socialism. Some part of the CPSU perhaps interpreted the events in the South-American country in the way that the leftist regimes needed the Soviet military equipment and assistance in order to survive. Bowker, "Brezhnev and Superpowers Relations," 101.

the main protagonists of this policy was initially Yuri Andropov.[870] But Soviet actions cannot be seen as an attempt to expand international communism, as Moscow got involved solely in the areas where the U.S. had abandoned their positions and had no interests there anymore. From this point of view, the USSR considered its activities neither dangerous nor escalating international tensions. The third hypothesis claims that the Soviet Union, in reality, was not more active in the Third World in the 1970s. Its actions just became more effective due to a certain paralysis of U.S. policy after the Watergate scandal and the Vietnam War.[871]

In fact, the Soviet policy remained largely defensive. The period between the invasions of Czechoslovakia and Afghanistan was the most peaceful era in the Soviet Union's existence. Despite this fact, the era is often called "a decade of stagnation"; in the context of Soviet history, the importance of a certain stability should not be underestimated. One of the key factors, which allowed such a state, was undoubtedly achieving a general strategic parity with the United States.[872] In reality, Moscow was losing its important allies in this period. It had to swallow the most bitter pill in 1976, when Egypt terminated the treaty on mutual friendship and cooperation. In addition, the country banned Soviet military vessels from using its ports. On the contrary, two years later, Anvar Sadat made a peace deal with Israel with American assistance. The Camp David agreement, in fact, only further confirmed the marginalization of the Soviet Union in the region of Middle East.[873] It was a definite retreat of Moscow from this strategic area. Till the end of the 1970s, Somalia and Sudan also escaped the Soviet influence. Therefore, considering the actions of the USSR in the Third World, one can speak about the deliberate "export of revolution." These anabases cannot be overestimated. Furthermore, the West remained dominant and the most active power in the Third World. Only one third of the developing countries were seeking an alternative for capitalism. Moreover, to proclaim Marxism-Leninism was for many regimes simply a tool for getting access to welcomed Soviet military assistance. After all,

---

[870] Westad, "The Fall of Détente," 19.
[871] Bowker, "Brezhnev and Superpowers Relations," 101-102.
[872] Volkogonov, *The Rise and Fall*, 276.
[873] Webber, "'Out of Area' Operations," 123.

this was also admitted by the official U.S. report published at the beginning of the 1980s. The Soviet global position was pronounced weaker than it had been during Khrushchev's era.[874]

The Soviet ideas about parameters of the military relaxation were clear. Moscow saw the center of this process in nuclear disarmament. The military elites informed the political leadership that despite significant advances in development of atomic weapons, the USSR would not achieve a desirable parity in this area until 1979.[875] As explained later, Moscow did not intend to weaken the Warsaw Pact's conventional strength until some clear disarmament agreements were concluded. The U.S. President Carter was inclined towards further disarmament at the time he joined the office. But simultaneously he strove for shifting existing boundaries of the Cold War reality. Among other things, Carter promised more transparent foreign policy.[876] This Kissinger-era approach, which substantially differed from Nixon, caught Moscow by surprise. As the Soviet ambassador to the United States Anatoly Dobrynin remembered, the Kremlin had not been able to understand why the new U.S. President repeatedly publicly attacked it when the USSR had been ready to make some guarantees on the issue of the arms race control. However, Carter believed that alongside further reduction of nuclear arsenals or deepening diplomatic and economic contacts with Moscow, it was also necessary to push for a change of the internal situation in the USSR, and by extension, the other Eastern Bloc countries. They should have converged with the American ideals. For this reason, he put a strong emphasis on respect for human rights.[877]

Furthermore, in the United States existed fears whether the Kremlin intended to test the new president in a similar way like it had done when J. F. Kennedy had come to power. However, Brezhnev assured Carter there was no such a risk. On January 18, 1977, symbolically on the inauguration day of the new White House head, he gave a breaking speech in Tula. For the first time the Soviet leader supported a strictly defensive

---

[874] Bowker, "Brezhnev and Superpowers Relations," 102.
[875] Mastny, "Imagining War in Europe," 35; Westad, "The Fall of Détente," 31.
[876] Zubok, *A Failed Empire,* 254.
[877] Westad, "The Fall of Détente," 9, 17.

strategy. The USSR did not seek a global supremacy that would have allowed it to strike the first nuclear blow to the United States, he declared; the Soviet aim was solely to keep defense capabilities strong enough to deter aggressive intentions of a potential enemy. The General Secretary probably hoped that the speech would calm down the ongoing campaign in the U.S. warning against the Soviet military threat and would help the new President to repulse the American hawks.[878] Thus, the Soviet policy officially aimed to reduce the arms race and confrontation, but the Eastern defense capabilities should have remained intact. Carter welcomed Brezhnev's message. Nevertheless, he immediately submitted his own, more ambitious alternative: He proposed to move from SALT II negotiations to much more radical disarmament moves, especially to achieve a breakthrough in the Vienna talks on the reduction of conventional arms in Central Europe.[879] For the above-described reasons, Carter's conception was unacceptable for Soviet leadership. Therefore, it inevitably led to paralysis of the entire disarmament negotiations.

Given the situation, the Warsaw Pact's political activity focused on nuclear disarmament. Its perspectives were also the key item of the second Committee of Ministers of Foreign Affairs meeting in Sofia, April 24-25, 1978. The member-states agreed that at the scheduled UN General Assembly session, they would demand holding the worldwide conference on disarmament. The aim was to create a parallel platform to the blocked Vienna talks and the very slowly proceeding SALT II negotiations,[880]

---

[878] Zubok, *A Failed Empire,* 254-255.

[879] Durman, *Útěk od praporů,* 185, 190.

[880] The intention to conclude the SALT II agreement was already approved in 1974. It was one of Brezhnev's biggest personal achievements. The part of the politburo, especially Marshal Grechko, fiercely opposed such a move. The General Secretary was able to make a deal with maximum effort only. According to testimony of his coworkers, it is not surprising that soon after his health deteriorated rapidly. Nevertheless, the SALT II negotiations proceeded with serious troubles. In February 1977, Brezhnev, on Gromyko's advice, wrote to Carter he would meet him personally only if the new treaty was ready for signing. Due to this reason, both politicians did not meet until December 1979. At the time of this summit in Vienna, Brezhnev's health as well as his mental condition were already broken. This fact also limited the possibility of successful Soviet-U.S. dialogue. Zubok, *A Failed Empire,* 257; Westad, "The Fall of Détente," 12, 15.

where some sort of deal corresponding to the Soviet intentions would have been made. Fully in accordance with the traditional praxis, the Warsaw Pact's actual strategy was defined by Gromyko's speech. The alliance was supposed to get involved in support for Soviet initiatives on terminating production of atomic ammunition and gradual reduction of its stockpiles,[881] an effort to expand the global range of the Non-Proliferation Treaty - a general ban on military nuclear tests or on the development of new types of weapons of mass destruction. It was a direct reaction to the planned usage of a neutron bomb by the U.S. army to stop potential advancing of the Warsaw Pact's offensive.[882] The member-states' approach towards SALT II and the Vienna talks should have differed: In the first case, they were supposed to appeal for a successful conclusion of the negotiations. But in the second, they should have just propagandistically made assurances about the "constructive approach" of the socialist countries on that forum. This is more proof that according to Moscow, conventional disarmament was not on the table at the moment.[883]

In this situation, when it became clear that détente was stagnating at best, the Soviet satellites presented different views on how to face the problem. Traditionally, Poland tried to be very active. Emil Wojtaszek, the Minister of Foreign Affairs, appealed for improving relations between the socialist countries and the West, including NATO members. Also, more frequent consultations within the Warsaw Treaty Organization - on the

---

[881] This appeal was presented by Leonid Brezhnev during celebration of the 60th anniversary of the Bolshevik coup in Russia.

[882] *Prejav ministra zahraničných vecí ZSSR s. A.A. Gromyka*, 24. 4. 1978, 1261/0/7, sv. 73, a. j. 78/info2, NA, Prague. At the end of the 1970s, the U.S. armed forces came up with a concept of nuclear weapon marked as ERW (Enhanced Radiation Warhead), which soon became known as a neutron bomb. It was based on a special nuclear reaction minimizing the destructive strength of a thermonuclear explosion while causing extreme radioactive contamination which maximized its deadly impact on the human forces. The weapon was intended as an effective device against the fearful Warsaw Pact's attack panzer battalions. In addition, deep psychological impact on the enemy troops was expected. Walsh, David M., *The Military Balance in the Cold War. US perception and policy, 1976-1985* (Abingdon: Routledge 2008), 92-93.

[883] *Informace o 2. zasedání Výboru ministrů zahraničních věcí členských států Varšavské smlouvy*, 12. 5. 1978, 1261/0/7, sv. 73, a. j. 78/info2, NA, Prague.

level of either the Committee of Foreign Affairs or the Political Consultative Committee - should have contributed to preserving détente. His Hungarian counterpart, Frigyes Puja, rather urged the orientation on to the neutral states. The East-German regime acted more aggressively. Given the situation, Oskar Fischer once again warned against "reactive NATO circles", with obvious reference to the West-German policy.[884] The fact that this approach was outdated and excessive was proven by the Soviet unwillingness to include any verbal assaults on the West in the final communiqué of the Committee of Ministers of Foreign Affairs' meeting. Considering the existing situation, Moscow used rather moderate and accommodating rhetoric.[885]

The recent development deeply unsatisfied Bucharest. The Romanian Minister of Foreign Affairs, Ştefan Andrei, clearly stated that the détente process had stopped and international tensions and military instability were rising again. Using strictly ideological Marxist-Leninist language, he declared skepticism about the results in the field of political relaxation so far. Aspects of the third basket of the Final Act as well as minor progress in disarmament especially bothered the Ceausescu regime. At the meeting in Sofia, Romania therefore presented a quite radical proposal on freezing the military budgets at the level of the year 1978. Till 1985, they should have been gradually reduced by 15 percent. An interesting factor of this initiative was the distribution of saved monies: The countries themselves were supposed to keep only half of the amount and to give the rest to the brand-new UN fund for assistance to the developing countries with GDP less than 600 dollars per capita. The Romanian disarmament proposals did not contradict all the aspects of the Warsaw Pact's conception. However, they were more extensive. Therefore, Bucharest decided to act separately during the upcoming international talks. Thus, not only did the issue of disarmament negotiations being at a dead end deteriorate the inter-bloc relations, but it also led to another cooldown in Romania-Warsaw Pact cooperation. As a result, after almost two and half years, the country again

---

[884] Ibid.
[885] *Komuniké ze zasedání výboru ministrů zahraničních věcí členských států Varšavské smlouvy*, 25. 4. 1978, ibid.

significantly corrected the alliance session resolution. The Romanian delegation constantly tried to incorporate its maximalist demands there. They threatened refusal to sign the document if it did not contain these.[886]

The special meeting of the UN General Assembly on disarmament brought no encouraging result from the Eastern point of view. The fact was that the resolution only partially corresponded with the socialist countries' thesis; it did not reflect the real course of the session. Moscow's defensive strategy was underlined by Gromyko's pledge that the USSR would never use its nuclear weapons against the states which had no such an arsenal or arms located on their territories. In this way, the Soviet Union also indirectly urged European NATO members to reject potential American demands on dislocation of new nuclear missiles on the Old Continent. Washington did not reflect the Soviet proposals. Furthermore, it also refused the draft of a ban on neutron weapons. On the contrary, in case of attack on the United States or their allies it reserved its right to use these arms. Romania confirmed its distance from the Warsaw Treaty Organization. Not only did Ceausescu's regime not respect the previously-set joint line of the alliance, but it also indirectly criticized the Soviet and Cuban activities in Africa.[887]

Despite various fails and deteriorating international atmosphere, Moscow still believed that the Eastern Bloc could set the main tone in the détente process. The Kremlin decided to tighten its approach. After two years, the Political Committee was held and diction of the final declaration would be significantly sharper. Traditionally, a draft of the document was worked out in Moscow.[888] It was based on the Kremlin's claim that the process of relaxation was facing problems mostly due to poor results in

---

[886] *Informace o 2. zasedání Výboru ministrů zahraničních věcí členských států Varšavské smlouvy*, 12. 5. 1978, ibid.

[887] *Informace o průběhu a výsledcích zvláštního zasedání Valného shromáždění OSN o odzbrojení*, 28. 7. 1978, sv. 79, a. j. 82/info11, ibid.

[888] *Vermerk über ein Gespräch des Generalsekretärs des ZK der SED, Genossen E. Honecker, mit dem Mitlied des Kollegiums des Außenministeriums der UdSSR, Leiter der 3. europäischen Abteilung, Genossen Bodnarenko, am 25. 10. 1978*, DY 30/2351, BArch, Berlin.

the area of disarmament.[889] However, during redaction of the text, this time the USSR took more of a partnership approach towards its allies than in previous cases.[890]

At the meeting itself, which took place on November 22-23, 1978 in Moscow, Brezhnev's speech fully reflected the deteriorating relations between two superpowers. The General Secretary warned against the increasing American military budget reaching $130 billion per year. According to Brezhnev, growing military spending suggested that the U.S. was developing new sophisticated weapons.[891] In this atmosphere, the Soviet leader anticipated an assault of some Western politicians on détente as well as the socialist states' policy. Once again, he declaratively denounced the arms race and advocated dissolution of the military blocs. At the same time, however, he declared that in the situations where the West was not reflecting Soviet incentives, it was necessary to do all that was possible for maintaining the existing correlation of forces. Brezhnev

---

[889] *Zasedání politického poradního výboru Varšavské smlouvy*, 3. 11. 1978, 1261/0/7, sv. 85, a. j. 89/18, NA, Prague.

[890] However, also in this case, the note by J. Nowak, the Polish diplomat, is true that some sops and certain room for corrections of the Warsaw Pact's documents provided by the USSR should have only strengthened its perception as an older partner and a "brother" who also took into account another's will. Nowak, *Od hegemonii do agonii*, 53. The alliance member-states' declaration as well as the communiqué of the meeting was prepared by the Soviet leadership three weeks before the session. Moscow presented it as largely a working draft for formulating a final version of the documents. Initial corrections of the text took place at the bilateral talks between the Soviet Union and individual member-states of the alliance. Just ahead of the Political Consultative Committee session an additional multilateral meeting of deputy foreign ministers was held. But no significant adjustments were made there. The deputies called the text "good". *Report by Hungarian Deputy Foreign Minister István Roska on the Meeting of the Warsaw Pact Deputy Foreign Ministers*, November 16, 1978, Accessed October 16, 2019. http://www.php.isn.ethz.ch/kms2.isn.ethz.ch/serviceengine/Files/PHP/17383/ipu blicationdocument_singledocument/c60c974e-4935-4a6e-8187-d86993377706/en/781116_Report_E.pdf.

[891] In fact, the Carter administration increased defense spending. In August 1977, it blessed the development of a B-2 bomber invisible to radar as well as the programs of Peacekeeper intercontinental ballistic missile and submarine MIRVs Trident. However, the President internally admitted he could have stopped the latter two if some disarmament agreement with the USSR was concluded. Westad, "The Fall of Détente," 17.

informed the allies about significant progress in SALT II negotiations. He hoped optimistically that the treaty would be ready to sign by one month. On the contrary, he adversely commented on the Vienna talks. The Warsaw Treaty Organization had made many steps towards breakthrough in deadlocked negotiation, but there was no Western reaction, he complained. The Soviet leader revisited argumentation by NATO, which had rejected a symmetric troop reduction by pointing out the existing conventional advantage of the Warsaw Pact. Therefore, he was thinking whether concluding some non-aggression and no-first-use treaties between both alliances would remove such an obstacle. This tactic proves that in the first stage, the USSR planned to limit its actions to great gestures and pompous agreements which would not have weakened its strength in praxis. Despite this, NATO itself admitted the possibility of concluding some sort of mentioned agreement during the CSCE summit in Beograd in 1977,[892] one year later, such a conception of the military relaxation was no longer acceptable for the United States. Given the situation, it would have undermined credibility of the American guarantees to Western Europe, as nuclear deterrence remained the core of NATO strategy.[893]

It is likely that Brezhnev strongly underestimated a doggedness of the American side. This is proven also by his approach towards the Soviet mid- range missiles SS-20. After 1976, the USSR started to replace old types SS-4 and SS-5 with them. Neither SALT II negotiations nor the Vienna talks concerned this category of weapons.[894] The General Secretary was aware that these very accurate and destructive arms were raising fears among the top West-European politicians. But at the Political Consultative Committee session, he informed the Warsaw Pact member-states that the Soviet side intended to keep this weapon as an "ace in the sleeve." Unrealistically, it wanted to connect potential negotiation on withdrawal of these missiles with the issue of dismantling the American

---

[892] Wanner, *Brežněv*, 58

[893] Mastny, *A Cardboard Castle?*, 45.

[894] SS-20 missiles were supposed to replace about 500 mid-range rockets SS-4 and SS-5. Unlike their predecessors, these carried three warheads instead of one. Śluszarczyk, *Układ Warszawski,* 78.

military bases in Europe.[895] Given the existing situation, such a measure appeared to be totally utopian.

The USSR did not leave anything to chance. As already explained, even in the process of détente, the Soviet leadership put a strong emphasis on own military strength. It is not surprising that at the Political Consultative Committee in Moscow, in the situation when indications about increasing NATO military capabilities existed, it tried to prompt the rest of the Warsaw Pact to another extensive modernization of the Unified Armed Forces. In this way, the alliance should have faced the actual challenges, in praxis. In the light of international development, Brezhnev urged the Soviet satellites to significantly increase their military investments.[896] To some extent, the demand on intensified armament was linked to the more active Soviet approach towards the developing world, but was mostly a reaction of Moscow to a new generation of technologically sophisticated Western weapons.[897] The previous meeting of the North Atlantic Council, held in May 1978, served as a pretext. The new Warsaw Pact supreme commander, Marshal Victor Kulikov, warned the Political Consultative Committee that at the gathering, NATO had approved increases of its troop numbers and intensified preparations for an eventual armed conflict in Europe.[898] The Warsaw Pact's countermeasures should have been complex. These insisted not only on strengthening troops through weaponry modernization, but also making some changes in the command structures.[899] This paved the way for a very

---

[895] Speech by Brezhnev at the Political Consultative Committee Meeting in Moscow, November 22, 1978, in *A Cardboard Castle?* (doc. 84): 418-421.
[896] Munteanu, "The Beginning of the End for Détente"; Luňák, *Plánování nemyslitelného*, 63-64; Mastny, *A Cardboard Castle?*, 48.
[897] Jarząbek, *PRL w politycznych strukturach*, 78-79.
[898] NATO justified its intention to increase its defense budget by 3% by the fact that between 1968 and 1976, the "real" GDP-related funding of military activities decreased in the West, but in the Warsaw Pact member-states rose by 4-5 % annually. *Telegram number 1350 of 8 December*, FCO 28/3273, TNA, London.
[899] *TÉZE DOKLADU hlavního velitele Spojených ozbrojených sil maršála Sovětského svazu V. G. Kulikova na zasedání Politického poradního výboru "O stavu a rozvoji vojenské spolupráce Spojených ozbrojených sil členských států Varšavské smlouvy" /listopad 1978/*, FMO GŠ/OS, k. 44, p. č. 148, č. j. 36054, VÚA, Prague.

controversial wartime statute of the Unified Armed Forces which is analyzed in detail in a later chapter.

The Soviet intention to face the decline of détente by a new wave of arms race inevitably resulted in the escalation of tensions between the Six and Romania.[900] The growing irritation of Bucharest could have been seen even before the Moscow meeting. The Ceausescu regime watched the alteration of the international environment with worry. It agreed that the Political Consultative Committee should have tried to overturn the existing trend. But the opinions of Bucharest and the rest of the Warsaw Treaty Organization differed significantly on how to proceed.[901] Moscow was aware of this rift. It wanted to solve the situation traditionally, in a restrained way. Thus, the Soviet side ignored another call of Erich Honecker by using the Political Consultative Committee session to firmly streamline Romanian policy.[902] However, Ceausescu's actions were

---

[900] Amid the Political Consultative Committee in Moscow, the West generally speculated about the allegedly deepening discords between the Warsaw Pact member-states. The effort to strengthen the alliance's forces was linked mostly to the Chinese factor. Some of NATO's members presumed that the USSR, in fact, asked its satellites to increase their military spending in Europe and this would have allowed Moscow to detach more of its troops to defend the border with China. Report by the British diplomacy for the North Atlantic Council of NATO on the situation within the Warsaw Pact after its Political Consultative Committee recent session, December 21, 1978, FCO 28/3274, 021/2, TNA, London.

[901] *Report by Hungarian Deputy Foreign Minister István Roska on the Meeting of the Warsaw Pact Deputy Foreign Ministers*, November 16, 1978, Accessed October 16, 2019. http://www.php.isn.ethz.ch/kms2.isn.ethz.ch/serviceengine/Files/PHP/17383/ipublicationdocument_singledocument/c60c974e-4935-4a6e-8187-d86993377706/en/781116_Report_E.pdf.

[902] Considering the Romanian stances, the Kremlin was thinking about holding an internal meeting of deputy foreign ministers of the Six, which would have coordinated their attitude towards Bucharest ahead of the Political Consultative Committee session. In the end, a conciliatory approach was chosen again. The Soviet diplomats did not hide that during shaping the documents, the Kremlin had tried to take into account some Romanian-specific positions. On the contrary, the offended Honecker reminded them that four years earlier, he had been given no opportunity for a response to Ceausescu's insults which had been the Romanian leader's reaction to the Polish and East-German criticism at the Warsaw Pact supreme body session. *Vermerk über ein Gespräch des Generalsekretärs des ZK der SED, Genossen E. Honecker, mit dem Mitlied des Kollegiums des Außenministeriums der UdSSR, Leiter der 3. europäischen Abteilung, Genossen*

ruthless this time. Besides partial obstructions,[903] the Romanian delegation especially refused to sign the communiqué on the military issues. Ceausescu declared that the outlined plans were not compatible with the economic capacities of his country. In addition, he stated that enormous military spending made the Warsaw Pact resolutions calling for disarmament and continuation of détente worthless.[904] On the contrary, the Romanian leader appealed to the Pact to intensify its disarmament initiatives.[905] But other members ignored these proposals and the final protocol was approved regardless of Romania. This outraged the RCP Executive Committee General Secretary. Ceausescu announced that passing the documents without Romanian consent he considered illegal, as the Warsaw Treaty Organization had approved everything on the basis of unanimity so far.[906]

Thus, the Political Consultative Committee gathering in 1978 confirmed the long-term and essential disagreement on the Warsaw Pact's future existence between Romania and the rest of the alliance. Bucharest planned to focus on the Cold War demilitarization leading to a loosening of both blocs. The Six, on the contrary, strove for deepening cooperation

---

Bodnarenko, am 25. 10. 1978, DY 30/2351, BArch, Berlin*; Zasedání politického poradního výboru Varšavské smlouvy,* 3. 11. 1978, 1261/0/7, sv. 85, a. j. 89/18, NA, Prague.

[903] Romania opposed for example the declarations on Indochina and Israel-Egypt relations.

[904] At the time, the Warsaw Treaty Organization engaged mostly in a campaign for a neutron-bomb ban. At the same time, it initiated a treaty on a no-first-use policy within the CSCE or even a non-aggression agreement between NATO and the Warsaw Pact. Jarząbek, *PRL w politycznych strukturach,* 79.

[905] According to Ceausescu, the Warsaw Treaty Organization should have focused mostly on the nuclear weapons production ban approved by the five world nuclear powers in the UN as well as striven for a deal with NATO on the military budgets' reduction by 5 percent at least.

[906] *Informacja o naradzie Doradczego Komitetu Politycznego państw-stron Układu Warszawskiego,* 27. 11. 1978, PZPR KC, XI/807, AAN, Warsaw. At the time of its foundation, no voting mechanism was defined within the Warsaw Treaty Organization. A principle of unanimity was established in praxis. In the 1960s, given the Romanian rebellion, some attempts appeared to introduce a voting system based on the simple majority, which would have been always easily achievable for the USSR of course. However, Bucharest opposed these proposals and the rest of the alliance failed to implement their demands.

within the scope of the Pact.[907] Ceausescu's refusal to participate in the new round of arms race as well as the changes in the command system was seen by the other member-states as another weakening of Romanian collaboration in the Warsaw Pact's military framework. Moreover, there were raising fears among the Eastern diplomats whether Romania was planning to leave the alliance military structures for good. In this case, the country would have kept its political membership in the Pact only as it provided Bucharest with international anchoring and security.[908]

At the RCP Executive Committee session, which immediately followed the Moscow meeting of the Warsaw Treaty Organization's supreme body, the Prime Minister Manea Mănescu delivered harsh criticism of both scheduled armament programs and the planned wartime statute.[909] The RCP leadership approved an official statement for the rest of the Pact that the alliance resolution issued without Romanian consent violated the rules of the founding charter. The RCP General Secretary did not intend to block the plans of the Six. But he insisted such decisions not to be formally made by the Political Consultative Committee or the Committee of Defense Ministers, as conclusions of these forums were binding for Romania. This proves once again that Ceausescu never wanted to break the ties to the Warsaw Pact completely; he just strove for

---

[907] Munteanu, "The Beginning of the End for Détente".

[908] *Informacja o naradzie Doradczego Komitetu Politycznego państw-stron Układu Warszawskiego*, 27. 11. 1978, PZPR KC, XI/807, AAN, Warsaw; *Postavení RSR v diferencované politice VKS vůči ZSS*, 5. 5. 1979, DTO 1945-1989, i. č. 34, e. č. 89, č. j. 013.289/79, AMZV, Prague; The West also did not see the Romanian exit from the alliance as an immediate issue, although it considered the ongoing discords between the Six and Bucharest as the most serious since the 1968 invasion of Czechoslovakia. *European political coopertion: Ministeral meeting in Brussel: 4 December 1978*, December 5, 1978, FCO 28/3274, TNA, London.

[909] *Stenographic Transcript of the meeting of the Consultative Political Committee of the Central Committee of the Romanian Communist Party*, November 20, 1978, Accessed October 16, 2019. http://www.php.isn.ethz.ch/ kms2.isn.ethz.ch/serviceengine/Files/PHP/16634/ipublicationdocument_singled ocument/68371719-6b4e -454a-a438-d56048050ebe/ en/781120_stenographic _transcript.pdf.

restriction of the Kremlin's ability to directly affect Romanian affairs through the alliance.[910]

The RCP leadership decided to launch a public campaign for the reduction of armament spending; even here, a strong anti-Soviet undertone could be found.[911] Moreover, already during the Moscow meeting, Ceausescu had threatened he would demonstratively make public some details of the military measures proposed by the Political Consultative Committee in front of the Romanian political institutions. This would provide the West with a pretext to start a campaign marking the Warsaw Pact as an initiator of the arms race.[912] Also the propagandistic claims about absolute unity of the alliance - strongly emphasized by the Kremlin - would have been disproven. After all, the Romanian reaction to the Political Consultative Committee session in Moscow echoed loudly in the West. It is plausible that Western political officials naturally contemplated the ways to exploit Bucharest's stance in both a political and propagandistic dimension.[913] Therefore, the rest of the Warsaw Pact was afraid to criticize the Romanian stance publicly. In this regard, the Bulgarian leadership, whose positions mostly reflected the actual approach of Moscow, warned against any open criticism that would have only increased the existing disputes between Romania and the rest of the Eastern Bloc. They suggested influencing Ceausescu solely by

---

[910] Munteanu, "The Beginning of the End for Détente".

[911] Luňák, *Plánování nemyslitelného*, 64; *Zahraniční styky RSR /podkladový materiál k návštěvě ministra zahraničních věcí s.B.Chňoupka v RSR/*, 11. 7. 1979, DTO 1945-1989, i. č. 34, e. č. 96, č. j. 015062/79, AMZV, Prague.

[912] *Informacja o naradzie Doradczego Komitetu Politycznego państw-stron Układu Warszawskiego*, 27. 11. 1978, PZPR KC, XI/807, AAN, Warsaw. In this regard, the Warsaw Pact member-states' diplomats speculated that Ceausescu's leadership was well aware of how much its approach within the alliance was making Romania attractive for the West. They thought that this was one of the reasons why Romania did not want to leave the Warsaw Treaty Organization. Besides various international complications, such a step would have resulted also in reducing the value of the country from the Western powers' point of view. *Postavení RSR v diferencované politice VKS vůči ZSS*, 5. 5. 1979, DTO 1945-1989, i. č. 34, e. č. 89, č. j. 013.289/79, AMZV, Prague.

[913] *Suplementary notes for the Prime minister: Romania and the Warsaw Pact*, November 30, 1978, FCO 28/3273, TNA, London.

diplomatic means.[914] Poland acted in a similar way.[915]

Claiming that Romania was the only Soviet satellite interested in preserving détente and reduction of armament would be totally odd. But unlike the others, Bucharest was able to oppose the Soviet approach which it considered wrong. Once again, the Romanian attitude strictly contrasted with the Polish-one, although Gierek's leadership was one of the loudest supporters of the continuation of easing international tensions in the second half of the 1970s. Within the existing limits, they tried to initiate some actions through the Warsaw Treaty Organization which would have helped such a development. Poland especially advocated improving economic cooperation between the blocs.[916] However, at the end of 1978,

---

[914] Along these lines, the BCP leadership sent a polite letter to its Romanian counterpart, presenting a hope that the RCP bodies would reconsider their attitude towards the measures in the field of mutual military cooperation approved by the Warsaw Pact. However, Romanian unwillingness to participate in further military integration of the alliance was considered problematic. Therefore, influencing Romania through bilateral talks should have been henceforward better coordinated among all the Warsaw Pact member-states. *Zpráva o současném vývoji vztahů BLR s RSR, SFRJ, ASLR, Tureckem a Řeckem*, 24. 1. 1979, DTO 1945-1989, i. č. 4, e. č. 50, č. j. 010.686/79, AMZV, Prague; *Čs.-rumunské vztahy po zasedání PPV VS v Moskvě*, 23. 1. 1979, i. č. 34, e. č. 89, č. j. 010.506/79, ibid.

[915] In the letter, the PUWP CC urged its Romanian counterpart to join an implementation of the conclusions of the last Political Consultative Committee sessions aimed at preserving détente. Letter by the PUWP CC to the RCP EC regarding the Political Consultative Committee meeting held in December 1978, December 20, 1978, PZPR KC, XIB/172, AAN, Waraw; Moreover, in talks with the Western diplomats, the Polish side tried to blunt an interpretation of Ceausescu's recent tense behavior. The Poles stated that the reports about dispute between the Warsaw Pact members were exaggerated. Report by the British embassy on the discussion with the Polish deputy foreign minister Dobrosileski, December 1, 1978, FCO 28/3273, TNA, London.

[916] Gierek used claims about the need to improve contacts with the Western supporters of détente in order to justify attempts to strengthen the Polish economic ties with the U.S., France, and West Germany. According to his interpretation, through maintaining relations with these countries, Poland should have "contributed to relaxation"; at the same time, he assured the Kremlin that Poland was not forgetting to urge the West to adhere to the Helsinki Final Act. *Przemówienie I Sekretarza KC PZPR na naradzie Doradczego Komitetu Politycznego Państw - Stron Układu Warszawskiego w Bukareszcie w dniach 25 -26. XI. 1976 r.*, PZPR KC, XIA/589, AAN, Warsaw.

the PUWP CC First Secretary was right in his assumption that the détente's future depended mostly on development of the Moscow-Washington relationship. At the same time, he was aware that the key political forces in the Kremlin were interested in preserving détente as well.[917] Thus, the Warsaw Pact member-states' unified approach remained an axiom to Gierek's leadership. In addition, referring to well-known, long-term repeated claims about NATO armaments, the Polish politburo also supported another strengthening of the Warsaw Pact's defense capabilities.[918] They had to be aware that such measures would inevitably result in additional military spending which limited civil economic sectors. In the light of this, it is not surprising that in mid-1979, the USSR, speaking through diplomat Maltsev, considerably appreciated the way the Soviet-Polish cooperation within the alliance structures proceeded.[919] From Moscow's point of view, no doubt, it was almost an ideal state of things.

The Warsaw Treaty Organization political bodies' activity on the eve of the fall of détente

During the discussion at the Moscow meeting of the Political Consultative Committee in November 1978, appeals occurred again for regular annual gatherings of the body, more frequent sessions of the Committee of Ministers of Foreign Affairs, and specifying, in fact substantial improving of the Joint Secretariat's work.[920] These demands

---

[917] *Projekt przemówienie I Sekretarza KC PZPR na naradzie Doradczego Komitetu Politycznego Państw - Stron Układu Warszawskiego w Moskwie,* 18. 11. 1978, XI/807, ibid.

[918] *Protokół z posiedzenia Biura Politycznego wspólnie z Prezydium Rządu w dniu 24 listopad 1978 r.;,* V/151 (mkf. 2977), ibid; Statement of the PUWP CC to outcomes of the Political Consultative Committee session, November 25, 1978, XI/807, ibid.

[919] *Pilna notatka z konsultacji w MZ z I z-cą ministra SZ ZSRR W. MALCEWEM /26-27.04/,* May 2, 1979, Accessed October 16, 2019. http://www.php.isn. ethz.ch/kms2.isn.ethz.ch/serviceengine/Files/PHP/111558/ipublicationdocument _singledocument/ 84961d8c-ad3c - 4cc3-a665 - b8574ff1d1ac/pl / 1979_Bilateral _Consultations_Malcev.pdf.

[920] *Informacja o naradzie Doradczego Komitetu Politycznego państw-stron Układu Warszawskiego,* 27. 11. 1978, PZPR KC, XI/807, AAN, Warsaw.

were typical, but as already explained, usually had no practical impacts.[921] However, given the deteriorating international situation, in terms of an effort to face - from the Eastern Bloc point of view - undesirable development, the issue of further ways of political cooperation within the Warsaw Pact came up again. In the first months of 1979, the Soviet Minister of Foreign Affairs Gromyko made a trip to the alliance member-states. It directly followed the last Political Consultative Committee session. Given very limited multilateral consultations, the bilateral meetings of the top Soviet diplomats with their counterparts from the Warsaw Treaty Organization countries started to be instrumentally labeled as a part of strengthening the alliance's political mechanisms. The series of such talks should have later resulted in regular sessions of the Committee of Ministers of Foreign Affairs. It is good to add that even four years after its establishing, neither the position nor competences of this body were clarified. Gromyko discussed the role of this forum with his Czechoslovak counterpart Bohuslav Chňoupek. They largely agreed that it should have been strongly connected with the Political Consultative Committee's activities and prepared the room for those top summits. The USSR did not envisage the Committee of Ministers of Foreign Affairs exceeding the scope of a non-binding consultative forum, beyond issuing some collective declarations.[922] In the light of this, the claim by Vojtech Mastny and Anna Locher, that the body took a key role in shaping the foreign policy of the Pact in the second half of the 1970s, cannot be accepted.[923]

At the end of the 1970s, the idea of setting a concrete term for the Political Consultative Committee sessions - once a year, at least - was revived. The Polish Minister of Foreign Affairs, Emil Wojtaszek, was an author of this initiative. It is symptomatic that the Soviet Union formally

---

[921] Anna Locher correctly notes that proclamations voiced at the Political Consultative Committee meetings were rarely new. Locher, "*Shaping the Policies of the Alliance*".

[922] *Informace o pracovním setkání ministra zahraničních věcí ČSSR s. B. Chňoupka s členem politického byra ÚV KSSS a ministrem zahraničních věcí SSSR s. A.A. Gomykem*, 9. 3. 1979, 1261/0/7, sv. 102, a. j. 102/info1, NA, Prague.

[923] Mastny, *A Cardboard Castle?*, 45; Locher, "*Shaping the Policies of the Alliance*".

welcomed his proposal. As an excuse, Moscow blamed a complicated international situation for the irregular gatherings of the Warsaw Treaty Organization supreme body. But it failed to support the Polish initiative by any concrete steps. The unpredictable reaction of Romania served as a suitable pretext. A well-known situation recurred again: The Soviet leadership verbally backed the idea of improving cooperation within the Warsaw Pact,[924] but in reality it tried to keep the issue on ice, or used the Romanian obstructions as an excuse.[925] After all, the same stance took Moscow towards the Czechoslovak initiative on establishing the alliance sub-committee of deputy foreign ministers and a joint academic institute.[926] The Kremlin's unwillingness to create a system of real multilateral mechanisms is documented by its attitude to the Joint Secretariat's work. It proved to be absolutely ineffective. In the period between the meetings of the Political Consultative Committee and the Committee of Ministers of Foreign Affairs, the institution still did not work, in fact.[927] Its activity remained limited to one or two weeks at the

---

[924] In May 1979, Poland informed the Soviet diplomacy about its intention to again put forward a proposal on improving the Warsaw Pact's political mechanisms at the upcoming meeting of the Committee of Ministers of Foreign Affairs. Victor Maltsev, the Soviet deputy foreign minister, assured the Polish side that the USSR attached great importance to this issue. A group of Soviet experts were dealing with the question, he claimed. However, referring to the Romanian positions, he suggested not to open the issue directly at the Committee of Ministers of Foreign Affairs meeting. It should have been discussed in backrooms and during bilateral consultations only, especially with Romanian officials. *Pilna notatka z konsultacji w MZ z I z-cą ministra SZ ZSRR W. MALCEWEM /26-27.04/*, May 2, 1979, Accessed October 16, 2019. http://www.php.isn.ethz.ch/kms2.isn.ethz.ch/serviceengine/Files/PHP/111558/ip ublicationdocument_singledocument/84961d8c-ad3c-4cc3-a665-b8574ff1d1ac/pl/1979_Bilateral_Consultations_Malcev.pdf.

[925] During the meeting with his Czechoslovak counterpart, the Soviet Minister of Foreign Affairs absolutely instrumentally accused Romania of blocking the effort to make the Warsaw Pact's mechanisms more effective. In reality, the country remained passive and silent on this issue at the time.

[926] *Informace o pracovním setkání ministra zahraničních věcí ČSSR s. B. Chňoupka s členem politického byra ÚV KSSS a ministrem zahraničních věcí SSSR s. A.A. Gomykem*, 9. 3. 1979, 1261/0/7, sv. 102, a. j. 102/info1, NA, Prague.

[927] The poor effectiveness of the Joint Secretariat is also proven by the fact that the Political Consultative Committee and Committee of Ministers of Foreign

time of those sessions. But the Polish proposal to appoint the General Secretary's permanent deputies met only calm support by the Kremlin. On the contrary, at the end of the 1970s, the USSR slowly admitted to its satellites that it considered dealing with the issues on a multilateral basis unsuitable and preferred bilateral talks instead. Thus, the questions regarding the Warsaw Pact's further steps were regularly discussed outside its official structures. In Moscow's perception, this meant a meeting of the Soviet officials with the diplomats of individual alliance member-states. But the satellites were not supposed to negotiate mutually. The approach that Moscow took at the end of the 1970s towards the Polish and Czechoslovak incentives defined the borders for political cooperation of the Eastern Bloc countries through the Warsaw Treaty Organization. Outwardly, the alliance mechanisms really should have appeared as a multilateral partnership.[928] In fact, they did not exceed the scope of the superpower-satellite countries' relations.

Nor did another Committee of Ministers of Foreign Affairs' session bring any key stimulus. The meeting, held in Budapest on May 14-15, 1979, mostly provided a platform for repeated appeals for implementation of the previous Political Consultative Committee initiatives. Besides pushing a pledge of the Final Act signatories to non-use conventional as well as nuclear forces first, the Warsaw Pact should

---

Affairs' sessions were still prepared mostly by the deputy foreign ministers at their multilateral meetings. Thus, these gatherings remained a key mechanism for work of the alliance political structures. *Report by Hungarian Deputy Foreign Minister István Roska on the Meeting of the Warsaw Pact Deputy Foreign Ministers*, November 16, 1978, Accessed October 16, 2019. http://www.php.isn. ethz.ch/kms2.isn.ethz.ch/serviceengine/Files/PHP/17383/ipublicationdocument_ singledocument/c60c974e-4935-4a6e-8187-d86993377706/en/781116_Report_ E.pdf.

[928] *Výklad s. A.A. Gromyka na jednání ministrů zahraničních věcí SSSR a ČSSR v Moskvě dne 19. a 20. 2. 1979,* 1261/0/7, sv. 102, a. j. 102/info1, NA, Prague. Fully along these lines, Moscow also neglected the Polish appeals for holding not only one but two sessions of the Committee of Ministers of Foreign Affairs in 1979. *Pilna notatka z konsultacji w MZ z I z-cą ministra SZ ZSRR W. MALCEWEM /26-27.04/,* May 2, 1979, Accessed October 16, 2019. http://www.php.isn.ethz.ch/kms2.isn.ethz.ch/serviceengine/Files/PHP/111558/ip ublicationdocument_singledocument / 84961d8c-ad3c-4cc3-a665-b8574ff1d1ac/ pl/1979_Bilateral_Consultations_Malcev.pdf.

have focused on launching the international talks on reduction of military spending in the upcoming three years.[929] This was just another proclamation without any practical impact. As explained later, the plans on development of the armies in the five-year period of 1981-1985 were finalized in the alliance military structures at the time. On the contrary, these envisaged a significant increasing of monies for armament purposes. On the other hand, Moscow sill believed that the military relaxation was possible. It was still exploring the idea of holding the particular conference till the end of 1979 with the signatories of the Final Act as participants.[930] So the USSR still sought ways in which to move the development of this issue to a more favorable direction from its point of view.

The human rights factor gradually complicated the situation. Regarding the upcoming CSCE summit in Madrid, the Warsaw Pact countries were instructed at the Committee of Ministers of Foreign Affairs' meeting to react to anticipated Western criticism with notes that within the Eastern Bloc, the economic and social rights were more respected than in the capitalist states. Moscow felt that after the unsuccessful (from its point of view) CSCE summit in Beograd, more positive outcomes of the Madrid meeting were necessary for preserving détente. This is also proven by the fact that an imminent priority of the Warsaw Treaty Organization was to convene a preliminary meeting of the Final Act signatories where a scenario of the upcoming CSCE summit would have been agreed and the most contentious topics blunted.[931]

---

[929] *Informace o zasedání Výboru ministrů zahraničních věcí členských států Varšavské smlouvy v Budapešti*, 31. 5. 1979, 1261/0/7, sv. 108, a. j. 109/info8, NA, Prague.

[930] *Pilna notatka z konsultacji w MZ z I z-cą ministra SZ ZSRR W. MALCEWEM /26-27.04/*, May 2, 1979, Accessed October 16, 2019. http://www.php.isn.ethz. ch/kms2.isn.ethz.ch/serviceengine/Files/PHP/111558/ipublicationdocument_sin gledocument/84961d8c-ad3c-4cc3-a665-b8574ff1d1ac/pl/1979_Bilateral_ Consultations_Malcev.pdf.

[931] *Informace o zasedání Výboru ministrů zahraničních věcí členských států Varšavské smlouvy v Budapešti*, 31. 5. 1979, 1261/0/7, sv. 108, a. j. 109/info8, NA, Prague; *Zpráva o možnostech aktivního československého příspěvku k uskutečňování politiky uvolňování napětí v Evropě na základě výsledků zasedání Výboru ministrů zahraničních věcí členských států Varšavské smlouvy a*

Given the existing circumstances, the Soviet Union planned to allow its loyal satellites to act at the international stage in favor of disarmament. Nevertheless, their opportunities remained strictly limited. In 1979, for example, up to this point the very passive Czechoslovak normalization regime put forward its own initiative. Czechoslovakia wanted to present it in the UN General Assembly. However, the content of the document had been in the stage of preparation consistently consulted with the Soviet top diplomatic institutions. Also, the Committee of Ministers of Foreign Affairs dealt with the Czechoslovak intention in passing. In addition, the material was based on the last Political Consultative Committee resolution. Thus, logically, it brought nothing path-breaking. It remained limited to general and vague calls for taking part in disarmament talks and appeals for successful outcomes.[932]

The key effort remained directed by the Soviet Union. This fact was fully proven in the moment when a new international crisis appeared on horizon. As mentioned before, after 1976, the USSR dispatched SS-20 mid-range missiles on its European territory. The step was justified by the note that it was no deployment of additional nuclear capacities, but just their modernization. However, the new generation of the missiles had incomparably better parameters. At first, the United States intended to answer the Soviet conduct with dislocation of neutron weapons in Europe. But such a scenario was scrapped in April 1978 due to reluctance in West Germany.[933] Therefore, the American plans on deployment of additional

---

z hlediska čs. přípravy na madridskou schůzku, 8. 6. 1979, sv. 109, a. j. 110/6, ibid.

[932] *Návrh iniciativy ČSSR v oblasti odzbrojení*, 8. 6. 1979; *Návrh deklarace o mezinárodní spolupráci pro dosažení cílů odzbrojení*, a. j. 110/7, ibid.

[933] Wanner, *Brežněv*, 33. In summer 1977, the American journal Washington Post published a series of articles explaining principles of a neutron bomb and its effect. A quick response by the anti-war movement in Western Europe followed with obviously strong anti-American features. The Soviet Union of course supported these tendencies and launched a massive campaign against a neutron weapon calling it a "capitalist bomb." Moscow found a strong ally especially in some part of the SPD, a government party in West Germany. In the end, the deployment of neutron weapons in Europe was shown to be way too risky for the future of U.S.-West Germany relations. Therefore, Washington abandoned the idea. Walsh, *The Military Balance in the Cold War,* 93-97.

nuclear missiles on the territories of NATO European members were intensified. It would be odd to say that Washington's decision strictly reflected fears of the SS-20 rockets. The United States had already begun to plan a modernization of its tactical missiles in Europe in the mid-1970s. This move should have compensated for the scheduled reduction of nuclear warheads and atomic carrier aircrafts on the continent. At the time, the Republican administration actually aimed to strengthen the internal cohesion of NATO and to increase its deterrent potential.[934]

The Soviet Union was seriously afraid of possible deployment of Pershing-II rockets and BGM-109G cruise missiles in Western Europe. The move was not seen as just a military threat. Moscow attached great political importance to this matter. It considered deployment of new destructive missiles as a big blow to the détente's future.[935] In autumn 1979, however, the USSR warned its satellites that such a move would have meant a significant violation of the balance of power and some reaction of the Warsaw Treaty Organization would be necessary. In addition, Brezhnev publicly announced that the Soviet Union had reduced numbers of its nuclear forces in recent years and had not been deploying these weapons on the territory of its allies.[936] He repeated that his country would not use atomic weapons against any country which possessed no such arms and did not have them deployed on its territory. He also vaguely outlined some willingness to reduce the numbers of nuclear carriers nearby

---

[934] In this regard, the United States put the emphasis on dialogue with their allies. In the mid-1970s, Kissinger thought that a modernization of nuclear forces in Europe would not increase its deterrent potential if it weakened the U.S. credibility among European members of NATO at the same time or significantly undermined the Vienna talks. On the contrary, in front of the allies, the American representatives should have underlined the importance of the Vienna disarmament negotiations and the U.S. willingness to make a deal. Action Memorandum, Hartman and Stern to Kissinger, "The DOD Modernization Program for Tactical Nuclear Forces in Europe", September 11, 1975, in *The Euromissiles Crisis and the End of the Cold War: 1977-1987*, vol. II (doc. 1), ed. Tomothy McDonnell (Washington: Wilson Center, 2009): 7.
[935] Stenographic Minutes of Meeting, Erich Honecker and Leonid Ilyich Brezhnev, October 4, 1979, in *The Euromissiles Crisis* (doc. 26): 470.
[936] *Dopis ÚV KSSS*, undated, October 1979, 1261/0/7, sv. 119, a. j. 119/14, NA, Prague.

the Soviet western borders. As a goodwill gesture, the General Secretary announced the unilateral reduction of Soviet troops and military equipment in East Germany.[937] In reality, such a step had no impact on combat capabilities of this contingent.[938] Also the Warsaw Treaty Organization should have engaged in the peaceful campaign. Brezhnev intended to use the initiatives under the auspices of the alliance to escalate pressure on progress in the Vienna talks. The propagandistic statement was supposed to present the Warsaw Pact as a strictly defense alliance which had never threatened another countries' security.[939] Furthermore, in October 1979, the Soviet leader sent an open letter to NATO countries. It stressed that the Soviet Union kept in Europe just the military strength needed to maintain its effective defense. On the contrary, potential deployment of the American "Pershings" was called an attempt to overturn the existing balance of forces. The Soviet general secretary's defense rhetoric was underlined by the note that the previous Warsaw Pact's initiatives on reduction of international tensions and disarmament remained valid.[940] The USSR of course did not envisage affecting the West solely through the Warsaw Treaty Organization resolutions. Brezhnev, for instance, instructed Erich Honecker to open the issue of deployment of the new American missiles during his bilateral talks with Helmudt Schmidt, the West-German chancellor.[941]

---

[937] In 1980, disregarding the outcomes of the Vienna talks, the USSR unilaterally withdrew 20 thousand troops from East Germany. Śluszarczyk, *Układ Warszawski*, 82. This step was greatly appreciated mostly by Romania. *20. zasedání vojenské rady Spojených ozbrojených sil*. 2. 11. 1979, FMO GŠ/OS, k. 45, p. č. 153, č. j. 36511, VÚA, Prague.

[938] In October 1979, Brezhnev assured Erich Honecker about this. Stenographic Minutes of Meeting, Erich Honecker and Leonid Ilyich Brezhnev, October 4, 1979, in *The Euromissiles Crisis,* (doc. 26): 470.

[939] *Návrh opatření na podporu nové sovětské mírové iniciativy v oblasti vojenského uvolnění v Evropě*, 11. 10. 1979, 1261/0/7, sv. 119, a. j. 119/14, NA, Prague.

[940] *Informace ÚV KSSS o dopisech s. L.I. Brežněva hlavám států členských zemí NATO v souvislosti s novou mírovou iniciativou SSSR*, 26. 10. 1979, sv. 121, a. j. 120/info7, ibid.

[941] Stenographic Minutes of Meeting, Erich Honecker and Leonid Ilyich Brezhnev, October 4, 1979, in *The Euromissiles Crisis,* (doc. 26): 470.

Considering the existing situation, the Carter administration was very interested in deployment of the American missiles. Therefore, it followed with concerns on the Warsaw Pact's effort to hold a conference on military relaxation. Washington suspected that the Eastern Bloc countries planned to use the potential summit to put forward a proposal which would have frozen any deployment of missiles with a range of more than 1000 kilometers in Europe. Regardless of the eventual success or failure of this tactic, the U.S. considered it dangerous and a threat to its own interests. Some NATO members actually could have gotten an impression that there was an alternative to the unpopular and controversial dispatching of new nuclear missiles on the Old Continent. Thus, Washington suspected that Moscow's real aim was not to make a deal but rather to slowdown or block deployment of new modern American weapons in Europe through disarmament initiatives. The Americans pointed to Brezhnev's smarmy rhetoric. Since March 1979, the CPSU CC General Secretary was striving to instigate new disarmament talks by noting that a potential agreement could also have actually included the dreaded Soviet SS-20 missiles and Tu-22M supersonic nuclear bombers.[942]

The last public political meeting of the Warsaw Treaty Organization before the tumultuous events of December 1979 was the gathering of the parliament's representatives held in Prague. It took place just two months earlier, October16-17. Of the logic beyond it, the session could have had no significant impact. Its real purpose was just to support the last initiatives of Brezhnev and strengthen the impression that even in the existing international atmosphere, authentic multilateral cooperation continued within the alliance. Traditionally, the meeting followed the line defined by the Political Consultative Committee and the Committee of Ministers of Foreign Affairs. The final documents urged the Final Act signatories to support the previous initiatives of the Warsaw Treaty

---

[942] State Department INR Publication, Theater Nuclear Force Negotiations: The Initial Soviet Approach, August 10, 1979, in *The Euromissiles Crisis,* (doc. 19): 357.

Organization.[943] Brezhnev's recent proposals and an appeal for ratification of the SALT II treaty by the American side[944] stood on the frontline, of course.[945]

Despite its marginal importance, the gathering of the parliament representatives revealed one significant aspect: The Soviet leadership's intention to involve other socialist countries in the alliance cooperation did not meet with success. Cuba, Laos, Mongolia, Vietnam, and North Korea were invited to the meeting. But only the delegation of first three states arrived. The Vietnamese communists excused, noting their participation in the Warsaw Treaty Organization session could have provided a pretext for other Chinese military attacks.[946] North Korea did not respond at all.[947]

---

[943] The final documents were prepared by the Czechoslovak side in advance, but it had them corrected by the Supreme Soviet of the USSR. Only after that were the texts sent to other member-states for commentary. *Komuniké z konzultativního zasedání představitelů parlamentů členských států Varšavské smlouvy*, 17. 10. 1979; *Výzva představitelů parlamentů členských zemí Varšavské smlouvy k parlamentům členských zemí Severoatlantického paktu*, 17. 10. 1979, 1261/0/7, sv. 122, a. j. 122/10, NA, Prague.

[944] SALT II was signed in June 1979 in Vienna. However, the ratification of the treaty dragged on in the U.S.

[945] Pointless adoration of the aging Soviet leader was also a significant part of the speech by Alois Indra, the chairman of the Czechoslovak Federal Assembly. *Projev vedoucího československé delegace, člena předsednictva ÚV KSČ a předsedy Federálního shromáždění ČSSR Aloise Indry na pražském konzultativním setkání dne 16. října 1979*, ibid.

[946] On February 17th 1979, China attacked Vietnam. It was largely revenge for previous Vietnamese military intervention in Cambodia which had toppled the Red Khmers brutal regime led by Pol Pot. The fights took 18 days. This clash strengthened the Kremlin's suspicions that the U.S.-China coalition was emerging at the international stage. In 1979, Deng Xiaoping visited the United States. This, among the others, prevented Brezhnev from his trip to Washington scheduled the following month. The Chinese attack on Vietnam came too soon after Deng had returned from America. The USSR therefore assumed that the operation had actually been agreed upon during his talks in the U.S. Washington undoubtedly considered reconciliation with China as an anti-Soviet tool. Mutual cooperation should have stopped another Soviet global expansion. Westad, "The Fall of Détente," 21-22.

[947] *Zpráva o konzultativním setkání představitelů parlamentů členských států Varšavské smlouvy v Praze*, 5. 11. 1979, 1261/0/7, sv. 122, a. j. 122/10, NA, Prague.

Additionally, the gap between the Six and Romania was not closing. Bucharest continued in various obstructions. Taking into account China, it objected to the invitation of the Laotian and Vietnamese officials.[948] In his speech, the Romanian representative, Nicolae Giosan, evaluated the Warsaw Pact's activities in similarly negative terms as those of NATO. Also, Ceausescu's constantly growing personal cult could be seen. The Romanian leader was adored by the delegation of his country and provocatively placed on the same level as Brezhnev. But the disputes still did not exceed the existing degree. Thus, after a phone consultation with Ceausescu, the final documents, in reality prepared by the USSR, were approved smoothly.[949]

In this situation, repeating the already issued initiatives had no potential to change the international atmosphere substantially. The issue of the Soviet SS-20 missiles was especially becoming more and more burning.[950] In December 1979, the North Atlantic Council met to discuss the problem. Just ahead of this session, a consultation at the level of deputy foreign ministers within the Warsaw Treaty Organization took place. This is another proof that that forum was the only operatively working political platform of the alliance at the time. The Soviet Union anticipated some harsh countermeasures of NATO. Therefore, Victor Maltsev instructed the member-states that in its following statements, the Warsaw Pact should

---

[948] In fact, Ceausescu generally condemned the Chinese actions against Vietnam. He urged Beijing to withdraw its troops and to solve the dispute by political means based on respect for national sovereignty and the territorial integrity of two socialist countries. *N. Ceausescu o konflikte chińsko-wietnamskim i polityce zagranicznej Rumunii – Po naradzie kierowniczych kadr przemysłu, budownictwa, transportu o rolnictva*, 8. 3. 1979, PZPR KC, XII 3056, 937/8, nr. 10197, AAN, Warsaw.

[949] Ibid; *Uspořádání II. konzultativního setkání nejvyšších představitelů parlamentů členských států Varšavské smlouvy v Praze ve dnech 16. a 17. října 1979*, 19. 6. 1979, 1261/0/7, sv. 111, a. j. 111/25, NA, Prague.

[950] Even the strong détente supporters among the Western politicians were not able to accept buildup of the mobile and sophisticated SS-20 missiles system. The new rockets overtopped their predecessors in every parameter. With a range of 5000 kilometers, they were able to hit targets in Western Europe very accurately even from the bases in Ural. Every missile carried three warheads capable of striking different targets. In praxis, these had the parameters of strategic missiles and decreased the effect of SALT II. Durman, *Útěk od praporů*, 196-97.

have emphasized willingness to reduce its troop numbers, if NATO took the same step. Since spring 1980, the Eastern Bloc countries were supposed to start an energetic diplomatic activity aiming at holding an all-European conference on military relaxation and disarmament.[951] At the end, on December 12, 1979, the North Atlantic Council passed so called Double-Track measures: It called Moscow to talks on withdrawal of the Soviet mid-range nuclear missiles from Europe. If this did not happen until 1982, the new modern American Pershing-II rockets and cruise missiles would be deployed.[952]

Moscow was not able to understand this move. It considered NATO's sense of insecurity due to the Warsaw Pact's military potential absolutely baseless. Despite that there was no Western equivalent to the SS-20, the USSR regarded the announced intentions of NATO as totally unjust.[953] In the political dimension, the Kremlin planned to face the situation through other appeals for launching disarmament talks.[954] In this regard, the fact that the deployment of new American nuclear weapons was very unpopular among the West-European society, played into the Soviet hands. Thus, an onset of the Euromissiles Crisis strongly mobilized the Western peace movement.[955] The public in Western Europe could have hardly believed the claims that the deployment of new nuclear missiles on the continent would in reality decrease the risk of an outburst of destructive war.[956] On the contrary, the Romanian attitude complicated the Warsaw Pact's position. Given the situation, the Ceausescu regime asserted unilateral cuts of troops and armament as well as refused to condemn NATO intentions on behalf of the alliance.[957]

---

[951] *Report by Hungarian Deputy Foreign Minister István Roska on the Meeting of the Warsaw Pact Deputy Foreign Ministers*, November 16, 1978, Accessed October 16, 2019. http://www.php.isn.ethz.ch/kms2.isn.ethz.ch/serviceengine/Files/PHP/17383/ipublicationdocument_singledocument/c60c974e-4935-4a6e-8187-d86993377706/en/781116_Report_E.pdf.

[952] Wanner, *Brežněv*, 33.

[953] Nünlist, "*Cold War Generals*".

[954] Jarząbek, *PRL w politycznych strukturach*, 80.

[955] Bowker, "Brezhnev and Superpowers Relations," 99.

[956] Holloway, "Jaderné zbraně," 46.

[957] *Report by Hungarian Deputy Foreign Minister István Roska on the Meeting of the Warsaw Pact Deputy Foreign Ministers*, November 16, 1978, Accessed

But the key moment which significantly reduced the USSR's ability to operatively act in the Euromissiles Crisis and at the same time definitively terminated détente, was the Soviet military intervention in Afghanistan launched just a few days after NATO had announced its Double-Track decision. If the Warsaw Pact's political structures were unable to prevent deterioration of the international situation in the second half of the 1970s, the Soviet action in the "land of Mujahideens" put a final stop to this effort. The alliance that was on its political climax in 1975, only five years later would take the path to its definite decline.

October 16, 2019. http://www.php.isn.ethz.ch/kms2.isn.ethz.ch/serviceengine/ Files/PHP/17383/ipublicationdocument_singledocument/c60c974e-4935-4a6e-8 187-d86993377706/en/781116_Report_E.pdf. During the Crimean meeting in August 1979, Brezhnev reproached Ceausescu on the Romanian attitude towards the Warsaw Treaty Organization, China, disarmament, respectively, the arms race, European security, the implementation of the Final Act, the situation in the Middle East and in Africa, and Romanian attempts for closer cooperation with the Balkan countries. The General Secretary stated that Romania continually followed its "separate course." He reminded Ceausescu of the recent Romanian behavior at the Political Consultative Committee and other Warsaw Pact sessions. Brezhnev gave assurances that he respected Romanian state sovereignty. According to his interpretation, national interests could have been effectively protected through joint approach only. Ceausescu chose a strictly defensive tactic once again and tried to assure him that Romania in reality fully cooperated with the Warsaw Pact countries on the international issues. The Soviet leadership later agreed it would solely focus on maximum elimination of the negative impacts resulting from Romanian policy. But it did not plan any decisive actions. ИНФОРМАЦИЯ о крымской встрече и беседе Л.И. Брежнева с Н. Чаушеску I августа 1979 года, August 20, 1979, Accessed October 16, 2019. http://digitalarchive.wilsoncenter.org/document/114160.

The Warsaw Pact's armament programs and the fall of détente

At the time of signing the Helsinki Final Act, the Warsaw Treaty Organization experienced its political climax. However, the trend to strengthen the Unified Armed Forces' combat capabilities continued. As already explained, after 1975, the member-states' political officials stressed further easing international tensions. Nevertheless, regarding modernization of NATO armaments and increasing of its military budget the claim remained valid that certain risk of an outburst of war in Europe persisted. For this reason, deepening of détente should have been accompanied by additional strengthening of the Warsaw Pact in parallel.[958] The Eastern Bloc's appeals for dissolution of both military-political alliances, in fact, did not exceed the scope of a pleasing phrase. Moreover, according to Vojtech Mastny, nothing suggests that the Soviet top military was really preparing for the military relaxation proclaimed by the Kremlin in the second half of the 1970s.[959]

Stalemate in disarmament talks

In praxis, the Warsaw Treaty Organization did not really take any step that aimed to ease military confrontation between both blocs. This was especially apparent at the Vienna disarmament talks. The Eastern rigorous position was based on a proclamation that an equal balance of forces existed between NATO and the Warsaw Pact in Europe. On the contrary, the West pointed to the conventional superiority of its potential enemy. Therefore, it still asserted asymmetric reduction of troops that would have eliminated such an advantage. According to Western thoughts,

---

[958] *Analýza výsledků plnění vojenskopolitické linie XIV. sjezdu KSČ a hlavní směry upevňování obrany ČSSR po XV. sjezdu strany*, 31. 10. 1975, 1261/0/6, sv. 171, a. j. 174/4, NA, Prague.

[959] Mastny, "Imagining War in Europe," 35.

a definite outcome of the disarmament process should have been the "common ceiling" of ground forces for both alliances dispatched in the region of Central Europe. Along these lines, the Soviet Union, which kept huge numbers of troops especially in East Germany, was supposed to make a three-fold reduction in the first stage, in comparison to the American side. The situation was illustrated by the course of the 8th round of the Vienna negotiations during the first half of 1976. The Warsaw Pact countries proposed purely symbolic measures: The USSR and USA should freeze and later cut their military presence in the region by 2-3 percent of the total numbers of both alliance armed forces dislocated in Central Europe. Although this move was supposed to include tanks, tactical bombers, missile units, and command structures of the army corps, given the proposed numbers it did not exceed the scope of a gesture. Reduction of other NATO and Warsaw Pact member-states' armed forces in the region should have taken place in the following stage.[960] In this way, the Soviet leaders perhaps wanted to reach some tangible result that could have been used in appeals for continuation of détente. Along these lines, in November 1976, the Political Consultative Committee session urged the Warsaw Pact countries for propagandistic actions which would have won the world public's support for a disarmament proposal put forward by the socialist states at the time.[961] However, in the atmosphere of persisting suspicions, the Eastern proposals at the Vienna talks were not accepted by the NATO countries, of course. Similarly, the Warsaw Treaty Organization did not accede to the Western incentive which, in exchange for an asymmetric reduction of troop numbers, even offered withdrawal of some American nuclear weapons and their carriers from Europe.[962]

---

[960] *Informace o posledním průběhu a dosavadních výsledcích jednání o snížení ozbrojených sil a výzbroje ve střední Evropě*, 21. 5. 1976, 1261/0/7, sv. 6, a. j. 4/info, NA, Prague.
[961] *Zpráva o zasedání politického poradního výboru států Varšavské smlouvy konaném v Bukurešti 25. – 26. 11. 1976*, 1. 12. 1976, sv. 23, a. j. 26/1, ibid.
[962] In December 1975, NATO offered withdrawal of 1,000 U.S. nuclear warheads and some of their carriers, 36 Pershing guided missiles, and 54 slowly obsoleting F-4 Phantom aircrafts from Europe. Letter by B. Chňoupek, the Czechoslovak minister of foreign affairs, to M. Dzúr, the Czechoslovak defense minister, February 12, 1976, FMO GŠ/OS, k. 7, p. č. 39, č. j. 39031, VÚA, Prague.

Paradoxically, Moscow's mistrust was partially based on the signature of the Final Act. The Kremlin considered successful conclusion of the Helsinki process as its own victory and presented it publicly in this way. For this reason, the USSR feared that the West would try to compensate for this "defeat" and was determined to "beat" the East in the field of disarmament talks through an asymmetric reduction of troops. The Warsaw Treaty Organization saw this concept as a direct effort to disrupt its defense system.[963] In this regard, it is important to keep in mind that détente was a process of better management of mutual rivalry between the superpowers, not a process of its total abandonment. It was not the end of the Cold War, but rather its defusing only.[964] The arms race and development of technologies continued on both sides and this fact, alongside the other aspects, unambiguously slowed down the tempo of negotiations on armaments control.[965]

All the Eastern proposals at the Vienna talks were still formulated strictly by the Soviet side. The Warsaw Treaty Organization countries' ability to influence them was much more limited than in case of political initiatives issued on behalf of the alliance. In February 1976, for instance, the Warsaw Pact direct participants in the Vienna negotiations, the USSR, PPR, CSSR, and GDR, put forward collectively the new, in fact only slightly adjusted, proposals.[966] These were purely the Soviet materials. The leadership of the satellites only formally confirmed they agreed with their proclaimed co-authorship.[967] The Warsaw Pact countries were not

---

[963] *Informace o posledním průběhu a dosavadních výsledcích jednání o snížení ozbrojených sil a výzbroje ve střední Evropě*, 21. 5. 1976, 1261/0/7, sv. 6, a. j. 4/info, NA, Prague; *Informace o průběhu jednání o snížení ozbrojených sil a výzbroje ve střední Evropě v 2. pololetí 1976*, 11. 2. 1977, sv. 30, a. j. 33/info1, ibid.
[964] Bowker, "Brezhnev and Superpowers Relations," 104, 106.
[965] Zubok, *A Failed Empire,* 242.
[966] In fact, this was just a slightly modified proposal of November 1973, adapted to the Western methods of determining troop numbers. The aim was to retake an initiative at the Vienna talks which was being gained by the West. This is proven also by the Soviet note that the guidelines for the Warsaw Pact member-states on the Vienna talks of 1973 remained valid.
[967] Letter by B. Chňoupek, the Czechoslovak minister of foreign affairs, to M. Dzúr, the Czechoslovak defense minister, February 12, 1976, FMO GŠ/OS, k. 7, p. č. 39, č. j. 39031, VÚA, Prague.

even able to influence which information about their armed forces would be presented in Vienna, as the Czechoslovak case suggests. In March 1976, the chief of the Soviet general staff Marshal Kulikov directly ordered the delegation of Czechoslovakia to admit about 170 thousand men in arms. The real peacetime numbers of the CSPA were at least 25 thousand troops higher at the time. Moreover, the Czechoslovak diplomats in the Austrian capital were not given these numbers directly from the army command. The stats went through the Soviet general staff and the head of the Soviet negotiators in the Vienna talks, Oleg Khlestov. In addition, this information had to be approved by the military institutions of the USSR.[968] Thus, an alleged parity showed by the reports on troop numbers in Central Europe which NATO and the Warsaw Treaty Organization exchanged in 1976, was obviously just an illusion.[969] Moreover, considering the geographical factor and dislocation of the armed forces across the entire continent, not only in the region of Central Europe, these stats were misleading.

Unification and modernization of the Unified Armed Forces' armaments

The preparation of the military five-year plan of 1976-1980 was under way already before the conclusion of the Final Act. But even at the time of climaxing détente, the intentions were not reviewed at all. The recommendation of the Unified Armed Forces' supreme commander set the modernization of weaponry and improvement of the command structures as the main tasks in the above-mentioned period.[970] Afterwards, the priorities defined by the Unified Command in particular bilateral

---

[968] *Ústní informace ministra národní obrany ČSSR o zveřejnění číselných údajů ČSLA při vídeňských jednáních o snížení ozbrojených sil a výzbroje ve střední Evropě*, March 19, 1976, ibid.

[969] The Warsaw Treaty Organization acknowledged 805 thousand troops, 14 thousand more than NATO. *Informace o průběhu jednání o snížení ozbrojených sil a výzbroje ve střední Evropě v 2. pololetí 1976*, 11. 2. 1977, 1261/0/7, sv. 30, a. j. 33/info1, NA, Prague.

[970] *DOKLAD náčelníka operační správy - zástupce náčelníka generálního štábu ČSLA k otázkám "Protokolu"*, undated, 1976, FMO GŠ/OS, k. 7, p. č. 38, č. j. 39030, VÚA, Prague.

protocols were approved by all the member-states, except for Romania, by January 1, 1976.[971] Mostly, a continuation of unification and standardization of the armament under the supervision of the Technical Committee was envisaged.[972] In the fall of the previous year, in his report on the state of the Unified Armed Forces, the supreme commander Yakubovski commended that in the majority of the alliance armies the tasks set for the previous five-year plan had been largely accomplished. He noted, however, that regardless of successful realization of the protocols, many units were still equipped with obsolete arms. This included even the key weapons like missiles, tanks, aircrafts, or vessels. Also, the buildup of protected hangars for aircrafts and permanent military airfields proceeded very slowly.[973]

At the turn of June and July 1976, further improvement of mutual military cooperation within the Warsaw Treaty Organization was discussed at an extraordinary meeting of individual member-states' general staffs in Poland. The military officials were happy to say that the establishing of collective bodies in 1969 had enhanced the alliance military collaboration significantly and made it more transparent. However, there still existed some inconsistency in reports to the Staff of the Unified Armed Forces. This should have been solved by unification of their format. Therefore, a great emphasis was put on the improvement of

---

[971] Through proper formulations, Romania continually tried to weaken attempts to present the Warsaw Treaty Organization as a decisive instance. In 1976, during the approving of a report on the development of the Unified Armed Forces in the years 1969-1974, the Romanian defense ministry opposed the text "to approve the measures" and rather insisted on the formulation "to acknowledge the measures." The country still referred to the text of military statutes approved by the Political Consultative Committee in 1969. Romania considered it as the only valid directive defining cooperation within the alliance. This attitude was largely ignored by the Six. Letter by M. Dzúr, the Czechoslovak defense minister, to G. Husák, the CPCz CC General Secretary, November 22, 1976, k. 8, p. č. 41, č. j. 39033, ibid.

[972] In May 1975, the Soviet General I. A. Fabrikov was appointed a head of this body. He replaced I. V. Schtepanyuk. *Porada vedoucích funkcionářů generálních /hlavního/ štábů armád Varšavské smlouvy*, 6. 7. 1976, k. 19, p. č. 58, č. j. 39051, ibid.

[973] Iakubovskii Report on the State of the Unified Armed Forces, December 30, 1975, in *A Cardboard Castle?* (doc. 77): 395-396.

Russian language skills within the officer corps of individual armies. For this purpose, the alliance drills should have been commanded in Russian only.[974] Another key task became surveillance of the estimated enemy, NATO. Information of individual member-states in this area should have been generalized and then discussed collectively at the Military Council sessions.[975] Regarding the intention to better analyze NATO military preparations, improving cooperation between the general staff's intelligence departments of the Warsaw Pact member-states as well as new collective meetings of their chief were envisaged.[976]

At the 9th Committee of Defense Ministers gathering, held in November 1976 in Sofia, the rhetoric known up to this point was slightly and temporarily modified.[977] It was stated that given the international situation, strengthening of NATO had new purpose: According to the Eastern claims, the West no longer directly strove for unleashing a war, but rather just for getting military superiority over the Warsaw Treaty Organization. This would have allowed it to negotiate with the socialist

---

[974] However, the achieved results were questionable. A report on the "Soyuz-81" exercise, which took place almost six years later, suggests that still not all the commanders spoke Russian at the required level. This was especially evident in the case of the officers who did not study at the Soviet military schools. *Rozbor operačně strategického velitelsko-štábního cvičení "Sojuz81"*, 8. 4. 1981, FMO GŠ/OS, kr. 48, p. č. 180, č. j. 37032, VÚA, Prague.

[975] *Porada vedoucích funkcionářů generálních /hlavního/ štábů armád Varšavské smlouvy*, 8. 6. 1976; *Porada vedoucích funkcionářů generálních /hlavního/ štábů armád Varšavské smlouvy*, 6. 7. 1976, k. 19, p. č. 58, č. j. 39051, ibid. In the mid-1970s, the military analytics of individual member-states focused on NATO maneuvers. Their analyses were collectively discussed at the Military Council sessions, where the Warsaw Pact countries checked their partial information obtained by their intelligence. *14. zasedání Vojenské rady Spojených ozbrojených sil*, 1. 6. 1976, k. 39, p. č. 134, č. j. 36037, ibid.

[976] *Plán realizace Protokolu čís.:009 ze 9. zasedání Výboru ministrů obrany členských států Varšavské smlouvy*, wrongly dated 1974, most likely 1976, k. 8, p. č. 42, č. j. 39035, ibid.

[977] This shows also an incident when the Czechoslovak defense minister, Martin Dzúr, wanted to go back to the Prague Spring events in his report. He had to be streamlined by the Soviet general Viktorov. He conveyed to Dzúr the Kremlin's opinion that in the light of actual issues of the CSCE there was no need to mention the event of the years 1968-1969 anymore. *Vystoupení ministra národní obrany na 9. zasedání Výboru ministrů obrany*, 8. 12. 1976, k. 42, p. č. 140, č. j. 36043, ibid.

countries from a position of strength. Nevertheless, even the Warsaw Pact's top military suspicions towards détente had to admit at the time, that measures of the Final Act really contributed to creating an atmosphere of confidence in the international environment. The first deputy chief of the Soviet general staff, General Mikhail Kozlov, stated that the Helsinki conference finally confirmed World War II results and therefore meant a beginning of the new age of easing tensions. It must be pointed out that the military officials fully respected that the further fate of détente would be decided by the political power. However, they remained very skeptical about some aspects of the Final Act. This mostly included the exchanges of military observers at drills taking place nearby borders of both alliances. The Warsaw Treaty Organization announced such exercises as well as invited the Western observers there,[978] but did not send its own representatives to similar NATO actions. The East acted in this issue ostentatiously, noting that NATO maneuvers still meant an unacceptable demonstration of force.[979] Thus, Kozlov warned that these measures were an innovative moment for the international situation in Europe and had the potential to affect the Warsaw Pact's defense capabilities in an undesirable way.[980]

After 1976, there were increasing talks about a need for unification of the Unified Armed Forces' weaponry and equipment. This term, in fact, covered the introduction of new Soviet arms into service. They could have also been produced in license in other Warsaw Pact countries; however, the following distribution proceeded according to the Unified Command's demands. It was envisaged that the process would

---

[978] The Warsaw Pact member-states should have consulted an announcement of drill with the supreme commander. However, political power still played a decisive role, including the question of the invitation of Western observers.

[979] Moreover, the Warsaw Treaty Organization still suspected that NATO's annual maneuvers held in West Germany could have been used to launch a surprise attack, regardless of existing balance of forces. Mastny, *A Cardboard Castle?*, 47.

[980] *9. zasedání výboru ministrů obrany členských států Varšavské smlouvy*, 13. 12. 1976, FMO GŠ/OS, k. 42, p. č. 140, č. j. 36043, VÚA, Prague; *Plán realizace Protokolu čís.:009 ze 9. zasedání Výboru ministrů obrany členských států Varšavské smlouvy*, wrongly dated 1974, most likely 1976, k. 8, p. č. 42, č. j. 39035, ibid.

peak in the five-year plan 1981-1985. But some projects would have been launched sooner.[981]

These intentions were accelerated by the arrival of the Carter administration in the United States. After a few months, the Warsaw Pact countries realized that the new political garniture in Washington opposed détente being transferred into the military's area. The East estimated that the U.S. government's priority was to prevent any weakening of NATO.[982] Thus, at a meeting of top Soviet military officials in 1977, Dmitri Ustinov, the new Soviet defense minister,[983] called NATO the most dangerous in its history so far, especially due to its allegedly increasing technological cooperation with China.[984] In fact, since the mid-1970s, the new highly-

---

[981] The first huge armament project which envisaged really deep cooperation of the member-states, was a licensed-production of the T-72 tank. The components should have been manufactured in the USSR, CSSR, PPR and PRB. The aim was to maximally accelerate the supplies of this weapon. Due to the complexity of the issue, it was discussed not only in the Warsaw Pact Technical Committee but also within the Comecon. *9. zasedání výboru ministrů obrany členských států Varšavské smlouvy*, 13. 12. 1976, k. 42, p. č. 140, č. j. 36043, ibid; *Plán realizace Protokolu čís.:009 ze 9. zasedání Výboru ministrů obrany členských států Varšavské smlouvy*, wrongly dated 1974, most likely 1976, k. 8, p. č. 42, č. j. 39035, ibid.

[982] *Informace o průběhu vídeňských jednání o snížení ozbrojených sil a výzbroje ve střední Evropě v 1. pololetí 1977*, 15. 8. 1977, 1261/0/7, sv. 47, a. j. 51/info2, NA, Prague.

[983] Dmitri Ustinov became the defense minister after the death of Grechko in 1976. He had served previously as the minister of weapons industry during Stalin's and Khrushchev's rule. His position in Brezhnev's system was based on his indisputable technocratic competences, explicit admiration for Stalin, and derogation of Khrushchev as a result. This was later supported by Ustinov's good personal relations with Brezhnev, Grechko, and his marshals. After he became the politburo member in 1976, Ustinov remained one of the five key persons in the Soviet leadership until his death in 1984. Durman, *Útěk od praporů*, 184.

[984] Mastny, "Imagining War in Europe," 35. Since 1976, in their internal discussions, Soviet officials began to talk about a risk of war with China again, like six years earlier. Chinese policy and propaganda at the time, however, greatly contributed to this. Some sources even say that at a certain point, Moscow was considering a preemptive strike aimed at elimination of the Chinese nuclear arsenal. In 1977, the USSR invested in the defense of its 12.5 thousand kilometers long border with China, where 3 million Chinese troops were dispatched, more money than the entire Warsaw Pact in general, as Moscow was afraid of an outburst of total war there. Schaefer, "The Sino-Soviet Conflict," 213.

sophisticated American weapons originally developed for the Vietnam battlefield, began to be introduced in NATO. Considering this, there was a real chance the West would gain superiority in the field where the East traditionally had kept advantage.[985]

In the second half of the 1970s, the Western lead in technologies as well as the improvements in NATO's organizational structure made the Warsaw Treaty Organization reconsider its potential enemy. Rather than emphasizing the Western alliance's weakness, its strength began to be stressed. At the Political Consultative Committee session in November 1976, the new chief of Staff of the Unified Armed Forces, Anatoly Gribkov,[986] warned that the West would be able to catch up with the Warsaw Pact's existing superiority in the upcoming three years. In this way, the Unified Command actually suggested to the member-states' leaders that the existing military advantage achieved during the 1970s was now in jeopardy.[987] As a result, the Warsaw Treaty Organization's supreme body formally decided that an expensive rearmament of the Unified Armed Forces would be the key task in the upcoming period. This caused various economic troubles to the member-states, as it was clear that increasing the import of Soviet military materials would be necessary to implement the required measures.[988]

---

[985] Luňák, *Plánování nemyslitelného*, 62.

[986] General Anatoly Gribkov replaced general Schtemenko, who died on April 26, 1976. For the first time, the chief of the alliance staff was appointed according to Article 14 of the statute of the Unified Armed Forces of 1969. This made filling the post conditional to consent of the member-states' governments. However, it was just a formality again. The Soviet satellites did not choose between any candidates. The supreme commander Yakubovski just confronted them with Moscow's decision and asked for their formal blessing. Symptomatically, he did not apply the political institutions directly, but rather instructed individual defense ministers to get needed consent. *KÁDROVÉ ZMĚNY VE ŠTÁBU SPOJENÝCH OZBROJENÝCH SIL*, October 1976, FMO GŠ/OS, k. 44, p. č. 146, č. j. 36052, VÚA, Prague; Letter by marshal I. Yakubovski to W. Jaruzelski on proposed appointing A. Gribkov the chief of Staff of the Unified Armed Forces, Setember 29m 1976, PZPR KC, XIA/583, AAN, Warsaw.

[987] Mastny, *A Cardboard Castle?*, 44-45

[988] Report by M. Dzúr, the Czechoslovak defense minister, to G. Husák, the CPCz CC General Secretary, January 18, 1977, FMO GŠ/OS, k. 8, p. č. 41, č. j. 39033, VÚA, Prague.

The problem was directly linked to intended unification and standardization of the armaments and equipment through which the Warsaw Treaty Organization was supposed to strengthen its combat capabilities. The Unified Command justified the move by these arguments: In case of combat operation, using the same weapons should have facilitated cooperation of individual member-states' armies. Also, service of damaged equipment and supplying would be made easier. Furthermore, the alliance command claimed that the time and money needed for development and production of the military materials would be reduced. However, also under the motto of unification, obsolete weapons were decommissioned and replaced by their modern equivalents which made the project even more expensive. The ambitious plans assumed a significant extension of licensed production and improving cooperation of the Warsaw Pact member-states' weapons industries. The related questions were not discussed between the Unified Command and national military institutions only, but also in the Comecon Permanent Commission for Defense Industry.[989] By the end of 1977, a brand-new Department of Unification and Standardization of Armaments and Combat Equipment began to work within the Technical Committee.[990] But the effectiveness of this body remains disputable. In 1981, for instance, the supreme commander had to admit that the Technical Committee functionaries' competences and results were incomparably worse than those of the deputy commanders for armaments in the Soviet army.[991]

---

[989] *Základní ustanovení o unifikaci a standardizaci výzbroje a vojenské techniky ve Spojených ozbrojených silách*, undated, perhaps 1977, k. 46, p. č. 163, č. j. 36519, ibid.

[990] In 1976, in order to make this process more effective, the Committee of Defense Ministers decided to establish a special Department of unification and standardization of armaments and combat equipment within the Technical Committee. Thus, the body was enlarged by 45 officials. As this was a modification of the alliance institution whose activities had been defined by the Statute of 1969, Romania demanded the changes to be approved by the national bodies first, not just by the Committee of Defense Ministers. *9. zasedání výboru ministrů obrany členských států Varšavské smlouvy*, 13. 12. 1976, k. 42, p. č. 140, č. j. 36043, ibid.

[991] *Účast ČSLA na cvičení „Sojuz-81"a výměna jednotek s PLA ve VVP*, 2. 1. 1981, k. 48, p. č. 180, č. j. 37032, ibid.

The costs of rearmament were also increased by the fact that slow lagging of the air forces and air defense system began to be considered a burning issue after 1977. But especially because the reinforcement of these units was enormously expensive. The Warsaw Pact's analytics concluded that the air forces should have played the main role in the NATO strategy of ground-forces support. The key modernization of the air defense system was expected in the years 1981-1985.[992] However, Romania soon started hindering these plans and refused to take part in a new round of improvement of cooperation of the air defense system units. In fact, this question was very acute, given the introduction of the a new generation of NATO combat aircraft and cruise missiles.[993]

In 1977, the Technical Committee instrumentally stated that unification was meeting the plans.[994] In reality, the first substantial cracks were appearing in the project. Poland, for example, experienced problems with its supply contracts of the military equipment for the Warsaw Pact's partners.[995] At the 10th session of the Committee of Defense Ministers, held in December 1977 in Budapest, Wojciech Jaruzelski had to inform

---

[992] In May 1977, at its 15th meeting, held in Prague, the Military Council stated that the Warsaw Pact's air-defense air force units - fighter jets - were on the fair level. The issue mostly concerned surface-to-air missiles. *15. zasedání Vojenské rady Spojených ozbrojených sil*, 23. 5. 1977, k. 39, p. č. 134, č. j. 36037, ibid.

[993] At the time, the air-defense system of GDR, PPL, and PRB showed the best results. These countries were also quickest in introducing the automatic command system ALMAZ-2 into service. On the other hand, the Unified Command considered the Hungarian and Romanian systems as a weakness. Because of this, Romania should have received 24 MiG-23 MF fighters by 1980, two times more than the rest of the Warsaw Pact countries. At the same time, Romania rejected the Unified Command's request to assign all the units equipped by modern KRUG and KUB missiles to the alliance emergency system. The country perhaps wanted to use these effective weapons regardless of the Warsaw Pact command. *16. zasedání Vojenské rady Spojených ozbrojených sil*, 24. 10. 1977, ibid; *10. zasedání Výboru ministrů obrany*, 5. 12. 1977, k. 42, p. č. 141, e. č. 36044, ibid.

[994] *16. zasedání Vojenské rady Spojených ozbrojených sil*, 24. 10. 1977, k. 39, p. č. 134, č. j. 36037, ibid.

[995] Intensified pressure on armament in the second half of the 1970s occurred in the moment when Poland began facing an economic crisis. In this regard, Wanda Jarząbek notes it is not possible to answer with certainty a question whether and how Polish officials presented this problem to Moscow. Jarząbek, *PRL w politycznych strukturach*, 78

his Czechoslovak counterpart that supplies of "Strela-1A" anti-aircraft system scheduled to conclude in 1979 would be delayed. It became clear that the unification and licensed production of weapons brought a serious problem: If one Warsaw Pact member-state was not able to fulfill its commitments in military production, the rest of the alliance armies were affected as well.[996] Despite this fact, some Soviet satellites still supported deepening mutual cooperation in military production. This is proven, for instance, by the planned massive production of the T-72 tank. In this regard, Bulgaria was not satisfied that only the wheels would have been produced in the country as their sales could not have covered the costs of purchasing numbers of this weapon required by the Unified Command.[997]

Regardless of the aforementioned complications, the Warsaw Treaty Organization managed to increase the combat capabilities of its troops significantly. After the death of Ivan Yakubovski in November 1976, Marshal Victor Kulikov, acting chief of the Soviet general staff, moved to the post of the supreme commander of the Pact.[998] In his first report on the state of the alliance formation, he was pleased to declare that in 1977 missile units rearmed by 8%, tank divisions by 4%, infantry fighting vehicles by 37%, and anti-tank guided missiles by 39%. He also appreciated the key strengthening of the anti-aircraft defense as well as establishing elite units equipped solely by the most modern armaments, the yacheykas.[999] However, it is clear that the armament was creating more and more difficulties for the Soviet satellites. During his meeting with the

---

[996] Czechoslovakia, for example, in basic parameters, met its commitments of the Protocol on assignment of troops to the Unified Armed Forces and their development in the years 1976-1980. What remained lacking was the introduction of new weapons in artillery and air-defense units. The reason was delayed arms supplies from Poland and Bulgaria. *Podkladové materiály pro projednání o rozvoji ČSLA v letech 1981-1985*, undated, 1980, FMO GŠ/OS, k. 21, p. č. 64, č. j. 40001, VÚA, Prague.

[997] *10. zasedání Výboru ministrů obrany*, 5. 12. 1977, k. 42, p. č. 141, e. č. 36044, ibid.

[998] Kulikov was proposed by the CPSU CC. The leadership of the rest of the Warsaw Pact countries just formally approved this choice. *Politbüro des ZK Reinschriftenprotokoll nr. 2*, 11. 1. 1977, DY 30/ J IV 2/2/1652, BArch, Berlin.

[999] Report by Marshal Kulikov on the State of the Unified Armed Force, January 30, 1978, in *A Cardboard Castle?* (doc. 82): 413-414.

Committee of Defense Ministers' members in December 1977, Janos Kadar complained that military production was hindering his plans to increase living standards in Hungary. But at the same time, he solemnly swore: *"We will give the armed forces everything, even if we wear just slippers."*[1000]

The arms race escalation and the fall of détente

It seemed more and more impossible to overturn the imposed trends. The year 1978 was clearly characterized by deteriorating relations between both Cold War blocs. It was obvious at the Vienna disarmament talks. The Soviet actions in Africa, an atmosphere of the unsuccessful CSCE summit in Beograd, and complications in the SALT II negotiations were reflected even at this forum. In addition, NATO began to modernize and strengthen its armed forces and increase their combat readiness. These moves were justified by similar measures of the Warsaw Treaty Organization and the alleged Eastern effort to extend its already existing superiority. Furthermore, in backrooms, Western representatives warned their Eastern counterparts[1001] that taking into account the Warsaw Pact's advantage in tanks and recent deployment of the Soviet SS-20 modern missiles, NATO was considering dislocation of neutron weapons in Europe. In the light of this, the Eastern Bloc countries abandoned their beliefs that some foreseeable breakthrough at the Vienna talks was possible. Their claims and initiatives officially presented in the Austrian capital did not reflect reality and were nothing more than instrumental appeals.[1002] The main set of tactics was to criticize the allegedly

---

[1000] *10. zasedání Výboru ministrů obrany*, 5. 12. 1977, FMO GŠ/OS, k. 42, p. č. 141, e. č. 36044, VÚA, Prague.

[1001] The Warsaw Pact countries' representatives in Vienna had several bilateral talks in backrooms with NATO states' delegations. Most likely, it was at the Kremlin's directive. In this regard, Poland was especially active and, of course, informed the USSR in detail about the negotiations. *Szyfrogram Nr. 3769/IV z Wiednia*, December 18, 1978, Accessed October 16, 2019. http://www.php.isn. ethz.ch/kms2.isn.ethz.ch/serviceengine/Files/PHP/111435/ipublicationdocument _singledocument / 7e163ed8-6f1a-4a9e-8bee-240e06e3edb3 / pl/1978_Talks_ Dean.pdf.

[1002] *SPRAWOZDANIE delegacji PRL na XIV rundę rokowań wiedeńskich w sprawie wzajemnej redukcji sił zbrojnych i zbrojeń w Europie Srodkowaj /31*

mendacious claims by NATO about the conventional superiority of the East.[1003] Moreover, the Soviet satellites still just seconded Moscow. The USSR at least waited for a formal approval of its allies - the direct participants of the talks - when presenting its own stances on behalf of the Warsaw Treaty Organization. Intensive consultations during the negotiations and the Soviet proposals to put the situation at the Vienna talks into the agenda of the alliance bodies' meetings only reflected the Kremlin's effort to secure maximum unified acting of the Pact member-states.[1004]

The troops reduction, admitted by the Soviet Union,[1005] in fact, did not distort the long-term plans of the Unified Armed Forces' development. The upcoming five-year plan actually envisaged no increasing of troop numbers assigned to the alliance formation. The

---

stycznia - 19 kwietnia 1978 r./, April 29, 1978, Accessed October 16, 2019. http://www.php.isn.ethz.ch/kms2.isn.ethz.ch/serviceengine/Files/PHP/111439/ip ublicationdocument_singledocument/758212af-25b1-40e8-8d73-c1955c8e0197/pl/1978_Note_SALT_II.pdf.

[1003] *Informácia o priebehu a výsledkoch viedenských rokovaní o vzájomnom znížení ozbrojených síl a výzbroje v strednej Európe v 2. polroku 1978 a dalšom postupe čs. delegágie v roku 1979*, 24. 1. 1979, 1261/0/7, sv. 96, a. j. 96/info11, NA, Prague.

[1004] In the end, the strategy of the Warsaw Pact countries was not discussed at the meetings of the alliance collective institutions but rather bilaterally with individual satellites' representatives of the Soviet general staff, under the direction of the Soviet top military. *Ujednocení dalšího postupu při vídeňských jednáních*, 26. 3. 1980, FMO GŠ/OS, k. 21, p. č. 65, č. j. 40002, VÚA, Prague; *Zpráva o průběhu 14. etapy vídeňských jednání*, 26. 4. 1978; *Zpráva o novém návrhu k snížení ozbrojených sil a výzbroje ve střední Evropě*, 1. 6. 1978; *Zpráva o novém návrhu socialistických států - přímých účastníků vídeňských jednání na vzájemné snížení ozbrojených sil a výzbroje ve střední Evropě*, 1. 6. 1978, k. 1, p. č. 23, e. č. 39006, ibid.

[1005] In April 1978, NATO proposed to impose ceilings for the numbers of military personnel in the region of Central Europe to 900 thousand men on both sides. According to Western estimation, the numbers of the Unified Armed Forces exceeded this quota by 268 thousand. The Warsaw treaty Organization officially admitted 987 thousand soldiers dispatched in Central Europe, NATO about 66 thousand less. The USSR put forward an alternative: to reduce NATO and Warsaw Pact troops in the region by 91 and 105 thousand men, respectively, in the upcoming 3-4 years, but stipulated a condition of withdrawal of 1,000 U.S. nuclear warheads, 36 Pershing missiles, and 54 F-4 Phantom aircrafts. Ibid.

emphasis was put on modernization - qualitative, rather than quantitative strengthening of the units.[1006] Nevertheless, the scheduled increase of weapons industry production was enormous. During the years 1980-1985, it was supposed to grow by approximately 45 percent. Recommendations for individual Warsaw Pact member-states in this matter was not issued solely by the alliance institutions. The guidelines for planned reorganization of communication troops of the CSPA, for instance, was sent by the Soviet general staff instead of the Staff of the Unified Armed Forces. The plans for development of the Soviet satellites' armed forces were naturally mostly based on the actual trends in the Soviet army.[1007] Officially, however, the armed forces buildup was still strictly directed by the supreme commander's recommendations. He discussed these issues with the military institutions of individual countries bilaterally. However, there is evidence that the Soviet satellites, at least in some cases, informed each other about the talks with Kulikov and the other representatives of the Unified Command.[1008]

Economic complications

The ambitious armament plans absolutely did not reflect the real economic capacities of the Eastern Bloc countries. Even though the main phase of the Unified Armed Forces' modernization should have come in the mid-1980s, at the 18th gathering of the Military Council in October 1978, Marshal Kulikov complained that no Warsaw Pact state was meeting

---

[1006] Any potential reduction of troop numbers was envisaged. They should remain at least on the same level as in 1980. *Zámysl výstavby ČSLA*, 28. 2. 1978, p. č. 24, e. č. 39007, ibid; Information by M. Dzúr, the Czechoslovak defense minister, for G. Husák, the CPCz CC General Secretary, November 16, 1978, k. 35, p. č. 127, e. č. 36027, ibid.

[1007] *Podklady pro projednání soudruha ministra národní obrany s hlavním velitelem Spojených ozbrojených sil*, 20. 1. 1978, k. 1, p. č. 28, č. j. 39012, ibid.

[1008] *Sdělení NGŠ PLA generála broni F. SIVICKÉHO*, 22. 1. 1978, k. 13, p. č. 50, č. j. 39043, ibid; Писмо главнокомандующему обьедиными вооруженными силами государств-участникоб варшавского договора маршалу советскофо союза товаришу Куликову В. Г., 6. 3. 1978, k. 1, p. č. 24, e. č. 39007, ibid; *Upřesnění návštěvy náčelníka štábu Spojených ozbrojených sil v ČSSR*, 13. 2. 1979, k. 44, p. č. 154, č. j. 36504, ibid.

100 percent of the scheduled supplies of new weapons and equipment.[1009] Therefore, he asked individual member-states' military officials to urge their national political institutions to accelerate the military production.[1010] The Czechoslovak example shows that the armies really began to provide the communist parties' leadership with the analyses warning against development within NATO.[1011] The West was allegedly preparing for a long war, including conventional as well as nuclear operations. The reports pointed to the trend of using recent scientific discoveries in the development of sophisticated arms to potentially win a conflict. These efforts should have climaxed in the upcoming decade, whereas NATO armed forces were supposed to reach a significantly higher qualitative level. This of course justified the demands for massive investments in the military area in the years 1981-1985. In this period, the Warsaw Treaty Organization was envisaged to rearm 75% of ground forces, 85% of air force units and air-defense system, as well as 70% of navy. The question remains of how much the national military institution followed Kulikov's order to urge a massive increase in defense spending. The Czechoslovak documentation suggests that CSPA acted to a very limited extent only. On the contrary, according to the directives of the defense ministry,

---

[1009] The rearmament process faced delays. There was the lack of certain weapons and equipment and a lot of obsolete arms were still in service. Terms for supplies of requested replacements were gradually prolonged. This especially concerned yacheykas, which should have been equipped with "Sani", "Podnos" and "Vasiljok" mortars, "Konkurs", "Metis" and Žalo-B" anti-tank weapons, "Osa-AK" and "Strela-10 sv" anti-aircraft missiles and MiG-25 fighters as well as their recon version RB. *TEZE DOKLADU zástupce hlavního velitele Spojených ozbrojených sil pro výzbroj - náčelníka Technického výboru na 11. zasedání Výboru ministrů obrany ke třetí otázce programu*, undated 1978, k. 35, p. č. 127, e. č. 36027, ibid.

[1010] *18. zasedání Vojenské rady Spojených ozbrojených sil*, 23. 10. 1978, k. 34, p. č. 125, č. j. 36025, ibid.

[1011] Former Czechoslovak general Mojmír Zachariáš retrospectively stated that the analysis talking about increasing military strength of NATO in the second half of the 1970s had been based on real intelligence reports and reflected neither political nor military order. Interviews with the Czechoslovak Generals, 2002-2003, in *Plánování nemyslitelného* (ad. 2): 420.

rearmament should have proceeded in line with the country's economic capacities.[1012]

In late 1978, the USSR tried to force its satellites to implement the set armament plans. It used the Political Consultative Committee session and also the following Committee of Defense Ministers meeting for this purpose. The Soviet defense resort stated that the main tendency in international relations was still easing tensions. However, it warned that this process was facing "significant obstacles" and accused the imperialists, led by the U.S., of policy aiming at a return of the Cold War. Accelerated NATO militarization, characterized by using science for military purposes, and an increase in military budgets, was marked as the result. At the 11th Committee of Defense Ministers' gathering, held in East Berlin on December 4-7, the Soviet general Petr Ivaschutin warned against missing the "technological moment of surprise". He appealed for development of new, modern conventional weapons systems, which - according to his opinion - would win a future conflict.[1013] The rousing reports claimed that NATO military spending could have risen by as much as 54% in the years 1978-1982. The supreme commander, Victor Kulikov, declared that the military's five-year plan of 1981-1985 had to be based mostly on these assumptions.[1014] The main emphasis was given to the introduction of modern weaponry of all kinds - R-17 (Scud) and Luna-M tactical missiles, T-72 tanks, Gvozdika, Akaciya and Dana self-propelled

---

[1012] Information by M. Dzúr, the Czechoslovak defense minister, for G. Husák, the CPCz CC General Secretary, November 16, 1978; *USNESENÍ Výboru ministrů obrany členských států Varšavské smlouvy k otázce "Obecné směry rozvoje Spojených ozbrojených sil členských států Varšavské smlouvy na léta 1981-1985"*, undated, 1978FMO GŠ/OS, k. 35, p. č. 127, e. č. 36027, VÚA, Prague.

[1013] Mastny, "Imagining War in Europe," 35. It is good to add that Ivashutin was already known for his extreme military prognosis in the first half of the 1960s.

[1014] *TÉZE DOKLADU představitelů ministerstva obrany SSSR na 11. zasedání Výboru ministrů členských států Varšavské smlouvy k první otázce programu "O stavu a perspektivách rozvoje sil NATO"*, undated 1978; *TEZE DOKLADU hlavního velitele Spojených ozbrojených sil členských států Varšavské smlouvy na 11. zasedání Výboru ministrů obrany ke druhé otázce programu - "Obecné směry rozvoje Spojených ozbrojených sil členských států Varšavské smlouvy na léta 1981-1985"*, undated 1978, FMO GŠ/OS, k. 35, p. č. 127, e. č. 36027, VÚA, Prague.

howitzers, guided anti-tank missiles, recent versions of rocket launchers, KRUG, KUB, and Strela anti-aircraft systems, MiG-23, MiG-25, Su-22 and Su-25 combat aircrafts, or Mi-24D helicopter gunships.[1015]

It cannot be said that only the Soviet and Unified Command's officials warned against the course of events in the West at the Warsaw Treaty Organization meetings. For example, the Polish defense minister, Wojciech Jaruzelski, identified himself with the analyses of possible development in NATO. But his speech, unlike the Soviet-ones, presented fears about the future of détente, rather than a call for intensification of the Warsaw Pact's armament programs.[1016] Thus, some rhetorical shift occurred. As mostly warnings against too much openness towards détente could have been heard at the alliance military bodies meetings in the first half of the 1970s, now concerns about its future fate were expressed instead. However, the conclusion was the same: To keep strengthening defense capabilities and not to allow NATO to gain superiority over the Warsaw Treaty Organization. In praxis, this meant continuing the intensive arms race.[1017]

That spirit was the context for the Political Consultative Committee session, held in Moscow, December 1978. Brezhnev ascribed the ongoing changes in NATO to the American hawks, whose influence had been decreasing in the previous period. He blamed them for "sabotaging" détente and called for common action to save it.[1018] However, the hardliners were also given plenty of room. In his report, Kulikov torpedoed the member-states because of "small" investments into the military area and delays in supplies by their weapons industry. Given the ongoing unification process, these had an impact on the entire Unified

---

[1015] *Usnesení Výboru ministrů obrany členských států Varšavské smlouvy k otázce "O vybavení vojsk a námořnictva vyčleněných do sestavy Spojených ozbrojených sil, výzbrojí a vojenskou technikou na léta 1981-1985"*, undated 1978, ibid.

[1016] *WSTĄPIENIE Ministra Obrony Narodowej PRL, gen.armii Wojciecha Jaruzelskiego w dyskusji nad I punktem porządku dziannego XI posiedzenia Komitetu Ministrów Obrony - Berlin, grudzień 1978r.*, PZPR KC, XIB/172, AAN, Warsaw.

[1017] *UCHWAŁA Komitetu Ministrów Obrony państw Układu Warszawskiego na temat: "O stanie i perspektywach rozwoju sił zbrojnych NATO"*, 27. 11. 1978, ibid.

[1018] Mastny, *A Cardboard Castle?*, 45.

Armed Forces.[1019] He pointed to the last meeting of the North Atlantic Council in which the intention was declared to significantly improve capabilities of the Western alliance. This provided the Warsaw Pact supreme body with an excuse for formal approval of substantially strengthening its own military capacities.[1020] However, this was just the confirmation of an already set course.

As mentioned in the previous chapter, Romania was the only alliance member-state that openly opposed these plans. Ceausescu refused to sign the Political Consultative Committee protocol.[1021] Romanian representatives also harshly condemned Soviet intentions at the following 11th meeting of the Committee of Defense Ministers. Furthermore, they tried to include in the resolution that the armament was causing "*more and more damages to the social progress of all the countries and is gradually complicating realization of disarmament.*" Facing the risk of escalation of the arms race, Bucharest tried once again to challenge the Warsaw Pact bodies' authority.[1022] Ion Coman, the defense minister, stated the

---

[1019] For instance, in December 1978, at the Committee of Defense Ministers, Jaruzelski informed his Czechoslovak counterpart that the Polish side would not start testing the already delayed supplies of "Strela-1M" missiles for the CSPA before 1980. He advised Dzúr to ask the USSR for these weapons. Nevertheless, Czechoslovakia remained interested in cooperation with Poland on the field of military production. Both ministers also discussed an opportunity of exchanging drills on the polygons in their countries. *INFORMACE o průběhu a výsledcích 11. zasedání Výboru ministrů obrany, konaného ve dnech 4.-7. prosince 1978 v BERLÍNĚ*, undated, 1978, FMO GŠ/OS, k. 35, p. č. 127, e. č. 36027, VÚA, Prague.

[1020] *TÉZE DOKLADU hlavního velitele Spojených ozbrojených sil maršála Sovětského svazu V. G. Kulikova na zasedání Politického poradního výboru "O stavu a rozvoji vojenské spolupráce Spojených ozbrojených sil členských států Varšavské smlouvy" /listopad 1978/; USNESENÍ* členských států Varšavské smlouvy, přijaté na zasedání Politického poradního výboru ke zprávě hlavního velitele Spojených ozbrojených sil, 23. 11. 1978, k. 44, p. č. 148, č. j. 36054, ibid.

[1021] Ibid.

[1022] The Romanian delegation proposed to issue the final documents on behalf of the member-states, rather than the entire alliance. Moreover, it also tried to enforce formulation in the text outlining the development of the armies in the five-year plan of 1981-1985 that, besides the Political Consultative Committee and the Committee of Defense Ministers, the RCP leadership would also have the authority to set out the measures. In fact, unlike the rest of the countries, Romania

Romanian intention to cooperate with all the countries regardless of their belonging to NATO or the Warsaw Treaty Organization. He also stepped up against the claim that there was a threat to peace and to the socialist states' security in the existing international situation. According to his words, Romania planned to act mostly in accordance with the text of the Final Act in future. In the following discussion, these statements were dismissed and condemned by the ministers of the Six, led by Dmitri Ustinov.[1023] On the contrary, during the final meeting of the Committee of Defense Ministers' members with the East-German leader, Erich Honecker, the head of the Soviet defense resort, roared about a strengthening of NATO that had to be responded to by development of better weapons in the Warsaw Pact countries. In this regard, he instrumentally said that the pact of the socialist countries had been established in reaction to the existence of the western alliance.[1024]

Undoubtedly, the massive military production did not trouble the Soviet satellites only. It also significantly harmed the economy of the USSR. It is an undeniable fact that during Brezhnev's rule, the Soviet armed forces were substantially strengthened, especially the air forces and

---

did not consider the plans outlined by the Unified Command as a directive, but rather just a non-binding recommendation which could have been taken into account by the Romanian armed forces. Report by M. Dzúr, the Czechoslovak defense minister, for G. Husák, the CPCz CC General Secretary, December 8, 1978, k. 35, p. č. 127, e. č. 36027, ibid.

[1023] For example, in an internal discussion, Kulikov and Ustinov commended the Czechoslovak defense minister Dzúr for his speech against stances of his Romanian counterpart. The supreme commander also criticized some Romanian measures which, in his opinion, harmed the air-defense system at the Warsaw Pact's southwestern flank. Due to stances of the Ceausescu regime, it was impossible to organize it through unified plans. Similar to the praxis of the first half of the 1970s, Bucharest rejected solving the issue through a single document under the auspices of the Warsaw Treaty Organization and instead proposed the series of bilateral state conventions instead. However, Kulikov intended to stop some of the sort of benevolence shown by his predecessor Yakubovski in this regard. Therefore, he instrumentally called the Romanian attitude a dangerous precedent.

[1024] *INFORMACE o průběhu a výsledcích 11. zasedání Výboru ministrů obrany, konaného ve dnech 4.-7. prosince 1978 v BERLÍNĚ*, undated, 1978, FMO GŠ/OS, k. 35, p. č. 127, e. č. 36027, VÚA, Prague.

navy. Nevertheless, the West still kept certain advantages in these areas.[1025] The longstanding catching up of the United States in the production of nuclear weapons and development of their carriers, which culminated in the 1970s, brought a certain, although not fully stable, military balance. The Soviet Union paid for this achievement by another restriction of the already low living standards of its population. As the arms race exhausted the Soviet economy, they significantly accelerated arrival of the crisis of the entire communist system.[1026]

The Soviet military improvements had at least two negative impacts: These literally churned the economy out and at the same time escalated the arms race in a closed circuit. Continual massive investments into army projects in the 1970s actually raised doubts in the West about Moscow's real intentions.[1027] Stagnation of living standards of the Soviet people as well as the economy sharply contrasted with the extraordinary efforts in the military area. Achieving strategic parity with the U.S. the USSR marked as an historical task. But these claims failed to mention that the strength of the American economy was double in comparison to the Soviet one. Thus, Washington invested an incomparably smaller part of its means in military projects. Furthermore, the Americans never relied so much on quantity: They believed that there was no need for dozens of guns to win a pistol duel, but rather one absolutely reliable.[1028]

[1025] Bowker, "Brezhnev and Superpowers Relations," 91; Westad, "The Fall of Détente," 18. Theodor Hoffman, former East-German admiral, retrospectively remembered that in the 1960s, he still considered the new generation of the East-German military vessels better than that of NATO. In the second half of the next decade, these ships could not have been measured even against those of Denmark based on modern technologies. Mastny, *A Cardboard Castle?*, 47.

[1026] Volkogonov, *The Rise and Fall*, 276.

[1027] Bowker, "Brezhnev and Superpowers Relations," 104.

[1028] Volkogonov, *The Rise and Fall*, 208-281.

Comparison of military spending in the years 1976-1985 based on the CIA and NIE estimates (USD billions)											
alliance/year	1976	1977	1978	1979	1980	1981	1982	1983	1984	1985	Total
Warsaw Treaty Organization	263	265	269	274	279	280	281	286	289	293	*2778*
NATO	276	282	287	294	302	313	330	349	361	373	*3166*

Table II[1029]

The Soviet military plans faced another challenge: Even the official Soviet stats admitted that already at the time when Brezhnev had come to power, the Soviet economy had been experiencing significant slowdown in comparison to the previous post-war boom. Actually, this trend was gradually deepening until the General Secretary's death.[1030] In

---

[1029] The table based on Firth, Noel E. and Noren, James H., *Soviet Military Spending* (Texas: A&M University Press, 1998), 120.
[1030] In 1973, the USSR stopped catching up with the West economically. Although this year is generally considered as the end of strong post-war economic growth, the slowdown in the USSR and the entire Eastern Bloc was more striking than in Western Europe and the United States. Perhaps the issue had deeper roots. An extreme tempo of economic growth in the 1950s - a result of the post-war recovery

the light of this, Brezhnev's leadership would have undoubtedly welcomed the conclusion of some international agreements allowing for the reduction of military spending. In the second half of the 1970s in fact, there were rising fears that the country would not be able to keep pace with the U.S. in another round of the arms race. From the Kremlin's point of view, restrictions of living standards of the people in favor of military spending could have potentially weakened internal stability of the entire regime.[1031] However, the Soviet military-industrial complex still pushed forward more and more demands justified by achieving and maintaining a strategic parity with the USA. The situation was complicated by the fact that at the time of decline of the Soviet political leadership's ability to act and growth of general bureaucratization and corruption, apparently the most loyal coworkers of Brezhnev actually came from the military circles. Marshal Ustinov, strongly fawning to the General Secretary, was a typical example.[1032]

According to some testimonies, the Soviet general staff itself was well aware of the real situation. Former East-German colonel Joachim Schunke, whose job was to monitor NATO's technological innovations, claims that in reality the Soviet top military did not draw any radical conclusions of gradually growing Western advantage in modern conventional weapons. They knew that given its weak economy, the USSR simply could not afford this technological battle. Therefore, besides armament programs, a strong emphasis was given to other elements of improving defense capabilities, especially organizational changes in the command system and maintaining troops in high combat and mobilization readiness.[1033] These aspects are analyzed in detail in the chapter focused on doctrinal and strategic changes at the end of the 1970s.

The Soviet defense minister, Dmitri Ustinov, was one of the main initiators of another round of the arms race. Especially in 1979, when

---

- had been just unsustainable. Thus, its decline cannot be solely ascribed to wrong economic decisions in the USSR and the Eastern Bloc. Harrison, Mark, "Economic Growth and Slowdown," in *Brezhnev Reconsidered*, eds. Erwin Bacon and Mark Sandle (New York: Palgrave, 2002): 41-55.

[1031] Westad, "The Fall of Détente," 13.

[1032] Volkogonov, *The Rise and Fall*, 279, 319-320.

[1033] Mastny, *A Cardboard Castle?*, 48.

Brezhnev's health was deteriorating rapidly, he found himself in an "armament fever." He even tried to force those Warsaw Pact countries, which had engaged little in military production so far, to join the armament programs. Thus, in the same year, Ustinov told the East-German military delegation in Moscow that in the light of the U.S. arms projects, some significant increase of weapons production in the GDR was necessary during the upcoming years 1981-1985. He expected the country to start manufacturing some heavy weaponry like aircrafts, tanks, and ships. But even the very loyal Heinz Hoffmann, the East-German defense minister, noted that realization of the Soviet recommendations would have been very complicated. The negotiation was extremely tough. Hoffmann was well aware of the fact that the East-German people would not agree with reductions in consumer goods' supplies in favor of investments into the weapons industry. Due to absence of a language barrier and proximity to the FRG, the East-Germans actually had the absolutely best change of the entire Eastern Bloc in comparing living standards with the West directly. But the Soviet side had no understanding for the East-German unwillingness to build a weapons industry. Moscow pointed to the fact that the GDR had the least-developed military production of all the Warsaw Pact member-states. Ustinov therefore implored the East-German representatives that their country had to begin manufacturing at least one major weapon system. The situation, when the East-Germans, in terms of the Eastern Bloc's very developed industry, engaged simply in producing automatic rifles, he called no longer sufficient.[1034]

Also, the weapons industry centers of the alliance, like Czechoslovakia, faced more and more troubles. In spring 1979, the report by the defense ministry stated that the tasks based on the Protocol on the

---

[1034] On this issue, the visions of the Soviet and East-German leadership differed. The SED top officials thought that their task was to build the GDR as a "showcase" of socialism nearby Western borders, and which would also be able to compete with the West-German economy in terms of production of consumer goods. Thus, they did not understand the appeals for the transfer of some resources to the weapons industry. The NPA remained clearly the most dependent on supplies of the Soviet arms of all the Warsaw Pact armies. Especially during the 1980s, more troubles appeared because of their shortcuts. Schaefer, "The GDR in the Warsaw Pact".

assignment of the CSPA troops into the Unified Armed forces in the years 1976-1980 were being fulfilled, but only with enormous effort. At the time, Czechoslovakia subordinated 200 thousand soldiers to the Unified Command in peacetime. In case of war, it should have been 700 thousand. This was, in fact, almost the entire military potential of the country. Even though, the real number of backups was still about 50 thousand below the demand of the alliance command.[1035] Moreover, creating new elite units required monies exceeding the scope of the aforementioned Protocol.[1036] In the years 1979-1980, the Czechoslovak government had to find additional budgetary resources for introducing modern equipment. In addition, keeping with the conclusions of the meetings within the Warsaw Pact's military structures, the army command was preparing the political power for the risk of a need to increase investments into the military programs significantly in the upcoming five-year plan.[1037]

In the first months of 1979, the supreme commander Kulikov sent to individual Warsaw Pact member-states the recommendations on buildup of their armies in the five-year plan of 1981-1985.[1038] As the

---

[1035] *Zpráva o průběhu plnění "Protokolu o vyčlenění vojsk ČSSR do sestavy Spojených ozbrojených sil a jejich rozvoji na léta 1976-1980, vyhodnocení plnění protokolu vyčlenění vojsk ČSSR do sestavy SOS a jejich rozvoj na léta 1976-1980,* 26. 3. 1979, FMO GŠ/OS, k. 1, p. č. 27, č. j. 39011, VÚA, Prague.

[1036] In the CSPA, the yacheykas were called the units equipped by T-72 tanks, MiG-23 aircrafts, Mi-24 gunships, "Fagot" anti-tank missiles, and "Strela" anti-aircraft complexes. A report from late 1978 cautiously criticized the state of almost all kinds of troops - the anti-tank defense, missile units, and artillery. The situation in the air forces was considered the worst. The military was right that MiG-21 fighters and Su-7 tactical bombers, which were a backbone of the Czechoslovak air forces at the time, cannot compete with new U.S. combat aircrafts F-15, F-16, and F-18. However, by 1980, only one squadron was supposed to use new MiG-23 BN fighter-bombers. Keeping with the previous appeals of the supreme commander, the army broached to the political power that in order to meet the requested tasks, it would have been good to increase investments into military area. *Osnova k zpracování vojenskoekonomického rozboru výstavby ČSLA*, 31. 10. 1978, k. 44, p. č. 147, č. j. 36053, ibid.

[1037] *Osnova k zpracování vojenskoekonomického rozboru výstavby ČSLA*, 31. 10. 1978, k. 44, p. č. 147, č. j. 36053, ibid.

[1038] We are reminded that the .protocols between the member-states and the Unified Command were negotiated bilaterally; they were signed by the supreme

Czechoslovak case shows, these guidelines were not welcomed warmly. Even the general staff of the CSPA, usually absolutely loyal to the Unified Command, stated in advance that practical measures would strictly reflect the human and economic capacities of the country. The Czechoslovak side considered many recommendations of the supreme commander excessive. Therefore, it tried to mitigate them by its own compromise on the proposals. In the air forces, for example, less than 30% of the marshal's demands would have been realized.[1039] But the alliance command put an extreme emphasis on this area. In April 1979, at the 19th Military Council session in Warsaw, Kulikov once more complained about very slow rearmament of these units and set an improvement in this field as a priority.[1040] The deputy supreme commander for the air forces, the Soviet general A. N. Katrich, warned that combat potential of NATO air forces would increase by 150% until 1986. In this regard, he criticized that some of the Warsaw Pact countries still used the absolutely obsolete MiG-15, MiG-17, MiG-19, and L-29 aircrafts along with the Mi-1 and Mi-4 helicopters. Based on a recommendation by the Soviet marshal A. I. Koldunov, reorganization of the air force units was supposed to take place - the air regiments consisting of forty MiG-23MF or twenty-five MiG-25P fighters were preferred. There should have been three pilots on two combat aircrafts. Regarding this, Kulikov emphasized that the solution of actual complex problems needed an absolutely unified approach within all the

---

commander and individual defense ministers. The Unified Command officials, led by Marshal Kulikov, got involved in these talks.

[1039] Kulikov, for instance, advised the Czechoslovak air force to replace MiG-17 and MiG-21 aircrafts with the powerful interceptor MiG-25. In total, 24 of them should have been introduced into service. But the Czechoslovak side did not welcome this measure. It proposed to purchase only 4 pieces of a recon version of this jet. *Jednání s NŠ/SOS dne 26.2. 1979*, 27. 2. 1979, FMO GŠ/OS, k. 44, p. č. 154, č. j. 36504, ibid; *ÚVODNÍ SLOVO náčelníka generálního štábu ČSLA - 1. ZMNO k dokladům o výstavbě ČSLA v letech 1981-1985*, undated, 1979, p. č. 151, č. j. 36501, ibid.

[1040] According to the then calculations, at least 120 air raids of fighter-bombers per day were necessary to successful combat activities of a single motor-rifle division belonging to the Unified Armed Forces' first echelon. For this reason, the fighter-attack units for ground forces support should have formed up to two-thirds of the entire air forces numbers.

Warsaw Pact member-states' armies.[1041] Thus, in the light of challenges related to the fall of détente, alliance cooperation should have been tightened, with the existing reality in the Soviet armed forces as a model.[1042]

The political power of the loyal Soviet satellites objected to the above-mentioned principles only very cautiously. In April 1979, for example, during his meeting with the Military Council members, Edward Gierek explained in detail the economic problems which Poland began to experience at the time. Nevertheless, he called another strengthening of defense capabilities necessary. After all, in his speech at the Military Council meeting, the Polish deputy defense minister, Eugeniusz Molczyk, stated that attempts at military confrontation with the socialist countries remained a continual aspect of the imperialistic policy.[1043] Thus, in 1980, Poland planned to increase its military spending by 1.7% at least. This happened despite the fact that the authorities were anticipating annual downturn of the Polish GDP by 2.5%. The previous ambitious economic plans actually envisaged growth by 2.8%.[1044] In fact, facing economic troubles, the Polish industry was unable to fulfill the already signed contracts for the rest of the alliance armies. These included not only the aforementioned "Strela-1" anti-aircraft missiles, but also Su-25, the Soviet combat aircraft for ground forces support produced in license.[1045]

[1041] Romania formally agreed with the proposals. However, it emphasized that every country had to implement these measures in accordance with its capacities. *19. zasedání vojenské rady Spojených ozbrojených sil*, 27. 4. 1979, FMO GŠ/OS, k. 44, p. č. 153, č. j. 36503, VÚA, Prague.

[1042] Since the second half of the 1970s, for instance, so called reorganization of combined divisions in the CSPA took place. Besides introducing modern weaponry, this meant mostly organizational unification with the Soviet army. Bílek and Koller, "Československá armada," 204.

[1043] *19. zasedání vojenské rady Spojených ozbrojených sil*, 27. 4. 1979, FMO GŠ/OS, k. 44, p. č. 153, č. j. 36503, VÚA, Prague.

[1044] *Projekt budzetu państwa na rok 1980*, undated 1979; Ibid, *Notatka w sprawie podstawowych problemów planu na 1980 rok*, 19. 10. 1979, PZPR KC, XIA 1130, AAN, Warsaw.

[1045] *Příprava "Protokolu o vyčlenění vojsk ČSSR do sestavy Spojených ozbrojených sil"*, 6. 11. 1980, FMO GŠ/OS, k. 21, p. č. 64, č. j. 40001, VÚA, Prague.

Statute on the Warsaw Treaty Organization cooperation with the developing countries

The Soviet Union obviously sought for ways to solve the aforementioned problems with implementation of the armament programs. One of them was the "*Statute on coordination of the Warsaw Treaty Organization's activities in realization of cooperation with the developing countries.*" At first glance, the document had no direct relation to the issue of intensified armament. The question was opened in 1978. The Soviet side justified the draft of this document by pointing out increasing military assistance provided by the Warsaw Pact countries to the developing states which had proceeded largely uncoordinated. The USSR noted that in this situation, many Third World countries were negotiating on supplies of the same military equipment with various alliance member-states at the same time. In this manner, they strove for lower final prices. Signed contracts on service or training of personnel were often evaded by the developing countries in the same way. The Soviet defense ministry officially stated that this was a direct effort of some pro-Western representatives in the developing states to harm the Warsaw Treaty Organization, to provoke undesirable competition between its members and to instigate discord between their military specialists. But in reality, taking into account the scheduled rearmament and beginning problems in supplies of modern military equipment, Moscow suspected of some of its European satellites that due to clear financial reasons they preferred export of weapons to the developing countries rather than fulfilling the supplies for the Warsaw Pact armies. Furthermore, licensed-produced arms were often sold without consent of the original manufacturer. Therefore, from the Soviet point of view, the aim of the statute was to reduce the alliance member-states' autonomy in terms of military cooperation with the developing world. Henceforward, it should have reflected not only the national interests of individual countries, but mostly the interests of the entire socialist camp. A weapons industry was supposed to primarily ensure the Warsaw Pact's requirements.[1046]

[1046] *TEZE DOKLADU představitelů ministerstva obrany SSSR na 11. zasedání Výboru ministrů obrany členských států Varšavské smlouvy ke čtvrté otázce "O*

Naturally, everything was under the guise of claims about "proletarian internationalism." However, the Warsaw Pact's cooperation with the developing countries should not have been coordinated by any alliance body, but by the Soviet defense ministry and general staff. The states exporting licensed-produced weapons were supposed to get permission by the original manufacturer. This significantly bolstered Soviet supervision on arms supplies, as the country was the main developer of the weapons system in the Warsaw Treaty Organization. The state which signed a contract was at least granted the right to conduct service works and provide technical assistance. The statute also clearly declared that the alliance member-states' priority was to meet the Unified Armed Forces' needs.[1047]

In late 1978, the supreme commander Kulikov considered the issue of the statute solved. He expected no serious discussion about it.[1048] But the initial draft of the document, which significantly reduced the individual government's ability to export military materials, aroused a relatively unusual outcry within the Warsaw Pact's military structures. For example, in internal documentation, Czechoslovakia pointed out that decisions about military contracts based on consultation between the member-state defense ministry and the Soviet institutions were not compatible with the Czechoslovak laws, which grated those competences to the party and military bodies.[1049] Thus, in December 1978, at the 11th session of the Committee of Defense Ministers, the Statute on cooperation with the developing countries had to be written off the agenda, as individual national institutions of the member-states wanted to assess the

---

koordinaci činnosti členských států Varšavské smlouvy při uskutečňování spolupráce s rozvojovými zeměmi", undated 1978, k. 35, p. č. 127, e. č. 36027, ibid.

[1047] *STATUT o koordinaci činnosti členských států Varšavské smlouvy při uskutečňování spolupráce s rozvojovými zeměmi*, undated 1978, , e. č. 36027, ibid.

[1048] Letter by F. Siwicki, the chief of the Polish general staff, to Mieczysław Jagielski, the deputy prime minister of Poland, December 3, 1978, URM-KT, s. 98/77, AAN, Warsaw.

[1049] *Změny v materiálech k 11. zasedání Výboru ministrů obrany*, 24. 11. 1978, FMO GŠ/OS, k. 35, p. č. 127, e. č. 36027, VÚA, Prague.

issue in detail.[1050] It is also not surprising that Romania strongly objected the Soviet intentions. Ion Coman noted that the Warsaw Pact's regulations contained no provision justifying such a document. In addition, he pointed out the strictly European range of the alliance. According to his interpretation, this as well as other foreign policy issues fell within the remits of individual countries only.[1051]

Symptomatically, the top national political bodies were not supposed send their comments on the statute to the Unified Command, but rather to the Soviet defense ministry. This institution intended to engage very actively in the negotiations.[1052] Later, the Soviet general staff began to work on the compromised text. Its representatives had a series of bilateral meetings with individual member-states' military officials. In October 1979, the final document was then discussed at a multilateral gathering of the representatives of the general staff's and the special foreign trade institutions. One month later, the material was finally approved by the 12th Committee of Defense Ministers' session in Warsaw. A formal blessing by the governments of individual member-states

---

[1050] In Czechoslovakia, for example, a special commission led by the deputy prime minister and the economic ministries, minister of foreign affairs, and the chief of general staff was established. The Czechoslovak documentation suggests that the political institutions paid much more attention to the preparation of the statute on coordination of military cooperation with the developing world than simultaneously preparing guidelines for wartime. This proves that at the end of the 1970s, the Husák normalization regime abandoned strictly passive positions within the alliance framework. It cautiously stepped up mostly against those intentions which had a potential to affect the economy of the country in a negative way, but not against the general line of foreign and military policy of the USSR. Report by M. Dzúr, the Czechoslovak defense minister, to G. Husák, the CPCz CC General Secretary, 8. 12. 1978, ibid; *PLÁN REALIZACE USNESENÍ z 11. zasedání Výboru ministrů obrany členských států Varšavské smlouvy*, undated, 1979, k. 8, p. č. 43, č. j. 39035, ibid.

[1051] *USNESENÍ Výboru ministrů obrany členských států Varšavské smlouvy k otázce "Statutu o koordinaci činnosti členských států Varšavské smlouvy při uskutečňování spolupráce s rozvojovými zeměmi"*, undated, 1978, FMO GŠ/OS, k. 35, p. č. 127, e. č. 36027, ibid.

[1052] *Uchwała Komitetu Ministrów Obrony państw Układu Warszawskiego dotycząca projektu "Statutu o koordynacji działań państw-stron Układu Warszawskiego w ramach współpracy wojskowj z państwamo rozwijającymi się"*, 2. 12. 1978, URM-KT, s. 98/77, AAN, Warsaw.

followed. This procedure proves that within the Warsaw Treaty Organization, the key armament issues were still mostly decided by the political power, not by the military itself.

The compromised version of the Statute left the Warsaw Treaty Organization countries significant room for military-technology cooperation with the developing world. The national bodies maintained their decisive competences in the issue. In case of a request by some developing state for assistance, they were supposed to work out their own proposal and put it forward to the allies. Nevertheless, the military-technology assistance should have taken into account the Unified Armed Forces' needs at first and only after that the "proletarian internationalism." Everything was top secret of course, in order to avoid revelation of combat capabilities and characteristics of the Warsaw Pact's weaponry. Providing assistance in the production of military equipment in license was the responsibility of that alliance member-state which relayed the license. However, participation of the competent authorities of the other alliance countries was not excluded. Their involvement should have been based on proper bilateral agreements.[1053]

The situation shows that at the turn of the 1970s and 1980s, the USSR did not engage with its satellites in strictly uncompromising and firm positions: unlike in the past, some proposals of Moscow were not taken as absolutely binding directives. They were subjects of some, however limited discussions. In this regard, the Kremlin could have undoubtedly relied on the high loyalty level of the party leadership in the countries of its sphere of influence. Except for Romania, which sometimes showed almost ostentatious independence and opposition, other Warsaw Pact states' party leaders gave Moscow no cause for concern. Reliability of the old chiefs in Hungary and Bulgaria had been known for a long time. The new-ones, installed at the turn of the 1960s and 1970s in Czechoslovakia, Poland, and East Germany, also dispelled all doubts about their loyalty to the USSR during the decade.[1054]

[1053] *Statut o koordinaci činnosti členských států Varšavské smlouvy při uskutečňování vojenskotechnické spolupráce s rozvojovými zeměmi*, 11. 1. 1980, 1261/0/7, sv. P 128/80, b.14, NA, Prague.

[1054] Tejchman and Litera, *Moskva a socialistické země na Balkáně*, 93.

The supreme commander Kulikov urged continual armament even on the eve of the fall of détente. In November 1979, only one month before the outburst of the Euromissile Crisis and the Soviet invasion of Afghanistan, the 20th meeting of the Military Council was held in Bucharest. There, the marshal declared once again that the Warsaw Treaty Organization had to respond to all the measures realized by NATO armed forces. In this regard, the supreme commander repeatedly criticized the slow introduction of modern weaponry.[1055] The Warsaw Pact's top military unambiguously blamed the Carter administration for the deteriorating international situation. In the context of the Soviet actions in Afghanistan, it is not surprising that the West did not plan to change the existing military trends. In 1980, the NATO military budget was increased by 13.5%. The largest part of this amount flew into new military equipment. For the following year, the Pentagon demanded another growth of military spending by 12.7%.[1056]

Even in this situation, planned countermeasures by the Warsaw Treaty Organization, however, faced the economic limits of its members. During his talks with the Unified Command's representatives, the chief of the Czechoslovak general staff, Miloslav Blahník, declared that even though the CSPA was fully trying to implement the supreme commander's recommendation for the upcoming five-year plan for 1981-1985, it was strictly limited by the country's economic capacities and budget allocations. The Staff of the Unified Armed Forces actually reflected most of the Czechoslovak remarks and incorporated them into the plan.[1057] This

---

[1055] *20. zasedání vojenské rady Spojených ozbrojených sil.* 2. 11. 1979, FMO GŠ/OS, k. 45, p. č. 153, č. j. 36511, VÚA, Prague.

[1056] *NÁVRH hlavních směrů rozvoje ČSLA na léta 1981-1985*, undated, 1980, k. 20, p. č. 61, č. j. 35502, ibid.

[1057] It is good to point out that information about scheduled development of individual armies of the Warsaw Pact member-states still had the Soviet generality only. This is suggested by composition of the Unified Command's delegation which discussed the details of the Protocol for the years 1981-1985 with the Czechoslovak side. It consisted solely of the Soviet and Czechoslovak functionaries of the Staff of the Unified Armed Forces - the representatives of other countries in this institution did not participate in the negotiation at all. *ZÁPIS z konzultace k návrhu "Protokolu o vyčlenění vojsk ČSSR do sestavy Spojených*

was a significant shift in comparison to the 1960s and even the first half of the following decade. In fact, many of the rejected supreme commander's recommendations concerned the key area of the air forces and anti-aircraft missiles.[1058] Also the troop numbers assigned by Czechoslovakia to the Unified Armed Forces were reduced.[1059]

Until the death of Brezhnev in November 1982, there were no important corrections in the Warsaw Treaty Organization armament plans. At the meeting of the alliance military bodies, usually just quarrels between the representatives of the Six and the Romanian delegations took place. In December 1980, for instance, Ceausescu used his meeting with the Committee of Defense Ministers' members to call once again for a solution to economic troubles and reduction of armament spending. He marked as the main priority the increase of living standards in the socialist countries. He reminded the members that GDP in the U.S. per capita is more than 250% higher than in Romania. Ustinov tried to refuse Ceausescu's notes politely, referring to the need to ensure defense capabilities at first. At the end, the situation resulted in an argument between the Romanian dictator and the Soviet defense minister, who was supported by his Bulgarian counterpart Dobri Dzhurov and the supreme

---

ozbrojených sil a jejich rozvoji na léta 1981-1985", provedené v PRAZE ve dnech 17.-20. dubna 1980, k. 21, p. č. 64, č. j. 40001, ibid.

[1058] The supreme commander suggested introducing into the armaments of the Czechoslovak air forces 8 MiG-25P and 4 recon MiG-25R. The CSSR refused to create these yacheykas, referring to the lack of foreign currency. Moreover, the Czechoslovak air forces intended to run 12 MiG-21 and 18 MiG-23 less than Kulikov had recommended. In case of helicopters, the Unified Command initially demanded 87 machines more than the Czechoslovak side wanted to purchase. *Přípravy jednání k návrhu "Protokolu"*, 12. 8. 1980; *Příprava "Protokolu o vyčlenění vojsk ČSSR do sestavy Spojených ozbrojených sil"*, 6. 11. 1980, ibid.

[1059] After 1980, the CSPA assigned to the alliance formation 142-145 thousand troops in peacetime and 430-450 thousand in wartime. However, its entire numbers remained unchanged. They were supposed to stay on the level of 190-200 thousand men in peace and 700 thousand in war. Material support, including weapons and equipment, should have cost 69 billion crowns. In fact, the supreme commander initially had asked for 15 billion more. *Návrh protokolu o vyčlenění vojsk ČSSR do sestavy Spojených ozbrojených sil a jejich rozvoji na léta 1981-1985*, undated, 1980, ibid,; *Koncepce výstavby Československé lidové armády na léta 1981-1985*, undated, 1980, k. 22, p. č. 66, č. j. 40004, ibid.

commander Kulikov.[1060] Only four months later, at the 14th session of the Committee of Defense Ministers in Moscow, Ion Coman stated once more that Romania asserted the freezing of existing military budgets followed by some cutbacks. His speech was again criticized by Ustinov who rejected any reduction in armament spending given the existing international situation. On the contrary, he asked the Warsaw Pact members to seek additional financial resources needed to secure the "balance of forces."[1061]

Victor Kulikov, one of the prominent Soviet hawks and Unified Command officials who thought in narrow ideological templates,[1062] enjoyed his notes about the ongoing deteriorating international situation. He used these claims to make repeated appeals for the socialist countries' unity and improving the Unified Armed Forces' combat capabilities.[1063] However, the marshal never went so far as to challenge the actual political line of the Soviet leadership. He always formally supported it publicly. For instance, at the ceremonial gathering of the Military Council, which took place in Moscow on the 25th Warsaw Treaty Organization anniversary, Kulikov emphasized the peace initiatives issued during the past years by the Political Consultative Committee. On behalf of the alliance military structures, he also supported deepening détente which the supreme body of the Pact was pushing forward at the time. Moreover, he talked about peaceful coexistence of the countries with different socio-economic systems. He just verbally assaulted the policy of the Carter administration. Also, a special statement for media later declared that the Military Council

---

[1060] *Informace o průběhu a výsledcích 13. zasedání výboru ministrů obrany konaného ve dnech 1.-3. prosince 1980 v Bukurešti*, 4. 12. 1980, k. 47, p. č. 176, e. č. 37021, ibid.

[1061] *14. zasedání výboru ministrů obrany členských států Varšavské smlouvy*, 7. 4. 1981, k. 49, p. č. 184, e. č. 37503, ibid.

[1062] Kulikov's behavior suggests that his thinking was very ideological. Anatoly Gribkov, the chief of Staff of the Unified Armed Forces, for example, showed much more pragmatism.

[1063] *26. zasedání vojenské rady Spojených ozbrojených sil*, 25. 10. 1982, k. 53, p. č. 204, e. č. 38013, ibid.

fully backed the last peace initiatives presented by the Warsaw Pact's political forums.[1064]

"The Euromissiles" and correlation of forces at the beginning of the 1980s

At the first glance, the strength of the Warsaw Pact's formation was enormous. At the beginning of the 1980s, allegedly its 245 divisions and 60,000 tanks faced the 76 divisions and 22,600 tanks of NATO.[1065] However, despite massive investments, quantity did not always meet quality. In 1981, in some Warsaw Pact member-states' armies still served the T-34 tanks from the time of World War II. Also, the air forces even used the totally obsolete MiG-15 and MiG-17 fighters from the 1950s.[1066] The modern generation of the Soviet fighter-bombers was lighter, better maneuverable in low altitudes, and significantly cheaper in comparison to NATO aircrafts, but their weaknesses were lower loads, shorter range, and insufficient radio-electronic equipment. Even for MiG-23 aircraft, a major drawback was its less effective destruction of the enemy than in the case of the modern NATO war planes. On the contrary, the new combat helicopter Mi-24 was evaluated extremely positively. The Warsaw Pact's analyses assumed that the enemy had no weapon on a similar level. In terms of core parameters, the T-72 tank was fully comparable with the enemy armor. The East also maintained a slight advantage in artillery. On

---

[1064] *Slavnostní zasedání vojenské rady Varšavské smlouvy v Moskvě*, 24. 5. 1980, k. 46, p. č. 161, č. j. 36518, ibid; *Slavnostní zasedání vojenské rady Varšavské smlouvy v Moskvě*, 24. 5. 1980; *Sdělení pro tisk, rozhlas a televizi o provedení 21. zasedání vojenské rady Spojených ozbrojených sil*, undated, May 1980, k. 47, p. č. 169, e. č. 37007, ibid.

[1065] Durman, *Útěk od praporů*, 131.

[1066] *Teze informace náčelníka štábu Spojených ozbrojených sil armádního generála A.I. Gribkova na 14. zasedání výboru ministrů obrany*, undated, 1981, FMO GŠ/OS, kr. 49, p. č. 184, e. č. 37503, ibid; In 1980, in the CSPA, there still operated two combined army squadrons equipped only with obsolete T-34 tanks. It was envisaged that these would serve even in the upcoming years. *NÁVRH hlavních směrů rozvoje ČSLA na léta 1981-1985*, undated, 1980; Ibid, k. 21, p. č. 64, č. j. 40001, *Příprava "Protokolu o vyčlenění vojsk ČSSR do sestavy Spojených ozbrojených sil"*, 6. 11. 1980, k. 20, p. č. 61, č. j. 35502, ibid.

the other hand, NATO had much better guided surface-to-surface missiles.[1067]

Taking into account the latest factor, among others, the Warsaw Pact's top military were scared of the prospective deployment of new American Pershing-II rocket and cruise missiles in Europe. In comparison to their predecessors, these systems were almost ten times more accurate.[1068] The Eastern analytics were afraid that the U.S. could have dispatched up to 460 pieces of these weapons on the territories of the European NATO members. Therefore, even the Warsaw Pact's top military, usually very skeptical about disarmament agreements, pinned their hopes on the potential SALT III deal. They saw it as a welcomed instrument to reduce the advantage of the enemy. Within the Unified Armed Forces, a strong emphasis was put on seeking an effective defense against the cruise missiles, however unsuccessful. An inability of the Eastern Bloc armies to cope with this threat was illustrated by the CSPA's effort to simulate elimination of these weapons by training L-29 Dolphin planes during the "Start-80" drills.[1069] Pershing-II missiles launched from West Germany were able to hit the Soviet territory very quickly. Moreover, the North Atlantic Council decision on December 1979 envisaged a larger deployment of these arms than the Eastern strategists had previously expected.[1070]

This begs the question why the USSR did not accept NATO's proposal and stop deployment of the feared American weapons by a withdrawal of its own SS-20 missiles. We need to understand that at the turn of the 1970s and 1980s, mid-range missiles were one of the core

---

[1067] *Srovnávací tabulky hlavních druhů výzbroje ČSLA a vojsk NATO*, undated 1982, k. 23, p. č. 74, e. č. 41001, ibid.

[1068] *NÁVRH hlavních směrů rozvoje ČSLA na léta 1981-1985*, undated, 1980, k. 20, p. č. 61, č. j. 35502, ibid.

[1069] *20. zasedání vojenské rady Spojených ozbrojených sil. 2. 11. 1979*, k. 45, p. č. 153, č. j. 36511, ibid; *PLÁN REALIZACE Protokolu čís.: 0020 z 20. zasedání Vojenské rady Spojených ozbrojených sil*, undated 1979, k. 46, p. č. 165, č. j. 36522, ibid.

[1070] NATO planned to deploy 108 Pershing-II rockets and 464 cruise missiles. The step should have strengthened deterrent potential and also improved a link between NATO conventional forces in Europe and the U.S. intercontinental ballistic missiles. Holloway, "Jaderné zbraně," 45.

Soviet offensive means. They would have been launched during an initial nuclear strike. SS-20 were supposed to destroy military and strategic objects on the territories of all NATO members as well as in nearby seawaters. Their positives included long range, high flexibility of potential targets, good accuracy, and quick deployment to battle stations. The USSR also acknowledged that NATO would not have equivalent weapons in the foreseeable future.[1071] According to Wojciech Jaruzelski's memories, the deployment of SS-20 should have compensated for the gradually obvious Western technological superiority.[1072] Brezhnev's former assistant Andrei Alexander-Agentov claims that the CPSU CC General Secretary just followed the military command, led by Grechko and Ustinov, on this issue. The army circles were very proud of their modern, mobile, and very accurate missiles. They considered them as a response to short and mid-range missiles of NATO deployed in Europe and waters around the USSR. The Soviet military command believed they confronted not only the U.S. nuclear forces in Europe, but also the atomic arsenals of Great Britain and France. At the same time, the country had to deploy some of its nuclear as well as conventional forces at the borders with China. Moreover, the Soviet military-industrial complex felt that the strategic forces still lagged behind the American ones in terms of quality. For this reason, it tried to achieve at least quantitative superiority. Ustinov's former assistant, Victor Starodubov, retrospectively explained that the USSR had built so many heavy missiles because they had been one of few things the Soviet industry could build well. In hindsight, despite massive buildup of its strategic forces in the 1970s, the Soviet Union did not achieve substantial superiority, as the U.S. neoconservatives warned at the time. The USSR still did not have the capability to launch a surprise disarming strike against the United States. In general, the Americans still maintained the

---

[1071] Information about parameters and planned use of SS-20 missiles were top secret. Ustinov urged the Warsaw Pact member-states' military command to keep it within a small circle of necessary personnel. This was similar in case of development of the Soviet Kiev class aircraft carriers in the Project 1143. Those four vessels were supposed to at least partially weaken NATO naval supremacy. Information by Marshal Ustinov on Soviet Strategic Offensive Forces, September 1981, in: *A Cardboard Castle?*, doc. 92, pp. 449-450.

[1072] Mastny, *A Cardboard Castle?*, 47.

advantage in many military aspects, however, the gap was no longer as wide as it had been earlier.[1073]

Armaments in the era of Brezhnev's successors

These trends did not change significantly after Leonid Brezhnev's death. Initially, the Soviet leader's demise aroused great concern in the representatives to the military-industrial complex, who had relied on some support by the General Secretary so far.[1074] Their fears vanished quite soon. The power of Brezhnev's successor, Yuri Andropov, was actually based on the army circles, especially the old guardians of the World War II results, among others. Thus, the political leadership of the CPSU still reflected their demands for an all-round strengthening of the Warsaw Treaty Organization.[1075]

As explained in detail later, it seems that Andropov himself was not a big supporter of further escalation of the arms race. On the other hand, he still believed in a chimera of the acute military threat by the West. In January 1983, at the Political Consultative Committee session in Prague, he warned that the risk of war had increased rapidly.[1076] The supreme commander Kulikov spoke in the same spirit during the meeting. He noted that NATO was maintaining its troops at high combat readiness which allowed them to launch an attack on the Warsaw Pact under the guise of holding a massive exercise without any additional preparations. With a well-known link to the strengthening of the enemy, the marshal asked the member-states for reconsidering an option to introduce modern equipment and weaponry over the schedule already in the years 1983-1985, not just in the following five-year plan. He pointed to the fact that the Unified Armed Forces were not supplied by the recent modern arms in numbers enough to replace obsolete equipment, often out of service life.[1077]

[1073] Zubok, *A Failed Empire,* 231, 243.

[1074] Volkogonov, *The Rise and Fall,* 326.

[1075] Durman, *Útěk od praporů,* 46.

[1076] Mastny, *A Cardboard Castle?,* 56.

[1077] This included mostly MiG-15, MiG-17 and Su-7b aircrafts, and T-34 tanks. *TÉZE REFERÁTU hlavního velitele Spojených ozbrojených sil na zasedání Politického poradního výboru "O Stavu Spojených ozbrojených sil členských států*

The continuity of the military policy of Brezhnev's era was secured, among other ways, by reconfirmation of Marshal Kulikov in the post of the Warsaw Pact supreme commander at the Prague session of the Political Consultative Committee. Other key-personnel of the Unified Command - the chiefs of the alliance Staff and the Technical Committee, Gribkov and Fabrikov - remained in their positions. For the sake of completeness, it should be pointed out that the prolongation of their terms was once again fully directed by Moscow. The rest of the member-states only formally approved the Soviet decision.[1078]

The intensified arms race dominated the agenda of the 14th Committee of Defense Ministers' session, held in Prague, January 11-13, 1983. Ustinov declared that despite significant improvement, the Warsaw Pact member-states still were not using their industrial potential enough for the purposes of military production. He admitted that introduction of new arms was gradually more expensive, but at the same time warned against the lagging of the Unified Armed Forces' weaponry.[1079] The Committee also appealed to individual alliance members for a massive introducing of the equipment previously used by yacheykas only.[1080] In this regard, the chief of Staff of the Unified Armed Forces, Gribkov, stated

---

Varšavské smlouvy a opatřeních k dalšímu zvyšování jejich bojeschopnosti", undated, 1983; *USNESENÍ členských států Varšavské smlouvy přijaté na poradě politického poradního výboru v referátu hlavního velitele Spojených ozbrojených sil 5. ledna 1983 (překlad z ruského znění)*, undated, 1983, FMO GŠ/OS, k. 54, p. č. 205, č. j. 38015, VÚA, Prague.

[1078] In December 1982, Andropov proposed that Kulikov hold the post of supreme commander even in the following term. After being approved as a commander in chief of the alliance troops, in January 1983, the marshal asked the member-states for consent to a mandate of Gribkov and Fabrikov. Letter by Y. Andropov, the CPSU CC General Secretary, to E. Honecker, the SED CC General Secretary, December 11, 1982, DY 30/ J IV 2/2/1982, BArch, Berlin; *USNESENÍ členských států Varšavské smlouvy, undated*, leden 1983; Ibid, p. č. 207, č. j. 38021, *Návrh na prodloužení funkčního období ve funkcích Spojeného velení*, 18. 1. 1983, FMO GŠ/OS, kr. 54, p. č. 205, č. j. 38015, VÚA, Prague.

[1079] *Informace o výsledcích 15. zasedání výboru ministrů obrany*, 14. 1. 1983, k. 52, p. č. 197, č. j. 38001, ibid.

[1080] It included the especially modernized Su-22 and MiG-23 jets, Mi-24 gunships, and OSA-AKM anti-aircraft systems. *Podkladový materiál k 1. bodu jednání 15. zasedání Výboru ministrů obrany*, undated, 1982, ibid.

that the efficiency of modern Soviet weapons was clearly proven by the lessons from local wars. If used properly, they should have overtopped the Western-ones, according to the likely Soviet-biased analysis.[1081] The Soviet satellites definitely did not welcome the appeals for further armament. For example, at the same meeting, the Czechoslovak defense minister, Martin Dzúr, demanded to start planning the upcoming military five-year plan as soon as possible. From the beginning, the issue should have been solved in complexity while also taking into account the needs of inhabitants as well as the economic capacities of the Warsaw Pact member-states.[1082]

Accelerated rearmament in the years 1983-1985 largely eliminated previous corrections achieved by the Soviet satellites during negotiations about military five-year plans. The issue was discussed by the Unified Command and the Soviet general staff officials with individual member-states traditionally on a bilateral basis. They tried to prompt them to create new yacheykas. Thus, the Warsaw Pact countries had to invest more monies into the military area than the original five-year protocols had set. But these were actually being fulfilled only with serious difficulty. They were undoubtedly limited also by the continual effort of some member-states to prefer the export of military equipment in order to get financial resources.[1083] A huge complication was also the catastrophic condition of the Polish economy hit by crisis. Poland was not able to meet

---

[1081] Gribkov's analysis was based on the course of the Israeli-Lebanese war, Iraqi-Iranian conflict, and British-Argentinean clash over the Falkland Islands. The Soviet side ascribed a failure of the Syrian forces facing Israel, especially the T-72 tanks and MiG-23 aircrafts, to poor training only. On the contrary, it exaggerated their successes. Attention was gradually paid to the air forces. The report stated that combat jets caused up to 75% of all damages in mentioned conflicts. Artillery accounted for another 15%; tanks and other weapons for only 10%. *Zkušenosti z lokálních válek*, undated, 1983, ibid.

[1082] *Vystoupení ministra národní obrany ČSSR armádního generála M. Dzúra k otázce: "Zvládnutí nové výzbroje a vojenské techniky v armádách členských států Varšavské smlouvy"*, undated, 1982, ibid.

[1083] At the beginning of 1983, for instance, the chief of general staff of the CSPA, General Blahník, informed the Unified Command about the failure of effort to rearm motor-rifle divisions by BVP-1 infantry fighting vehicles, as 120 pieces had been released for export.

its commitments in armament contracts, which especially affected the joint production of a T-72 tank.[1084] Moreover, the introduction of modern equipment inevitably forced the Soviet satellites to increase their import of Soviet military materials and weapons.[1085] Kulikov advocated for these measures among others by the fact that Spain had become a member of NATO. He noted that this step had increased significantly the depth of the European battlefield.[1086]

In June 1983, NATO officially refused negotiations on the previous Warsaw Pact's peace initiatives. This step undoubtedly contributed to further escalation of international tensions and the Cold War in general. The exaggerated claim that since the end of the World War II there had not been such a threat to global peace was coming in the forefront

---

[1084] *Porada k plnění plánů rozvoje spojeneckých armád*, 23. 1. 1983; *VYSTOUPENÍ náčelníka generálního štábu ČSLA - 1. ZMNO na poradě k vyhodnocení plnění PROTOKOLU na roky 1981 a 1982*, undated, 1983; *Výsledky jednání v SSSR ve dnech 9.-10. 2. 1983*, 14. 2. 1983, FMO GŠ/OS, kr. 23, p. č. 77, č. j. 41006, VÚA, Prague.

[1085] Czechoslovakia, for example, had to set aside an additional 4 billion crowns in the years 1983-1985 in order to purchase the modern Soviet weapons. The supplies of missiles, tanks, aircraft, guns, anti-aircraft weapons, radio-electronic warfare devices, and ammunition were expected. Prague did not oppose the intention to accelerate introduction of modern equipment in the years 1983-1985. However, it proposed to pay the costs of these arms imported from the USSR by increasing its own military exports to the Soviet Union, Syria, and Libya. This included mostly L-39 and L-410 aircraft and BVP-1 vehicle. The Czechoslovak government also wanted to supervise conditions of the agreement. At the end of January, the defense minister, Martin Dzúr, firmly opposed the possibility that the chief of general staff of the CSPA, Blahník, had the authority to sign a revision of the development plan for the CSPA in the five-year plan of 1981-1985 during his talk with the Unified Command officials. The document was approved after careful consideration by the federal government presidium on July 7, 1983. Letter by M. Dzúr, the Czechoslovak defense minister, to M. Blahník, the chief of general staff of the CSPA, January 25, 1983, ibid; *DOKLAD náčelníka operační správy - ZNGŠ k plnění "PROTOKOLU" za léta 1981-1982*, undated, 1983, k. 24, p. č. 81, č. j. 41104, ibid; *PODKLAD pro vystoupení ministra národní obrany ČSSR u příležitosti jeho návštěvy v SSSR*, undated, 1983, p. č. 79, č. j. 41102, ibid; *Informace pro schůzi ROS o 15. zasedání výboru ministrů obrany členských států Varšavské smlouvy*, undated, 1982-1983, k. 52, p. č. 197, č. j. 38001, ibid.

[1086] *Výsledky jednání v SSSR ve dnech 9.-10. 2. 1983*, 14. 2. 1983, k. 23, p. č. 77, č. j. 41006, ibid.

of the rhetoric of the Eastern Bloc countries' officials.[1087]

Moreover, the Warsaw Pact's military analytics were aware of the continual strengthening of U.S. nuclear forces, especially the introduction of new and much more accurate carriers. They also followed with displeasure the Strategic Defense Initiative (SDI), the so-called "Star Wars" project,[1088] and the development of radar-undetectable aircrafts.[1089] The USSR realized it was hardly able to keep up with the U.S. in the new round of arms race. Therefore, it began to put forward more constructive proposals during disarmament talks. However, Moscow found itself more and more in a defensive position against the firm approach of the Reagan administration. At the beginning of 1983, during the Vienna talks, the Warsaw Treaty Organization even admitted an up-to-this-point unthinkable asymmetric reduction of troops in Central Europe. This should have been the first phase, followed by setting an equal quota of 900 thousand soldiers for units of both alliances dispatched in the region. According to this proposal, the troop numbers should have been frozen there during the negotiations.[1090] But given the situation, the United States was not interested in a deal with the USSR. Thus, talks about the further reduction of strategic nuclear arsenals ground to a halt. Washington intentionally put forward such a proposal there, which solely limited those sorts of weapons where the USSR maintained superiority. At the same time, the U.S. refused the Soviet effort to freeze the nuclear warhead and

---

[1087] *PODKLADY pro jednání se soudruhem ANDROPOVEM*, undated, 1983, k. 24, p. č. 79, č. j. 41102, ibid.

[1088] The Strategic Defense Initiative assumed deployment of an anti-missile system in outer space. The aim was to make nuclear weapons useless and surpass combat means. From the Soviet political as well as military leadership, potential elimination of atomic arsenals would make the country vulnerable to the first American nuclear strike. At home, Reagan advocated for the SDI project by claiming it had the potential to force the USSR to negotiate on nuclear disarmament under U.S. conditions. Zubok, *A Failed Empire*, 272-273.

[1089] *Základní tendence rozvoje a výstavby ozbrojených sil hlavních států Severoatlantického paktu s důrazem na ozbrojené síly USA, NSR a Francie*, undated, 1983, FMO GŠ/OS, k. 24, p. č. 79, č. j. 41102, VÚA, Prague.

[1090] The USSR should have reduced its military presence by 20 thousand troops, the U.S. by 13 thousand only. *Jednání o snížení počtu ozbrojených sil a výzbroje ve střední Evropě*, undated, 1983, ibid.

its carrier numbers during the negotiation.[1091] The Geneva talks on the Euromissiles issue brought similarly joyless results. The United States made stopping deployment of Pershing-II and BGM-109G missiles conditional on withdrawal of all the Soviet mid-range rockets from the European part of the USSR. But they also planned to keep on the Old Continent so-called forward means - nuclear carriers launched from aircrafts, ships, and submarines. The deal would also have excluded British and French weapons. In 1983, the USSR agreed with withdrawal of its missiles but under the conditions of removal of all the mid-range nuclear weapons from Europe and stipulating quotas for carrier numbers for NATO and the Warsaw Treaty Organization. Moscow was right in its assumption that the United States was not interested in any constructive agreement and was determined to deploy their weapons at all costs.[1092]

Reagan was against agreements like SALT. He considered them a product of the previous, from his point of view, wrong policy of concessions to Moscow. He asserted a theory that Soviet ambitions had to be faced from a position of strength and by the method of shocks. After all, the continual Soviet armaments project reflected the general staff's attitudes of calling for a permanent increase of own security; this entrenched Reagan's mind.[1093] In reality, the arms race escalation in the first half of the 1980s harmed the Soviet economy much more: Although the U.S. military spending per capita was higher than in the USSR, the power of the American economy was more than double. Despite this fact, the Soviet Union outclassed the United States in total military spending, its account of GDP, as well as the armed forces' numbers.

[1091] A center of the Soviet strategic forces lay in intercontinental ballistic nuclear surface-to-surface missiles. On the contrary, the Americans relied mostly on weapons launched from submarines and vessels. Therefore, Moscow proposed a reduction of all kind of carriers until both sides were able to decide which means would be restricted. Ibid.

[1092] Ibid.

[1093] At the beginning of 1983, the U.S. realized that the USSR was developing a hybrid of SS-16 and SS-20 missiles. A stationary intercontinental ballistic rocket launched from silo was being rebuilt into a mobile system. The American intelligence satellites also discovered a huge radar station in Krasnoyarsk region used for missiles' navigation, which was not compatible with the SALT treaty. Durman, *Útěk od praporů*, 258-259.

County /parameter	Ground domestic product USD billions	Military spending USD billions	Military spending per capita USD	Military spending account of GDP %	Number of troops Thous.	Number of inhabitants Thous.	Number of inhabitants per one soldier Inhabitants/ one soldier
USA	2924.8	180	782	6.1	2116.8	230049	107
West Germany	687.12	20.17	328	4.3	495	61440	124
France	570	21.23	391	4.1	493	54257	110
Great Britain	449.85	27.77	497	5.4	327	55830	171
Italy	350.154	7.2	126	2.5	370	57300	155
USSR	934.9-	192	711	12.4-15	3673	270000	74
Poland	88.1-	5.41	150	4.3	317	36000	114
East Germany	96.8-	6,96	408	7,7	166	17068	103
Hungary	37.7-52.8	1.24	115	3.0	106	10750	101
Bulgaria	30.2-39.1	1.34	150	4.2	148	8930	60
Czechoslovakia	73	3.396	248	5.2	194	15314	78

Table III[1094]

The USSR tried to solve this issue by making the ongoing rearmament fully effective. Such a strategy was explained to the Warsaw Pact members by Marshal Kulikov at the 27th meeting of the Military Council, held in Bucharest, April 1983. The modernization should have in particular included the most effective and - from an economic point of view - most optimal weapons. The supreme commander also criticized the

---

[1094] *Srovnání počtů a finančních nákladů na armády některých evropských států NATO a Varšavské smlouvy v roce 1981*, undated, 1983, FMO GŠ/OS, kr. 24, p. č. 79, č. j. 41102, VÚA, Prague.

fact that the military industry in various Warsaw Pact countries produced civil materials as well and thus did not exploit its full capacities.[1095] In the autumn of the same year, Ustinov added that modern equipment had to replace the especially obsolete weapons from the 1960s. In the upcoming five-year plan of 1986-1990, a qualitative improvement of almost all manner of troops was expected. Their numbers should have remained at least at the level of the year 1985.[1096] In December 1983, the Soviet defense minister urged his alliance counterparts that given the existing military-political situation, the Warsaw Treaty Organization could not catch up with the enemy, but must still keep a certain advantage. He noted that actual events forced even the USSR to review previously set economic tasks like improvement of living standards. "*The Soviet people have always consciously understood*," Ustinov stated.[1097] We are reminded that at the time, the head of the Soviet defense resort was one of two key persons who defined the basic course of the Kremlin's policy, as Yuri Andropov's health condition was already very serious.

Despite various complications, the ambitious armaments plans were largely filled by the Warsaw Pact members. In December 1983, at the 16th Committee of Defense Ministers session in Sofia, Kulikov commended that combat capabilities of the ground forces had increased by 30%, the air-defense system by 10%, and radio-electronic troops by 60%. Until 1985, the air forces should have had at least 18% of its planes be the most recent types. However, Poland, Bulgaria, and Romania all experienced serious troubles with realization of the set tasks.[1098] The latter

---

[1095] *27. zasedání vojenské rady Spojených ozbrojených sil*, 29. 4. 1983, k. 54, p. č. 206, č. j. 38016, ibid.

[1096] *VYSTOUPENÍ člena předsednictva ÚV KSSS, ministra obrany SSSR, maršála Sovětského svazu D. F. USTINOVA na mimořádném zasedání výboru ministrů obrany členských států Varšavské smlouvy 20. října 1983*, k. 55, p. č. 215, č. j. 38108, ibid; *USNESENÍ výboru ministrů obrany členských států Varšavské smlouvy k 1. otázce programu 16. zasedání "Všeobecné směry rozvoje a vybavení výzbrojí a bojovou technikou Spojených ozbrojených sil členských států Varšavské smlouvy v letech 1986-1990"*, undated, 1983, k. 56, p. č. 216, č. j. 38109, ibid.

[1097] *INFORMACE o průběhu a výsledcích 16. zasedání výboru ministrů obrany konaného ve dnech 5.-7. prosince 1983 v SOFII*, 12. 12. 1983, ibid.

[1098] Ibid.

country solved the issue in the way typical for the Ceausescu regime. Regardless of recommendations by Moscow and the Unified Command, Bucharest announced that its priority was to stop frantic armament. Ceausescu rejected any increase of military spending, at least till 1985. He intended to use the resources to improve the living standards of the Romanian population. In reality, this meant trying to at least basically fix the catastrophic conditions of a large number of people in the country.[1099] Nevertheless, in economically beneficial cases, the Romanian dictator did not hesitate to support some armament programs. For instance, he tried to secure production of the T-80 tank in Romania.[1100] Such a step would have reduced the costs of its following purchase; moreover, the Romanian industry would have gotten a lucrative contract. On the contrary, Bucharest did not intend to participate in any programs which caused an increase of military spending. These especially included the expensive construction of a tropospheric communication network for the Unified Armed Forces. In this regard, the joint alliance budget began to rise again.[1101]

[1099] A crisis fully hit the Romanian economy at the end of the 1970s. The country faced its debt, non-functional agriculture, poor supplies, and stagnation of GDP. Since 1977, spontaneous protests of starving people took place. In 1981, the agricultural programs totally collapsed. As the result, food rationing was introduced under the guise of "rational nutrition according to scientific norms." Three years later, a heating norm was established for apartments. It was set up to 16°C, but in reality, often felt to 10°C only. This firm restriction of living standards of the Romanian inhabitants was caused mostly by Ceausescu's effort to quickly repay all the foreign loans and to strengthen the country's sovereignty this way. Tejchman, *Nicolae Ceausescu*, 116–121.

[1100] *27. zasedání vojenské rady Spojených ozbrojených sil*, 29. 4. 1983, FMO GŠ/OS, k. 54, p. č. 206, č. j. 38016, VÚA, Prague.

[1101] The project was approved in 1982. It was fully directed by the Soviet defense ministry. Individual member-states had to increase their contributions to the Unified Command's budget again. The problem appeared especially in 1985, when the monies for the Warsaw Pact command rose by 26.6%, to 17.16 billion rubles. This was sharp annual growth in comparison to the previous year when the Unified Command's spending had lowered after a long time, as the scheduled buildup of a protected command post in the USSR had been scrapped. *Podpis dohody o zřízení soustavy troposférického spojení SOS a schématu organizační struktury orgánů velení SOS*, 13. 7. 1982, k. 52, p. č. 197, č. j. 38001, ibid; *Návrh rozpočtu Spojeného velení na rok 1984 a vyhodnocení rozpočtu za rok 1982*, 20.

The Romanian appeals for unilateral reduction of military spending directly opposed the intentions of the Soviet top military. Therefore, the USSR sought to secure that Bucharest remained isolated within the Warsaw Treaty Organization on this issue. For example, at the extraordinary meeting of Committee of Defense Ministers, held in October 1983 in East Berlin due to escalation of the Euromissiles Crisis, Ustinov instructed his Czechoslovak counterpart[1102] in backrooms to oppose the Romanian attitudes openly. In accordance with the Soviet stances, Dzúr later declared that the Romanian proposals meant a "passive observation" and just encouraged the aggressor to intensify its preparation for war. Also along these lines, the Unified Command officials, Kulikov and Gribkov, as well as Ustinov himself and the Bulgarian defense minister Dobri Dzhurov, criticized the Romanian policy. They all similarly said that it was impossible to reduce the military budgets of the Warsaw Pact countries when the annual U.S. investments into armed forces would have constantly risen till the end of decade.[1103]

No significant correction of the armament programs in the Warsaw Pact countries occurred even during the short rule of Andropov's successor, Konstantin Chernenko. Thus, at the meetings of the Warsaw Pact military bodies, Marshal Kulikov constantly appealed for better exploitation of industry capacities for military purposes. He criticized the especially slow rearmament of the air forces.[1104] Since the end of the

---

9. 1983, k. 55, p. č. 214, č. j. 38107, ibid; *Návrh rozpočtu Spojeného velení na rok 1985 a vyhodnocení rozpočtu za rok 1983*, 22. 9. 1984, k. 58, p. č. 224, č. j. 38208, ibid.

[1102] At the time, the Czechoslovak delegations acted with an extreme loyalty at the alliance military bodies' meetings. Czechoslovakia often asked the USSR for sending the thesis of prepared speeches of the Soviet military officials in advance, in order to harmonize the reports of the Czechoslovak representatives with them. This concerned in particular the evaluation of the international and military-political situation. In the issues which affected the country's economy, the Husák regime at least partially pursued the national interests. *Vystoupení s. ministra národní obrany*, 31. 10. 1984, k. 59, p. č. 229, č. j. 38217, ibid.

[1103] *INFORMACE o průběhu a výsledcích mimořádného zasedání výboru ministrů obrany konaného dne 20. října 1983 v Berlíně*, k. 55, p. č. 215, č. j. 38108, ibid.

[1104] *29. zasedání vojenské rady Spojených ozbrojených sil*, 28. 4. 1984, k. 57, p. č. 218, č. j. 38117, ibid; *30. zasedání vojenské rady Spojených ozbrojených sil*, 23. 10. 1984, k. 58, p. č. 224, č. j. 38208, ibid.

1970s, NATO air forces had been reinforced rapidly. The Western units were given more and more modern aircrafts like F-15, F-16, Tornado, and Mirage 2000, or Patriot surface-to-air missiles.[1105]

On the other hand, nothing suggests that the CPSU leadership was really thinking about increasing military spending in this period. However, at the meetings of the Council of Defense, such an option was asserted by the chief of the Soviet general staff, Marshal Nikolai Ogarkov, who criticized stagnation of the military-industrial complex. According to his opinion, many projects were ineffective and too pompous. At the same time, he called the effort to directly compete with the U.S. in the arms race suicidal. This was in fact a direct attack on Dmitri Ustinov's policy, who was in charge of the military production. In September 1984, the defense minister responded by dismissing the chief of general staff as their mutual relations had been very strained since the outburst of war in Afghanistan. Ogarkov's successor, Marshal Sergei Akhromeyev, brought new stimulus. He was a modernist, did not hesitate to criticize the state of things, and at the same time realistically considered the Soviet options. The defeat of advocates of extreme armament was suggested also by the articles in both the Soviet daily and military press which occurred in October. These stated that despite the unfavorable international situation requiring huge resources for the country's defense, it was unacceptable to think about a reduction of social programs. Soon after that, in December 1984, the last important protagonist of the Cold War escalation, Dmitri Ustinov, died. Thus, the conception by some members of the general staff who asserted responding to Reagan's policy by increasing the military budget by a breathtaking 14% was definitely buried. At the time, the USSR actually invested a larger proportion of its resources into defense capacities than in 1940, when the country was preparing for potential engagement in World War II. After Gorbachev came to power, he realized with horror that the

---

[1105] In 1984, the Warsaw Pact's analyses stated that NATO air forces had been reinforced by 25% during the last five years. *Vystoupení s. ministra národní obrany*, 31. 10. 1984, k. 59, p. č. 229, č. j. 38217, ibid.

total monies flowing into the military area were not equal to the officially declared 16%, but rather almost two fifths of the country's budget.[1106]

[1106] Zubok, *A Failed Empire,* 277; Durman, *Útěk od praporů,* 268, 295. Dmitri Volkogonov noted that indirect military spending during Andropov's rule approached 70% of the state's budget and that this fact had been tabooed. However, this claim is most likely exaggerated. Volkogonov, *The Rise and Fall,* 362.

Approval of the wartime statute of the Warsaw Pact

Firm Soviet control of the Eastern Bloc countries' armed forces was one of the core aspects of the Warsaw Pact's existence. At the same time, it is symptomatic, that for a long time of the alliance's existence, military cooperation between the member-states was defined very vaguely. This included mostly features of collaboration of the armies as well as their command structures in time of war. This essential question was not clarified until the early 1980s, when the organization approved the *Statute of the Unified Armed Forces of the Warsaw Treaty Organization member-states and their command institutions in wartime*. The following chapter analyzes the origins and approval of this controversial document, which meant not just a significant milestone in the alliance's development, but also substantial interference with the autonomy and national sovereignty of the countries of the Soviet sphere of interest in Europe.

Outlines of the Warsaw Pact military structures' development until 1969

To understand the context of the wartime statute origins, it is necessary to remind the reader, at least in short, of the development of features of military cooperation within the Warsaw Treaty Organization in the period which is not a subject of this work. The alliance's establishment in 1955 terminated the formal autonomy of the Soviet European satellites' armed forces. The USSR had controlled the situation in the East-European armies already before, through a large network of its military advisers. However, despite Soviet supervision, the Eastern Bloc states' armed forces were independent entities and subject to the national institutions up to this point. Even though the military framework of the established alliance was initially rudimentary, the *Protocol on creation of the Unified Command of the treaty member-states' armed forces*, approved at the Warsaw conference, assigned a vast majority of the signatory-countries' military units to the incipient Unified Armed Forces. The new supranational institutions - the Political Consultative Committee and the Unified Command - were given decisive authority over them.[1107]

In September 1955, the alliance command - ruled by the Soviet generality - put forward the secret *Statute of the Unified Command of the Member States of the Warsaw Treaty*. In its basic dimension, the document defined powers and obligations of the supreme commander, his deputies as well as the Staff of the Unified Armed Forces. However, it consisted of just brief formulations and the practical implementation of its content was not specified in detail. The material also lacked any mention of the Unified Command's accountability to political institutions.[1108] The supreme commander was only obligated to realize the Political Consultative Committee resolutions when they concerned the Unified Armed Forces.[1109] This vague document, approved at the first meeting of the alliance's supreme body in Prague at the beginning of the next year, became for 14 long years the only official directive defining military cooperation within the Warsaw Pact. Although the Unified Command had no clear structures until the 1960s, it served as finalization of the process of subordinating the East-European armed forces to Moscow. Hence, the Soviet generality was able to plan using the satellites' armies under the auspices of the alliance without any interference from foreign military representatives.

In praxis, such a state remained unchanged until the political fall of the CPSU CC First Secretary, Nikita Khrushchev, in 1964. An attempt of the Polish generality to review the cooperation within the Warsaw

---

[1107] *Protokol o vytvoření spojeného velení ozbrojených sil států-účastníků smlouvy o přátelství, spolupráci a vzájemné pomoci*, May 1955, 1261/0/44, i. č. 41, k. 20. NA, Prague.

[1108] Political control of the military bodies was just briefly mentioned in the "Protocol on Establishing the Unified Command of the Armed Forces" of May 1955. Its first provision stated that "*the Political Consultative Committee is in charge of decisions on issues concerning strengthening of defense capabilities and organization of the Unified Armed Forces of the Treaty member-states.*"

[1109] Statute of the Unified Command of the Member States of the Warsaw Treaty, in *Cold War International History Project Bulletin*, no 11 (1998), 236-237.

Treaty Organization after the Eastern Bloc crisis in 1956 was just a unique initiative without any real effect.[1110] After Leonid Brezhnev came to power, the Kremlin decided to partially modify the ruling mechanisms of its sphere of influence in Europe. Among other things, this process started talks on a reform of the Warsaw Pact's military as well as political structures. One of the pursued goals was to work out a new statute which would have redefined the rules of military cooperation within the alliance and at the same time strengthen what had been up till now only a marginally effective structure of the Unified Command. The USSR and some alliance member-states differed in their opinions on just how deep the changes should have been implemented. During the talks in 1966, Moscow basically proposed to just codify its existing dominance. Demands of the loyal allies should have been satisfied by a sop in the form of bigger representation in brand-new or reformed military bodies.[1111] A very important aspect was the Soviet intention not to define, but rather to roughly outline the Unified Command's activities in a state of war.[1112] At the time, the Soviet military officials actually clearly admitted for the first time that in a potential conflict, the function of the Warsaw Pact Unified Command would be retaken by the general staff of the Soviet army.[1113]

The Warsaw Pact troops command in a wartime

The fundamental question, how would be the alliance troops be commanded in case of an outbreak of armed fighting, remained unanswered for most of the time of the Warsaw Pact's existence. Since the end of the World War II, the Soviet strategists most likely envisaged using the East-European armies in a new eventual conflict. They assumed that

---

[1110] For more see Mastny, "We are in a Bind," 231-232.

[1111] Later, regarding a reform of the alliance's staff, Brezhnev talked about the important propagandistic dimension. Extending this institution to representatives of all the Warsaw Pact countries could terminate accusations that the alliance was ruled by the Soviet generals only. L. Brezhnev's opinions on actual international issues, February 27, 1969, PZPR KC, XIA/87, AAN, Warsaw.

[1112] *Zpráva o jednání náčelníků generálních štábů armád členských států Varšavské smlouvy provedeném v Moskvě ve dnech 4.-9. února 1966,* 1261/0/44, i. č. 43, k. 24, NA, Prague.

[1113] Luňák, *Plánování nemyslitelného,* 45.

the satellites' armies would fall within the remit of the top Soviet command in this case.[1114] Establishing of the Warsaw Treaty Organization and the Unified Command perhaps did not change this intention at all. In this regard, Frank Umbach, the German security analytic, notes that at the time of its establishing, the latter institutions could not have been seen as a body for waging a war on a coalition basis. In the early stage of the alliance's existence, the Unified Command had mostly administrative-coordinative features. It was rather just a connection link in communication between the Soviet general staff and the Soviet satellites' military institutions. Until the 1960s, it definitely lacked the necessary personnel and structures for providing effective command to the troops at the battlefield.[1115]

In 1961, at the latest, after the changes in the Warsaw Pact's combat preparation and operation plans caused the Second Berlin Crisis, the Soviet command clearly envisaged waging a potential war on the basis of the entire alliance.[1116] Nevertheless, the Soviet generality still kept secret an expected scheme of command for such a scenario. A very paradoxical situation occurred, as the Czechoslovak case shows. On one hand, the Soviet military command assured its counterparts from the member-states about their sovereignty. At the same time, however, it refused to answer concrete and essential questions by the Czechoslovak generals about organization of the command in a potential conflict.[1117] The

---

[1114] Ross Johnson, Artur, *"Soviet Control of East European Military Forces: New Evidence on Imposition of the 1980 "Wartime Statute","* Accessed October 16, 2019.https://www.cia.gov/library/publications/cold-war/wartime-statutes/wartime-statutes.pdf.

[1115] Umbach, Frank, *Die Rotte Bündnis. Entwicklung und Zerfall des Warschauer Paktes 1955-1991* (Berlin: Links Verlag, 1996), 251.

[1116] *Informační zpráva o poradě náčelníků HPS spřátelených armád ve dnech 5.-7.června 1962,* 1261/0/44, i. č. 43, k. 24, NA, Prague.

[1117] In the mid-1960s, the Czechoslovak general staff assumed that in case of global armed conflict, the organizational scheme of the Warsaw Pact command would not significantly differ from that of World War II. The first variant envisaged that the Stavka, the highest military body for waging a global war, would be led by the Political Consultative Committee. The Unified Command would have been a part of the Stavka for combat operations in Europe. The second version anticipated that instead of the Warsaw Pact's supreme body, the Stavka would be chaired by the "top supreme commander," the Soviet leader, in practice.

queries about what would be the real composition of the general staff of the CSPA during war were rebuffed by the words of the Soviet generality: *"This is not your business."*[1118]

In the context of the Warsaw Treaty Organization's development so far, the reform documents, approved by the Political Consultative Committee in March 1969, undoubtedly meant a shift. But they failed to solve the core question of the member-states' military cooperation in a war. The comprehensive Statute of the alliance's military institutions defined the rules for peacetime only. The reason lay in very different visions of the USSR and its satellites on waging a war on the Warsaw Pact basis, especially on the autonomy of individual member-states' armed forces. Brezhnev's notes suggest that even the Soviet political leadership as well as the military command were not unanimous in the opinion on the need to push forward a controversial document regulating the activities during a war. Just a couple of weeks before the scheduled approval of the Warsaw Pact's military structures, for example, the Soviet leader told the Polish representatives that the alliance's supreme commander, Yakubovski, had not decided the fate of a wartime statute yet.[1119] Apparently, the CPSU leadership was aware that even in the atmosphere after the military suppression of the Prague Spring, it would not be easy to enforce within the Warsaw Treaty Organization the document revealing the real intentions for the period of war. At the talks in the Hungarian capital, discord emerged even during the discussion on a state of emergency that would have granted the supreme commander much more competences. Moscow and the member-states were unable to agree on who should have gained the right to declare such a state. Even this much less contentious issue, in comparison to the wartime statute, had to be left out of the new regulations in the end. The situation obviously bothered the military more than the political circles of the USSR. Dissatisfaction of the top Soviet generality with the alliance member-states' unwillingness to support its

---

Pravděpodobná funkce Spojeného velení za války, undated, 1966, FMO GŠ/OS, k. 12, p. č. 49, č. j. 39042, VÚA, Prague.
[1118] Interviews with the Czechoslovak Generals, 2002-2003, in *Plánování nemyslitelného* (ad. 2): 379-380.
[1119] L. Brezhnev's opinions on actual international issues, February 27, 1969, PZPR KC, XIA/87, AAN, Warsaw.

intentions was illustrated by the unusually sharp speech of the supreme commander Yakubovski.[1120]

The official claims assumed that the statute clarifying the Unified Command's competences during a war would be approved in the near future. This did not happen.[1121] In this regard, the HSWP CC First Secretary, Janos Kadar, was right in his note that although making the rules of military cooperation within the Warsaw Treaty Organization more precise in peacetime was necessary, there had been at least some basic provisions included in the founding charter as well as the Protocol on establishing the Unified Command and its Statute of 1955. To the contrary, there was no official document outlining the features of cooperation in case of armed conflict.[1122] However, after 1969, the issue took a back seat. The tendencies to complete the Warsaw Pact's military statutes for peacetime by those of war did not reemerge until mid-1973. At the alliance meeting of the chiefs of the general staffs in Sofia, the Soviet delegation announced that in the case of an outbreak of armed conflict, mutual relations between the Unified Command and the national military institutions would be organized neither on the basis of the Statute of 1969 nor other partial documents regulating military cooperation within the Warsaw Pact. Therefore, the Staff of the Armed Forces should have begun to work on regulations for a period of war. It was expected that the material would later be put forward to the alliance political bodies for approval.

[1120] *Minutes of the Hungarian Party Politburo Session - Report on the PCC Meeting by the First Secretary of the MSzMP (János Kádár)*, March 24, 1969, Accessed October 16, 2019. http://www.php.isn.ethz.ch/kms2.isn.ethz.ch/service engine/Files/PHP/18017/ipublicationdocument_singledocument/dff6430a-4402-4217-ba18-509256cb20cb/en/690324_Minutes_Eng.pdf.

[1121] Mastny, *A Cardboard Castle?*, 323.

[1122] *Minutes of the Hungarian Party Politburo Session - Report on the PCC Meeting by the First Secretary of the MSzMP (János Kádár)*, March 24, 1969, Accessed October 16, 2019. http://www.php.isn.ethz.ch/kms2.isn.ethz.ch/service engine/Files/PHP/18017/ipublicationdocument_singledocument/dff6430a-4402-4217-ba18-509256cb20cb/en/690324_Minutes_Eng.pdf.

This statement was supported by all the present delegations, including the Romanian.[1123]

At first glance, it could seem paradoxical that the Soviet generality initiated supplementation of the statutes for peacetime by their war equivalent at the time of climaxing détente.[1124] We are reminded that this period conveniently coincided with the effort to strictly define the Unified Command's activities through a series of additional statutory documents. In the years 1969-1973, at least fourteen of them were concluded within the alliance's military structures.[1125] Given its dominant position, the USSR maintained mechanisms for how to eventually assert its will to the allies regardless of the official procedures. Despite this fact, the military bodies' activities were largely given clearly defined rules. An absence of any guidelines for wartime strongly contrasted with this state. We are reminded that in the 1970s, the alliance military structures enjoyed a significant level of autonomy. The political power did not substantially interfere with their work. There is no evidence suggesting the political leadership of the Warsaw Pact countries dealt with the new initiatives to work out the wartime statute at all.

The reason why the issue was reopened in 1973 has to be seen mostly as an alternative to thinking about the features of potential conflict. As explained before, the Warsaw Pact generality at the time was clearly abandoning the rigid idea that since the beginning, a future war had to be waged on the basis of a total nuclear clash. A rise in consideration about the realness of a conventional war in the first half of the 1970s also required a revision of the Warsaw Treaty Organization command structures. The Soviet generality thought that the Unified Armed Forces' success in such a conflict depended on their strict combat cooperation. Therefore, an emphasis was placed on joint command. Despite significant

---

[1123] At the meeting, Romania was not represented by its chief of general staff, but by his deputy only. *Porada vedoucích funkcionářů generálních štábů*, 18. 6. 1973, FMO GŠ/OS, k. 15, p. č. 52, č. j. 39044, VÚA, Prague.

[1124] Mastny, Vojtech, *Learning from the Enemy: NATO as a Model for the Warsaw Pact* (Zürich: Forschungsstelle für Sicherheitspolitik und Konfliktanalyse der ETH, 2001), 34.

[1125] *Porada vedoucích funkcionářů generálních štábů*, 18. 6. 1973, FMO GŠ/OS, k. 15, p. č. 52, č. j. 39044, VÚA, Prague.

improvement of the alliance's command structures at the end of the 1960s, the USSR probably still envisaged that in case of war, their function would be retaken by the top Soviet command and the general staff of the Soviet army.[1126] From the point of view of the Soviet generality, the aim of the new statute was to strengthen competences of the supreme commander - in reality the Soviet marshal - towards the alliance troops.[1127]

The initiative of 1973 was just an ephemeral episode. Given the unavailable primary documents of the Unified Command activities, the question why this happened cannot be answered. It is also unclear whether the Staff of the Unified Armed Forces really began to work on the statute as was agreed in Sofia. However, the available documentation of the following meeting of the alliance bodies remains silent about any activities in this issue. Vojtech Mastny assumes that the USSR did not comprehensively push forward a definition of the Warsaw Pact's military cooperation during a war in the following years due to climaxing détente and some fears of an anticipated negative reaction by Romania. But it never abandoned these plans.[1128]

A slow fall of the relaxation process in the second half of the 1970s and probably also the changes at the Unified Command's key posts were the impulses for a shift in the issue of the wartime statute of the Warsaw Treaty Organization. In 1976, in a period of seven months, not only did the Soviet minister of defense Grechko die, but also the alliance's supreme commander Yakubovski and the chief of Staff of the Armed Forces Schtemenko. The latter two were replaced by Marshal Victor Kulikov and General Anatoly Gribkov. Kulikov especially considered finalization of the project, which had stagnated for many years, as one of his priorities.[1129] Thus, in the following years, he engaged significantly in working out and passing the Statute.[1130] However, it is unlikely he was able to act on his own desire by virtue of his function, without any previous

---

[1126] Umbach, *Die Rotte Bündnis,* 250-252.
[1127] Mastny, *Learning from the Enemy,* 34.
[1128] Ibid.
[1129] Mastny, *A Cardboard Castle?,* 48.
[1130] Letter by the chief of the CSPA general staff, K. Rusov, to the Warsaw Pact's supreme commander, V. Kulikov, March 5, 1979, FMO GŠ/OS, k. 44, p. č. 152, č. j. 36502, VÚA, Prague.

instruction of the Soviet defense ministry, where Marshal Dmitri Ustinov became the head after Grechko's death.

Negotiations on the Statute

The agenda of the 10th Committee of Defense Ministers' session, which took place in November 1977 in Sofia, contained the item "On improvements of the organizational structure of the Unified Armed Forces' command bodies." The new supreme commander was supposed to comment on it. The vague title did not allow for anticipating the real content of the Marshal's speech. The report of the previous meeting of the heads of defense resorts actually talked about possible enlargement of the Staff of the Unified Armed Forces' personnel. There was no mention of the wartime statute.[1131] Now, Kulikov warned in his speech that NATO had not abandoned its plans for a surprise attack on the Warsaw Pact countries. He appealed for the Unified Armed Forces and its command system to be ready even in peacetime to quickly repulse any aggression and then destroy the enemy. These claims were typical of the rhetoric of the Eastern generality at the time. Nevertheless, referring to the experience of recent maneuvers,[1132] the supreme commander announced that the existing structure of the alliance's command did not allow for a quick and easy transition from a state of peace to a state of war. He suggested two steps for a solution: Within the Unified Command, new posts should be created and personnel numbers increased.[1133] But above all, the *Statute of*

[1131] *10. zasedání Výboru ministrů obrany - návrh*, 1. 11. 1977, k. 42, p. č. 141, e. č. 36044, ibid; Report on the 9th meeting of the Committee of Defense Ministers of the Warsaw Pact in Sofia for the CPCz CC General secretary, G. Husák, November 5, 1976, p. č. 140, e. č. 36043, ibid.

[1132] Concretely, Kulikov mentioned maneuvers "West," "Soyuz," "Val," and "Transit" held in 1977.

[1133] The Unified Command should be extended to include the deputy supreme commanders for navy and air forces and the supreme commander assistant for back-territories. Also, the Staff of the Unified Armed Forces should be slightly modified. Another 60 officers and 50 civilians were supposed to join the Warsaw Pact command structures. The persisting Soviet dominance was illustrated by the fact that the Soviet generals A. P. Borisov, V. I. Maryasov, and M. V. Proskurin were installed to new functions in the alliance staff - chief of communications, chief of chemical troops, and chief of recon and informational department,

the Unified Armed Forces of the Warsaw Treaty member-states and their command institutions in wartime should be worked out in the following year. The Marshal was supported by all the delegations except for the Romanian. As the representatives of the USSR and five of its loyal satellites all agreed that Kulikov's proposals reflected the outcomes of the recent Warsaw Pact military as well as political bodies' meetings, the Romanian defense minister, Ion Coman, opposed and added his written objections to the final statement.[1134]

The intention to work out the wartime statute was presented as a part of broader considered steps towards the improvement of the Unified Armed Forces' control. The military five-year plan of 1981-1985 on the development of the alliance troops actually mostly stressed enhancement of the command system. A unified organizational command structure for all sorts of troops should have been implemented. At the same time, completion of the network of various command posts providing a permanent connection and coordination of activities of the Unified Armed Forces was expected.[1135]

The works on the Statute proceeded slower than Kulikov had suggested in Sofia. First, given the previous experience, it was obvious that this was both an essential as well as extremely sensitive issue. Therefore, the military alone was not able to decide it without being given authority and consent by the political power. The Political Consultative Committee session, held at the end of November 1978, was used for this purpose. The meeting of the Warsaw Pact's supreme body took place in the Soviet capital amid a new deteriorating phase of East-West relations.

---

respectively. The rest of the Warsaw Pact member-states were not given any direct command positions within the extended Staff of the Unified Armed Forces, but solely posts of senior officers and deputy commanders. *Doplnění štábu Spojených ozbrojených sil novými funkcemi*, 13. 5. 1978; *Ustanovení do nově vytvořených funkcí ve štábu Spojených ozbrojených sil*, 1. 7. 1978, FMO GŠ/OS, k. 44, p. č. 146, č. j. 36052, VÚA, Prague.

[1134] *10. zasedání Výboru ministrů obrany*, 5. 12. 1977, k. 42, p. č. 141, e. č. 36044, ibid.

[1135] *TEZE DOKLADU hlavního velitele Spojených ozbrojených sil členských států Varšavské smlouvy na 11. zasedání Výboru ministrů obrany ke druhé otázce programu - "Obecné směry rozvoje Spojených ozbrojených sil členských států Varšavské smlouvy na léta 1981-1985"*, undated 1978, k. 35, p. č. 127, e. č. 36027, ibid.

Thus, Marshal Kulikov presented the works on the Statute as just one of the many broader measures to strengthen the Warsaw Treaty Organization, proven to be necessary during the operational preparation in the recent years. In his claims, the Marshal referred also to stances of individual member-states' military bodies.[1136] Afterwards, the Political Consultative Committee authorized the supreme commander to work out a draft of the document in cooperation with the defense ministers. "*A single top main command*" was determined as its basic premise. New command structures of the Unified Armed Forces for the western and southeastern war theatres as well as Unified war navies of the Warsaw Pact in the Baltic Sea and the Black Sea should also have been created.[1137]

An important aspect was a preliminary measure issued by the Political Consultative Committee on the matter: In case of sudden escalation of a military-political situation before the new statute was approved, the command of the alliance forces would not fall within the remit of the existing Unified Command, but for this purpose, they established "*the High Main Command*" and the Soviet general staff at first.[1138] This was clear proof that the existing Warsaw Pact's military institutions were not considered for a potential conflict. In fact, the alliance's military structures served solely as a tool of control of the Eastern Bloc countries' armed forces. These were not designed for commanding a joint combat operation. Thus, in case of emergency or war, the general staff of the Soviet army was given absolute power over the Unified Armed Forces in praxis, because it remained questionable whether the Warsaw Pact political bodies, given their poor operational capabilities, would have been able to quickly establish the High Main Command if an unexpected international crisis broke out. Furthermore, even if created, it is unlikely that this institution lacking any specific definition could have

---

[1136] *TÉZE DOKLADU hlavního velitele Spojených ozbrojených sil maršála Sovětského svazu V. G. Kulikova na zasedání Politického poradního výboru "O stavu a rozvoji vojenské spolupráce Spojených ozbrojených sil členských států Varšavské smlouvy" /listopad 1978/,* k. 44, p. č. 148, č. j. 36054, ibid.

[1137] *USNESENÍ členských států Varšavské smlouvy, přijaté na zasedání Politického poradního výboru ke zprávě hlavního velitele Spojených ozbrojených sil,* 23. 11. 1978, ibid.

[1138] Ibid.

really taken command over the alliance troops. It is good to add, that no broad discussion on these measures took place in the Political Consultative Committee.

It was solely Romania who opposed the aforementioned intentions in the military area. The country refused to participate in both changes in the command system as well as the new round of intensified armament envisaged by the Warsaw Pact's supreme body.[1139] During the RCP Executive Committee session, held immediately after the Pact's meeting in Moscow, the planned armament programs and the statement on the Soviet general staff's authority towards the alliance troops were criticized harshly. Manea Mănescu, the Romanian Prime Minister, called the measures a violation of independence and sovereignty of the Warsaw Pact member-states and the Soviet attempt to interfere with their internal affairs.[1140] As already explained, Moscow intended to influence Ceausescu through diplomacy only, and this approach also concerned the current situation. Therefore, the rest of the alliance members were instructed by the Soviet side at the time to avoid any open arguments with the Romanians at the alliance meeting and not to challenge their attitudes excessively.[1141]

Works on the Statute

In December 1978, at the 11th Committee of Defense Ministers' meeting in Berlin, Kulikov suggested establishing the working groups within individual Warsaw Pact countries' general staffs which were supposed to cooperate with the Unified Command on the issue of the

---

[1139] *Informacja o naradzie Doradczego Komitetu Politycznego państw-stron Układu Warszawskiego*, 27. 11. 1978, PZPR KC, XI/807, AAN, Warsaw; *Postavení RSR v diferencované politice VKS vůči ZSS*, 5. 5. 1979, DTO 1945-1989, i. č. 34, e. č. 89, č. j. 013.289/79, AMZV, Prague.

[1140] *Stenographic Transcript of the meeting of the Consultative Political Committee of the Central Committee of the Romanian Communist Party*, November 20, 1978, Accessed October 16, 2019. http://www.php.isn.ethz.ch/ kms2.isn.ethz.ch/serviceengine/Files/PHP/16634/ipublicationdocument_singled ocument / 68371719-6b4e-454a-a438-d56048050ebe/en / 781120_stenographic_ transcript.pdf.

[1141] Watts, Larry "The Soviet-Romanian Clash over History, Identity and Dominion," *Cold War International History Project e-Dossier*, no. 29 (2010).

wartime statute. He expected that the material would be worked out during the upcoming year and then approved at the next gathering of the heads of defense resorts.[1142] Unlike in the 1960s, when the reforms of the military framework were negotiated on the basis of special meetings, now the Unified Command discussed the issue with individual member-states bilaterally.

At the beginning of 1979, the first draft of the document was sent by the Unified Command to the Soviet satellites. Communication between the alliance command and their political institutions did not proceed directly but rather through the general staffs and the defense ministries, as was typical in the Warsaw Pact after all. Finalization of the agreement was supervised by both of the nominally highest military officials of the alliance - the supreme commander and the chief of Staff of the Unified Armed Forces. The Soviet generality demanded the national political and military bodies approve the text already before the alliance meetings.[1143] According to a common practice, the pre-discussed material should have been just formally authorized there. Kulikov and Gribkov also urged meeting the outlined deadlines.[1144] Moscow obviously gave much importance to this matter. Taking into account previous experience, the USSR strove to prevent the issue of the wartime statute grounding to a halt again.

The document was most likely shaped by the top Soviet military circles. Only then the member-states were given an opportunity to comment on it. However, the initial draft was also carefully criticized by the loyal Soviet satellites. The Czechoslovak representatives, for instance,

---

[1142] *Informace o průběhu a výsledcích 11. zasedání výboru ministrů obrany, konaného ve dnech 4.-7. prosince 1978 v Berlíně*, undated; *NÁVRH programu 12. zasedání Výboru ministrů obrany členských států Varšavské smlouvy*, undated 1978, FMO GŠ/OS, k. 35, p. č. 127, e. č. 36027, VÚA, Prague.

[1143] During his visit to Czechoslovakia in February, for example, the chief of Staff of the Unified Armed Forces Gribkov urged the Czechoslovak Council of the State Defense to approve not only the preliminary, but also the final text of the statute. In this way, he tried to prevent any, however unlikely, complications.

[1144] *Statut Spojených ozbrojených sil členských států Varšavské smlouvy a orgány jejich řízení za války*, 28. 2. 1979, FMO GŠ/OS, k. 44, p. č. 152, č. j. 36502, VÚA, Prague; Letter by the Czechoslovak defense minister, M. Dúzr, to the CPCz CC General Secretary, G. Husák, September 5, 1979, k. 46, p. č. 164, č. j. 36520, ibid.

realized that the material contained plenty of "white areas." The ambiguous formulations included among other things very important aspects like the role and competences of the Warsaw Treaty Organization's military as well as political bodies in case of war. In this regard, Gribkov assured that the document would be mostly of operational features and therefore did not have to be all encompassing. But he admitted some Czechoslovak proposals to be relevant and promised to include them in the final text.[1145] In this regard, the Czechoslovak military institutions ensured the alliance's military structures that the country agreed with the proposed principles of command of the Unified Armed Forces during a war and the CSSR was just aiming to maximize cooperation in a potential conflict. Indeed, the recommended adjustments did not substantially change the essence of the document. The CSPA generality reconciled the measures. They only tried to clarify the most obvious discrepancies and ambiguous formulations rather than attempting to impede an approval of the statute in any way.[1146] Even the usually loyal Polish generality made many comments.[1147] They demanded around six dozen various

---

[1145] *Statut Spojených ozbrojených sil členských států Varšavské smlouvy a orgány jejich řízení za války*, 28. 2. 1979, k. 44, p. č. 152, č. j. 36502, ibid.

[1146] Letter by the chief of the CSPA general staff, K. Rusov, to the Warsaw Pact's supreme commander, V. Kulikov, March 5, 1979, ibid.

[1147] The Polish generals played a non-negligible role in the Warsaw Pact. Formally, they commanded the second largest army of the alliance located in the strategic area between the USSR and Germany. Their loyalty towards the Pact was essential for combat capabilities of the armed forces of the country, whose population kept strong anti-Soviet feelings. The memories of former Polish generals gave the impression that they primarily tried to extricate the PPA units from direct influence of the Soviet military command. However, some documents prove this claim wrong. Nevertheless, the Polish generality undoubtedly strove for bigger maneuvering space within the Unified Command, as they considered their real competences unsatisfactory. But they did not dare to cross the line of loyalty towards the PUWP leadership as well as the Soviet ally. Mastny, Vojtech, "*Warsaw Pact Generals in Polish Uniforms*, " Accessed October 16, 2019. http://www.php.isn.ethz.ch/lory1.ethz.ch/collections/coll_polgen/introduction8f 94.html?navinfo=15708; Paczkowski, Andrzej, "Die Polnische Volksarmee im Warschauer Pakt, " in *Der Warschauer Pakt. Von der Gründung bis zum Zusammenbruch 1955 bis 1991,* eds. Torsten Diedrich, Winfried Heinemann, and Christian Ostermann (Berlin: Christoph Links Verlag, 2009): 129.

adjustments and amendments.[1148] But the military failed to get support by the political power of the PPR. Objections by the Polish general staff concerning restrictions on national sovereignty were dismissed by the PUWP CC First Secretary, Edward Gierek.[1149] Also, the head of the Polish defense resort, Wojciech Jaruzelski, appreciated the scheduled changes in the alliance command system at the aforementioned session of the Committee of Defense Ministers in late 1978.[1150]

In September 1979, the partially modified document was submitted by the Unified Command again. By virtue of it, the Warsaw Pact member-states' armies were at the disposal of the Soviet command for a total war. Moscow's idea was the subordination of all troops to the alliance supreme commander. The Soviet generality argued from the World War II experience when the Czechoslovak and Polish units at the eastern front fought within the Belarus and Ukraine Fronts of the Red Army. According to this scheme, the Warsaw Pact command was supposed to directly order the units regardless of the country that had assigned them to the alliance formation.[1151] The command structure of the Unified Armed Forces at the battlefield largely corresponded to peacetime. It was just duplicated for the western and southwestern war theatres. Henceforward, the Warsaw Treaty Organization would have an individual supreme commander as well as a chief of staff for each of those. However, the Top High Command was established as a superstructure where Soviet dominance was strengthened again in comparison to the existing Unified Command. The commanders of Baltic and Black Sea fleets of the USSR - in fact, the Soviet admirals - were authorized to command the alliance Unified Navies. The design of the joint air-defense system remained essentially the same as in peace.[1152] The Warsaw Pact member-states even

---

[1148] Mastny, *A Cardboard Castle?*, 49.

[1149] Ross Johnson, *"Soviet Control of East European Military Forces"*.

[1150] *WSTĄPIENIE Ministra Obrony Narodowej PRL, gen.armii Wojciecha Jaruzelskiego w dyskusji nad 1 punktem porządku dziannego XI posiedzenia Komitetu Ministrów Obrony - Berlin, grudzień 1978r.*, PZPR KC, XIB/172, AAN, Warsaw.

[1151] Luňák, *Plánování nemyslitelného*, 64-65.

[1152] The integration of the Warsaw Pact member-states' air-defense system had already begun in the first half of the 1960s. Officially, the unified air-defense

lost their authority over the rest of their units which had not belonged to the alliance forces so far. The statute called them backups of the Top High Command intended to replenish the Unified Armed Forces. Despite this, their assignment to the Pact's formation at a certain stage of conflict was made conditional on "*a mutual agreement between the national military command and the Top High Command*;" this was taken for granted in the Eastern Bloc's reality.[1153] Moreover, the national military institutions were explicitly subordinated to the Top High Command according to the statute. The member-states were at least granted representation in the Military Council at the war theatre that was supposed to supervise the alliance troops' activities. This body had the right to issue statements which should have been implemented through orders of the supreme commander at the war theatre. Transition of the Unified Armed Forces to submission to the brand-new Top High Command should have been triggered by the member-states' decision reflecting an actual military-political situation or in case of an attack on the Warsaw Treaty Organization.[1154]

It also soon became clear that this reviewed draft was unacceptable. Besides rebellious Romania, which demanded the armed forces be under exclusive control of the national institutions even in wartime,[1155] the other alliance countries made new criticisms as well.[1156]

---

system of the alliance was established in 1969 and its structures were fully controlled by the Soviet officers. Therefore, it was not necessary to make any substantial changes in this area, despite the fact that serious deficiencies in coordination of the member-states' air-defenses still existed in practice.

[1153] The initial draft of the statute even assumed that in case of war, troop assignment to the Unified Armed Forces would not be decided by the political power of the member-states like up to this point, but rather by the national military bodies only.

[1154] *STATUT SPOJENÝCH OZBROJENÝCH SIL ČLENSKÝCH STÁTŮ VARŠAVSKÉ SMLOUVY A JEJICH ORGÁNŮ VELENÍ /v době války/ /návrh/*, 5. 9. 1979, FMO GŠ/OS, k. 46, p. č. 164, č. j. 36520, VÚA, Prague.

[1155] Luňák, *Plánování nemyslitelného*, 64.

[1156] Correspondence of the Czechoslovak military institutions suggest that these comments were really subjected to negotiations with the alliance staff. The Unified Command incorporated most of the proposals into the document. In some issues, the satellites were able to assert their will relatively straightforwardly. Czechoslovakia, for instance, firmly rejected the Staff of the Unified Armed Forces' claim that there was no need to include an amendment ordering

In terms of relations between the Top High Command and individual Warsaw Pact members, for example, the document totally ignored their political power. It referred solely to their military bodies. After objections of some Soviet satellites, the initial formulations had to be widened to "military-political leadership." However, as the statute obligated these institutions to act fully in accordance with the decisions of the entire Warsaw Treaty Organization and its military command, the member-states, in fact, lost all their sovereignty in the case of war, rather than strengthening their political representatives' role. On the contrary, some partial sops in reviewed documents gave the impression that some of the alliance members' powers were improved.[1157] The Czechoslovak objections that establishing the Top High Command and its composition was described just roughly in the statute were refused by the alliance command with a note that the issue would be additionally clarified by the Political Consultative Committee.[1158]

---

commanders of national units to submit their reports to the national leadership. *Statut SOS v době války*, 24. 10. 1979, FMO GŠ/OS, k. 46, p. č. 164, č. j. 36520, VÚA, Prague.

[1157] The initial draft did not envisage that the top supreme commander and composition of the Top High Command would be decided by the alliance member-states. However, such a formulation was pointless. In the following passage, the latter institution was practically identified with the Soviet army's general staff, whose composition could not be decided by the satellites of course. The national political and military bodies' position was also formally strengthened on the issue of dislocation of parts of the air-defense system outside the country's territory. The member-states also got bigger competences in the area of political agitation in their armies. In case of war, the established Political Administration of the Unified Armed Forces should have respected individual armies' specifics. The final document also contained assurances that the cadre's questions in individual units fell within the remits of national command bodies. On the contrary, a passage talking about the presence of the representatives of every member-state in the Top High Command, who would have served as a link between individual armies and the Warsaw Pact command, was deleted. *STATUT SPOJENÝCH OZBROJENÝCH SIL ČLENSKÝCH STÁTŮ VARŠAVSKÉ SMLOUVY A JEJICH ORGÁNŮ VELENÍ /v době války/ /návrh/*, 5. 9. 1979, ibid; *Statut Spojených ozbrojených sil členských států Varšavské smlouvy a jejich orgánů velení v době války*, 18. 1. 1980, 1261/0/7, sv. P 128/80, b. 13, NA, Prague.

[1158] *Statut SOS v době války*, 24. 10. 1979, FMO GŠ/OS, k. 46, p. č. 164, č. j. 36520, VÚA, Prague.

In reality, some proposed measures remained hardly acceptable even for the loyal Soviet satellites. This is proven by the requirement of Czechoslovakia that the document should have explicitly stated that it was valid for the time of the Warsaw Pact's existence only.[1159] The Czechoslovak documentation generally reveals that even servile commanders of the CSPA did not welcome the Soviet intentions. However, they were frankly not in a position to oppose. Therefore, there was no criticism even in internal materials. Nevertheless, during negotiations with the Unified Command officials, they tried to adjust the most burning aspects of the statute and sought potential ways to withdraw its remit in the future.

In November 1979, the reviewed document was approved by the working group in the Unified Command despite the Romanian disagreement.[1160] One month later, the final version was discussed at the 12th meeting of the Committee of Defense Ministers in Warsaw. Kulikov emphasized that the material had been worked out in accordance with the resolution of the last Political Consultative Committee session. He presented it as a result of a joint effort of the defense ministers, general staffs, and the Unified Command. The supreme commander did not hide any longer that in the case of war, *all* the Warsaw Pact countries' armed forces would be commanded by a single body. Then, the member-states' delegations, except for the Romanian one, flattered the Marshal in their speeches. They called the statute a significant measure to improve defense capabilities of the alliance and stated their unanimous consent to it.[1161] Given the situation, the Romanian defense minister Coman tried to use an

---

[1159] *STANOVISKO k návrhu "Statutu Spojených ozbrojených sil členských států Varšavské smlouvy a jejich orgánů velení /v době války/"*, undated 1979, ibid.

[1160] Mastny, *A Cardboard Castle?*, 49.

[1161] For example, the Czechoslovak defense minister, Martin Dzúr, said that the statute fully met the interests of the nations of Czechoslovakia and it was a significant contribution to deepening international unity and strengthening defense capabilities of the Warsaw Pact member-states. He emphasized the need for centralization of command which he called a basic "Leninist principle" of buildup and administration of armed forces. Besides these ideological clichés, he also correctly added that within the Unified Armed Forces, the Soviet army played a decisive role which represented undoubtedly the mightiest force of the Warsaw Pact.

obstructive strategy: He intentionally did not challenge the existence of the wartime statute, but demanded substantial adjustments. First of all, the Political Consultative Committee should have served as the Top High Command. Its advisory body should have become the existing collective Unified Command, rather than the General Staff of the Soviet Army. Bucharest agreed with establishing alliance commands at the war theatres. However, they were supposed to play a solely coordinative role, without the power to give direct orders to individual armies. Thus, the command over the Eastern Bloc' armed forces would have still fallen within the remits of the national military bodies.[1162] At the same time, Romania rejected integration of the Baltic and Black Sea navies. The country proposed just establishing their command center for coordination of combat activities instead. Also, no army political administration on the basis of the entire Warsaw Pact should have been created under any circumstances.[1163] These Romanian proposals were swept off the table by Kulikov. He said they contradicted the last Political Consultative Committee resolution. The Six was aware that the Romanian appeal for other adjustments of the statute, which should have been assessed one year later by the defense ministers again, in reality targeted the intended centralization of the command. Thus, the Committee of Defense Ministers, despite the Romanian protests,[1164] advised the appropriate Warsaw Pact member-states' institutions to approve the document. According to

---

[1162] Romania proposed individual member-states' armies to be commanded by the deputy supreme commanders, in practice deputy defense ministers, and national general staffs. They were supposed to mostly respect the orders of the commander in chief of their country.

[1163] As previously explained, during the entire 1970s, the Ceausescu regime rejected all attempts of the Unified Command to increase its influence over the propaganda within the Romanian armed forces.

[1164] Initially, Coman was ready to approve the resolution if it contained an amendment with Romanian comments and objections. This was a common practice on how different Romanian stances were handled at the Warsaw Pact military bodies' meetings in the 1970s. However, he changed his position later. He absolutely rejected a statement on the statute with a note that the draft did not correspond to the principles of military-political cooperation described in the founding charter and other documents. Speaking through its defense minister, Romania requested a complete revision of the statute. The Romanian representatives later conveyed the concrete proposals to the Unified Command.

Kulikov's formal proposal, the party leaders and the prime ministers were supposed to sign it. A final blessing should have been given by the next session of the Political Consultative Committee.[1165]

In February 1980, the supreme commander accompanied by the chief of Staff of the Unified Armed Forces made a trip to the Warsaw Pact countries. Kulikov should have been present during the signing of the statute. He did not visit Romania only.[1166] The course of his talks in Czechoslovakia reveals that the Unified Command officials had reconciled with the fact that Romania would not join the agreement. The CPCz CC General Secretary, Gustav Husák, asked Kulikov for Bucharest's position. The Marshal replied that despite an alleged verbal agreement with the principles of the statute, Romania was not ready to sign the document officially. He also stated that "*a patient clarifying and fundamental policy,*" rather than harsh actions was needed in the approach towards the Ceausescu regime.[1167]

The Political Consultative Committee's consent to the wartime statute was just a formal matter. A secret statement of this body was worked out on March 18, 1980, almost two months before the meeting itself. Referring to the last session in 1978, the statute was called a joint measure to strengthen the Warsaw Pact's defense capabilities in reaction to the persisting "*imperialistic threat.*" Signatures of Ceausescu as well as the new Romanian Prime Minister, Ilie Verdeţ, were missing. Afterwards, Kulikov stated that despite the absence of Romanian consent, the document was considered approved and valid.[1168]

No discussion at the Political Consultative Committee session about the supreme commander's report containing the issue of the wartime statute was expected. After all, the Marshal did not prepare any important

[1165] *Informace o průběhu a výsledcích 12. zasedání výboru ministrů obrany, konaného ve dnech 3.-6. prosince 1979 ve Varšavě,* undated 1979, FMO GŠ/OS, k. 45, p. č. 160, e. č. 36516, VÚA, Prague.
[1166] *Návrh schválení Statutů,* 7. 1. 1980; *Návštěva hlavního velitele Spojených ozbrojených sil v ČSSR,* 15. 2. 1980, k. 46, p. č. 164, č. j. 36520, ibid.
[1167] *Návštěva hlavního velitele SOS v ČSSR,* 22. 2. 1980, ibid.
[1168] *USNESENÍ členských států Varšavské smlouvy 18. března 1980;* Letter by the Czechoslovak defense minister, M. Dúzr, to the CPCz CC General Secretary, G. Husák, March 31, 1980, ibid.

speech. He just laconically said that henceforward, it was necessary to focus on implementation of the measures included in the document.[1169] Furthermore, the wartime statute was just one item on the agenda of the alliance's supreme body which took place on May 14-15 in the Polish capital. It primarily sought to define a new tactic for how to face the fall of détente and risk of escalation of the arms race, which the Eastern Bloc could not afford any more.[1170] Taking into account the recent development, a clash with the Romanian delegation could be anticipated. According to Gribkov's testimony, an angry Ceausescu quickly left Warsaw right after the meeting had ended.[1171] Although the Six considered his attitudes as directly opposing policy of Moscow and the entire Warsaw Treaty Organization, it appreciated that the Romanian leader had presented his stances at the internal talks only.[1172] At the end, the Romanian delegation also made a certain concession. It agreed with the resolution referring to, among other things, another strengthening of the

---

[1169] *Předložení materiálů k jednání Politického poradního výboru*, 12. 4. 1980, FMO GŠ/OS, k. 47, p. č. 173, č. j. 37012, VÚA, Prague; *Důvodová zpráva o nadcházejícím zasedání politického poradního výboru Varšavské smlouvy*, 29. 4. 1980, 1261/0/7, sv. P 138/80, perr 8, NA, Prague.

[1170] *Zpráva o průběhu a výsledcích zasedání politického poradního výboru Varšavské smlouvy ve dnech 14.-15. května 1980 ve Varšavě*, 27. 5. 1980, sv. P139/80, b. 7, ibid.

[1171] Luňák, *Plánování nemyslitelného*, 65.

[1172] *Zpráva o průběhu a výsledcích zasedání politického poradního výboru Varšavské smlouvy ve dnech 14.-15. května 1980 ve Varšavě*, 27. 5. 1980, 1261/0/7, sv. P139/80, b. 7, NA, Prague. After the last Political Consultative Committee session, Ceausescu made public some secret intentions of the Warsaw Treaty Organization. The East-German defense minister, Heinz Hoffmann, criticized Romania for this already in 1978 at the Committee of Defense Ministers' meeting. He pointed to the fact that the West had had the results of the recent Political Consultative Committee gathering concerning the military issues and warningly emphasized that this situation did not contribute to a deepening of mutual trust between the alliance members. *Informace o poznatcích BLR k současné politice ČLR v oblasti Balkánského poloostrova*, 29. 1. 1979, DTO 1945-1989, i. č. 4, e. č. 50, č. j. 010.780/79, AMZV, Prague. *Informace o průběhu a výsledcích 11. zasedání výboru ministrů obrany, konaného ve dnech 4.-7. prosince 1978 v Berlíně*, undated, FMO GŠ/OS, k. 35, p. č. 127, e. č. 36027, VÚA, Prague.

Unified Armed Forces.[1173] The statute, approved in absolute secrecy, remained non-binding for Romania.[1174]

Consequences of the Statute's approval

Adoption of the statute for a time of war largely reflected changes which the Warsaw Treaty Organization went through during the 25 years of its existence. The USSR always kept clear dominance within the alliance. However, in the early period, it was not based on any statutory documents and this fact had increased room for Moscow's wantonness. We are reminded that the original statute of the Unified Command of 1955 was an extremely brief document, compared to the regulations approved 25 years later. Brezhnev's modification of cooperation within the alliance's framework did not aim to weaken the Soviet position, but rather to codify the existing state of things. Hence, it was straightforwardly declared that in the case of armed conflict the Soviet general staff would become a part of the command structures which would be superior to the member-states' national institutions.

At first, the statute defined in detail the composition and activities of the Unified Armed Forces and their command in war. It became a guideline for technical and logistic support, political agitation as well as funding combat operations at each war theater.[1175] The document also

---

[1173] *Překlad - USNESENÍ členských států Varšavské smlouvy, přijaté na zasedání Politického poradního výboru k vystoupení hlavního velitele Spojených ozbrojených sil*, 21. 5. 1980, k. 47, p. č. 173, č. j. 37012, ibid.

[1174] Retrospectively, Gribkov criticized this approach. He thought that in the case of bigger respect for the allies, Romanian consent could also have been obtained. Mastny, *A Cardboard Castle?*, 49; Mastny, *Learning from the Enemy*, 35.

[1175] In the end, the funding proportion of the high command at war theater was based on the ratio of contributions to the alliance's budget set up by the Political Consultative Committee in 1969. At the western war theater, the USSR should have paid 44.5%, Poland 23.1%, East Germany and Czechoslovakia 16.2% equally. At the southwestern war theater, the USSR would have covered 44.5%, Romania 24.1%, Bulgaria 16.9%, and Hungary 14.5%. The initial draft became a bone of contention. Czechoslovakia was supposed to pay 22.2%, while East Germany just 10.2%. Referring to the population number of individual Warsaw Pact member-states, Czechoslovakia declared that its proportional contribution to the western war theater should have been around 12.4% only. Afterwards, the USSR decided that national income would be used as a criterion. Therefore, the

stated once again that the eventual "destruction of an aggressor" would not be limited by state borderlines at all. Thus, it formally confirmed a strategy of the Warsaw Pact troops' massive counterattack after initial repulsion of the enemy and following invasion of Western Europe. The command structures for the case of war were split into two independent branches for the western and southwestern war theaters. The posts of the supreme commanders would be held by Kulikov and Gribkov, respectively.[1176] A brand-new Top High Command was intended as their coordinative center.[1177] The symbolic appointment of Brezhnev - old and in ill-health - as the head of this body, reflected an initial crisis of the Soviet leadership as well as beginning decay of the Warsaw Treaty Organization itself.[1178] Such a step was again taken without Romanian consent. It is good to add that the other Soviet satellites' representatives accepted it with "sarcastic smiles."[1179] After the death of the CPSU CC General Secretary in 1982, the position remained vacant till the alliance's dissolution.[1180]

The document revealed that in case of war, the USSR planned to restrict almost the entire decisive autonomy of its satellites. On one hand, combat operations should have been conducted in close collaboration with the Warsaw Pact member-states' military as well as political bodies, but

---

CSSR and NDR were supposed to pay the same amount. It is good to add that in the case of an outbreak of global total war, this issue, as well as many others, would have been most likely irrelevant. *Výše podílových vkladů členských států Varšavské smlouvy do rozpočtu finančního zabezpečení činnosti a orgánů Spojeného velení,* undated 1979; *STATUT SPOJENÝCH OZBROJENÝCH SIL ČLENSKÝCH STÁTŮ VARŠAVSKÉ SMLOUVY A JEJICH ORGÁNŮ VELENÍ /v době války/ /návrh/,* 5. 9. 1979, FMO GŠ/OS, k. 46, p. č. 164, č. j. 36520, VÚA, Prague.

[1176] Memorandum of conversation with Marshals Ustinov and Kulikov concerning a Soviet War Games, June 14, 1982, in *A Cardboard Castle?* (doc. 95): 462-465.

[1177] *Statut Spojených ozbrojených sil členských států Varšavské smlouvy a jejich orgánů velení v době války,* 18. 1. 1980, 1261/0/7, sv. P 128/80, b. 13, NA, Prague.

[1178] Luňák, *Plánování nemyslitelného,* 65.

[1179] Mastny, *A Cardboard Castle?,* 49. Brezhnev was obsessed with decorations and medals in general. He was given around two hundred of those during his life. Already in 1976, he had been appointed the Marshal of the Soviet Union, the highest rank in the Soviet army. Bacon, "Reconsidering Brezhnev," 9-10.

[1180] Umbach, *Die Rotte Bündnis,* 260.

solely orders by the alliance command were binding for the troops. The Pact members nominated their representatives there, but they were given strictly defined, second-rank functions only. Moreover, these officers were authorized to participate only in planning which directly concerned units of the country of their origin. Therefore, it cannot be considered as alliance cooperation in the proper sense of the word.[1181] The most significant representation of the Soviet satellites' military officials should have been participation of the Military Council at war theater. Nevertheless, the supreme instance remained the Top High Command which also held wide competences in the issue of logistic support at the member-states' territories. The satellites were left with just a symbolic responsibility for propaganda in their armies. On the other hand, the top supreme commander was granted the right to interfere with this activity in order to *"strengthen willingness to achieve a decisive victory over the imperialistic aggressor."* The member-states retained certain - but in practice obviously rudimentary - decisive power through the Political Consultative Committee which was supposed to deal with military issues even in a period of war.[1182]

Paradoxically, the wartime statute was approved at the time, when the East-European armed forces were probably losing their importance from the Soviet point of view. At the beginning of the 1980s, a term *"war within the coalition"* was disappearing from works by the Soviet military

---

[1181] For example, Polish representation in the command structures at war theater should have only included 70 people. In practice, they were supposed to command the Polish Front and participate in planning the operation of this echelon. The former general of the PPA, Antoni Jasiński, who served as the Polish deputy supreme commander of the Warsaw Pact, claims that command of individual member-states' troops through their own officers was assured this way. He adds that the Polish generality in the alliance command was responsible for the PPA units only. The non-Soviet deputy supreme commanders had no decisive authority over the other member-states' armies as a title of their function suggested. *Rozmowa z generałem Antonim Jasińskim 3 marca 2000*, Accessed October 16, 2019. http://www.php.isn.ethz.ch/kms2.isn.ethz.ch/serviceengine/Files/PHP/20 668/ipublicationdocument_singledocument/331a8bf0-5f20-4bf5-beca-9af3aa784 a6d/pl/3Jasinski.pdf.

[1182] *Statut Spojených ozbrojených sil členských států Varšavské smlouvy a jejich orgánů velení v době války*, 18. 1. 1980, 1261/0/7, sv. P 128/80, b. 13, NA, Prague.

theorists. According to their opinion, the Eastern Bloc countries' units lacked the operational capabilities of the Soviet army needed to succeed in war waged by modern conventional weapon systems. This conception was supported especially by Marshal Nikolai Ogarkov, the chief of the Soviet general staff.[1183] However, the loyalty level of the Warsaw Pact member-states' military officials towards Moscow at the time was the highest for the alliance's existence so far. This fact also largely contributed to approval of the statute. The measures which had been totally unacceptable at the end of the 1960s, the East-European military representatives now recommended their political counterparts to pass.[1184]

During evaluation of the real impacts of adoption of the Warsaw Pact's wartime statute, the fact must be taken into account that the events fortunately never reached the stage when the measures were implemented in practice. Petr Luňák notes that the document hardly improved mechanisms of the Warsaw Treaty Organization, as Moscow had initially expected. One of the reasons lay in relations between the alliance supreme commander and the chief of the Soviet general staff which was still unsolved. Most likely, the question remained open because of extreme personal antipathy between Marshals Kulikov and Ogarkov.[1185] Moreover, the statute brought no substantial changes in the alliance military structures in peacetime. In this regard, a huge gap between the Warsaw Treaty Organization and NATO persisted. The command structures of the latter would not have been significantly changed during a transition from a state of peace to a state of war, as the Warsaw Pact still envisaged. Thus,

---

[1183] Ross Johnson, "*Soviet Control of East European Military Forces*".

[1184] In 1978, for example, ahead of the Political Consultative Committee session, the Czechoslovak defense minister Dzúr advised Gustav Husák to support working-out the statute. He vaguely argued the need for conducting contemporary combat operations and assured that the document would clearly define relations between the party and state institutions and the Warsaw Pact command in case of war. In front of the political leadership, he also defended a demand that in case of a sudden escalation of situation, the up to this point non-existing Top High Command and the Soviet general staff would take command over the Warsaw Pact troops. Letter by the Czechoslovak defense minister, M. Dúzr, to the CPCz CC General Secretary, G. Husák, November 16, 1978, FMO GŠ/OS, k. 44, p. č. 148, č. j. 36054, VÚA, Prague.

[1185] Luňák, *Plánování nemyslitelného*, 65.

the wartime statute of 1980 just confirmed that the existing Unified Command of the alliance had a solely administrative and organizational role.[1186] Later, the USSR unveiled that in the case of war, this institution would be incorporated into the Top High Command represented mostly by the Soviet army's general staff.[1187]

Retrospective testimonies of some Polish officers suggest that the statute was a too controversial document interfering with the alliance member-states' sovereignty and therefore never enforced in practice.[1188] But implementation of its partial measures proves these claims wrong. The final plan for joint use of the Unified Armed Forces was finished in 1983.[1189] However, already in early June 1980, Kulikov informed the Warsaw Pact member-states' military officials that in connection with approval of the wartime statute, funding of construction of the protected command posts on the Soviet territory was canceled. Instead, the monies would be invested into the buildup of similar objects of western strategic direction, in Poland.[1190] Given the absence of key Soviet documents, we can only speculate about the reasons behind this decision. However, it is

---

[1186] *Rozmowa z generałem Antonim Jasińskim 3 marca 2000*, Accessed October 16, 2019. http://www.php.isn.ethz.ch/kms2.isn.ethz.ch/serviceengine/Files/PHP/20668/ipublicationdocument_singledocument/331a8bf0-5f20-4bf5-beca-9af3aa784a6d/pl/3Jasinski.pdf.

[1187] The Unified Command in peace and the command structures for individual war theaters in case of war were strictly separated. In March 1985, for example, the supreme commander informed the Czechoslovak side that given this reason, the CSPA was not able to nominate to positions at the western war theater those officers who already served in the Unified Command bodies. *Doplňování velení hlavního velitele vojsk na západním válčišti*, 22. 3. 1985; *Objasnění požadavku hlavního velitele SOS na doplňování orgánů velení*, 22. 4. 1985, FMO GŠ/OS, k. 60, p. č. 235, č. j. 38304, VÚA, Prague.

[1188] Tejchman an Litera, *Moskva a socialistické země na Balkáně*, 86.

[1189] *USNESENÍ členských států Varšavské smlouvy přijaté na poradě politického poradního výboru v referátu hlavního velitele Spojených ozbrojených sil 5. ledna 1983 (překlad z ruského znění)*, undated, 1983, FMO GŠ/OS, k. 54, p. č. 205, č. j. 38015, VÚA, Prague.

[1190] We are reminded that construction of the protected command posts on the Bulgarian and Polish territories had been proceeding since 1974. In the southwestern direction, in Bulgaria, the object was almost finished at the time. *Chráněná velitelská stanoviště Spojených ozbrojených sil*, 31. 7. 1980, k. 22, p. č. 67, č. j. 40004, ibid.

likely that the step was linked to tightening Soviet control over the Warsaw Pact member-states' armies through the recently adopted regulations. Since this point, the Soviet generality envisaged that the Unified Armed Forces would be commanded from outside the territory of the USSR, from the posts in Poland and Bulgaria.[1191] In late 1981, in terms of realization of the wartime statute's measures, the Unified Command put forward a draft of organizational changes in the Unified Armed Forces' command structures at the western war theater. The numbers of personnel in charge should have been increased, including the non-Soviet representatives.[1192] Their potential competences remained doubtful. The Czechoslovak defense minister, Martin Dzúr, noted that the real ranks of nominated Czechoslovak officers were higher than those indicated for in case of armed conflict. Although the sources available say that he made just a laconic comment on the issue,[1193] this was probably evidence that the non-Soviet armies' members should have played a secondary role in the Warsaw Pact command during a war. The fact that the wartime statute was considered valid within the alliance military framework, is also proven by the Czechoslovak military bodies' effort to choose those officers for service in the Staff of the Armed Forces who did not have to be replaced during the transition from peace to war.[1194] In 1983, establishing of the alliance command structures for the western war theater in wartime was

---

[1191] Letter by the Czechoslovak defense minister, M. Dúzr, to the CPCz CC General Secretary and the Czechoslovak president, G. Husák, August 18, 1980, ibid.

[1192] For example, the Czechoslovak institutions stated that in existing Staff of the Unified Armed Forces were serving 41 officers of the CSPA only and therefore it was necessary to select another 22 adepts. Composition of the command structures at the western war theater was discussed at the 24th meeting of the Warsaw Pact Military Council. In this regard, general A. G. Merezhko, deputy chief of Staff of the Unified Armed Forces, granted the right of the member-states to select and send representatives to the alliance command. *Návrh organizační struktury hlavního velení hlavního velitele SOS na západním válčišti*, 11. 11. 1981, k. 51, p. č. 189, č. j. 37519, ibid.

[1193] *Zařazení důstojníků ČSLA v HV SOS*, 22. 12. 1982, ibid.

[1194] *Návrh organizační struktury orgánů hlavního velení SOS na západním válčišti*, 3. 11. 1981, ibid.

simulated in various military drills.[1195] At the end of the 1980s, for example, the scenario of "Shield-88" maneuvers was also based on provisions of the wartime statute. It envisaged submission of the 8th Polish army to the Soviet command of the Front which was directly responsible to the highest command of the USSR.[1196]

The statute in the Western hands

As already explained, the process of preparation and approval of the statute proceeded in absolute secrecy. The document was so confidential that only a narrow circle of the top party and military officials in the Warsaw Pact countries knew about its existence.[1197] The statute was signed on March 18, 1980. However, on May 25, at the latest, some parts of the material were obtained by the CIA.[1198] The information leaked through Polish Colonel Ryszard Kukliński,[1199] a pro-Western spy. Therefore, the West had already received the first reports on the statute in March 1980. At that time, Kukliński informed that the PPA command had voluntarily agreed with subordination of the Polish troops to the Soviet general staff and joint command of the Warsaw Pact in case of war and representation of the Polish officers in this body would have been strictly rudimentary.[1200]

[1195] Report on Speech by Marshal Ogarkov at a Warsaw Pact Chiefs of Staff Meeting in Minsk, September 8-10, 1982, in *A Cardboard Castle?* (doc. 96): 466-468.

[1196] Ross Johnson, "*Soviet Control of East European Military Forces*".

[1197] Umbach, *Die Rotte Bündnis,* 260.

[1198] Ross Johnson, "*Soviet Control of East European Military Forces*".

[1199] Since the beginning of the 1970s till November 1981, Colonel Kukliński provided the United States with sensitive information on the Eastern Bloc military planning. He smuggled more than 30 thousand top-secret documents of the Warsaw Treaty Organization and the Soviet armed forces, including war plans, military maps, mobilization guidelines, decision-making procedures within the Unified Command, and the alliance nuclear doctrine. He served as the supreme commander Kulikov's liaison in the Unified Command. In November 1981, he was forced to flee Poland in order to avoid arrest. Kramer, "Colonel Kuklinski and Polish Crisis," 48-59; Kukliński, Ryszard, *Wojna z narodem. Rozmowa z byłym płk. dypl. Ryszardem J. Kuklińskim* (Warszawa: Kurs, 1987).

[1200] Umbach, *Die Rotte Bündnis,* 260.

Les Griggs, a former U.S. intelligence officer and an Eastern Europe expert, notes that the U.S. Army Psychological Operations command tried to use this fragmented information to compile a report on the loss of sovereignty of the Soviet satellites which could have been used for propagandistic purposes. The Pentagon wanted to make the information about the statute public as soon as possible, as a diplomatic weapon against the USSR. But the CIA headquarters quickly refused such a scenario, referring to potential exposure of their intelligence sources in the Warsaw Pact command. Nevertheless, the information was used by NATO to specify its strike plans of which were supposed to totally eliminate the Warsaw Pact command structures right after an outbreak of eventual conflict.[1201]

* * *

The acceptance of the wartime statute meant the last step on the Warsaw Pact's path from Khrushchev's organization on paper to a real alliance. Once again, a huge gap between its military and political structures was proven. Despite some significant modifications, the Pact's political dimension operated continually without any statutory framework. This, among other things, made it easier for Moscow to shape the alliance's meetings according to its imminent needs and to define their agenda. In contrast, the functioning of the Warsaw Pact's military structures was defined more clearly. If the organization in the late 1980s still remained from the political perspective more or less a "cardboard castle," as Vojtech Mastny calls it, from the military point of view, it was a fully operational alliance. Putting aside features of this cooperation, namely the absolutely obvious dominance of the Soviet Union, clear and sophisticated mechanisms for collaboration of the member-states' armed forces in both peace and war were established in the Warsaw Pact during the 1970s.

Therefore, the Warsaw Pact's importance in terms of a tool for the Eastern Bloc management must be seen in the military dimension. From that perspective, at the turn of the 1970s and 1980s, the Warsaw Pact

---

[1201] Griggs, Les, "*Origins and Significance of the Warsaw Pact Wartime Statute Documents,*" Accessed October 16, 2019. https:// www.cia.gov / library / publications/cold-war/wartime-statutes/wartime-statutes.pdf.

represented an organization with a system of regular consultations and interaction, largely comprehensive rules, and finally, strong loyalty of the military officials of most member-states to both the alliance itself and its leading power. That reality, in any way, cannot be underestimated and overshadowed by the fact that in the political dimension the Warsaw Pact remained an eggshell, although much stronger than in 1955. At the time, in fact, a powerful, well-organized military machinery waited in the "cardboard castle."

Changes in the Warsaw Pact
military doctrine after 1975

Alteration of the international situation after 1975 also resulted in another shift in thinking of the Eastern military strategists and the entire doctrine of the Warsaw Treaty Organization. Their basic assumptions remained the same, only the estimation of the possible course of future war was changing. Thus, till the mid-1980s, the Warsaw Pact strategic doctrine was based on belief in its own strong offensive capabilities. The ideological precepts labeled capitalism aggressive and West Germany revanchist. If the Warsaw Treaty Organization was supposed to defend socialism effectively, it inevitably had to be able not only to repulse an aggression, but also move combat quickly to territories of NATO. This next war should have been a decisive clash between two different systems.[1202] The highest strategic priority was still given to the region of Central Europe. There was the biggest concentration of NATO as well as the Warsaw Pact troops. The Eastern analytics noted with concern that in this area, the Western alliance dispatched the largest, best-trained and equipped military group in the world.[1203]

Given the absence of the Unified Command's key documentation, the following conclusions cannot be considered definite and ultimate. It is good to add that even frequently mentioned drill scenarios must not be confused with the real war plans. In the second half of the 1970s, the Warsaw Pact's overall operational plans still remained in the Moscow-based military institutions. These would have been given to other countries

---

[1202] Bange, Oliver, "Comments on and Contextualization of Polish Documents related to SOYUZ 75 and SHCHIT 88," *Cold War International History Project e-Dossier*, no. 20 (2010).
[1203] *NÁVRH hlavních směrů rozvoje ČSLA na léta 1981-1985*, undated, 1980, FMO GŠ/OS, k. 20, p. č. 61, č. j. 35502, VÚA, Prague.

solely in case of an outbreak of war.[1204] However, command-staff exercises provided regular opportunities to test the effectiveness of actual strategic concepts, eventually to try various innovations. Within the Unified Armed Forces, three main sorts of maneuvers were held at the time: strategic-operational "Soyuz," tactical "Druzhba," and naval "Val."[1205] They aimed to test the limits and possible extent of combat operations.[1206] In the years following the Helsinki conference, especially the drills "Shield-76," "Soyuz-77," and "Val-77" took place in order to check the latest strategic plans. Afterwards, the lessons learned were used to make additional adjustments in these plans and provided strategists with some feedback.[1207]

As mentioned before, the Helsinki process caused an overall decrease in massive military maneuvers held in Europe. This concerned NATO as well as the Warsaw Treaty Organization.[1208] However, it did not mean that preparations for war were definitely stopped. In 1975 alone, four dozen joint actions under the auspices of the Unified Command took place, including twenty drills of various extents.[1209] Nevertheless, the trend to reduce maneuvers with deployment of large numbers of troops continued.

[1204] Kramer, Mark, "Commentary on Soyuz-75 and Shchit-88 Military Exercise Documents," *Cold War International History Project e-Dossier*, no. 20 (2010).

[1205] Bange, "Comments on and Contextualization of Polish Documents".

[1206] Heuser, "Warsaw Pact Military Doctrine," 438.

[1207] Soviet Statement at the Chiefs of General Staff Meeting in Sofia, 12.-14. 6. 1978, in *A Cardboard Castle?* (doc. 83): 415-417.

[1208] Overall, however temporary, the Warsaw Pact's withdrawal from large maneuvers was documented even by the fact, that unlike in the past, the CSPA units did not take part in the annual alliance "Shield" drills during the years 1977-1979. In 1978, the Czechoslovak military officials were surprised that the chief of Staff of the Unified Armed forces Gribkov asked the country to symbolically send at least one motor rifle battalion to the scheduled exercise "Shield-79" in Hungary. According to the original intentions, Soviet, Romanian, Bulgarian, and Hungarian troops should have solely participated in the maneuvers. The significance of the event was illustrated by the fact that this was the first ever Warsaw Pact maneuvers covered by Bulgarian press. *Plán společných opatření Spojených ozbrojených sil - doplněk*, 12. 10. 1978, FMO GŠ/OS, k. 34, p. č. 125, č. j. 36052, VÚA, Prague; *Warsaw Pact Exercise (Shield 79)*, May 28, 1979, FCO 28/3746, TNA, London.

[1209] In 1975, only Romanian troops were absent in testing the new operational plans of joint combat action of the Unified Armed Forces.

The most significant training action in that year was the operational-staff exercise "Soyuz-75." However, it played out on the maps only.[1210] In context of previous training measures, this was almost an imperceptible action.[1211] The scenario brought no substantial innovations: Even at the time of signing the CSCE Final Act, the Warsaw Pact's strategists anticipated massive use of nuclear weapons on both sides in case of war. A bigger emphasis was put on the use of radio-electronic warfare devices during an offensive operation in the western direction. Furthermore, computers were significantly engaged for the first time. During analysis of the exercise, their benefits were praised.[1212]

Repulsion of NATO's aggression

The Warsaw Treaty Organization still envisaged that in case of war, its troops would launch an incursion into Western Europe. It has to be stressed once more that such a scenario would have been realized in the second stage of conflict, after the Unified Armed Forces repulsed the initial enemy aggression. However, the Eastern military intelligence's reports clearly stated that NATO also did not operate with a possibility of starting a war. According to the Western plans, the Warsaw Pact would strike first. In this regard, there was almost no difference in both alliances' assumptions.[1213] This fully reflected the Cold War mistrust. Nevertheless, training of the Unified Armed Forces in the years 1975-1976 was clearly based on the fact that the main tasks of the alliance troops was to cover the member-states' borderlines, repulse an air aggression, and only then launch a "decisive attack." Individual armies should have been prepared to fight in these operations as parts of the alliance forces, rather than

---

[1210] Iakubovskii Report on the State of the Unified Armed Forces, December 31, 1975, in *A Cardboard Castle?* (doc. 77): 395-396.

[1211] "Soyuz-75" was incomparably smaller than the Unified Armed Forces' maneuvers before 1973. We are reminded that about 100 thousand troops had taken part in the "Shield-72" drills, for example. Kramer, "Commentary on Soyuz-75 and Shchit-88".

[1212] Bange, "Comments on and Contextualization of Polish Documents".

[1213] *PODKLADY k informaci a závěrům o cvičeních NATO v r. 1975,* undated, most likely 1976, FMO GŠ/OS, k. 39, p. č. 134, č. j. 36037, VÚA, Prague.

individual units.[1214] For this reason, a strong emphasis was put on unification of the organizational structure of the Warsaw Pact member-states' armed forces.[1215]

At the beginning of the 1980s, cooperation in the field of military science was also increased significantly within the Unified Armed Forces. On the level of alliance, the issues of operational and combat deployment of troops or a theory of conducting a joint front operation were discussed, for example, even without Romania.[1216] In 1981, a statute on military-political cooperation within the Unified Armed Forces was worked out. According to General Gribkov, it was supposed to unify the member-states' activities in this area.[1217] A strong emphasis was put on exploration of possible trends in NATO's operational and combat preparation. Conclusions began to be published in a special bulletin issued by the Staff of the Unified Armed Forces for the needs of military elites.[1218] In 1982, cooperation in this field reached a level where it was necessary for better coordination to establish the function of the chief of alliance staff for military-science. The post was held, of course, by the Soviet general, M.

[1214] *Zpráva pro PÚV KSČ o splnění úkolů ČSLA ve výcvikovém roce 1975-1976 a hlavních úkolech pro výcvikový rok 1976-1977*, 15. 10. 1976, 1261/0/7, sv. 17, a. j. 19/3, NA, Prague.

[1215] Czechoslovakia, for example, correctly stated that the country had no potential to build an army capable of destroying NATO units dispatched in West Germany on its own. For this reason, it should have focused mostly on defensive tasks - repulsion of a surprise air strike, defense of state borders, and eventually to deal a necessary nuclear blow to the enemy in time. The rest of the objectives should have been solved in the scope of the Warsaw Treaty Organization. *Základní záměry koncepce výstavby ČSLA na léta 1981-1985 s výhledem do roku 1990 (podklady)*, undated 1976-1977, FMO GŠ/OS, k. 7, p. č. 38, č. j. 39030, VÚA, Prague.

[1216] *Slavnostní zasedání vojenské rady Varšavské smlouvy v Moskvě*, 24. 5. 1980, k. 46, p. č. 161, č. j. 36518, ibid.

[1217] At the 23rd meeting of the Military Council, the material was rejected by the Romanian general M. Nicolescu. He stated that the field of military science fell within the remits of the national leaderships and Romania considered the existing level of alliance cooperation in this area sufficient. *Výsledky 23. zasedání vojenské rady Spojených ozbrojených sil*, 23. 4. 1981, k. 49, p. č. 182, e. č. 37501, ibid.

[1218] *PLÁN REALIZACE Protokolu čís.:023 z 23. zasedání Vojenské rady Spojených ozbrojených sil*, undated 1981, ibid.

G. Titov. His task was to oversee development of new theoretical bases for the Warsaw Pact's operational and combat preparation.[1219]

In the second half of the 1970s, not only the Warsaw Pact's war games were based on the assumption that a war would be launched by NATO. According to the 1977 modified operational plan of the Czechoslovak Front, a Western aggression and its termination by the Eastern troops would also have preceded an advance into the enemy's territories. Therefore, emphasis was placed on defense of borderlines with West Germany and Austria. As a result, the Czechoslovak Front as well as the Soviet troops would have gotten an opportunity to regroup and then start an offensive operation themselves.[1220] To stop NATO's initial offensive remained unambiguously the main declared goal of the Unified Armed Forces even at the time of slowly falling détente. In December 1978, the 11th meeting of the Committee of Defense Ministers stated that rise of offensive features in NATO military doctrine could be seen.[1221] Six months earlier, at a meeting of the chief of general staffs of the Warsaw Pact member-states, the Soviet general S. F. Romanov repeated that a future war would be a class-conflict. However, this would happen only if the imperialistic aggressor started it, he added. Although destruction of the enemy forces and a "total victory" should have been achieved in the following operation - which was based on the assumption of a campaign into Western Europe - defense of its own territory was considered essential.[1222] This conception persisted even at the beginning of the

---

[1219] *Zavedení funkce zástupce NŠ SOS pro vojenskovědeckou práci*, 10. 8. 1982, k. 53, p. č. 202, e. č. 38009, ibid.

[1220] Operační plán Československé lidové armády, December 28, 1977, in *Plánování nemyslitelného* (doc. 11): 249-276.

[1221] *TÉZE DOKLADU představitelů ministerstva obrany SSSR na 11. zasedání Výboru ministrů členských států Varšavské smlouvy k první otázce programu "O stavu a perspektivách rozvoje sil NATO"*, undated 1978, FMO GŠ/OS, k. 35, p. č. 127, e. č. 36027, VÚA, Prague.

[1222] Report by Marshal Kulikov on the State of the Unified Armed Force, January 30, 1978, in *A Cardboard Castle?* (doc. 82): 413-414; Soviet Statement at the Chiefs of General Staff Meeting in Sofia, June 12-14, 1978, ibid, (doc. 83): 415-417.

1980s.[1223] At the time, an eventuality emerged in drill scenarios that a global war was being started by China. Combat operations in the Far East would have engaged the non-negligible forces of the Soviet army and after 40 days, this situation would have been used by NATO to launch a surprise attack on the Warsaw Treaty Organization.[1224]

Some historians dealing with the Warsaw Pact note that despite the aforementioned proclamations, preparation for repulsion of aggression remained very superficial. In this regard, Vojtech Mastny correctly points to the Eastern strategists' assumption that NATO would start a war with significant, numerous disadvantages. Therefore, to stop the enemy at some point in the fighting would have been no serious problem for the Unified Armed Forces. Thus, a frequently practiced launch of a counteroffensive was considered as the key phase.[1225] On the other hand, it would be odd to say that the Warsaw Pact troops did not train defensively at all in the second half of the 1970s. At the time of falling détente, the supreme commander Kulikov gave individual armies specific tasks which should have been tested in the upcoming drills. During the "Neutron-79" maneuvers, for example, the CSPA trained simultaneously in the stopping of enemy aggression as well as forming the offensive corps to destroy those units which had penetrated Czechoslovak borders.[1226] In 1979, the Unified Armed Forces were supposed to practice various stages of combat operations, including a breakthrough in the enemy's defense as well as

---

[1223] *PLÁN REALIZACE Protokolu čís.:023 z 23. zasedání Vojenské rady Spojených ozbrojených sil*, undated 1981, FMO GŠ/OS, k. 49, p. č. 182, e. č. 37501, VÚA, Prague.

[1224] The Soviet war games held in mid-1982 presumed that the Chinese attack on the USSR would also be supported by Japan and South Korea. The Soviet army would have been engaged by 290 enemy divisions. After the NATO strike on the Warsaw Treaty Organization, Iran and Saudi Arabia were expected to join the fight against the USSR as well. Memorandum of conversation with Marshals Ustinov and Kulikov concerning a Soviet War Games, June 14, 1982, in *A Cardboard Castle?* (doc. 95): 462-465.

[1225] Mastny, *A Cardboard Castle?*, 46.

[1226] *PLÁN REALIZACE Protokolu čís.: 0018 z 18. zasedání Vojenské rady Spojených ozbrojených sil*, undated, 1978, FMO GŠ/OS, k. 8, p. č. 42, č. j., 390034, VÚA, Prague.

defensive tasks.[1227] However, in October 1982, at the 26th meeting of the Military Council in Warsaw, Kulikov complained that insufficient units and means within the Unified Armed Forces were assigned to protection of state borders.[1228] The documents of this body's meetings prove that in the second half of the previous decade, the drills were focused mostly on a breakthrough in the enemy's defense which was necessary in order to launch a march into Western Europe.[1229]

Some inevitable penetration of NATO troops into the Warsaw Pact member-states' territories in the early stage of conflict was largely anticipated. The initial period of war was simulated, for example, by the already mentioned "West-77" maneuvers. The Eastern strategists put forward a scenario that 3-4 days before the outbreak of war, NATO had mobilized and deployed its troops and fleets to battle stations. The objective of their first operation was to seize the whole GDR and the western regions of CSSR and PPR. NATO strove for crushing the main

---

[1227] The former presumption was valid mostly for the East-German armed forces, the latter for the Czechoslovak army. *18. zasedání Vojenské rady Spojených ozbrojených sil*, 23. 10. 1978, k. 34, p. č. 125, č. j. 36025, ibid.

[1228] *26. zasedání vojenské rady Spojených ozbrojených sil*, 25. 10. 1982, k. 53, p. č. 204, e. č. 38013, ibid.

[1229] In 1977, the drills included especially a breakthrough in the enemy's defense, overcoming water obstacles and missile units' shooting. In this regard, an emphasis was put on cooperation of staffs. The main actions, "Soyuz-77" and "West-77," were just smaller command-staff exercises. An offensive operation remained the back-bone of drills in 1979. The supreme commander tasked individual armies with test-objectives: Bulgaria should have practiced repulsion of a large wedging group of the enemy's troops and anti-tank combat; Hungary, breakthrough in fortified areas and activities of airborne strike forces during seizure of mountain cols; East Germany, cooperation of artillery and helicopters during a breakthrough in enemy defense; Poland, command to the troops during transition from peace to a state of war; Romania, combat support for the troops during their transport to a battlefield; the USSR organization of fire during a breakthrough in enemy's defense; and Czechoslovakia, command to the air forces during an offensive operation. The Staff of the Unified Armed Forces should have received detailed information on the drills' results in order to draw proper operational conclusions from them. *16. zasedání Vojenské rady Spojených ozbrojených sil*, 24. 10. 1977, k. 39, p. č. 134, č. j. 36037, ibid; *20. zasedání vojenské rady Spojených ozbrojených sil*. 2. 11. 1979, k. 45, p. č. 153, č. j. 36511, ibid.

formation of the Unified Armed Forces in direction of the line Lviv-Kaliningrad and reaching the Soviet western border in 6-7 days. This would have created favorable conditions for the Western air forces and paratrooper units' attack on the second echelon of the Warsaw Pact troops.[1230] Also, the "Soyuz-77" maneuvers expected a similar development. Its scenario assumed the central group of NATO armies' attacking in the direction of Regensburg-Tábor-Katowice were tasked with destruction of the main formation of the Warsaw Pact troops located in nearby Prague, Karlovy Vary and Strakonice. According to NATO's alleged intention, its advancing forces were supposed to reach the Wroclaw-Katowice-Lučenec line between the sixth and eighth day of the operation. The Warsaw Pact strategists anticipated that the enemy would almost surely violate Austrian neutrality and attack Czechoslovakia from the south. The Unified Armed Forces' task remained the same: to repulse an aggression and to launch a devastating counteroffensive in the direction of Prague-Nurnberg-Stuttgart. An initial stopping of the enemy in the depth of 35-50 kilometers of the Czechoslovak territory took two days. During the following offensive operations, the main NATO forces should have been destroyed and German-French borders reached.[1231]

These assumptions were permanently present in the Eastern strategist's thinking, despite the fact they were informed about the defensive nature of NATO's military doctrine.[1232] It is good to point out that the reports were absolutely relevant. In the second half of the 1970s, the Warsaw Pact countries' spies, especially those of the GDR, managed

---

[1230] Marshal Ogarkov Analysis of the "Zapad" Exercise, May 30 – June 9, 1977, in *A Cardboard Castle?* (doc. 81): 406-412.

[1231] Czechoslovak Analysis of the "Soyuz 77" Exercise, March 21-29, 1977, ibid, (doc. 79): 402-403.

[1232] Mastny, *A Cardboard Castle?*, 46. In the early 1980s, for example, the East-German spies presented convincing reports that NATO operational plans envisaged an outbreak of war after a fast, conventional attack by the Warsaw Pact. The Western alliance assumed that it would be able to notice the Eastern armies' preparation for such an assault about 48 hours in advance. East German Intelligence Report on the Operational Plan of the U.S. 5th Army Corps in War Time, December 16, 1982, ibid (doc. 97): 469-471.

to penetrate the top structures of NATO and obtained valued documents there.[1233]

Rationalization of offensive operation planning

In the mid-1970s, large command exercises of the Warsaw Pact anticipated increasingly faster advances of the Unified Armed Forces' offensive. The troops were supposed to reach the English Channel coast in 6 days and approach Lyon in France just 3 days later.[1234] In case of war, the Eastern Bloc should have still striven for defeating the enemy. Therefore, during a conventional stage of conflict, the Unified Armed Forces' task was to create favorable conditions for a final victory. This meant to take a strategic initiative and move the combat operations deep into the enemy territories before the U.S. reinforcements arrived to the Old Continent or nuclear weapons were used.[1235]

In the second half of the 1970s, these thoughts about an extremely fast offensive operation were certainly reconsidered. Based on reports by the East-German spies, the Warsaw Pact came to the conclusion that NATO's ability to face the Unified Armed Forces' attack was increasing constantly. This is also proven by the fact that during training of the alliance offensive operations, the depth that the attacking should have reached was reduced significantly.[1236] Although the 1977 operational plan

---

[1233] In the second half of the 1970s, the East-German intelligence HVA managed to infiltrate an agent, code-named "Topas," in NATO central in Brussels. His British wife also performed espionage in favor of the East. A female agent, code-named "Michelle," obtained a large number of top secret documents in NATO headquarters as well. Sometimes, undetected she even brought them home, where her husband, an agent with the code name "Bordeaux," took pictures of them. Then, she put the documents back in her boss' safe. Thus, the Warsaw Treaty Organization had NATO materials concerning situations with armies, armament, strategy in disarmament talks, internal problems and relations, but also nuclear planning. In 1977, the almost complete documentation to "Hilex-7," "Hilex-8," and "Wintex-77" drills as well as conclusions of an expert group on the issue of defense and the Committee of Defense Planning of December 1976 were obtained. Description of Activities of an East German Spy inside NATO, April-May 1977, ibid (doc. 80): 404-405.

[1234] Mastny, *A Cardboard Castle?*, 47.

[1235] Luňák, *Plánování nemyslitelného*, 62-63.

[1236] Mastny, "Imagining War in Europe," 36.

of the CSPA still envisaged reaching French borders within 9 days, regardless of using nuclear weapons, much more sober scenarios were tested during the exercises. The "West-77" maneuvers, for example, simulated a situation in which NATO surprise-attacked the Warsaw Pact under the guise of large military drills in the eastern regions of West Germany.[1237] It took the Unified Armed Forces 8 days to launch a counterattack, even after the second echelon had joined the fight. The expected depth of their advance after 4 days was no more than 150 kilometers. This shift in military planning must be ascribed to the Eastern generality's fears of Western technological innovations.[1238] As the result, the first echelon of the Unified Armed Forces consisted of just 40% of their total numbers. The rest were located 300-1200 kilometers east of the opening phase of combat, in order to avoid the initial devastating enemy strike. The Soviet 45th army even waited 1500 kilometers away from the

---

[1237] Not only for this reason, the Warsaw Treaty Organization observed NATO maneuvers with fear. Already in 1976, the Unified Command stated with displeasure that the number of NATO joint exercises had increased by 32%. In the following years, especially the long-term, one to three month-long maneuvers under the alliance command simulating transition to war and its initial stages, were considered alarming. Therefore, in late 1978, Erich Honecker urged large drills of the Warsaw Pact to be held also in the GDR, given the situation. The measure should have demonstrated the alliance's determination to defend the East-German state and react to NATO maneuvers in West Germany at the same time. In 1981, the chief of Staff of the Unified Armed Forces Gribkov warned against increasing intensity and extent of NATO exercises nearby the Warsaw Pact countries' borders that could have been not only an indicator of growing military preparations by the West, but also a pretext for reinforcing the troops by transport of soldiers across the Atlantic Ocean or mobilization in West Germany. *PODKLADY k informaci a závěrům o cvičeních NATO v r. 1975,* undated, most likely 1976, FMO GŠ/OS, k. 39, p. č. 134, č. j. 36037, ibid; *TÉZE DOKLADU představitelů ministerstva obrany SSSR na 11. zasedání Výboru ministrů členských států Varšavské smlouvy k první otázce programu "O stavu a perspektivách rozvoje sil NATO",* undated 1978; *INFORMACE o průběhu a výsledcích 11. zasedání Výboru ministrů obrany, konaného ve dnech 4.-7. prosince 1978 v BERLÍNĚ,* undated, 1978, k. 35, p. č. 127, e. č. 36027, ibid; *Výsledky 23. zasedání vojenské rady Spojených ozbrojených sil,* 23. 4. 1981, k. 49, p. č. 182, e. č. 37501, ibid.

[1238] Luňák, "War Plans from Stalin to Brezhnev," 63.

place of the first clashes. Given these circumstances, NATO had a numerous advantage up to 1.5:1 on the frontlines.[1239]

Former Czechoslovak general, Mojmír Zachariáš, retrospectively said that at the turn of the 1970s and 1980s, the Warsaw Pact's strategists had been aware of NATO's increasing superiority in air-to-air and especially air-to-surface missiles.[1240] Allegedly, the problem was considered so serious that even the biggest optimists within the Unified Command started to see the existing strategic offensive plans as unrealistic.[1241] At the end of the 1970s, assumptions of a possibility to reach the Atlantic coast during 12-16 days were reviewed, taking into account political, economic and technological development. The operational planning of the Warsaw Pact became much more rational. In the following decade, the alliance military doctrine focused mostly on quick elimination of NATO's eastern countries from war operations, rather than seizure of most of the continent.[1242] Thus, an intended advance slowed down significantly. The "Shield-84" maneuvers' scenario, for example, envisaged that the Unified Armed Forces would reach the eastern French borders no sooner than in 10-17 days and those of Spain even in 30-35 days.[1243]

[1239] Marshal Ogarkov Analysis of the "Zapad" Exercise, May 30 – June 9, 1977, in *A Cardboard Castle?* (doc. 81): 406-412.

[1240] Not only more and more devastating conventional weapons was the issue. The Warsaw Pact's strategists were also afraid of the increasing quality of the U.S. nuclear arsenal. As any extensive development of nuclear forces was restricted by SALT agreements, the United States resorted to their modernization and improvements. *NÁVRH hlavních směrů rozvoje ČSLA na léta 1981-1985*, undated, 1980, FMO GŠ/OS, k. 20, p. č. 61, č. j. 35502, VÚA, Prague.

[1241] Interviews with the Czechoslovak Generals, 2002-2003, in *Plánování nemyslitelného* (ad. 2): 420-421.

[1242] Uhl, Mattias, "Storming on to Paris. The 1961 Buria Exercise and the Planned Solution of the Berlin Crisis," in *War Plans and Alliances in the Cold War*, eds. Vojtech Mastny, Sven Holtsmark and Andreas Wenger (London: Routledge, 2006): 65.

[1243] Luňák, *Plánování nemyslitelného*, 69.

Nuclear or conventional war?

Since the mid-1970s, the Eastern military strategists faced these key questions: Will the future war be waged by nuclear means? If so, who and in which phase of the conflict would use them first? As already explained, the possibility of avoiding the use of atomic weapons was already considered within the Warsaw Treaty Organization in the first half of the seventh decade. However, a conventional stage of conflict remained very limited in these thoughts. After all, this concerned NATO as well. A scenario of "Wintex-75" maneuvers of the Western alliance assumed limited use of tactical nuclear ammunition after only two and half days of fight. A slow transition to total nuclear war would have followed. Weapons of mass destruction were used by the West first, as it was not able to withstand the Unified Armed Forces' conventional advantage. The clash came to the total nuclear stage by the sixth day and the first strategic strikes were again delivered by NATO.[1244] However, the Western alliance slowly realized that due to a substantial shift in weapon technologies, there was an increasing ability to stop an offensive of its potential enemy solely by conventional means. In 1975, for example, NATO stated that given an improvement in the area of anti-tank weapons, a successful attack of Soviet tank battalions through the territory of West Germany was no longer real.[1245]

The Eastern strategists' considerations of the possibilities of avoiding nuclear conflict were full of doubts and often even contradictions. One of the results was a return to the idea that during a conflict, the Warsaw Treaty Organization would be able to estimate precisely the moment of the first use of atomic weapons by NATO. Therefore, the East could have launched its nuclear arsenals on the enemy

---

[1244] *PODKLADY k informaci a závěrům o cvičeních NATO v r. 1975,* undated, most likely 1976, FMO GŠ/OS, k. 39, p. č. 134, č. j. 36037, VÚA, Prague.

[1245] The character of tanks was subjected to long disputes between the Warsaw Pact and NATO at the Vienna disarmament talks. The Eastern Bloc countries loudly opposed the claims by the Western media that the Warsaw Pact's tanks were strictly an offensive weapon. *Informace o průběhu jednání o snížení ozbrojených sil a výzbroje ve střední Evropě v 2. pololetí 1976,* 11. 2. 1977. 1261/0/7, sv. 30, a. j. 33/info1, NA, Prague; *Telegram z Vídně,* 30. 1. 1978, FMO GŠ/OS, k. 1, p. č. 23, e. č. 39006, VÚA, Prague.

with a slight advantage. According to drill scenarios in the second half of the 1970s, both alliances reached for nuclear weapons at almost the same moment. From the point of the Soviet military thinking, this approach met the non-use first principle announced by Brezhnev in his Tula speech in 1977. Paradoxically, he did that at the moment when NATO strengthened its conventional capacities substantially.[1246] However, such a conception was allowed by a new generation of very accurate Soviet surface-to-surface missiles. It was not just SS-20 mobile mid-range rockets, which started the Euro-missiles Crisis, but also the replacement of obsolete carriers Scud, Frog, and Scaleboard by modern SS-21 and SS-23 missiles. These were capable of hitting targets in Western Europe effectively as well.[1247] Furthermore, in 1975, the Soviet strategic units were reinforced by more than 300 intercontinental ballistic missiles SS-18 "Satan," each of them carrying up to 10 warheads.[1248]

The aforementioned scenario of the "simultaneous" strike was tested during the "West-77" exercise. According to its premises, the Warsaw Pact troops cut off and destroyed advancing enemy units and started retaking previously lost territories in the GDR. In this situation, NATO reached for nuclear weapons on the tenth day of the conflict. The Warsaw Pact command noticed Western preparations for a massive atomic attack. Therefore, it got ready for alteration of the conflict's character and began to plan its own nuclear strike, at the same time. At the moment when 30-40 minutes remained until launch of the Western arsenal, the Warsaw Pact's nuclear weapons were already in full combat readiness.[1249] Their task was to totally weaken the enemy's nuclear forces in order to reduce loss of its own units. One of the considered and preferred alternatives was such use of nuclear weapons which would have paralyzed the enemy's administrative centers. Both sides launched their nuclear weapons almost simultaneously; the Warsaw Pact just under a minute ahead. However, this

---

[1246] Luňák, *Plánování nemyslitelného*, 62-63.
[1247] Heuser, "Warsaw Pact Military Doctrine," 438-441.
[1248] Zubok, *A Failed Empire*, 243.
[1249] The "Soyuz-77" exercise even presumed that despite a war being launched by conventional means, nuclear weapons would always be in battle stations, ready for prompt use. Czechoslovak Analysis of the "Soyuz-77" Exercise, March 21-29, 1977, in *A Cardboard Castle?* (doc. 79): 402-403.

was largely considered a counterattack. NATO would have already delivered 680 nuclear strikes onto the advancing Eastern troops. Another 400 targeted the western parts of the USSR. The losses were estimated much more rationally than in the 1960s: After the first wave of the attack, only 36% of the Unified Armed Forces remained ready to fight. The air forces suffered losses of 70%. The battlefield was contaminated by radioactivity and hit by widespread fires. The Warsaw Pact's strike met similar success; it killed almost 250 thousand enemy troops and caused massive material damages. But the East lost almost two times more soldiers. Although the overall potential of both alliances was approximately on the same level, the enemy's tactical missiles, which were supposed to eliminate the advancing troops, were still considered more effective by the Warsaw Pact's analytics.[1250]

The Eastern strategists were abandoning their deeply rooted assumptions of nuclear war, but only very slowly. In the years 1977-1979, the combat preparation of the Unified Armed Forces should have focused on a combined war, waged by both nuclear as well as conventional weapons. One of the main tasks still envisaged was the training of the advancing troops through areas devastated by nuclear explosions. Thus, even in the late 1970s, the Warsaw Pact armies still primarily drilled for a nuclear war.[1251] In June 1978, at a meeting of the alliance member-states' general staff, the Soviet general S. F. Romanov emphasized the importance of the "West-77" maneuvers described above. He stated that its basic points had simulated the actually anticipated features of a conflict in Europe. The General reiterated that nuclear weapons still remained the most effective tool to destroy the enemy.[1252] In autumn of the same year, a report for the CPCz CC Presidium also warned against the

---

[1250] Marshal Ogarkov Analysis of the "Zapad" Exercise, May 30 – June 9, 1977, in *A Cardboard Castle?* (doc. 81): 406-412.
[1251] *Rozkaz ministra národní obrany ČSSR pro přípravu vojsk ČSLA ve výcvikovém roce 1977-1978*, 24. 10. 1977, 1261/0/7, sv. 51, a. j. 57/1, NA, Prague.
[1252] Soviet Statement at the Chiefs of General Staff Meeting in Sofia, June 12-14, 1978, in *A Cardboard Castle?* (doc. 83): 415-417.

undervaluation of a surprise nuclear attack in the actual deteriorating international situation.[1253]

Despite the idea of a "simultaneous" nuclear strike, the Warsaw Treaty Organization gradually put the emphasis on the scenario of a conventional phase of war. As already mentioned, in the ideal situation, NATO should have been defeated before the U.S. managed to transport reinforcements to Europe.[1254] Therefore, the Warsaw Pact's priority was to keep the conflict in the conventional stage as long as possible; until the enemy decided to use nuclear weapons.[1255] According to the 1977 operational plan of the CPSA, a potential atomic strike of the Warsaw Pact should have been launched on the Soviet leadership's direct order.[1256] The East-German documentation reveals that ideas on conducting an offensive operation without nuclear weapons began to emerge in the years 1977-1979. However, it does not mean these were ultimately considered.[1257] A specific result of the Warsaw Pact strategists' withdrawal from inevitability of nuclear war was the establishing of so-called "operational maneuverable groups". The Political Consultative Committee decided on their creation in 1978; these were special units equipped with recent military technology. They were supposed to advance west ahead of the main formation of the Unified Armed Forces and literally fight their way through the technologically superior corps of NATO to strategic centers in Western Europe.[1258] A strong emphasis was also put on airdrops of the special units in the depth of enemy's territory. Since the beginning of the 1980s, the Warsaw Treaty Organization also envisaged operations of special marauding groups. Those strike commandos would have undermined the Western troops' capabilities to attack the Unified Armed Forces' main formation effectively. Since the late 1970s, their actions became an integral part of the Warsaw Pact's maneuvers as well as its

---

[1253] *Zpráva pro předsednictvo ÚV KSČ o splnění úkolů ČSLA ve výcvikovém roce 1977-1978 a hlavních úkolech ČSLA pro výcvikový rok 1978-1979*, 10. 10. 1978, 1261/0/7, sv. 8, a. j. 8/1, NA, Prague.

[1254] Mastny, *A Cardboard Castle?*, 45.

[1255] Heuser, "Warsaw Pact Military Doctrine," 444.

[1256] Luňák, *Plánování nemyslitelného*, 63.

[1257] Kramer, "Warsaw Pact Military Planning," 14-15.

[1258] Mastny, *A Cardboard Castle?*, 48; Mastny, "Imagining War in Europe," 36.

entire strategy.[1259] These units of strength of one division were supposed to fight on the Western territories on their own, without direct contact with the main Eastern military echelons, in practice. At the turn of the 1970s and 1980s, massive airdrops on the enemy's territory became an alternative to nuclear strikes in the drill scenarios. This included, for example, the "South-78," "North-79," or "Brotherhood in Arms-80" exercises.[1260] At the time, during the alliance's military bodies' meetings, Kulikov also urged better preparation for a breakthrough in the enemy's defense by conventional means only, which envisaged bigger use of the gunships and airdrops of the special forces.[1261]

As repeatedly explained, Brezhnev considered a possible nuclear war as a disaster. Therefore, he made at least political gestures aimed to reduce risk of its outbreak. In his talk with the West-German chancellor, Helmut Schmidt, just ahead of the beginning of the Euro-missile Crisis, he summed up his opinions on the issue: *"If somebody wanted strike us with the first blow, or if somebody in the Soviet Union wanted to commit such a madness, all the inhabitants in Europe would die. I do not want to allow the question of effectiveness of nuclear war to be discussed at all."*[1262]

[1259] Luňák, *Plánování nemyslitelného*, 67.

[1260] Heuser, "Warsaw Pact Military Doctrine," 442-443. The "Brotherhood in Arms-80" maneuvers were based on the three-stage operation: In the first, the enemy's defense should have been broken; in the second, his entire defense perimeter overcome and the second echelon troops deployed; in the final stage, an offensive combined with airborne proceeded. During analysis of the drills, the Soviet, Polish, and East-German commanders had to explain their attitude towards transition to a nuclear phase of the conflict. These scenarios were discussed: The first echelon, consisting of the Soviet troops and the NPA, had 840 tactical nuclear warheads, including surface-to-surface missiles as well as air bombs. The Warsaw Pact's nuclear strikes should have targeted NATO's atomic arsenals, airfields and air-defense systems, naval bases, command structures and communication lines, and large groups of troops. Those assaults were expected to eliminate resistance in the anticipated corridor of the attacking front. For additional support, other nuclear backups would have been earmarked. Kramer, "Warsaw Pact Military Planning," 14-15.

[1261] *20. zasedání vojenské rady Spojených ozbrojených sil.* 2. 11. 1979, FMO GŠ/OS, k. 45, p. č. 153, č. j. 36511, VÚA, Prague; *22. zasedání vojenské rady Spojených ozbrojených sil*, October 1980, k. 48, p. č. 178, č. j. 37023, ibid.

[1262] Quoted in Durman, *Útěk od praporů*, 126.

However, in the second half of the 1970s, the American right wing enjoyed its warnings that the USSR had never accepted the concept of mutual assured destruction. Allegedly, the country still built its armed forces in order to win in nuclear war. According to the above-mentioned, it is obvious that these tendencies were really persisting in at least some part of the Soviet military circles. But at the same time, it is necessary to point out that these did not represent the stance of the majority in the CPSU political leadership. On the contrary, after the death of Grechko in 1976 and placement of Dmitri Ustinov to the post of defense minister, the Soviet military policy went through certain moderation.[1263]

In this regard, the situation was probably complicated by the fact that in Brezhnev's ruling system, a principle of stability of cadres concerned not only political functions, but also to a similar extent the army. Most of the key positions in the armed forces were still held by the old generation of commanders, born in the World War II fights. In term of age, they were pensioners. Their stereotypical, black-and-white perspective and lack of education did not meet the needs of the time. Marshal Nikolai Vasilevich Ogarkov, who served as the chief of the Soviet general staff in the years 1977-1984, became a certain speaker of the military "modernists." He was considered one of the main military theorists of his era. His thinking was not bound by narrow military schemes; in many issues he took similar views to some pro-reform economists, technocrats, and apparatchiks. But unlike these and even Defense Minister Ustinov, the Marshal was ready to take much higher risks. For this reason, Yuri Andropov called him "little Napoleon." Ogarkov did not deem the possibility of victory in nuclear war as absolutely odd. But more openly than the others, he declared his skepticism about the ability of the aging Brezhnev leadership to prepare the country for such a conflict. Therefore, in accordance with changing military doctrine, the general staff under Ogarkov's command preferred the eventual World War III to be waged mostly on the basis of conventional weapons. On paper, given its quantitative advantage, the

---

[1263] Bowker, "Brezhnev and Superpowers Relations," 96-97.

Warsaw Treaty Organization had better prospects for victory in that sort of conflict.[1264]

The Ogarkov-Ustinov duo undoubtedly had some influence on the Soviet political elites' decisions. More precisely, while shaping the security policy, the key persons within the CPSU leadership partially listened to their advice. In the SALT II negotiations,[1265] for example, the American side tried to include a note in the final communiqué that an eventual nuclear conflict would have had no winner. Although Brezhnev himself most likely agreed with this claim, the two aforementioned marshals prevented this statement from being incorporated into the text. The Soviet military elites played a similarly negative role in the efforts to mitigate the Euromissile Crisis. Ustinov firmly rejected Helmut Schmidt's offer to solve the problem by a pledge that the number of warheads carried by SS-20 MIRVs would not exceed the number of those installed on their predecessors, SS-4 and SS-5. Brezhnev already lacked the strength to effectively oppose the defense minister.[1266] In this regard, the situation was also complicated by the fact that Ustinov together with Andrei Gromyko and Yuri Andropov formed a triumvirate that was getting more and more political power at the time of the General Secretary's deteriorating physical condition.

Nevertheless, at the end of the 1970s, the Soviet leadership firmly opposed weapons of mass destruction, at least verbally. It put strong

---

[1264] Durman, *Útěk od praporů*, 186, 188.

[1265] The SALT II treaty was signed in June 1979 in Vienna. Compared to its predecessor, the treaty meant a huge leap. It reduced the Soviet advantage in overall firing strength and the American superiority in numbers of warheads as well as missiles launched from submarines and forward bases, which the original agreement failed to address. In the years 1981-1985, the total number of carriers of each side was expected not to exceed 2,250. Just 1,320 of these could carry more than one warhead. The new treaty also concerned bombers and tightened up a definition of existing strategic arsenals' modernization. However, given the development of MIRV technology, the number of warheads was continually rising, in fact. Therefore, it is not surprising that at the end of the 1970s, some critical voices could be heard in the West calling the SALT agreements absolutely ineffective in terms of arms race control. Bowker, "Brezhnev and Superpowers Relations," 95.

[1266] Durman, *Útěk od praporů*, 195, 200.

emphasis on the campaign against a neutron bomb, initiative on a no-first-use treaty, and even a nonaggression agreement between NATO and the Warsaw Treaty Organization.[1267] These tendencies climaxed in June 1982, when Moscow pledged not to use nuclear ammunition first, regardless of the Western stance. Such a commitment went far beyond the scope of Brezhnev's speech in Tula in 1977.[1268] In this regard, Beatrice Heuser notes that at this time, a certain breakthrough really occurred. Despite the political declarations, an eventuality of initiating a preemptive nuclear strike on the enemy's atomic arsenals still persisted in the Warsaw Pact's drill scenarios up to this point.[1269]

One of the first exercises which tested "selective" and gradual use of nuclear weapons as well as operational maneuverable groups was "Soyuz-81." It incorporated the findings of Eastern observations of NATO's "Wintex-81" drills, held practically simultaneously. Its course reveals that considerations about a possible victory in the nuclear stage of conflict remained present in the Warsaw Pact's strategists' thinking. An atomic arsenal was even repeatedly marked as a suitable means to accelerate the troops' advance.[1270] In this regard, the term "accommodating-retaliation nuclear strike" occurred for the first time. The Warsaw Pact would have reached for it soon after the enemy had used its nuclear weapons.[1271] Nevertheless, the priority was to postpone the stage of total nuclear war as long as possible and use atomic ammunition limitedly and cautiously. During the maneuvers' analysis, the supreme commander Kulikov criticized the fact that existing military procedures insufficiently exploited the World War II experience.[1272] In practice, this meant a return to military thinking known from the Stalin era. According to those assumptions, a new global conflict would have just been a

---

[1267] Jarząbek, *PRL w politycznych strukturach*, 79.
[1268] Luňák, *Plánování nemyslitelného*, 68.
[1269] Heuser, "Warsaw Pact Military Doctrine," 444.
[1270] Bange, "Comments on and Contextualization of Polish Documents".
[1271] *Účast ČSLA na cvičení "Sojuz-81" a výměna jednotek s PLA ve VVP*, 2. 1. 1981, FMO GŠ/OS, kr. 48, p. č. 180, č. j. 37032, VÚA, Prague.
[1272] Later, Kulikov officially presented these theses at the Military Council session. *28. zasedání vojenské rady Spojených ozbrojených sil*, 31. 10. 1983, k. 55, p. č. 214, č. j. 38107, ibid.

repetition of the previous war and nuclear weapons were expected to affect its outcomes marginally. Thus, in Kulikov's view, the Warsaw Pact's first echelon were supposed to quickly penetrate deep into the enemy's defense, eliminate an atomic artillery by conventional means, and conduct reconnaissance for its own limited nuclear strikes at the same time. However, compared to the other commanders, Kulikov still showed at least some level of rational thinking. During evaluation of the drills, for example, he attacked an approach of some staff which underestimated the devastating strength of NATO nuclear assaults and envisaged fast unrealistic restoration of the combat readiness of hit units.[1273] At the beginning of the 1980s, the Warsaw Pact's strategists were well aware that in comparison to the previous period, the enemy reinforced substantially. Therefore, the exercises should have focused mostly on simulation of active and continual resistance, even in the initial stage of fighting.[1274]

Despite revision of thoughts about an inevitable nuclear war, more and more attention was paid to the elimination of damages caused by NATO's nuclear strikes. The Unified Command tasked the alliance member-states' civil defenses with this.[1275] The western countries of the Warsaw Treaty Organization would have primarily focused on cleaning contaminated areas, where an advance of the second echelon of Soviet troops was planned. After all, the satellites were generally expected to provide them with maximum possible support.[1276] In May 1982, the Staff

---

[1273] The Polish staff envisaged launch of the offensive just one day after being hit by a tactical nuclear strike. On the contrary, Kulikov anticipated at least 6 days needed for recovery of the Front. *Rozbor velitelsko-štábního cvičení "Sojuz81"*, 8. 4. 1981, k. 48, p. č. 180, č. j. 37032, ibid.

[1274] *PLÁN REALIZACE Protokolu čís.:023 z 23. zasedání Vojenské rady Spojených ozbrojených sil*, undated 1981, k. 49, p. č. 182, e. č. 37501, ibid.

[1275] Commanders of Warsaw Pact countries' civil defenses regularly met under the alliance auspices. Symbolically, their gatherings took place at the Soviet general staff or at the staff of civil defense of the USSR. Between the member-states' armies, bilateral protocols on cooperation in the field of civil defense were concluded. *16. zasedání Vojenské rady Spojených ozbrojených sil*, 24. 10. 1977, k. 39, p. č. 134, č. j. 36037, ibid; *PLÁN REALIZACE USNESENÍ z 10. zasedání Výboru ministrů obrany členských států Varšavské smlouvy*, undated, 1978, k. 8, p. č. 43, č. j., 39035, ibid.

[1276] Marshal Ogarkov Analysis of the "Zapad" Exercise, May 30 – June 9, 1977, in *A Cardboard Castle?* (doc. 81): 406-412.

of the Unified Armed Forces issued a preliminary guideline on the Warsaw Pact countries' notification of nuclear strikes. The aim of this system was mutual sharing of information about the situation on territories hit by atomic weapons. In this manner, the manpower and continuation of economic production should have been protected as much as possible. Nevertheless, the existing protection against weapons of mass destruction was largely correctly considered insufficient.[1277] The situation reveals that in the 1970s and 1980s, an integration of military cooperation within the Warsaw Treaty Organization concerned most areas and was not limited to the tasks resulting from the actual strategy, which was at the time focused on conventional combat.

A rise in thinking that a conventional phase of war will be its decisive stage also resulted in another strengthening of the air forces. An air war was gradually considered as a key to victory. At the 25th Military Council session, held in April 1982 in East Berlin, warnings of an effort invested by NATO to break through the Warsaw Pact's air-defenses resonated once again. Besides standard improvement of technological equipment of the alliance joint air-defense system and making its command more effective, the drills including live-ammunition shooting should also have been enhanced. Their scenarios should have been more difficult. Regarding escalation of the Euromissiles Crisis, strong emphasis was placed on the fight against cruise missiles.[1278] The air-defense troops were supposed to focus on elimination of bombers in low altitudes, AWACS aircrafts, or SR-71 and U-2 recon jets.[1279]

Increasing technological superiority by the West at the turn of the 1970s and 1980s posed a risk for the Warsaw Pact's strategy for another

---

[1277] Letter by M. Blahník, the chief of Czechoslovak general staff, to A. I. Gribkov, the chief of Staff of the Unified Armed Forces, August 25, 1982, FMO GŠ/OS, k. 53, p. č. 201, e. č. 38008, VÚA, Prague; *26. zasedání vojenské rady Spojených ozbrojených sil*, 25. 10. 1982, p. č. 204, e. č. 38013, ibid.

[1278] For example, the Technical Committee was supposed to consider the possibility of producing a universal, remote-controlled target for training shooting. *25. zasedání vojenské rady Spojených ozbrojených sil*, 30. 4. 1982, p. č. 198, e. č. 38004, ibid.

[1279] *PLÁN REALIZACE Protokolu čís.: 0014 z 14. zasedání Výboru ministrů obrany členských států Varšavské smlouvy*, undated 1982, k. 51, p. č. 193, e. č. 37525, ibid.

reason: NATO, for example, incorporated into its plans so called "deep fire attacks" against the Unified Armed Forces' second echelon. From this point, its advance could be stopped without nuclear weapons.[1280] This conception exceeded the scope of NATO's defensive strategy of territorial security of the alliance member-states. Using accurate artillery as well as air forces, strikes against an advancing second echelon of the Unified Armed Forces should have prevented it from moving to the frontlines. The Soviet military elites saw this strategic development as a proof of NATO's offensive intentions. They expected the Warsaw Pact to be exposed to massive pressure and the "deep fire attacks" had the potential to threaten realization of the march into Western Europe. Continual modernization of the U.S. ground forces fueled these fears. In the light of this development, Marshal Ogarkov began to compare modern conventional weapons to means of mass destruction. Also, Ustinov, the Soviet defense minister, made similar declarations.[1281]

Taking into account the anticipated technological advantage of the enemy, a great emphasis was put on quick mobilization and transition from peace to a state of war. At the end of the 1970s, Kulikov complained that the shortening of time norms for transition of some units to combat readiness was absolutely insufficient.[1282] These aspects were criticized even at the beginning of the following decade. During the already mentioned "Soyuz-81" maneuvers, for example, the procedure of transition to full combat readiness took the CSPA troops 14 minutes. The chief of Staff of the Unified Armed Forces Gribkov called this time "unacceptably long."[1283] Thus, required terms within the Warsaw Pact were constantly being reduced to the extremely short level. On the contrary, they became even laxer in NATO.[1284] But the Eastern strategists did not believe it. In 1981, for instance, they presented an assumption that

---

[1280] Mastny, *A Cardboard Castle?*, 46-47.
[1281] Luňák, *Plánování nemyslitelného*, 66-69.
[1282] *20. zasedání vojenské rady Spojených ozbrojených sil*. 2. 11. 1979, FMO GŠ/OS, k. 45, p. č. 153, č. j. 36511, VÚA, Prague.
[1283] *Rozbor velitelsko-štábního cvičení "Sojuz81"*, 8. 4. 1981; Ibid, kr. 53, p. č. 204, e. č. 38013, *26. zasedání vojenské rady Spojených ozbrojených sil*, 25. 10. 1982, k. 48, p. č. 180, č. j. 37032, ibid.
[1284] Mastny, *A Cardboard Castle?*, 47.

the aim of NATO was to reduce time of a complete transition of the troops to combat readiness from the actual 10-12 days to only 8 days.[1285] Given the alleged Western ability to strike faster than ever before, even the regular Eastern units would have been kept in permanent battle readiness in their barracks, in practice.[1286] Therefore, the norms for its achievement could be a matter of minutes.

In the 1980s, the Soviet military elites finally openly admitted that a war in Europe could have been waged without nuclear weapons. In the years 1982-1983, such scenarios began to dominate the Warsaw Pact drills. Nevertheless, atomic arsenals should have been ready for use in any stage of the conflict.[1287] Given the strengthening of the enemy, the Warsaw Pact's strategists considered an eventuality of defeating NATO with conventional means increasingly impossible.[1288] Thus, the drills still often simulated the possibility that war would reach some nuclear phase in the end. Both fighting sides were expected to use atomic ammunition.[1289] Despite this fact, Mark Kramer notes that the East-German military documentation contains almost no plans of massive nuclear strikes after 1981. They can only be found again from 1988 onwards. However, after Gorbachev's strategic changes, they were purely of a defensive character.[1290]

In the period of Brezhnev's successors, Yuri Andropov and Konstantin Chernenko, the risk of sudden attack by NATO gradually came forward in strategic thinking. A scenario of the "Soyuz-83" drills, for

---

[1285] *Výsledky 23. zasedání vojenské rady Spojených ozbrojených sil*, 23. 4. 1981, FMO GŠ/OS, kr. 49, p. č. 182, e. č. 37501, VÚA, Prague.

[1286] Bange, "Comments on and Contextualization of Polish Documents". The situation could be seen well in the East-German NPA. In the late 1970s and especially in the following decade, its units achieved standard mobilization terms which should have balanced NATO's technological advantage. As a potential attack by submarine missiles was considered extremely dangerous, the East-German troops should have been able to gather up to 30 minutes after receiving an order in mobilization locations at least 5 kilometers from their barracks. Schaefer, "*The GDR in the Warsaw Pact*".

[1287] Luňák, *Plánování nemyslitelného*, 69.

[1288] Mastny, *A Cardboard Castle?*, 56.

[1289] Heuser, "Warsaw Pact Military Doctrine," 443.

[1290] Kramer, "Warsaw Pact Military Planning," 15.

example, presumed that the Western alliance had increased its military capabilities to the point that it was able to strike in Central Europe from several directions with little advance notice.[1291] For this reason, the Eastern strategists were more and more afraid of large maneuvers held in Western Europe. During such actions, a huge number of troops gathered which allowed for the launching of a surprise attack even without any additional mobilization.[1292] Therefore, an increasingly strong emphasis was put on defensive tasks, a phase of repulsion of the enemy's aggression.[1293] In October 1984, at the 30th meeting of Military Council in Sofia, the supreme commander Kulikov complained that many commanders and staffs underestimated the risk of sudden attack by NATO, including both conventional and nuclear weapons. However, it was anticipated that at least in the initial stage of conflict, the West would not use atomic ammunition and would act with accurate conventional and radio-electronic warfare devices only.[1294]

The shift towards a defensive approach was also supported by a statement of Bernard Rogers, the commander in chief of NATO, that an eventual war would not be waged in the area of Western Europe, but the Warsaw Pact troops would be dealt devastating hits on their own territories instead.[1295] At the beginning of 1984, the Unified Command in its reaction ordered the frontal divisions to create stocks of fuel, ammunition, and food for at least 20 days of combat activities even in peacetime. In practice, this

---

[1291] Mastny, *A Cardboard Castle?*, 56.

[1292] *28. zasedání vojenské rady Spojených ozbrojených sil*, 31. 10. 1983, FMO GŠ/OS, kr. 55, p. č. 214, č. j. 38107, VÚA, Prague.

[1293] A defense against the enemy's attack was practiced during "Soyuz-83," "Druzhba-83," and "Granit-83" drills, for example. *USNESENÍ výboru ministrů obrany členských států Varšavské smlouvy k 1. otázce programu 16. zasedání "Všeobecné směry rozvoje a vybavení výzbrojí a bojovou technikou Spojených ozbrojených sil členských států Varšavské smlouvy v letech 1986-1990"*, undated, 1983; *INFORMACE o průběhu a výsledcích 16. zasedání výboru ministrů obrany konaného ve dnech 5.-7. prosince 1983 v SOFII*, 12. 12. 1983, k. 56, p. č. 216, č. j. 38109, ibid; *28. zasedání vojenské rady Spojených ozbrojených sil*, 31. 10. 1983, k. 55, p. č. 214, č. j. 38107, ibid.

[1294] *30. zasedání vojenské rady Spojených ozbrojených sil*, 23. 10. 1984, k. 58, p. č. 224, č. j. 38208, ibid.

[1295] *Vystoupení ministra národní obrany ČSSR na mimořádném zasedání výboru ministrů obrany*, undated, 1983, k. 55, p. č. 215, č. j. 38108, ibid.

meant being permanently ready for defense as well as the entire first stage of an offensive operation.[1296] This was undoubtedly supported by the Eastern military analytics' assumption, presented by the Soviet general Ivashutin in December 1984 at the 17th Committee of Defense Ministers' session in Budapest. He stated that the actual priority of U.S. policy was to "ensure readiness to destroy socialism as a social-political system by all means." In this regard, he warned against the strengthening of NATO's southern flank which would engage a significant part of the Warsaw Pact forces and weaken its main formation at the Central-European war theater in case of armed conflict. Therefore, the Unified Armed Forces should have been prepared for conducting combat operations at two relatively similar fronts.[1297]

***

The aforementioned shows that despite the fact that the Warsaw Pact's strategic doctrine went through obvious alterations from the mid-1970s to the arrival of Mikhail Gorbachev, its details remained unclear and swayed. It reacted to the fast development of new military technologies as well as dynamic changes of the international situation. This just proves the suggestion that in the case of an outbreak of a conflict, the existing strategic plans would have probably been modified and adjusted to the largely unpredictable development of a total war. Fortunately, such a scenario never happened. We are reminded that the changing consideration of combat features as well as analysis of military drills had at least one common aspect: They anticipated destruction and devastation at an unprecedented level with enormous losses of human lives. In the end, the war between the Warsaw Treaty Organization and NATO would have been waged on every continent, even in outer space; the results of this "life-or-death" struggle would have had devastating consequences for the whole planet, as Marshal Ogarkov once realistically stated.[1298]

[1296] *Dopis náčelníka štábu SOS k metodice vytváření a rozložení zásob*, 8. 2. 1984, k. 56, p. č. 216, č. j. 38109, ibid.

[1297] *INFORMACE o průběhu a výsledcích 17. zasedání výboru ministrů obrany, konaného ve dnech 3.-5. prosince 1984 v BUDAPEŠTI*, undated, 1984, k. 58, p. č. 224, č. j. 38208, ibid.

[1298] Report on Speech by Marshal Ogarkov at a Warsaw Pact Chiefs of Staff Meeting in Minsk, September 8-10, 1982, in *A Cardboard Castle?* (doc. 96): 466-

468. In this sense, former Soviet general Alexander Liakhovski retrospectively
testified. In the 1980s, the Soviet general staff allegedly increasingly thought that
there would not be any war in the end, as neither NATO nor the Warsaw Treaty
Organization wanted to start it. Even the military was aware that such a conflict
would have inevitably resulted in an absolute disaster. Thus, in the middle of the
decade, there was mostly an effort to preserve the balance of forces needed for
existence of the mutual deterrence conception. Hoffenaar and Findlay, *Military
planning for European Theatre Conflict*, 62-63.

The role of the Warsaw Treaty Organization in the Polish Crisis 1980-1981

In an outbreak of the communist regime's crisis in Poland at the beginning of the 1980s, the Warsaw Treaty Organization played no significant role. Despite this fact, strong speculations about its military intervention aiming to preserve the Soviet model of socialism in the country spread till December 13, 1981, when General Wojciech Jaruzelski, the Polish leader, declared martial law, a state of war. Such a step then allowed the regime to survive until 1989. He retrospectively called this move a "lesser evil" that, among other things, had allegedly eliminated a threat of seizure of Poland by the Warsaw Pact troops.[1299] Taking into account this claim, it is necessary for this work to contain a chapter that analyzes in detail the alliance's role in the Polish Crisis of 1980-1981.

It must be emphasized that the official structures of the Warsaw Treaty Organization did not in fact interfere with the development in the PPR. However, in some phases of the crisis, Moscow and especially some of its satellites tried to use the alliance to influence the situation in Poland in three ways, including politics, military, and propaganda. During the critical years, an appeal for holding the Warsaw Pact's extraordinary meeting of party leaders - the Political Consultative Committee session - could be heard many times. Such a gathering would have served as an appropriate forum to escalate duress on the Polish officials or even a direct dictate to the PUWP leadership. From the military point of view, the Unified Armed Forces' drills were a suitable instrument to interfere with the events in Poland. Increased activity of the armed forces in the region of Central Europe should have created a feeling of jeopardy, threatening

---

[1299] Jaruzelski, Wojciech, *Historia nie powinna dzielić* (Toruń : Adma Marszałek, 2006), 11, 38.

the Polish opposition, and in this way making a declaration of a state of emergency easier for the Polish authorities.[1300] Moreover, propagandistic claims about an alleged risk of Polish withdrawal from the Warsaw Pact were a useful euphemism for warning against losing the country from the Soviet sphere of influence. Such statements, however, have to be separated from accusations that instability in Poland jeopardized strategic military communications and the entire Warsaw Pact which had an absolutely real basis.

Nevertheless, the essence of the Warsaw Pact's mechanisms and the available evidence make it impossible to focus on the alliance itself only. The issue has to be explained in the context of the general development of consequences of the critical events in Poland and especially the changing policy of the Soviet leadership. Finally, the question to what extent the Warsaw Pact's intervention in Poland was imminent in 1981 cannot be neglected. Until his death in 2014, General Jaruzelski justified the action against the Polish opposition launched on December 13 as the step which had stopped the invasion in the last minute.

Strategic aspects of the strikes in Poland in summer 1980

At the beginning of the 1980s, Poland played an important role in the Warsaw Pact's structures and operational plans. The PPA was part of the Unified Armed Forces' first echelon. After the Soviet army, it

---

[1300] The following text uses a term "martial law," mostly because of the fact, that until December 13, 1981, it was not clear which sort of law would be implemented. Jaruzelski himself admitted big differences between a state of emergency ("stan wyjątkowy") and state of war ("stan wojenny"), the former of which he imposed in the end, as it formally granted him bigger rights. Ibid, p. 16. Those terms are often confused in literature, which is also the result of taking on an English term "martial law". Ambiguity of this term is proven even by the reaction of the Romanian leadership, which discussed at length on December 13 what sort of state Jaruzelski actually had imposed. *Stenographic Transcript of the Meeting of the Consultative Political Committee of the Central Committee of the Romanian Communist Party*, December 17, 1981, Accessed October 16, 2019. http://www.php.isn.ethz.ch/kms2.isn.ethz.ch/serviceengine/Files/PHP/16627/ipu blicationdocument_singledocument / 02670c0a-92c0-480f-811a-29950004329c/ en/811217_stenographic_transcript.pdf.

represented the alliance's largest military force. Additionally, various Soviet nuclear depots and key communication and logistic links between the USSR and its troops in the GDR were located in Poland.[1301] In July 1980, regarding an approval of the Warsaw Pact's wartime statute, the USSR decided to drop construction of a protected command post of the Unified Command on its own territory. The object should have been built in Poland instead, and this fact increased the significance of the Polish territory once again. From this point, the Soviet generality expected that in a potential conflict, the Warsaw Pact troops at the western war theater would be commanded from the posts in the PPR.[1302] Thus, during the crisis, not only were the Soviet officials concerned about the internal situation in Poland, but also about potential impact on foreign policy of the country and its role within the Warsaw Pact's system. An eventual arrival of a non-communist government and possible revision of the Polish international orientation Moscow considered dangerous even from the military-strategic point of view.[1303]

The Polish leadership soon realized that an eventual escalation of unrest in the country could jeopardize an unproblematic provision of the Soviet strategic interests. In July 1980, before he traveled to the Crimea meeting with Brezhnev, the PUWP CC First Secretary, Edward Gierek, at the Polish politburo session called the riots in Lublin[1304] a strategic attack on the supply lines of the Soviet troops in the GDR. He was afraid of the Kremlin's reaction unless the situation calmed down quickly.[1305] On August 20, just one week after the strikes in the Gdansk shipyards had broken out, the defense minister Jaruzelski, noted the need to prevent any

---

[1301] Kramer, "Poland 1980–81," 1, 118.

[1302] *Chráněná velitelská stanoviště Spojených ozbrojených sil,* 31. 7. 1980; Letter by M. Dzúr, the Czechoslovak defense minister, to G. Husák, the CPCz CC General Secretary and the president of the CSSR, August 18, 1980, FMO GŠ/OS, k. 22, p. č. 67, č. j. 40004, VÚA, Prague.

[1303] Kramer, "Poland 1980–81," 1, 118.

[1304] The wave of strikes hit Lublin in the period of July 11-20, 1980, Paczkowski, Andrzej and Byrne, Malcolm, eds., *From Solidarity to Martial Law: The Polish Crisis of 1980-1981: A Documentary History* (New York: CEU Press, 2007), xxix.

[1305] Protocol No. 13 of PUWP CC Politburo Meeting, July 18, 1980, in: ibid (doc. 1): 47.

harm to Soviet interests. The riots resulted in the cancelling of scheduled military drills of the Polish as well as the alliance's armed forces. Warsaw had to ask the Soviet military command for using alternative communications for access to the PPR. In this regard, Gierek appealed for public stressing of allegiance to the socialist camp and the USSR as the only guarantee of security and further development of Poland.[1306] There were relevant fears within the Polish politburo that the riots could have taken an anti-Soviet slant. Three days later, the head of the PUWP considered it useful to assure the party plenum that the situation in the country would not affect the Polish foreign policy that would still have been mostly based on the alliance with Moscow.[1307] Gierek's statement was also followed up by his successor, Stanisław Kania.[1308] At the beginning of September, in his speech delivered during inauguration as the first secretary, he gave reassurances that Poland still intended to participate in the strengthening of the Warsaw Treaty Organization.[1309]

Despite these declarations, the situation in Poland raised some fears in the Kremlin. The Soviet politburo reacted to the summer wave of strikes and unrests by establishing a top-secret commission, led by Mikhail Suslov.[1310] Besides Suslov, the ministers of foreign affairs and defense, Andrei Gromyko and Dmitri Ustinov, along with the head of the KGB, Yuri Andropov, were the members.[1311] Based on an official authorization, the main task of this working group was to inform the Soviet politburo

---

[1306] *Protokół Nr. 20 z posiedzenia Biura Politycznego KC PZPR 20 sierpnia 1980r*, PZPR KC, V/159, mkf. 2985, AAN, Warsaw.

[1307] *Projekt wstępny referatu I Sekretarza KC PZPR na IV Plenum*, 23. 8. 1980, ibid.

[1308] On September 5,1980, Gierek resigned as party leader under duress. Behind the pressure for such a step, he saw mostly a conspiracy of his closest coworkers, led by Kania, rather than real failures of his policy so far. Paczkowski, *From Solidarity*, 10.

[1309] *Projekt referatu na VI Plenum KC S. Kani*, 27. 8. 1980, PZPR KC, V/160, mkf. 2986, AAN, Warsaw.

[1310] Valery Musatov, a former Soviet diplomat, stated that the Soviet leadership had kept existence of the commission secret even from the diplomatic corps. *Wejdą nie wejdą: Polska 1980-1982 konferencja w Jachrance listopad 1997* (London: ANEKS, 1997), 149.

[1311] Mastny, "The Soviet Non-Invasion of Poland," 191.

about events in the PPR.[1312] However, one of its first steps was issuing a document recommending putting a large part of the Soviet troops at a state of full combat readiness for the case of providing Poland with "military assistance."[1313] Since the beginning of considerations of a military intervention, the Soviet leadership was aware of the real possibility that the Polish armed forces, or at least their part, would refuse obedience and join the side of "counterrevolution."[1314] Nevertheless, the character of these materials of Suslov's commission is questionable; in this regard, the Polish historian Łukasz Kamiński notes that the documents most likely were never the subject of discussion in the Soviet politburo.[1315]

The extent of preparations at the end of summer 1980 suggests that the Soviet defense ministry planned no operation comparable to the 1968 invasion of Czechoslovakia at the time. Taking into account the real number of mobilized divisions and an effort to involve mostly the second rank units, it is likely that at the turn of the summer and autumn, the Kremlin rather dealt with a theoretical opportunity to provide the Polish leadership with limited military assistance.[1316] Moreover, no suggestion can be found in the available documents that at this stage of thought about the military action in Poland, Moscow envisaged involvement of its European satellites' armed forces or even the entire Warsaw Treaty Organization.

As early as in August, the Western observers stated that the revolt in the PPR was similarly as important as the Eastern Bloc crises in 1956 and 1968. The press began to speculate about possible Soviet military

---

[1312] CPSU CC Politburo Decision Setting Up Suslov Commission, August 25, 1980, Cold War International History Project Bulletin, no 5 (1995): 116.

[1313] Acting on a resolution by the Suslov commission, the Soviet defense ministry mobilized about 100 thousand reservists and 15 thousand vehicles. Kramer, "In Case Military Assistance is Provided," 102.

[1314] CPSU CC Politburo Commision Order to Enhance Readiness of Military Units for Possible Use in Poland, August 28, 1980, in *From Solidarity to Martial Law* (doc. 5): 64. In this case, the Suslov commission recommended to increase initial numbers of the intervention troops by another 5-7 divisions.

[1315] Kamiński, Łukasz, *Przed i po 13 grudnia. Państwa bloku wschodniego wobec kryzysu w PRL, 1980 - 1982*, tom I, (Warszawa: IPN, 2006), XIX-XX.

[1316] Kramer, "In Case Military Assistance is Provided," 106-107.

intervention.[1317] On August 27, as one of the first in the Polish leadership, the PUWP CC Secretary and the politburo member, Stefan Olszowski,[1318] warned against the threat of outside intervention, actually pointing to the development in Hungary and Czechoslovakia.[1319] For this reason, he appealed to maintaining contacts and consultations with the Eastern Bloc party leaders while dealing with the situation.[1320] Such an opinion was not unique within the Polish politburo. Jaruzelski even called for imminent consultations on the highest level. He was supported by Jerzy Waszczuk, a candidate politburo member and CC member, who noted that the events in the PPR could merit a response from the entire "socialist camp."[1321] Thus, involvement of the Warsaw Pact countries' leadership in a solution of the crisis was not an outside initiative, but from the beginning this idea received strong support by a significant part of the PUWP top officials. In this manner, the Polish side probably wanted to prevent its isolation, in which Dubček's leadership had found itself in 1968.

The first reaction of the Soviet satellites

Not only Moscow, but also its satellites carefully monitored the situation in Poland. As early as September 1980, some officials of the ruling parties in the Eastern Bloc saw the event in the PPR as a serious threat to the whole socialist commonwealth. The party leaderships in neighboring countries, especially Czechoslovakia and East Germany, measured the unrests in Poland as extraordinarily sensitive. However, the Soviet satellites much more feared the spread of opposition activities to other Eastern Bloc states than the strategic aspect of the Polish position within the Warsaw Pact's system.

[1317] *Informace pro ÚV KSČ a vedení FMZV – Současná situace v Polsku*, 21. 8. 1980, DTO 1945-1989, i. č. 31, e. č. 50, č. j. 330/3, AMZV, Prague.

[1318] During the years 1971-1976 and 1982-1985, Olszowski also held the post of minister of foreign affairs of the PPR.

[1319] Protocol of PUWP CC Politburo Meeting, August 27, 1980, in *From Solidarity to Martial Law* (doc. 4): 57.

[1320] *Protokół Nr. 27 z posiedzenia Biura Politycznego KC PZPR 28 sierpnia 1980r.*, PZPR KC, V/159, mkf. 2985, AAN, Warsaw.

[1321] *Protokół Nr. 28 z posiedzenia Biura Politycznego KC PZPR 29 sierpnia 1980r*, ibid.

Based on the instruction of Erich Mielke, the minister of state security, the East-German security apparatus paid very close attention to the events by the eastern neighbors since August 12.[1322] After the Gdansk Agreement was signed,[1323] the SED CC documents began to frankly talk about counterrevolution in the PPR. The basic viewpoint of the East-German leadership in negotiations with the Eastern Bloc countries became a direct comparison to the Polish Crisis and the Prague Spring.[1324] The Czechoslovak diplomacy had also observed the situation in Poland very carefully since July 1980, taking into account the existence of the Polish minority on the territory of the CSSR.[1325] The CPCz CC Presidium compared the unrest in Poland to the Hungarian uprising in 1956 and the events of spring 1968. In this regard, it considered displays of discontent by the northern neighbor an even more serious threat to socialism, as the "class enemy" allegedly had learned a lesson and now planned its activities from the long-term view and broader perspective.[1326]

Husák's leadership correctly noticed that the Polish opposition was essentially challenging neither the leading role of the PUWP nor the foreign ties of the country.[1327] The agenda of the Solidarity movement focused mostly on economic issues which were slowly supplemented by political demand. But they lacked substantial foreign-political

---

[1322] Kamiński, *Przed i po 13 grudnia,* XIX.

[1323] The Gdansk agreement was signed on August 31 between the Polish government officials, the PUWP, and representatives of the Interfactory Strike Committee. Polish workers were granted especially unprecedented rights to establish trade unions independent to the party, organize strikes without fear of repressions, issue statements on the country's problems, and publish their own newsletters. The Gdańsk Agreement, August 31, 1980, in *From Solidarity to Martial Law* (doc. 7): 70.

[1324] Wilke, Manfred, Erier, Peter, Kubina, Michael, Laude, Horst and Muller, Hans-Peter, "The SED Politburo and the Polish Crisis, 1980/1982," *Cold War International History Project Bulletin,* no 5 (1995): 121, 127.

[1325] *Informace o současné vnitropolitické situaci v PLR,* July 1980, DTO 1945-1989, i. č. 31, e. č. 50, č. j. 330/1, AMZV, Prague.

[1326] *Předběžné poznatky o vnitropolitickém vývoji v Polsku,* 12. 9. 1980, 1261/0/7, sv. P147/80, b. info6, NA, Prague.

[1327] Ibid.

connotations.[1328] In September, Honecker's leadership also had to admit that there had been no outright protests against Polish international commitments during the strikes and demonstrations so far. Using firm ideological language, the East-German analysis of the situation, however, warned against further development. The SED pointed out that some parts of the opposition tended to criticize relations between Poland and the USSR as unequal and even humiliating. References to the obligation the army had to keep its size relative to the size of the country along with subsequent economic troubles also appeared. In terms of the comparison of "counterrevolution" in the CSSR and PPR, the SED noted similar calls for equalization of the country in economic, political, and military cooperation with the Soviet Union as well as achieving a neutral position in the international confrontation between the East and West. Such a construct later allowed the East-German leadership to accuse solidarity of an effort to withdraw Poland from commitments resulting from the Warsaw Treaty Organization.[1329]

The Warsaw Pact maneuvers - recurring of the Czechoslovak scenario?

At the turn of summer and autumn 1980, the tendencies to at least partially repeat the 1968 Czechoslovak scenario apparently existed in the Soviet leadership. The Kremlin intended to use the scheduled series of the Warsaw Pact drills on the territories of Poland and its neighbors to encourage the not very strong Polish leadership to a firm intervention against the opposition.[1330] After all, this had been also the original aim of the large "Šumava" exercise, which had taken place in Czechoslovakia soon before the invasion.

[1328] Задорожнюк, Элла Григорьевна, *Бласть - общество – реформы : Центральная и Юго-Восточная Европа. Вторая половина XX века*, Москва: Наука, 2006), 232.

[1329] Analiza Rozwoj wydarzeń w Polskiej Rzeczpospolitej Ludowej od VI Zjazdu Polskiej Zjednoczenej Partii Robotniczej rozesłana pierwszym sekretarzom komitetów okręgowych SED, kierownikom wydziałów KC oraz członkom i tym kandydatom na członków KC, którzy wchdzili w skład Rady Ministrów, September 1980, in *Przed i po 13 grudnia*, tom 1 (doc. 36): 81.

[1330] Mastny, *A Cardboard Castle?*, 50.

During the Polish Crisis, 14 drills of the Warsaw Treaty Organization of various extent and several bilateral Soviet-Polish military exercises were held.[1331] At the time, the analytics considered them mostly an outside pressure and a threatening campaign aiming, at least in its first stage, to force the Polish leadership to abolish the Gdansk Agreement.[1332] The first of a series of important maneuvers was the Unified Armed Forces' large drill "Brotherhood in Arms-80." It took place at the beginning of September on the territory of the GDR, with participation of 40 thousand troops of all the alliance member-states. No similar action had been held in the last 10 years.[1333] However, the exercise had been scheduled since 1979.[1334] This fact had to be admitted even by West-European press. However, newspapers pointed out that although the West had received an announcement 21 days in advance, in line with the Final Act, its observers had not been invited.[1335] We are reminded that this was a standard approach of the Warsaw Treaty Organization. Despite predominating beliefs among the Western political representation that the exercise was mostly a demonstration of the communist alliance's force in the situation of falling détente, the press began to link the drills in the GDR directly to the events in Poland. These assumptions were also supported by the reports that at the time, loyal East-German generals were complaining about the unusual level of arrogance of the Soviet military

---

[1331] Kramer, "Poland 1980–81," 120. In this regard, Mark Kramer speculates that during these exercises revisited invasion plans were tested and maneuvers, at the same time, provided the Soviet officers with opportunity to try to see how quickly the Polish armed forces could be neutralized by the Warsaw Pact troops in a case of their resistance.

[1332] Thee, Marek, "The Polish Drama: Its Meaning and International Impact " *Journal of Peace Research,* no. 1 (1982): 3.

[1333] Romania participated on the staff level only. The Warsaw Pact supreme commander Kulikov and the Soviet defense minister Ustinov agreed that absence of Romanian combat units was no essential problem. *Konzeption für die Vorbereitung und Durchführung der gemeinsamen operativ-strategischen Übung "WAFFENBRUDERSCHAFT-80",* 13. 5. 1980, DY 30/ J IV 2/2/1838, BArch, Berlin.

[1334] *Kalendářní plán opatření ČSLA ve výcvikovém roce 1979-1980,* 5. 11. 1979, 1261/0/7, sv. 122, a.j. 122/1, NA, Prague.

[1335] "Polish troops join Warsaw Pact forces in E Germany for excercises," *The Times,* August 30, 1980.

officials, despite the fact that the maneuvers were formally commanded by Heinz Hoffman, the defense minister of the GDR.[1336] The Eastern Bloc countries indeed called any connection between the drills and the unrest in Poland "disgusting speculations." Their propaganda focused primarily on praising the Unified Armed Forces' combat capabilities.[1337] During the years 1980-1981, Western media stories about a planned intervention of the Soviet army or the Warsaw Pact in Poland accompanied almost every Eastern military exercise. However, NATO countries' political representations saw them mostly as a demonstration of the strength of the communist alliance.[1338]

      The increased activities of the Warsaw Treaty Organization in Central Europe had to be seen in the broader context of Cold War development, rather than a solely Soviet reaction to the events in Poland. As already explained, at the turn of the 1970s and 1980s, détente was increasingly falling and the atmosphere of East-West confrontation returned. It resulted in rising military budgets and number of drills at the borders of NATO and the Warsaw Pact in Europe.[1339] Almost simultaneously with the "Brotherhood in Arms-80" exercise, for example, even more massive NATO maneuvers such as "Autumn Forge-80" were held in West Germany, involving 320 thousand troops, according to some

---

[1336] "Warsaw Pact Diversity: Cracks Appear in Monolith," *International Herald Tribune*, September 8, 1980.

[1337] "Ke společnému cvičení armád Varšavské smlouvy," *Rudé právo*, September 15, 1980.

[1338] On September 22, for example, International Herald Tribune published a story which warned against suspicious maneuvers of the Soviet troops at nearby Polish borders. However, the U.S. representatives in unofficial interviews confirmed they did not consider movement of the Soviet divisions as an imminent threat to Poland. "Unusual Soviet Maneuvers Are Spotted Near Poland," International Herald Tribune, September 22, 1980; The British side also informed NATO headquarters that in respect for the "Brotherhood in Arms-80" maneuvers, it detected no movement of the Soviet troops directly related to the events in Poland. *Telegram number 166 of 20 September*, FCO 28/4063, TNA, London.

[1339] *Zpráva pro předsednictvo ústředního výboru Komunistické strany Československa o splnění úkolů ČSLA ve výcvikovém roce 1979-1980 a hlavních úkolech ČSLA pro výcvikový rok 1980-1981*, 24. 10. 1980, 1261/0/7, sv. P152/80, b. 1, NA, Prague.

estimations. It was the largest Western exercise in a decade.[1340] Thus, the increased number of Warsaw Pact maneuvers in the region primarily reflected the international situation, even though the Soviet as well as Polish leadership were able to use it as an instrument of political duress and the spreading of an atmosphere of fear.

"Poland remains the Warsaw Pact member"

Given the growing activity of the opposition, at the politburo session held at the end of September, Józef Pińkowski, the Polish prime minister, did not rule out a need to use force, unless the situation calmed down.[1341] Up to this point, the Polish leadership probably had not really considered repression in terms of a broad use of violence, let alone foreign military action.[1342] The fact that some top PUWP officials were afraid of such a move is also proven by Olszowski's talks with Gustáv Husák which took place on September 15. The Polish representative reassured the CPCz CC General Secretary the situation was good within the security apparatus and army, especially in their command structures. Nevertheless, he suggested that in case of an order to use force against the Polish population, the reaction of regular soldiers was hardly predictable. In this regard, he was reminded of the events of December 1970.[1343] Husák was given at least verbal assurance that the Polish leadership would demand,

---

[1340] *Velkomanévry ozbrojených sil NATO v září 1980 v severní části Německé spolkové republiky*, 20. 2. 1980, DTO 1945-1989, i. č. 25, e. č. 1, č. j. 011.444/80, AMZV, Prague. Similarly to the "Brotherhood in Arms-80" drills, General Bernard Rogers, the Supreme Allied Commander of NATO, had decided on holding maneuvers long before the Polish Crisis began. Although the series of "Autumn Forge" exercises had started already in 1975, their extent increased annually.

[1341] *Protokół Nr. 34 z posiedzenia Biura Politycznego KC PZPR 29 września 1980r*, PZPR KC, V/160, mkf. 2986, AAN, Warsaw.

[1342] On August 16, 1980, the interior ministry established a special staff tasked with preparation of the "Summer-80" operation aiming to suppress the opposition. In practice, however, no massive repressions took place. Paczkowski, *From Solidarity*, 7.

[1343] During suppression of the workers' riots in December 1970, in a few cases, some members of security forces refused to follow orders. Eisler, Jerzy, *Grudzień 1970: geneza-przebieg-konsekwencje* (Warszawa: Sensacje XX Wieku, 2000).

among other things, clear support for existing international commitments and allied ties of the PPR by Solidarity.[1344]

Kania made a similar statement in the turbulent atmosphere caused by a one-hour warning general strike of Solidarity.[1345] On October 4, in his report on the situation in the country for the politburo, he stressed that Poland remain a Warsaw Pact member as well as the "country of the socialist commonwealth." He also stated that the party leadership was aware of the existence of enemies of socialism in the country and did not underestimate their activities. Kania's declaration should have primarily sent a signal to the Eastern Bloc states that the situation in the PPR differed from the Prague Spring in essential aspects. This was preemptive assurance, as neither important anti-Soviet acts, nor any questioning of the foreign orientation of the country had occurred in Poland so far. However, at the same meeting, Jaruzelski warned that supply cuts and decline of "social discipline" caused by the strikes could have resulted in a general weakening of the state's defense capabilities.[1346]

The resolution of the 6th PUWP CC plenum, which decided the important personal changes in CC as well as the politburo, reassured that alteration of the leadership would not affect Polish membership in the Warsaw Treaty Organization or the Comecon and the party would not allow any tendencies towards changing the orientation of the country after 1945. The Warsaw Pact members were also given a guarantee that the internal problems would not influence the Polish activities on the issue of holding a conference on military relaxation and disarmament in Europe. We are reminded that the entire alliance had initiated it back in 1979 and the Polish capital had been proposed as a hosting place.[1347]

[1344] Protokół spotkania Gustáva Husáka ze Stefanem Olszowskim 15. wrzesień 1980 r., September 16, 1980, in *Przed i po 13 grudnia*, tom I (doc. 31): 60.

[1345] On October 3, 1980, Solidarity held a one-hour warning strike, as a protest against non-compliance with the Gdansk agreement, especially in terms of the promised raise of wages.

[1346] Unspecific report on development in the PPR, October 8, 1980, DTO 1945-1989, i. č. 31, e. č. 50, č. j. 330/8, AMZV, Prague.

[1347] A hosting place was proposed on Gierek's initiative. *Analýza současného vnitropolitického vývoje PLR z hlediska závěrů 6. pléna ÚV PSDS*, 14. 10. 1980, č. j. 016.531/80, ibid.

In October 1980, given the real situation in Poland up to this point, even the awaiting Eastern Bloc countries began to engage in the Polish question. Despite the aforementioned assurances, they expressed some fears that the internal unrest, among other things, could have affected Polish foreign policy and, by extension, membership in the Warsaw Treaty Organization.

At the end of September, Janos Kadar in his talks with Nguyễn Cơ Thạch, the Vietnamese minister of foreign affairs,[1348] regarding the events in Poland, stressed non-interference with the internal affairs of the "fraternal country." He voiced an opinion that the Polish communists would be able to solve the crisis on their own. Honecker also officially presented Nguyễn an intention not to intervene in the Polish events. He just admitted some economic assistance.[1349] In his speech in the parliament on September 27 regarding the crisis in Poland, Kadar only vaguely mentioned the importance of the Warsaw Treaty Organization. He called for a solution on the socialist basis.[1350] However, in mid-October, he already stated that during conciliation with the Polish opposition, some lines could not be crossed, which were defined by the leading role of the PUWP and the country's membership in the alliance.[1351] In fact, Kadar's phrases about a "political and socialist way of solution" meant that Hungary simply demanded Poland remain a state of "real socialism" and Warsaw Pact member. Force would have been used if the essence of the communist regime was jeopardized or in case of a change in Polish foreign orientation.[1352]

[1348] We are reminded that at the turn of the 1970s and 1980s, the Eastern Bloc significantly widened contacts and cooperation with the socialist states outside Europe, especially Mongolia, Vietnam, and Cuba. Their officials were informed about the events in Poland in detail.

[1349] Szyfogram nf 172/IV z ambasády PLR do MSZ, tajne, October 3, 1980, in *Przed i po 13 grudnia*, tom I (doc. 39): 96.

[1350] *Reakce stranického a státního vedení NDR, MLR, SFRJ, BLR a THE SRR na vývoj situace v PLR*, undated, DTO 1945-1989, i. č. 31, e. č. 50, AMZV, Prague.

[1351] Protokól posiedzenia KC WSPR, October 16, 1980, in *Przed i po 13 grudnia*, tom I (doc. 44): 113.

[1352] Tischler, János, "The Hungarian Party Leadership and the Polish Crisis of 1980-1981," *Cold War International History Project Bulletin*, no. 11 (1998): 77.

A much more radical solution was urged by Petar Mladenov, the Bulgarian minister of foreign affairs. At a session of the politburo in October, he became one of the first representatives of the Warsaw Pact countries who outwardly talked about the need for outside military intervention if the Polish leadership failed to restore order by its own means. This vaguely defined measure was supposed to prevent the unrest from spreading across other Eastern Bloc countries. Some unspecific calls that Poland should have solved its situation in cooperation with other socialist states, especially the Warsaw Pact members, were also presented by Todor Zhivkov, the Bulgarian dictator. In this regard, he pointed out the important position of Poland in the alliance system.[1353] Zhivkov's appeals for a collective solution in the scope of the socialist countries were supported by his daughter Lyudmila, a politburo member. She even unsuccessfully urged her father to take an initiative and propose Brezhnev hold a party leaders' meeting or proper bilateral talks.[1354] At the beginning of November, Zhivkov already stated that the socialist states could not allow the fall of real socialism in Poland and withdrawal of the country from Comecon and the Warsaw Treaty Organization.[1355] This was just a propagandistic pressure statement - the Bulgarian regime was aware that the Polish opposition did not openly challenge socialism, the PUWP leading role, or Polish commitments in the Comecon and the Warsaw Pact.[1356]

Romania took a specific position towards the events in Poland. Ceausescu's leadership formulated its official statement on the unrest as early as late August. In keeping with its foreign policy up to this point, Bucharest stressed the independence of Poland in solving internal troubles

---

[1353] Informacja Todora Žiwkowa o niektórych aktualnych problemach związanych z przygotowaniem XII Zjazdu BKP, October 14, 1980, in *Przed i po 13 grudnia*, tom I (doc. 43): 109.

[1354] Transcript of Bulgarian (BCP CC) Politburo Meeting, October 25, 1980, in *From Solidarity to Martial Law* (doc. 14): 119.

[1355] Stenogram z plenum KC BKP, November 4-5, 1980, in *Przed i po 13 grudnia*, tom I (doc. 57): 155.

[1356] Informacia o wydarzeniach w PLR przyęta przes Biuro Politiczne KC BKP na posiedzeniu 21 i 25 paźdrenika 1980 r., October 25, 1980, ibid (doc. 50): 135.

and warned against any outside interference.[1357] In mid-September, Ceausescu began probing a topicality of foreign military intervention. In his private talk with Zdzisław Kurowski, the PUWP CC Organizational Committee secretary, he outlined concerns about such a step, as it paradoxically could have escalated the situation in Poland as well as international tension, according to Ceausescu's opinion. Although he admitted some unspecific possibility of assistance in solving the issue, he mostly meant improvement of economic cooperation without interfering with Polish internal affairs.[1358] Nevertheless, it is obvious that the Romanian dictator had absolutely no sympathy for the revolt in Poland. However, he was afraid of the spread of similar tendencies to Romania, though far less than Honecker or Husák. Therefore, he implemented some preventive measures, including strengthening the security apparatus protecting the Ceausescu and Petrescu family clan, for example.[1359] During the Polish Crisis, a specific position of Romania within the Warsaw Treaty Organization and the Eastern Bloc in general was even reflected by the fact that Ceausescu, as the only party leader of the alliance member-states, was not informed by Moscow about outcomes of negotiations between the Soviet politburo and the Polish leadership.[1360] Even Fidel Castro, the Cuban leader, was provided with more official information about some meetings.[1361] Moreover, Zbigniew Brzezinski, the national security adviser of the U.S. president Carter, even considered Romania as a neutral country in fact, like Yugoslavia or Finland. In the

---

[1357] *Reakce stranického a státního vedení NDR, MLR, SFRJ, BLR a THE SRR na vývoj situace v PLR*, undated, DTO 1945-1989, i. č. 31, e. č. 50, AMZV, Prague.

[1358] Notatka ze spotkania Zdzisława Kurowskiego z Nicolae Ceauşescu 12 wreśnia 1980 r., tajne, September 13, 1980, in *Przed i po 13 grudnia*, tom I (doc. 29): 55.

[1359] *Reakce stranického a státního vedení NDR, MLR, SFRJ, BLR a THE SRR na vývoj situace v PLR*, undated, DTO 1945-1989, i. č. 31, e. č. 50, AMZV, Prague.

[1360] Letter from Leonid Brezhnev to Erich Honecker, Novemebr 4, 1980, in *From Solidarity to Martial Law* (doc. 17): 132; Transcript of CPSU CC Politburo Meeting, April 16, 1981, ibid (doc. 44): 265.

[1361] Memorandum of Meeting between Leonid Brezhnev, Erich Honecker, Gustáv Husák et al, in Moscow, May 16, 1981, ibid (doc. 49): 280.

measures proposed in connection with the events in Poland, he also advised the White House to intensify consultations with these states.[1362]

Regular meetings of the Warsaw Pact's bodies in autumn 1980

In the second half of October 1980, two scheduled meetings of the Warsaw Pact's bodies - the Committee of Ministers of Foreign Affairs and the Military Council - took place. The Polish question was not put into the official agenda. The representatives of the PPR just touched the issue very quickly in their speeches. The reason lay in the essence and mechanism of both bodies. We are reminded that the Committee of Ministers of Foreign Affairs served mostly as a coordinative forum for unification of the alliance member-states' positions before negotiations with the West. The Military Council dealt with solely military matters during its sessions; it made neither political decisions nor statements.

The regular 22nd meeting of the Military Council was held on October 15-17 in Prague. The situation in Poland was mentioned in an additional part of the report by Eugenius Molczyk, the Polish deputy defense minister. His speech, however, just contained ideological clichés and only vaguely outlined the possibility of internal repressions. As the main reason of the crisis, he identified the economic and social policy of Gierek's leadership characterized by violation of the "Leninist's norms" that allegedly had allowed internal as well as foreign enemies become active in the period of internal political instability. Molczyk assured the Warsaw Pact's military officials that the Polish armed forces were ready to "protect socialism". He also reminded them that given the Warsaw Treaty Organization's existence, Poland had its borderlines secured and this fact gave the Polish authorities a relatively free hand in suppression of the opposition.[1363]

[1362] Special Coordination Committee, Summary of Conclusions, "Meeting on Poland," with Attachment, September 3, 1980, in *From Solidarity to Martial Law* (doc. 10): 87.

[1363] *22. zasedání vojenské rady Spojených ozbrojených sil – informační zpráva pro s. ministra národní obrany ČSSR*, 17. 10. 1980, FMO GŠ/OS, kr. 48, p. č. 178, č. j. 37023, VÚA, Prague.

From the military point of view, a much more important aspect of the Prague meeting was approval of continual construction of the protected command post for the alliance staff on the territory of the PPR. This activity was supposed to absorb even the majority of the Unified Armed Forces' budget in 1981.[1364] This reveals that despite the proclamations made by some political representatives of the Eastern Bloc countries that the Polish membership in the alliance was in jeopardy, the military institutions of the Warsaw Treaty Organization did not anticipate such a scenario at all.

The Committee of Ministers of Foreign Affairs' session was held on October 19-20 in Warsaw. It discussed the issues connected with the upcoming CSCE summit in Madrid and the initiative to convene a conference on military relaxation and disarmament. In keeping with this, the final communiqué strictly concerned the international issues of the East-West relation.[1365] However, the Polish side paid much attention to the preparation and course of the meeting. It tried to ensure that the gathering would be absolutely standard, unaffected by the unrests in the country. At the same time, the Polish leadership used the session to publicly declare the constancy of the basic principles of its foreign policy, including belonging to the Warsaw Treaty Organization and the Eastern Bloc in general. The speeches of Józef Czyrek, the minister of foreign affairs, and Józef Pińkowski, the prime minister, which were later spread in full by the Polish press, totally corresponded with this approach. In his speech, Czyrek stressed that the PPR absolutely reject any interference with its internal affairs by the West.[1366] Therefore, the Western representatives considered outcomes of the session as a success of the Kania leadership

---

[1364] *Návrh rozpočtu Spojených ozbrojených sil na rok 1981 a vyhodnocení rozpočtu Spojených ozbrojených sil za rok 1979*, 3. 10. 1980, ibid .

[1365] *Kommuniqué der Tagung des Komitees der Minister für Auswärtige Angelegenheiten der Teilnehmerstaaten des Warschauer Vertrages in Warschau*, October 20, 1980, Accessed October 16, 2019. http://www.php.isn.ethz.ch/kms2. isn.ethz.ch/serviceengine/Files/PHP/100371/ipublicationdocument_singledocum ent/4baa31c2-4419-4298-a440-2560a4917da3/de/201080_Communique.pdf.

[1366] *Informace o zasedání výboru ministrů zahraničních věcí ve Varšavě*, 6. 11. 1980, 1261/0/7, sv. P153/80, b. info3, NA, Prague.

which had mostly managed to assure anxious allies that Poland would remain a Warsaw Pact member.[1367]

Given the existing and very complex internal situation, the Polish leadership had to also take into account public opinion. Therefore, it tried to reassure the population through the press that the meeting was scheduled regularly and took place in the Polish capital simply because of the standard rotation of hosting countries and it being Poland's turn. At the same time, it was declared that the agenda had solely consisted of talks on the international situation and Moscow had no concerns about the foreign policy of Poland.[1368] In this way, the PUWP leadership sought to avoid spreading speculation about the involvement of the Warsaw Treaty Organization in solving the crisis in the country. The existence of these fears is also proven by the fact that during the first two months of the unrest, the Polish artist unions hesitated to support workers on strikes openly, afraid of possible intervention by the "allied troops."[1369]

The Polish question was not placed on the Warsaw Pact bodies' agenda in 1980. However, the Soviet leadership intensively dealt with it. At the politburo session on October 29, the defense minister Ustinov urged the imposition of martial law in the PPR. Referring to indecisiveness of the Polish army, he reminded the attendees that the Northern group of the Soviet troops was waiting in full combat readiness.[1370] However, one week earlier, a secretly established group of Polish officers, led by Jaruzelski, had begun to work on the imposition of martial law.[1371] The minister of foreign affairs Gromyko also stated that the USSR could not afford to lose Poland under any circumstances. In this regard, Konstantin Rusakov, the CPSU CC Secretary, asked a direct question of multilateral military assistance. He noted that Kania had opposed such a possibility up to this

---

[1367] "Kania beruhigt Verbündete: Polen bleibt bündnistreu," *Die Presse*, October 21, 1980.

[1368] *Trybuna Ludu o obsahu nadcházejícího zasedání ministrů zahraničních věcí Varšavské smlouvy*, 12. 10. 1980, 506/0/45, k. 5968, NA, Prague

[1369] *Současná situace v PLR a perspektivy dalšího vývoje*, November 1980, DTO 1945-1989, i. č. 31, e. č. 50, č. j. 3301/4, AMZV, Prague.

[1370] Transcript of CPSU CC Politburo Meeting, October 29, 1980, in *From Solidarity to Martial Law* (doc. 15): 123.

[1371] Paczkowski, *From Solidarity*, xxxiii.

point and the Polish officials had even denied that the country was in the same situation as Hungary or Czechoslovakia were in the past. Nevertheless, a report by Nikolai Baibakov, the Soviet deputy prime minister, openly mentioned an eventual intervention by the Warsaw Treaty Organization. But Rusakov pointed out the poor relations between Poland and its neighbors. He recommended the Polish leadership continue to solve the crisis on their own. Cooperation with the existing Polish leadership was also relatively appreciated by Brezhnev. Thus, at this time, Moscow was not interested in replacing Kania or Pińkowski with hardliners like Mieczysław Moczar, for instance. The head of the commission for supervising the events in Poland, Mikhail Suslov, even called him the "leftist" who could have caused some real damages.[1372]

In October, Kania tried to convince the Kremlin to postpone the long-planned Warsaw Pact "Soyuz-80" maneuvers which should have taken place on the Polish territory in December. According to the PUWP leader, given the existing situation, the opposition would have considered presence of the foreign troops as a provocation. Despite this, the Soviet leadership initially agreed, though it later changed its attitude. Vojtech Mastny notes that planning of the Warsaw Pact invasion of Poland started at this moment.[1373] Moscow saw the potential entrance of troops to the Polish territory in a different perspective than the suppression of the Prague Spring. The incursion into Czechoslovakia was motivated mostly by the effort to stop the reform process and remove its supporters within the CPCz leadership. Before the invasion, neither Dubček nor other reformists were informed about military intentions and the CSPA did not participate in the operation at all. On the contrary, the "Soyuz-80" drills were carefully coordinated with the Polish side; the PPA officers were supposed to cooperate with the Warsaw Pact forces. Thus, the Kremlin did not plan to take action against the Polish leadership but rather to bolster its position.[1374] Nevertheless, the Polish ruling representatives preferred a domestic solution at the time.

[1372] Transcript of CPSU CC Politburo Meeting, October 29, 1980, in *From Solidarity to Martial Law* (doc. 15): 123.

[1373] Mastny, "The Soviet Non-Invasion of Poland," 192.

[1374] Kramer, "Colonel Kuklinski and Polish Crisis," 52.

Despite the fact that during autumn 1980, the Western press frequently published stories about an impendent military invasion of Poland,[1375] it must be emphasized that the Soviet satellites were not unanimous whether the Polish Crisis had to be solved through outside intervention. Kadar did not even hide his opinion that the "imperialists," in fact, wished for the Soviet troops to invade Poland, as such a step would have discredited the USSR and communism in general for a long time.[1376] At an extraordinary meeting of the collegium of his office, the Czechoslovak minister of foreign affairs, Bohuslav Chňoupek, noted that given the existing situation, it was absolutely crucial for Poland to solve its problems on its own. The Eastern Bloc countries should have provided economic assistance only.[1377] Although Husák admitted the relevance of this step, the CPCz leadership obviously did not welcome the request by its Polish counterpart for economic salvation.[1378]

As already mentioned, the catastrophic condition of the Polish economy also affected the Warsaw Treaty Organization. In 1980, Poland asked for deferment of joining the series of bilateral agreements between the alliance member-states on mutual supplies of military equipment in the years 1981-1985,[1379] at least until the real perspectives of the Polish economy were clarified. Warsaw subsequently joined the agreements in January of 1981.[1380] It was mostly a political gesture, as the county's economic situation did not improve at all.

[1375] Polish diplomacy monitoring of West-European and North-American press, PZPR KC, XIA/1276, AAN, Warsaw.

[1376] Protokól posiedzenia KC WSPR, October 16, 1980, in *Przed i po 13 grudnia*, tom I (doc. 44,): 113.

[1377] *Zápis z mimořádné porady kolegia ministra, konaného dne 10. a 11. 10. 1980*, 15. 10. 1980, PK 1953-1989, kat. č. 807, k. č. 218, č. j. 016.648/80, AMZV, Prague.

[1378] Protokół spotkania Gustáva Husáka ze Stefanem Olszowskim 15. wrzesień 1980 r., September 16, 1980, in *Przed i po 13 grudnia*, tom I (doc. 31): 60.

[1379] The treaties were supposed to mostly deepen and widen coordination and specialization in production of military equipment by individual alliance member-states.

[1380] *Zpráva o uzavření dlouhodobých dohod o vzájemných dodávkách vojenské techniky v letech 1981-1985 mezi vládou Československé socialistické republiky a vládami členských států Varšavské smlouvy a o uzavření významných*

Multilateral solution of the crisis

The first real attempt to solve the crisis in the PPR on the basis of the entire Warsaw Treaty Organization occurred at the end of November 1980. During October and November, the Eastern Bloc countries' representatives discussed the Polish question at several bilateral meetings. But every time, they confined themselves to evaluation of the situation and outlines of expected development, without any significant effort to really engage in a solution. In October, during the Bulgarian Prime Minister Stanko Todorov's visit in the GDR, Erich Honecker and Willi Stoph pointed out the troubles caused to the East-German economy and political stability by the Polish Crisis.[1381] In his talk with Gustáv Husák, Janos Kadar only stressed the strategic importance of Poland in the Warsaw Pact's system.[1382] Actually, at the time, the U.S. press speculated that an aspect of strategic communications could have forced Moscow to intervene, in the end. The columnists advised the American diplomacy to send a clear message to the Kremlin that in such a case there would be no chance to establish correct relations with the new administration of President Ronald Reagan.[1383]

At the end of November, the initiative was not taken by Moscow, but quite surprisingly by Erich Honecker. Similar to 1968, some of its satellites pushed the Soviet Union towards a firm solution of the crisis in the Eastern Bloc. The East-German leader obviously supported the use of force, however not necessarily in the form of outside intervention. In his November 20th meeting with Stefan Olszowski, Honecker emphasized that revolutions did not disperse without violence. He referred to the solution of the situation in the GDR in 1953 and later in Hungary as well as in

---

dokumentů s nesocialistickými zeměmi v roce 1980, 25. 8. 1980, 1261/0/8, sv. P16/81, b. 2, NA, Prague.

[1381] *Návštěva předsedy vlády s. Todorova v NDR*, 14. 11. 1980, DTO 1945-1989, i. č. 25, e. č. 2, č. j. 017.358/80, AMZV, Prague.

[1382] Protókol spotkania Gustáva Husáka z Jánosem Kádárem, November 12, 1980, in *Przed i po 13 grudnia*, tom I (doc. 65): 178.

[1383] *Claris nr 5211/827 z Waszyngtonu*, 24. 11. 1980, PZPR KC, XIX/1276, AAN, Warsaw.

Czechoslovakia.[1384] On November 26, Honecker sent a letter to Brezhnev. He urged an emergency meeting of the party leaders where the possibility of collective measures and assistance to the Polish leadership in dealing with the crisis would have been discussed with Kania. This brief message neither contained an explicit call for the intervention, nor mentioned the Warsaw Pact's framework.[1385] Regarding Honecker's appeal, some historians note that the East-German leader tried to start the same consultative mechanism of the Warsaw Treaty Organization like in 1968.[1386] However, it has to be stressed that similar to the Czechoslovak case, these were not talks within the alliance's official structures, but solely informal multilateral consultations.[1387] Unlike the negotiations on the Prague Spring, participation of Romania as well as Poland itself was expected this time.

The question why Honecker did not tried to convene a regular session of the Political Consultative Committee of the Warsaw Pact rather than an unanchored meeting cannot be answered unambiguously. As explained in the previous chapters, during Brezhnev's rule the scheme of the alliance supreme body sessions had features which probably did not correspond to ideas of the East-German leader about a pressure-character of the talks. We are reminded that in the 1970s, the Political Consultative

---

[1384] Notatka z rozmowy Ericha Honeckera ze Stefanem Olszowskim przeprowadzonej 20. listopada 1980 t. 21. listopadu 1980, in: Przed i po 13 grudnia, tom I, doc. 74, s. 190. In November, an opinion was held by a non-negligible part of mostly lower PUWP functionaries that the situation could be solved by domestic, however harsh means, including use of security and armed forces or installing hardliners like M. Moczar as the head of the party. At the same time, they strictly rejected comparing the extent of the crisis with the Hungarian as well as Czechoslovak events, as such an attitude largely legitimated thoughts about military invasion from orthodox communist's point of view. *Situace v PSDS po 6. plénu*, 24. 11. 1980, DTO 1945-1989, i. č. 31, e. č. 50, č. j. 330/2, AMZV, Prague.

[1385] Enclosure # 2 to Protocol #49 from November 28, 1980, *Cold War International History Project Bulletin*, no. 5 (1995): 124.

[1386] Wilke, Erier, Kubina, Laude and Muller, "The SED Politburo and the Polish Crisis," 127.

[1387] Some historians wrongly call the meeting the Political Consultative Committee session. For example: Vykoukal and Litera and Tejchman, *Východ*, 520; Wanner, *Brežněv*, 64.

Committee gatherings were held in two-year intervals, with one exception, and its last meeting took place in May 1980. Moreover, the sessions had increasingly taken on a mostly propagandistic dimension and their holdings were accompanied by long media campaigns. Afterwards, the public resolutions of the body were given wide publicity and at the same time became key guidelines defining the Eastern Bloc countries' foreign policy. Domestic issues were almost not discussed at the meeting at all. Furthermore, it is very likely that Romania would have strongly opposed the tendency to connect the Warsaw Pact's official structures with internal affairs of its member-states. Honecker's approach actually proves that Moscow, and its satellites, by extension, adhered to the official alliance framework only if it fit the actual situation and pursued goals.

Regarding Honecker's initiative, it has to be stressed that despite the fact that the SED CC General Secretary undoubtedly supported an intervention, most likely on the basis of the Warsaw Treaty Organization, he never formulated this demand openly.[1388] Vojtech Mastny speculates that Honecker's enormous interest in the East-German participation in the eventual invasion was motivated by competition between the leaderships of the GDR and the PPR for the position of "the first among the Soviet allies".[1389] At the time, the East-German army received orders to get into a state of high combat readiness. This meant nothing else than in the case of a political decision, the troops of the GDR would join a joint action of the Warsaw Treaty Organization.[1390]

In front of the East-German politburo, Honecker justified a need for the meeting of the Eastern Bloc party leaders by acknowledging the long-term character of the Polish Crisis and its deteriorating tendency. He complained about the recent anti-Soviet demonstrations as well as the insults against the GDR.[1391] However, at the end of November, the East-

---

[1388] Kamiński, *Przed i po 13 grudnia,* XX.
[1389] Mastny, *A Cardboard Castle?,* 50.
[1390] Kramer, "The Warsaw Pact and the Polish Crisis," 124.
[1391] The anti-Russian acts did not aim for the Soviet policy at the time, but to some tense moments from the Russia-Poland history instead. Also, Honecker incorrectly - ideologically – explained anti-German feelings by the close partnership between East Berlin and Moscow. Sprawozdanie Biura Politycznego

German authorities thought that the Polish leadership soon planned to take firm action against the opposition on their own. On November 30, the defense minister Hoffmann informed Honecker that martial law would most likely be declared by the 7th plenum of the PUWP CC scheduled on December 1.[1392] But this report was misleading. In reality, on November 26, Kania asked the Polish politburo to deliberate strategy based on a real estimation of forces. He warned that otherwise there was a risk of recurrence of the 1956 Hungarian scenario. Rash actions were also opposed by Jaruzelski, who admitted that given the existing situation, the imposition of martial law was impossible. The same view was taken even by the other members of the Polish politburo.[1393]

"Soyuz-80" maneuvers

At the turn of November, simultaneously with Honecker's call for multilateral talks, the Soviet leadership decided that the Warsaw Pact military exercise "Soyuz-80," which should have been postponed to the following year by Kania's previous request, would take place on the Polish territory in December 1980. Some historians consider this move as the beginning of real preparation for the intervention.[1394] Given the absence of key Soviet documentation, the real purpose for increased activity of the Soviet, Czechoslovak, and East-German troops in the first weeks of December can be estimated just on the basis of preserved documents in the former Soviet satellites' archives[1395] and testimonies of participants in

---

na XIII Plenum KC SED, November 28, 1980, in *Przed i po 13 grudnia*, tom I (doc. 80): 210.

[1392] List gen. Heinza Hoffmanna do Ericha Honeckera o sytuacji w PRL, November 30, 1980, in *Przed i po 13 grudnia*, tom I (doc. 86): 223.

[1393] *Protokół Nr. 51 z posiedzenia Biura Politycznego KC PZPR 26. XI. 1980r*, PZPR KC, V/161, mkf. 2987, AAN, Warsaw.

[1394] For example Luňák, *Plánování nemyslitelného*, 75; Mastny, "The Soviet Non-Invasion of Poland," 192.

[1395] For example, a report by a commission of the Chamber of Deputies of the Parliament of the Czech Republic tasked with investigation of circumstances of "Norbert," "Zásah," and "Vlna" actions stated that a majority of documents concerning the "Giant Mountains" military operation, an involvement of the CSPA in the "Soyuz-80" maneuvers in fact, had been shredded in the years 1982-

these events. Nevertheless, it is clear that the first half of December 1980, when the "Soyuz-80" maneuvers and the meeting of the Warsaw Pact countries' party leaders in Moscow were held, was the period when the danger of outside intervention in Poland was the most imminent.[1396]

The "Soyuz-80" drills obviously had a direct connection with the development in Poland. After Moscow reviewed its initial decision to postpone the exercise, an emergency meeting of Nikolai Ogarkov, the chief of the Soviet general staff, with the military officials of the CSSR, GDR and PPR took place in the Soviet capital on December 1. Ogarkov was informed about the plans of the upcoming maneuvers.[1397] These had been worked out by the Soviet general staff with assistance of the Soviet officers from the Staff of the Unified Armed Forces.[1398] In this regard, it is good to point out that despite the "Soyuz" drills being one of the key exercises of the Warsaw Treaty Organization, the action was directed by the Chief of General Staff of the Soviet army, rather than the Unified Command's personnel. This fact again proves the rudimentary influence of the alliance headquarters on planning important military operations.

1984, Accessed October 16, 2019. http://www.psp.cz/eknih/1993ps/stenprot/028 schuz/s028011.htm.

[1396] Victor Sebestyen, a British publicist of Hungarian origin, in his book of questionable quality "*Revolution 1989. The Fall of the Soviet Empire,*" in the part dealing with the Polish Crisis regarding the "Soyuz-80" maneuvers, notes an important document from the Russian defense ministry archive unavailable to the other scholars. It allegedly consists of a few minutes of talks between Ustinov, Andropov and Brezhnev revealing that since the beginning, the real Soviet intention was just to extend the drills in a way which would have given an impression that the Warsaw Pact's troops were ready for an invasion. Taking into account Sebestyen's superficial method of work with sources and often questionable dating of the documents he refers to, this claim cannot be taken seriously. Moreover, the outlined approach corresponds to the early months of 1981, when the key Soviet political officials most likely abandoned the military intervention altogether. In addition, Sebestyen did not conduct complex archival research, but due to his connections in Russia, he was able to get some fragmented documentation. Sebestyen, Victor, *1989. Pád východního bloku* (Brno: Computer Press, 2011), 61-62.

[1397] Tůma, Oldřich, "The Czechoslovak Communist Regime and the Polish Crisis 1980-1981," *Cold War International History Project Bulletin*, no. 11 (1998): 60-76.

[1398] Kramer, "Poland 1980–81," 120.

Based on the available document, an assumption can be made that the Kremlin had no clear plan initially. It intended to adjust the course of the maneuvers to actual development of the situation, especially to outcomes of the upcoming talks of the party leaders in Moscow. In total, 18 divisions were supposed to take part in the action - fifteen of the USSR, two of Czechoslovakia, and one of East Germany. From the beginning, involvement of the Polish armed forces was planned.[1399] Participation of the Soviet satellites' armies was just formal and mostly of political importance, similar to the operation "Danube." The role of the Polish armed forces was primarily to remain passive during the deployment of the foreign divisions nearby big cities and key centers in the PPR. Most likely, such a move would have created room for a solution of the Polish Crisis according to the Kremlin's demands.[1400] It is possible that Moscow did not inform representatives of the satellites' armies about the entire plan of action, but rather about those parts concerning them directly.[1401] Keep in mind that this was a standard praxis in operational planning of the Warsaw Treaty Organization.

The strictly formal role of the PPA in this plan is also supported by a claim of Stanisław Kania that he saw a map with a large mass of the Warsaw Pact troops dispatched on the territory of the PPR without any specific task of the Polish army.[1402] Such an outlined plan shocked the defense minister Jaruzelski, who was informed about the conclusions of the meeting of the Soviet general staff by generals Tadeusz Hupałowski and Franciszek Puchała. After long negotiations with the Soviet side, Jaruzelski unsuccessfully demanded the East-German troops not to enter the territory of the PPR. He did manage at least to make a deal on active participation of two Polish divisions. However, they were supposed to ensure smooth passage for the Czechoslovak and East-German units and

---

[1399] Notatka Objaśnenia z instruktażu przeprowadzonego w Stazbie Generalnym Armii Radzieckiej na temat wspólnych ćwiczeń Zjednoczonych Sił Zbrojnych państw stron Układu Warszawskiego, tajny dokument sztabowy, December 1-3, 1980, in *Przed i po 13 grudnia*, tom I (doc. 88): 225.
[1400] Mastny, "The Soviet Non-Invasion of Poland," 193.
[1401] Tůma, "The Czechoslovak Communist Regime and the Polish Crisis," 63.
[1402] *Wejdą nie wejdą*, 163. This map was most likely shredded, similarly to a majority of materials about the "Giant Mountains" action.

to prevent eventual acts of resistance by the Polish population. On December 3, the supreme commander of the Unified Armed Forces, Victor Kulikov, asked Jaruzelski for formal permission to enter Polish territory by the Warsaw Pact troops in the night from December 7 to 8. In his memoirs, Jaruzelski retrospectively stated he had rejected Kulikov's demand, but there is no evidence for this claim.[1403]

The "Soyuz-80" maneuvers envisaged two phases: In the first, the troops would have collected outside the Polish territory. The second was supposed to start on the Soviet general staff's order to cross the state borders of Poland. This was not the direct military intervention, but rather continuation of the drills on Polish territory scheduled for approximately four days. After the end of the exercise, the troops would have retreated to the nearby big cities. Most likely, deployment of the "fraternal armies" would have helped the Polish leadership in its actions against the opposition.[1404] Thus, presence of the Warsaw Pact troops was not the real invasion, but according to the original plan it played a mostly supportive, psychological role.

The official plan of the December maneuvers corresponded to typical practice of the Warsaw Pact's operation at the time. According to the available materials on the CSPA participation, code-named "the Giant Mountain," in the first stage, regrouping of the troops would have taken place before their joint combat operations. The second stage was supposed to simulate a time lag in the scenario to days 20-30 after an outbreak of war, when the troops of the Alliance Front crossed the Rhine and launched an offensive operation in the western direction.[1405] However, some aspects of the maneuvers were nonstandard which allowed speculation about the real purpose of the exercise. In accordance with the CSCE Final Act, "Soyuz-80" was announced in advance, but in a shorter term than the originally stipulated 21 days. At least in case of the CSPA, it was the action of a long-term schedule of combat drills. Preparations proceeded in

---

[1403] Mastny, "The Soviet Non-Invasion of Poland," 193-194.

[1404] Luňák, *Plánování nemyslitelného*, 75.

[1405] Notatka informacyjna gen. Miroslava Blahnika dla gen. Martina Dzúra na temat zamierzeń i przebiegu wspólnych ćwiczeń Zjednoczonych Sił Zbrojnych państw stron Układu Warszawskiego, December 3, 1980, in *Przed i po 13 grudnia*, tom I (doc. 90): 230.

unusually short time and contradicted guidelines for holding such maneuvers. The exercise was not even discussed by the Council of State Defense. The 1st and 9th tank division took part - these were in terms of professionalism, organization, and from the communist regime's point of view, the political and ideological elite of the CSPA. Special criteria were established for selection of the soldiers: integrity, professional skills, and "ideological competences" were decisive factors.[1406] Also, the NPA command put a similar emphasis on moral quality of the units and thorough control of the political situation in areas of their deployment on the territory of the PPR.[1407] Moreover, the divisions were equipped with nonstandard weapons and large provisions of ammunition.[1408]

However, it is very unlikely that the meeting at the General Staff of the Soviet Army and the following maneuvers were the real preparations for the invasion. A relatively small number of involved troops is the main argument: For the invasion of much smaller Czechoslovakia, Moscow had used ten more divisions, and unlike in the Polish case, no active resistance of the population as well as the army had been anticipated. According to former estimates of the Soviet intelligence, at least 45 divisions would have been needed in case of organized resistance by the Polish side.[1409] Therefore, the troops' movement must be seen as an act of pressure aiming to push the Polish leadership to the imposition of martial law.[1410]

In his report for the Western intelligence, Ryszard Kukliński, the Polish colonel,[1411] stated that only a small percentage of Polish officers,

---

[1406] *Zpráva vyšetřovací komise Poslanecké sněmovny pro vyšetření okolností souvisejících s akcemi Norbert, Vlna a Zásah, případně obdobnými akcemi,* February 2, 1995, Accessed October 16, 2019. http://www.psp.cz/eknih/1993ps/stenprot/028schuz/s028010.htm.

[1407] Rozkaz nr 118/80 ministra obrony narodowej NRD o przygotowaniu i przeprowadzeniu wspólnych ćwiczeń Zjednoczonych Sił Zbrojnych państw stron Układu Warszawskiego, December 6, 1980, in: Przed i po 13 grudnia, tom I (doc. 95): 286.

[1408] Mastny, "The Soviet Non-Invasion of Poland," 194.

[1409] Ibid, 193.

[1410] Kamiński, *Przed i po 13 grudnia,* XXV.

[1411] We are reminded that since the beginning of the 1970s till November 1981, Colonel Kukliński provided the U.S. with extremely sensitive information about

who participated in planning the exercise, were concerned about its holding. Based on this, he concluded that in the case of the Warsaw Pact's invasion any resistance by the Polish army was unlikely. The PPA officer corps allegedly agreed with the assumption that the presence of a large number of foreign troops would calm down the situation in the country.[1412] However, this claim was largely misleading. On the contrary, regarding the "Soyuz-80" maneuvers, the Polish political as well as military elites were afraid of escalation of the opposition's resistance after the Warsaw Pact troops entered the country, and especially because of historical reminiscences connected to involvement of the East-German units.[1413] Even the hardliners, including the Polish pro-Moscow officers, were aware of this and therefore tried to prevent the NPA participation. In their talks with the East-German officers, the Polish generals stressed how complicated the situation in the PPA was and expressed their hopes that the massive maneuvers would not be held on Polish territory, in the end. In this regard, they presented an opinion that the situation in the PPR was not as serious as the Eastern Bloc countries' representatives thought.[1414]

To a lesser extent, this also concerned the participation of the CSPA.[1415] On December 4, during preparation of the maneuvers, Jaroslav Gottwald, the Czechoslovak general, met on the Polish territory General Józef Wilczyński, the deputy commander of the Silesian military district,

---

the Eastern Bloc's military planning. During the Polish Crisis, he worked on a staff which worked on plans of imposition of martial law and at the same time he served as a liaison officer of Supreme Commander Kulikov in the Unified Command. Kramer, "Colonel Kuklinski and Polish Crisis," 49; Kukliński, *Wojna z narodem.*

[1412] Message from Ryszard Kukliński on Impending Warsaw Pact Invasion, December 4, 1980, in *From Solidarity to Martial Law* (doc. 21): 139.

[1413] It cannot be ruled out that the Polish political as well as military leadership tried to maximally limit the entrance of foreign troops to the Polish territory and reduce a risk of an outside military intervention in this way. In the context of the events, the issue of participation of a relatively small number of the East-German troops seems to be secondary, at least.

[1414] Raport gen Manfreda Gehmerta o wynikach rozpoznania dokonanego na terytoirium PRL w celu przeprowadzenia wspólnych ćwiczeń Zjednoczonych Sił Zbrojnych państw stron Układu Warszawskiego, December 16, 1980, in: *Przed i po 13 grudnia*, tom I (doc. 111): 317.

[1415] Kramer, "Colonel Kuklinski and Polish Crisis," 53.

and unnamed representative of the general staff of the PPA. The Polish side asked the CSPA to send a group of officers before the exercise began, which would have specified the expected movement of the Czechoslovak troops on the territory of the PPR. The Poles also warned that the arrival of the Czechoslovak units could have been seen by the Polish population not only as a provocation but also as a prologue of the outside intervention.[1416] This supports claims by some historians that Moscow never planned an invasion on the basis of the Warsaw Treaty Organization and only the unilateral Soviet action was eligible.[1417]

Non-topicality of the intervention under the auspices of the entire alliance is also proven by the fact that during the regular session of the Committee of Defense Ministers of the Warsaw Pact, held December 1-3 in Bucharest, the situation in Poland was discussed in backrooms only. The report by Dobri Dzhurov, the head of Bulgarian defense resort, suggests that the Polish representatives, led by general Molczyk,[1418] managed to convince the Soviet side that the PUWP still controlled the army as well as the power apparatus and just hesitated to take decisive measures. The Polish army officials were aware that only two ways out of the crises existed: political negotiations and finding *modus vivendi* with the opposition, or use of force. Dzhurov, without naming his source, reported to the Bulgarian leadership that unlike Kania, the Polish generality leaned towards the latter option.[1419] The only item indirectly connected with Poland on the official agenda was an appeal by Anatoly Gribkov, the chief of Staff of the Unified Armed Forces, for another improvement of strategic communication and logistic lines. The fact that the crisis in Poland did not substantially affect relations between the

---

[1416] Informacja gen. Jaroslava Gottwalda o sytuacji w PRL, December 6 1980, in: *Przed i po 13 grudnia*, tom I (doc. 97): 290.

[1417] For example Kamiński, *Przed i po 13 grudnia,* XXV.

[1418] Molczyk led the Polish delegation from a position of Jaruzelski's deputy. The defense minister of Poland was busy with domestic issues and therefore did not take part in the meeting. This was a very rare situation. *Informace o průběhu a výsledcích 13. zasedání výboru ministrů obrany konaného ve dnech 1.-3. prosince v Bukurešti,* 4. 12. 1980, FMO GŠ/OS, kr. 47, p. č. 176, e. č. 37021, VÚA, Prague.

[1419] Bulgarian Report on the Defense Ministers' meeting in Bucharest, December 8, 1980, in *A Cardboard Castle?* (doc. 89): 441.

Warsaw Pact member-states' armies is also proven by the speech of Josef Mikulec, the Czechoslovak chief of the Department of Higher Military Education, who expressed appreciation for the ongoing cooperation between Polish and Czechoslovakian military academies regarding the exchange of students.[1420]

The meeting of the party leaders in Moscow

On December 5, as the Soviet, Czechoslovak, and East-German troops were gathering on the Polish borders, the two-day meeting of the Warsaw Pact member-states' party leaders initiated by Honecker had begun in Moscow. At the request of the Soviet leader, the talks were secret. No information should have been published, except for an official communiqué.[1421] By telephone, Brezhnev informed the Polish leadership that the meeting would concern the crisis in the PPR and publicly declared analysis of the international situation would be just a "masking" topic.[1422] The preparations stayed in secret. The meeting obviously caught the West by surprise. The Western press reminded its readers that similar negotiations had preceded the 1968 military intervention in Czechoslovakia. However, it was presumed that there was no imminent risk of invasion and Moscow as well as its most critical satellites of the Polish events - the GDR and CSSR - would give Kania more time to consolidate the situation.[1423] The official communiqué presented the Polish Crisis as a secondary item on an agenda which allegedly primarily

---

[1420] *Informace o průběhu a výsledcích 13. zasedání výboru ministrů obrany konaného ve dnech 1.-3. prosince v Bukurešti*, 4. 12. 1980, FMO GŠ/OS, k. 47, p. č. 176, e. č. 37021, VÚA, Prague. An exchange study program of the PPA, CPSA, and NPA officers at military academies, as well as fellowships of selected Soviet officers in CSSR actually began in 1980. *Vystoupení náčelníka ideologické správy – zástupce náčelníka HPS ČSLA genmjr. J. Klíchy na vojensko-teoretické konferenci k 25. výročí podepsání Smlouvy o přátelství, spolupráci a vzájemné pomoci ve Varšavě konané v Moskvě 22. 5. 1980*, p. č. 169, e. č. 37007, ibid.

[1421] *Záznam ze setkání vedoucích stranických a státních představitelů členských států Varšavské smlouvy*, 11. 12. 1980, 1261/0/7, sv. P157/80, b. info2, NA, Prague.

[1422] *Wejdą nie wejdą*, 163.

[1423] *Ohlasy moskevského setkání nejvyšších představitelů socialistických zemí*, 5. 12. 1980, 506/0/45, k. 5968, NA, Prague.

included a perspective of a restart of the relaxation process between the East and West.[1424]

Stanisław Kania, who took part in the similar Warsaw Pact's session for the first time, tried to calm down the Eastern Bloc leaders. He used relatively radical language and did not hesitate to admit that some elements of anarchy and counterrevolution could be found in Poland. But he gave assurances that the Polish armed forces remained in order and that key cadres respected the party line.[1425] He assessed the riots as the biggest crisis to ever hit Poland. At the same time, Kania supported a solution on the basis of the entire Eastern Bloc - he rejected, for example, a separate accession of Poland to the International Monetary Fund, even though he admitted that such a move would have helped economic stabilization. Kania tried to satisfy the hardliners with a claim that the PUWP would not hesitate to take repressive measures to defend socialism in Poland.[1426]

There are records of the official speeches delivered at the Moscow meeting. In this regard, it must be noted that we do not know which talks took place in backrooms. The Prague Spring experience shows that many negotiations proceeded in secret, face to face. Nevertheless, the documents say that only Ceausescu handed over the entire responsibility for a solution to the Polish communists.[1427] Taking into account a long-term and

---

[1424] "Встреча руководящих деятелей государств-участников Баршавского договора," *Правда*, December 6, 1980; *Известия*, December 6, 1980; "Za jednotu a semknutost socialistických zemí," *Rudé právo*, December 6, 1980.

[1425] He expressed some concerns about political work in the army and a new round of conscription, when about one quarter of the Polish soldiers had switched. Sprawozdanie ze spotkanie przywódców państw członkowskich Układu Warszawskiego w Moskwie 5 grudnia 1980 r. opracowane w Wydziale Zagranicznym KC WSPR, December 8, 1980, in *Przed i po 13 grudnia*, tom I (doc. 101): 300.

[1426] *Záznam ze setkání vedoucích stranických a státních představitelů členských států Varšavské smlouvy*, 11. 12. 1980, 1261/0/7, sv. P157/80, b. info2, NA, Prague.

[1427] Ceausescu's participation caught the West by surprise. It was generally presumed that his stance on Polish events would be similar to the invasion of Czechoslovakia, which he had condemned. *Ohlasy moskevského setkání nejvyšších představitelů socialistických zemí*, 5. 12. 1980, 506/0/45, k. 5968, NA, Prague. Later in December, Ceausescu discussed the Moscow meeting with A. Grlićkov, the Yugoslav official. In this regard, he assured the Yugoslav side that

permanent accent of state sovereignty and principle of non-interference with internal affairs, it was a logical move. However, it is interesting that he largely connected the whole crisis with economic troubles. In this regard, the Romanian dictator called for a coordinated solution within the Comecon.[1428] Erich Honecker, an initiator of the meeting, surprisingly expressed his beliefs that the PUWP was still strong enough to solve the situation on their own. But unlike Ceausescu, he preferred a tough and violent approach.[1429]

Honecker's speech was full of ideological clichés and did not link the crisis in Poland to the Warsaw Pact's interests at all. Also, Todor Zhivkov, despite his previous statements, did not consider Polish withdrawal from the alliance real. He emphasized that the West saw the situation in similar terms. According to the Bulgarian dictator, there still remained the risk of another model of leftist regime being imposed in Poland. "A dictatorship of the proletariat" would have been replaced by a "democratic socialism." Then, the regimes in the rest of the Eastern Bloc

---

Romania still insisted that the Polish leadership had to take the action against the opposition on its own. *Belegrad – Szyfrogram Nr 3697/IV*, 19. 12. 1980, PZPR KC, XIX/1276, AAN, Warsaw; In his talks with the Western diplomats, the Romanian foreign minister Andrei later stated that the Romanian participation had significantly affected the decision to postpone an invasion of Poland for the time being. It is likely that he just tried to increase the importance of his country in the Western eyes by such a statement. Report by the British embassy in Bonn on the talks with the Romanian foreign minister S. Andrei regarding the Warsaw Pact member-states' leaders in Moscow, December 15, 1980, FCO 28/4063, TNA, London.

[1428] *Záznam ze setkání vedoucích stranických a státních představitelů členských států Varšavské smlouvy*, 11. 12. 1980, 1261/0/7, sv. P157/80, b. info2, NA, Prague. This was an instrumental step. So far, Ceausescu strictly opposed any integration of the Eastern Bloc, including substantial improvement of cooperation within the Comecon. At the beginning of the 1980s, however, the Romanian economy began to struggle and the Romanian leader, also taking into account the development in Poland, was afraid of a potential outburst of unrest within the Romanian population. Therefore, he was probably compliant to support some multilateral stabilizing measures.

[1429] Sprawozdanie ze spotkanie przywódców państw członkowskich Układu Warszawskiego w Moskwie 5 grudnia 1980 r. opracowane w Wydziale Zagranicznym KC WSPR, December 8, 1980, in *Przed i po 13 grudnia*, tom I (doc. 101): 300.

countries could have been jeopardized by pressure for similar changes. However, Zhivkov agreed that the Polish leadership had enough power to act on their own. Also, Kadar identified himself with this evaluation. He just repeated that the red line for leaving initiative to the Polish leadership was actually preserving the country's membership in the Warsaw Treaty Organization.[1430] Only the speech by Gustáv Husák could be seen as an appeal for military invasion. His detailed explanation of the events in Czechoslovakia in 1968 and appreciation of then-military intervention suggested that the CPCz CC General Secretary also leaned towards similar solution this time. But he explicitly denied it later, calling the invasion a last resort.[1431]

Some parts of Brezhnev's speech could also be understood in the sense that in certain situations, an outside military action could have been taken. At first, the Soviet leader stressed that Moscow would not allow the PPR to leave political, economic, and military structures of the Eastern Bloc. At the same time, he warned that the USSR would not tolerate jeopardizing the Warsaw Pact member-states' security interests. He called the unrests in Poland an attack on the whole "socialist commonwealth."[1432] One week later, the Soviet politburo decided that these parts of Brezhnev's speech had to become a starting point for further activities of the Polish leadership.[1433]

Kania tried to pre-discuss the course of the meeting with Brezhnev, but he failed. However, he at least cooperated with the Soviet representatives on preparation of the text of the final communiqué. Then, the PUWP CC First Secretary tried to convince the Soviet leader that if foreign troops entered Poland, an uprising would break out. In the end,

---

[1430] Hungarian diplomacy actually informed the Yugoslav side, the British minister of foreign affairs Carrington as well as the SPD deputy chairman Hans-Jürgen Wischnewski that Polish withdrawal from the Warsaw Treaty Organization was out of the question. Ibid.

[1431] *Záznam ze setkání vedoucích stranických a státních představitelů členských států Varšavské smlouvy*, 11. 12. 1980, 1261/0/7, sv. P157/80, b. info2, NA, Prague.

[1432] Ibid.

[1433] Transcript of CPSU CC Politburo Meeting, December 11, 1980, in *From Solidarity to Martial Law* (doc. 25): 167.

Brezhnev promised him that the intervention would take place only under the Polish leadership's support and in case of another escalation of the situation in the country. At the same time, Kania reassured the Kremlin that the Polish top military alone had begun to work on the suppression of Solidarity.[1434]

After returning from Moscow, the head of the PUWP informed the Polish politburo that the meeting had introduced a new dimension into the situation. In this regard, he respected the Warsaw Pact member-states' right to interfere with a solution to the crisis as it was also affecting their internal political situation. The fact that the party leaders remained convinced of the Polish leadership's ability to solve the problem by its own means, Kania considered as a key finding. Therefore, he urged the politburo to act quickly, especially to work out a conception of intervention by the interior ministry and prosecution against the most radical parts of the opposition. At the same time, an attitude towards informing the Warsaw Pact member-states' ambassadors to Poland should have been changed. Henceforward, they were supposed to receive not only reports on negative things, but also reports about a constructive approach of the party to a solution of the situation.[1435]

A question whether the Polish side just tried to get more room for its own decisions or the Polish security forces intended to take action against the opposition regardless of the Soviet military plans cannot be unequivocally answered. Given the absence of key Soviet documents, it also remains unclear whether the original intention of the "Soyuz-80" maneuvers was modified because of Kania's personal impression on Brezhnev or the CPSU CC General Secretary had given the orders stopping arrival of a large number of troops to the Polish territory already before the Moscow meeting.[1436] Nevertheless, Kania's actions in Moscow were largely appreciated by the Soviet Politburo, including Mikhail

---

[1434] Mastny, "The Soviet Non-Invasion of Poland," 194-195.
[1435] *Protokół Nr. 53 z posiedzenia Biura Politycznego KC PZPR 6. XII. 1980r*, PZPR KC, V/161, mkf. 2987, AAN, Warsaw.
[1436] Mastny, *A Cardboard Castle?*, 51.

Suslov, the head of the commission for supervision of the events in Poland.[1437]

The Soviet, Czechoslovak as well as East-German troops actually entered the Polish territory in the end. But it was a gesture of support for Kania's leadership, rather than pressure for an imminent elimination of the opposition.[1438] Between December 8-10, a command-staff exercise took place in Wroclaw, including staffs of the USSR, CSSR, GDR, and PPR.[1439] Contrary to the initial plans, this stage of drills did not consist of a massive participation of the troops.[1440] Based on the Unified Command's instructions, the drills on the territory of the PPR then continued. Officers of the Soviet, Polish, East-German, and Czechoslovak armed forces also participated in its command. The East-German military communication reveals that the key orders were those by the chief of the Soviet general staff Ogarkov and the Warsaw Pact's supreme commander Kulikov.[1441] Officially, the "Soyuz-80" maneuvers should have ended on December 21. But the chief of Staff of the Unified Armed forces Gribkov received an unexpected order by Ustinov to prolong it. According to Kulikov's testimony, this step was taken on Jaruzelski's request. The Polish leadership was not concerned about the presence of the Warsaw Pact troops on the Polish territory. They were probably aware of the supportive purpose of the exercise mostly including the staffs.[1442] Moreover,

---

[1437] Transcript of CPSU CC Politburo Meeting, December 11, 1980, in *From Solidarity to Martial Law* (doc. 25): 167.

[1438] Luňák, *Plánování nemyslitelného*, 76.

[1439] Pismo gen. Martina Dzúra do Gustáva Husáka dotyczące wspólnych ćwiczeń Zjednoczonych Sił Zbrojnych państw stron Układu Warszawskiego, before December 8, 1980, in: *Przed i po 13 grudnia*, tom I (doc. 100): 298.

[1440] Notatka informacyjna gen. Miroslava Blahnika dla gen. Martina Dzúra na temat zamierzeń i przebiegu wspólnych ćwiczeń Zjednoczonych Sił Zbrojnych państw stron Układu Warszawskiego, December 3, 1980, in *Przed i po 13 grudnia*, tom I (doc. 90): 230.

[1441] Rozkaz nr 15/80 przewodniczącego Narodowej Rady Obrony NRD o przeprowadzeniu wspólnych ćwiczeń Zjednoczonych Sił Zbrojnych państw stron Układu Warszawskiego, December 10, 1980, in *Przed i po 13 grudnia*, tom I (doc. 104): 309.

[1442] This was another significant step back from the 1968 Czechoslovak scenario. The "Šumava" drills had been initially intended solely as a command-staff

continuing, however limited maneuvers raised a sense of insecurity among the Polish population and threatened the opposition.[1443] Regarding the events of December, even the West did not miss that Solidarity had called for stopping "irresponsible strikes."[1444] Also the Catholic Church realized that further escalation of the revolt could have led to loss of positions gained so far and urged Solidarity to restraint.[1445]

The Western reaction to the December events

At the beginning, the CIA deemed that the Warsaw Pact's maneuvers in December were not necessarily a presage to an invasion of Poland, although it increased its possibility.[1446] The West chose a careful approach. In November, NATO diplomats still firmly objected to an eventuality of providing "military assistance" to Poland. But according to the East-German sources, they were instructed to engage in no such official discussions on the situation in the PPR with their Eastern Bloc counterparts.[1447] On December 7, the leaving Carter administration concluded that the Kremlin had presumably finished the preparations needed for the military intervention in Poland. The American side actually anticipated an incursion of the troops under the pretext of joint maneuvers.[1448] The CIA reported that the Warsaw Pact's military preparation reached the stage when 15-25 divisions were ready to invade Poland. At the same time, it noted that unlike the invasion of Czechoslovakia, neither transport of the Soviet airborne units to western

---

exercise too, but the Soviet side had used it to dispatch a large military contingent on the Czechoslovak territory.

[1443] Mastny, "The Soviet Non-Invasion of Poland," 196.

[1444] "Union Plans No Strikes," *International Herald Tribune*, December 7, 1980.

[1445] *K současné situaci v Polské lidové republice*, 20. 1. 1981, DTO 1945-1989, i. č. 31, e. č. 50, č. j. 330/12, AMZV, Prague.

[1446] CIA Alert Memorandum, "Poland", December 3, 1980, in *From Solidarity to Martial Law* (doc. 20): 138.

[1447] Informacja nr 19/80 wywiadu wojskowego NRD o sytuacji w PLR, December 3, 1980, in *Przed i po 13 grudnia*, tom I (doc. 89): 227.

[1448] Minutes of U.S. Special Coordination Committee Meeting, December 7, 1980, in *From Solidarity to Martial Law* (doc. 23): 162.

airfields in the USSR, nor increase of readiness of the Soviet strategic forces and navy had proceeded so far.[1449]

Immediately after the Moscow meeting, Peter Carrington, the British minister of foreign affairs, warned against serious consequences for the USSR in case of an invasion of Poland.[1450] On the contrary, the U.S. officially took a neutral stance through Edmund Muskie, the secretary of state.[1451] NATO issued its joint statement on the Polish Crisis on December 12. On behalf of the ministers of foreign affairs, the alliance urged Moscow not to interfere with the Polish events. But at the same time, the West stressed its passive position.[1452]

Leaving the thoughts of intervention

At the end of 1980, Moscow most likely reviewed its thoughts of an eventual military action in the PPR. The Kremlin began to use the top representatives of the Warsaw Pact Unified Command to put pressure on the Polish leadership to solve the problem quickly and on their own. At the same time, there was a significant growth in the activities of Marshal Kulikov, the supreme commander of the Unified Armed Forces, who tried to establish cooperation with the Polish hardliners and ensure participation of the Polish troops in a potential emergency intervention.[1453] The key Soviet institutions - ministries of foreign affairs and defense and the KGB - identified from Moscow's point of view trustworthy Polish representatives with whom the situation in the country could be permanently consulted.[1454] The fact that at the turn of 1980 and 1981, despite no improvements in the situation, Moscow abandoned a possibility

---

[1449] CIA Situation Report, "Poland", December 8, 1980, in *From Solidarity to Martial Law* (doc. 24): 165.

[1450] *Reakce Západu na moskevské setkání představitelů Varšavské smlouvy*, 7. 12. 1980, 506/0/45, k. 5968, NA, Prague.

[1451] *Edmund Muskie a další američtí činitelé o výsledcích moskevské schůzky představitelů Varšavské smlouvy*, 7. 12. 1980, ibid.

[1452] Mastny, "The Soviet Non-Invasion of Poland," 196.

[1453] Mastny, *A Cardboard Castle?*, 51-52. Kulikov himself admitted later that he had played an important role in the Polish Crisis. *Wejdą nie wejdą*, 154.

[1454] Transcript of CPSU CC Politburo Meeting, January 22, 1981, in *From Solidarity to Martial Law* (doc. 29): 184.

to militarily intervene in Poland - alone as well as through the Warsaw Treaty Organization. This was confirmed in January 1981 by the unnamed senior officer of the Soviet army group in Hungary. The Kremlin was mostly afraid of Western political and economic sanctions. Furthermore, there were some fears that an entrance of troops into Poland would make the opposition stick together and result in a long-term armed resistance. During 1981, the USSR therefore just urged the Polish leadership to solve the situation on their own.[1455] There are suggestions that the representatives of the Warsaw Pact's military structures played a very important role in Moscow's policy towards the PPR.

At the beginning of January, a Soviet delegation arrived in Poland tasked with an assessment of the situation in the country after the events of December of the last year. Its head, the chief of the CPSU CC International Department, Leonid Zamyatin, once again raised the accusation that Solidarity's activities were weakening not only Poland, but also the entire socialist commonwealth and the Warsaw Pact's defense capabilities. In this regard, the delegation attacked the Polish side for a lax attitude towards worker's strikes in the weapons industry. Zamyatin especially criticized an effort to solve the crisis by political means and appealed for using a repressive apparatus.[1456] At the time, the PUPW top echelons again declared Polish belonging to the Warsaw Treaty Organization and awareness of the important role that the PPA played in the alliance's defense system. However, the international ties of Poland began to be cautiously challenged by the opposition. The Polish leadership tried to blunt these critics with the claim that only this alliance could guarantee the peace and independence of the country.[1457]

[1455] Budapeszt- Szyfrogram nr 975/I z ambasady PRL do MSZ, January 27, 1981, in *Przed i po 13 grudnia*, tom I (doc. 127): 341.

[1456] *Notatka informacyjna*, undated, most likely February 1981, PZPR KC, XIA/1258, AAN, Warsaw; *Notatka informacyjna z pobytu w Polsce delegacji KC KPZR z L.Zamiatinem w dniach 13.-21.I. 1981 r.*, 2. 2. 1981, ibid; PUWP CC Report on Leonid Zamyatin's Visit to Katowice, January 16, 1981, in *From Solidarity to Martial Law* (doc. 28): 180.

[1457] *Polska Zjednoczona partia robotnicza przewodnia siła odnowy*, 2. 1. 1981, PZPR KC, V/163, mkf. 2989, AAN, Warsaw; *Dalekopis 558 dla I sekretarza KW*

After Zamyatin returned to Moscow, at the politburo session on January 22, he recommended just to continue putting pressure on the Polish side to take firmer steps in suppression of the opposition. Such an approach was approved by almost the entire Soviet leadership. They did so despite a report by Supreme Commander Kulikov confirming that the December events had not caused any real turn in Poland. For this purpose, Defense Minister Dmitri Ustinov suggested enlarging the extent of the Warsaw Pact's "Soyuz-81" maneuvers scheduled in March. In this manner, Moscow planned to increase an impression that the alliance troops were ready to intervene.[1458] The KGB and ministry of foreign affairs were tasked with working out a specific draft of procedure and then submit it for discussion to the Suslov commission for supervision of the events in the PPR.[1459]

Regarding an action against Solidarity, the Polish as well as top Soviet officials faced the question of how much the Polish security forces remained loyal to the communist regime.[1460] In mid-January, Marshal Kulikov visited two divisions of the PPA. The command cadres left a good impression on him. At the same time, the talks with Stanisław Kania

---

PZPR – Propozycje aktualnych haseł politycznych, undated, early 1981, XII 3120, 945/11, ibid.

[1458] In fact, the Warsaw Pact troops' activity, including a series of small, limited maneuvers, continued around the PPR from the official end of the "Soyuz-80" drills until the beginning of March 1981. Ouimet, *The Rise and Fall of the Brezhnev Doctrine,* 180.

[1459] Transcript of CPSU CC Politburo Meeting, January 22, 1981, in *From Solidarity to Martial Law* (doc. 29): 184.

[1460] At the politburo session, Gromyko pointed to emerging cases when some Polish voivodeship military districts refused to follow the party's orders and clean out the trade union buildings seized by Solidarity. Zamyatin opposed that these were two isolated excesses. Transcript of CPSU CC Politburo Meeting, January 22, 1981, in *From Solidarity to Martial Law* (doc. 29): 184. A general deterioration of the situation within the Polish armed forces in the early months of 1981 was also documented by the Polish politburo's concerns that the opposition would try neutralize the army by appealing to "national feelings" and instigate recruits sympathizing with Solidarity to spread the movement's ideas among professional soldiers. Supplement No. 1 to PUWP CC Politburo Protocol No. 657 Analyzing the Intentions of Solidarity, January 26, 1981, ibid (doc. 30): 188.

reassured the supreme commander in his view that the party leadership, in absolute contradiction to his idea, preferred long-term political measures instead of the harsh use of violence. The Marshal actually believed that the Polish army was ready to follow an order to suppress the opposition.[1461]

The primary goal of Kulikov's visit in Poland was to sign a *Protocol on buildup of the PPA* in the years 1981-1985. The Polish side raised many objections to the document.[1462] A military budget, in fact, already affected the catastrophic condition of the Polish economy in a very negative way. In 1980, defense spending exceeded the planned amount by more than 1.2 billion zlotys. The prospects for 1981 envisaged the defense ministry's budget to rise by another 7.4%.[1463] Therefore, Kulikov was concerned that in this situation the Polish armed forces would be able to meet the Unified Command's directives for the years 1981-1985, just up to 50%.[1464] Thus, the economic aspects of the Polish Crisis increasingly influenced practical military cooperation within the Warsaw Treaty Organization.

During his visit in the PPR, the supreme commander also asked Kania for consent to enter the Unified Armed Forces into Polish territory during the "Soyuz-81" maneuvers. The PUWP CC First Secretary did not object to the exercise itself. However, he was reluctant to give a blessing to the arrival of a large number of foreign troops to the country.[1465] Similar

---

[1461] Letter from East German Defense Minister Heinz Hoffmann to Erich Honecker regarding Conversation with Marshal Kulikov, January 19, 1981, in *A Cardboard Castle?* (doc. 90a): 444.

[1462] Ibid.

[1463] *Przedłożenie rządowe sprawozdanie z wykonania budżetu państwa za okres od 1 stycznia do 31 grudnia 1980 r.,* PZPR KC, XIB/490, AAN, Warsaw; *Informacja o projekcie budżetu panstva na 1981 rok,* 23. 12. 1980, V/161, mkf. 2987, ibid.

[1464] Bericht des IM "Birnbaum" über das Zusammentreffen mit dem Mitglied des ZK der Polischen Vereinigsten Arbeiterpartei und Vizeminister für Nationale Verteidigung und Chef des Generalstabes der Polischen Armee Waffengeneral Siwicki, January 27, 1981, Accessed October 16, 2019. http://www.php.isn. ethz.ch/kms2.isn.ethz.ch/serviceengine/Files/PHP/44998/ipublicationdocument_ singledocument / 6f22e930-84b4-4d85-8146-8aea128c2449/de/810127_siwicki. pdf.

[1465] Mastny, *A Cardboard Castle?,* 51.

to December 1980, he initially did not agree with the entrance of the East-German units.[1466] The scheduled maneuvers provided Moscow with an opportunity to establish the Warsaw Pact command post at the Legnica base, where the Soviet troops had been dispatched since the end of World War II. During 1981, the officers of the Staff of the Unified Armed Forces operated there, totally isolated from the potentially undesirable influence of the Polish population and media.[1467] It is possible that the Soviet top military wanted to ensure a command center on the Polish territory for the eventual military intervention.

At the turn of January and February 1981, the Eastern Bloc military officials had not abandoned a possibility of intervention in Poland yet. Along these lines, the Czechoslovak security analytics were given an order to pay strong attention to the presumed reaction of the West to a potential invasion.[1468] However, it is likely that an action on the basis of the whole Warsaw Treaty Organization was not considered at all. In January, even the hardliners, like the East-German defense minister Heinz Hoffmann or the chief of the Polish general staff Florian Siwicki, talked about potential separate "assistance to the Polish army" provided solely by the Soviet, East-German, and Czechoslovak troops. Siwicki again reassured the defense minister of the GDR about the reliability of the PPA, informing him that the Polish general staff had several plans of using the armed forces against the opposition.[1469] It remains unclear how relevant

---

[1466] Letter from East German Defense Minister Heinz Hoffmann to Erich Honecker regarding Conversation with Marshal Kulikov, January 19, 1981, in *A Cardboard Castle?* (doc. 90a): 444.

[1467] Mastny, "The Soviet Non-Invasion of Poland," 196.

[1468] In February, the Czechoslovak military representatives still considered the most essential question to be whether the military intervention by the USSR or the other Warsaw Pact member-states in Poland would or would not take place. Informacja o wpływie wydarzeń polskich na sytuację w CSR przesłana gen. Jánowi Kováčowi przes gen. Vladimíra Stárka, February 12, 1981, in *Przed i po 13 grudnia*, tom I (doc. 138): 360.

[1469] Bericht des IM "Birnbaum" über das Zusammentreffen mit dem Mitglied des ZK der Polischen Vereinigsten Arbeiterpartei und Vizeminister für Nationale Verteidigung und Chef des Generalstabes der Polischen Armee Waffengeneral Siwicki, January 27, 1981, Accessed October 16, 2019. http://www.php.isn. ethz.ch/kms2.isn.ethz.ch/serviceengine/Files/PHP/44998/ipublicationdocument_

Siwicki's claims were. They suggest, however, that not only the Polish political leadership, but also a radically oriented part of the generality preferred to solve the crisis on the domestic basis.

U.S. intelligence saw the main reason of unwillingness to use the Polish army against the opposition in the Kania leadership's fears that an eventual failure of such an operation would have significantly increased the risk of an outside invasion. In this case, the *National Intelligence Estimate* deemed it more likely an individual action by the Soviet troops[1470] than an intervention on the multilateral basis or even an invasion of the entire Warsaw Treaty Organization. Involvement of the Romanian troops seemed extremely unlikely, almost ruled out by the West. At the same time, the Americans correctly assumed that the Kremlin hesitated with the military intervention in Poland because of three reasons: the USSR was afraid of the political consequences of relations with the West, a potential clash with the Polish army, and weakening of its position in the world communist movement. But the U.S. envisaged that Moscow would not allow its strategic interests to be jeopardized. These also included the Polish leadership's willingness and ability to guarantee fulfilling the commitments in the Warsaw Pact's system.[1471] After all, in his conversation with Kania and Jaruzelski at the 26th CPSU Congress, Brezhnev actually appealed for better securing of those communications in the PPR that were crucial for the Warsaw Treaty Organization.[1472]

Concerns about further deterioration of relations with the West played an important role in consideration of the Eastern Bloc officials about the military intervention in Poland. Soon before becoming a prime

---

singledocument / 6f22e930-84b4-4d85-8146-8aea128c2449 / de/810127_siwicki. pdf.

[1470] The U.S. intelligence admitted that also the Soviet divisions located in the GRD and CSSR would have joined the operation.

[1471] National Intelligence Estimate (NIE 12.6-81), "Poland's Prospects over the Next Six Months", January 27, 1981, in *From Solidarity to Martial Law* (doc. 31): 193.

[1472] Informacja przekazana Erichowi Honeckerowi o rozmowie Leonida Breżniewa ze Stanisławem Kanią i Wojtiechem Jaruzelskim przeprowadzonej 4 marca 1981 r., w trakcie XXVI Zjazdu KPZR, before March 9, 1981, in *Przed i po 13 grudnia*, tom I (doc. 157): 408.

minister, Jaruzelski speculated that escalation of the crisis, especially its transition to the key sphere of alliance between Warsaw and Moscow, could have been an intention of the new U.S. administration of Ronald Reagan.[1473] The PUWP leadership even suspected of the Polish emigration and the West that they wanted to escalate the situation to the point when the USSR would be forced into military action. Such a move would have definitely terminated détente and created room for uprisings in other Soviet satellites following the disintegration of the Eastern Bloc.[1474] However, William Dyess, an American diplomat, told the Polish representatives that unlike the outside intervention, Washington would have seen the suppression of unrests by domestic forces as an internal affair of Poland.[1475]

An unsuccessful attempt to impose martial law during "Soyuz-81" maneuvers

In February, some alteration of the Soviet tactic so far was caused by the appointment of Wojciech Jaruzelski to the post of prime minister.[1476] Along with the Kremlin's expectations, the new head of government substantially intensified preparations for the imposition of martial law by domestic authorities. Just one month later, during his visit to the 26th CPSU Congress, Jaruzelski informed Moscow about the finalization of plans to suppress Solidarity. The Soviet leadership welcomed this Polish activity. At the same time, it wanted to support the success of a strike against the opposition as much as possible. Therefore, Jaruzelski was asked to take his action during the upcoming "Soyuz-81" maneuvers of the Warsaw Pact. For this purpose, even the alliance's

---

[1473] *Protokół Nr. 70 z posiedzenia Biura Politycznego KC PZPR w dniu 7 lutego 1981r.,* PZPR KC, V/164, mkf. 2990, AAN, Warsaw.

[1474] Informacja o działalnosti organizacji polskich emigrantów záchodních organizacji rewanżystowskich o ziomkostw, partii politycznych oraz innych wrogich ośrodków skierowanej przeciwko PRL, February 13, 1981, in *Przed i po 13 grudnia*, tom I (doc. 139): 364.

[1475] Mastny, "The Soviet Non-Invasion of Poland," 197.

[1476] Jaruzelski became the Prime Minster on February 10, 1981, with the significant help of Moscow. He continued to hold the position of defense minister. Paczkowski, *From Solidarity*, xxxvi.

communication network on the territory of the PPR was provided to the Polish armed forces. A group of Soviet officers, led by Afanasy Scheglov, a representative of the supreme commander in the PPA,[1477] traveled across the country and checked the Polish military commanders' loyalty. Based on their reports, Kulikov was reassured in his opinion that the Polish armed and security forces were ready to follow any orders by the party bodies and state authorities.[1478]

According to the supreme commander's order, the "Soyuz-81" exercise should have taken place between March 17-25, 1981, in the territories of the PPR and its neighboring states.[1479] Despite the Soviet leadership's intention being to improve conditions for the imposition of martial law by the Polish authorities, the course of maneuvers itself did not go beyond the typical scope of the Unified Armed Forces' drills at the time.[1480] The Soviet officers serving at the aforementioned Legnica base were even temporarily joined by the military officials of the GDR, CSSR,

---

[1477] We are reminded that the representatives of the supreme commander had extensive power over the Soviet satellites' military units assigned to the Unified Armed Forces. After 1969, Moscow controlled the situation within the Warsaw Pact countries' armies through these officers.

[1478] Mastny, "The Soviet Non-Invasion of Poland," 198.

[1479] The Warsaw Pact's maneuvers were preceded by a series of bilateral drills, including the CSPA units' entrance into the territory of the PPR based on a bilateral agreement between the Czechoslovak and Polish defense ministries on the mutual use of military proving grounds. Afterwards, the troops drilled in the Wendrzyn area on February 2-19. Also, entrance of the Polish units into the Czechoslovak territory was expected during the maneuvers, specifically into the Dobrá voda area. In this regard, the Czechoslovak side paid great attention to the political indoctrination of sent troops as well as to surveillance of the Polish units operating in the CSSR. *Účast ČSLA na cvičení "Sojuz-81" a výměna jednotek s PLA ve VVP*, 2. 1. 1981, FMO GŠ/OS, k. 48, p. č. 180, č. j. 37032, VÚA, Prague.

[1480] "Soyuz-81" envisaged practice of a three-stage operation typical of the Warsaw Treaty Organization: In the first, a combat readiness was ordered. Three days later, the enemy's attack started the second phase, with objectives to repulse the aggression and regroup the alliance's formation to launch a counterattack. The offensive operation took place in the third phase, when the Warsaw Pact was supposed to resort to fights with nuclear weapons. *Operačně-strategické VŠC "SOJUZ-81" – informace o instruktáži*, 9. 3. 1981, ibid.

and PPR. Their active participation in analysis of the exercise[1481] proves that "Soyuz-81" was not primarily a preparation for the invasion of Poland, as the Western media speculated.[1482] Activity of 18 of the Warsaw Pact's divisions in the region would have only maximized the sense of insecurity, thus making action against the opposition easier for the Polish leadership.[1483] The not-fully-standard purpose of the exercise was also proven also by increased attention paid to it by the Eastern Bloc countries' media. This corresponded to the drills' intimidating role. The East-German defense minister Hoffmann also informed the SED politburo in detail about the maneuvers, which definitely was not a common praxis.[1484]

During meetings in March, just ahead of the "Soyuz-81" maneuvers, the Eastern Bloc political representatives *de facto* ruled out the intervention. At the 26th CPSU Congress, Brezhnev personally instructed Honecker to put pressure on Kania continually, but in a sense that the Polish leadership has to restore order in the country on its own.[1485] The PUWP CC First Secretary heard the appeals for firm defense of the regime by its own forces also during his visit in Hungary. Janos Kadar urged the Polish side to reach for repressions if political means failed to solve the crisis quickly. He expressed his hope that the PUWP was still strong

---

[1481] *Složení skupin a zabezpečení příslušníků MNO zařazených na cvičení "SOJUZ-81"*, undated; *Rozbor velitelsko-štábního cvičení "Sojuz-81"*, 8. 4. 1981, ibid.

[1482] Also, Antonín Kříž considers the "Soyuz-81" and "Druzhba-81" drills as a rehearsal of an invasion of Poland. His conclusions, however, miss the political development and incorrectly note that the PPA did not take part in the maneuvers. Kříž, Antonín, *Krkonoše 80. Cvičení nebo...?* (s.n., after 1995), 47.

[1483] Despite massive participation of troops, the Eastern press only emphasized the command-staff scope of the exercise and pointed out the much larger NATO maneuvers "Autumn Forge-80" and "Cold Winter-81." "Komentátor TASS o cvičení SOJUZ 81: Zběsilost západní propagandy," *Rudé právo*, March 30, 1981.
[1484] *Politbüro des ZK Reinschriftenprotokoll nr. 12*, 24. 3. 1981, DY 30/ J IV 2/2/1886, BArch, Berlin.

[1485] Transcript of CPSU CC Politburo Meeting, March 12, 1981, in *From Solidarity to Martial Law* (doc. 35): 221.

enough to take such an action.[1486] At the same time, the HSWP CC First Secretary repeated his opinion that the limit for leaving initiative on the Polish leadership was the country's belonging to the Warsaw Treaty Organization. A membership in the alliance, frequently stressed by the Eastern Bloc officials regarding the Polish Crisis, in reality meant to stay within the Soviet sphere of influence. Kadar reminded them that in 1956, the Hungarian leadership must have asked Moscow for help restoring order in the country for similar reasons.[1487] However, he emphasized a need to avoid recurrence of this scenario, as the Polish opposition as well as some groups in NATO countries would have welcomed the invasion.[1488]

Nevertheless, the "Soyuz-81" drills provoked a tumultuous reaction from the West. The U.S. officially asked Moscow to reveal details about the Warsaw Pact maneuvers.[1489] Reagan even accused the Kremlin of violation of the CSCE Final Act, as more than 25 thousand troops took part in the exercise, according to American estimation. The new U.S. president put the crisis in Poland on the same level as the conflict in Afghanistan or the Cuban engagement in Africa - he called it one of the reasons why there was no chance to think about a restoration of the USA-USSR dialogue.[1490]

Regarding the maneuvers, the Western press also emphasized the propagandistic dimension of the exercise. In the moment when the situation in the PPR partially stabilized, the drills should have publicly demonstrated an alliance between Warsaw and Moscow. But Kulikov refused all the Western objections, noting that the action had been

---

[1486] Sprawozdanie z wizyty Stanisława Kani w Budapeszcie opracowane dla Komitetu Politycznego KC WSPR, March 20, 1981, in *Przed i po 13 grudnia*, tom I (doc. 161): 417.

[1487] Kadar, of course, described the Soviet invasion of Hungary in an extremely simplified and ideologically biased way.

[1488] The HSWP CC First Secretary noted to statement by Joseph Luns, NATO general secretary, who had expressed his belief in September 1980 that the Soviet intervention in Poland would be inevitable. *Informacja z rozmów tow. Stanisława Kani z tow. Janosem Kadarem, Budapeszt, 19. III. 1981 r.*, PZPR KC, V/165, mkf. 2991, AAN, Warsaw.

[1489] *USA požaduje na SSSR zveřejnění podrobností o manévrech Varšavské smlouvy*, 16. 3. 1981, 506/0/45, k. 5968, NA, Prague.

[1490] "Mr. Reagan chips Russians," *The Times*, March 16, 1981.

scheduled already before the outbreak of the Polish Crisis.[1491] On the contrary, in front of the Warsaw Pact armies' officials, the supreme commander used the exercise to accuse the West of "fueling counterrevolution" in the PPR, destroying the socialist countries' unity, and several other provocations during the maneuvers. In this regard, he quoted Brezhnev's words delivered at the 26th CPSU Congress: "...*Polish communists, Polish working class and people at work in that country can firmly rely on their friends and allies; we shall not abandon socialist Poland, fraternal Poland...*" The Marshal, who had supported tough approaches in suppressing the opposition, even in this situation still acknowledged the responsibility for solution lay primarily with the PUWP and Polish government.[1492]

Overall, the behavior of the supreme commander Kulikov during the years 1980-1981 proves that he did not hesitate to support an outside military intervention if the Polish leadership was not able to handle the crisis by its own means.[1493] Regarding the Warsaw Treaty Organization, it is important to note that the Unified Command most likely was not unanimous on this matter. Kulikov's language of an orthodox communist, for example, differed from the relatively pragmatic acting of Anatoly Gribkov, the chief of Staff of the Unified Armed Forces. The latter correctly realized the situation in Poland varied significantly from the 1968 events. In front of his superiors, the General allegedly noted that the reforms in Czechoslovakia had been started by the top power echelons, while Poland experienced a revolt of people who had lost their belief in the party. Furthermore, he expected that the Polish armed forces were so

---

[1491] "Soviet-Polish Maneuvers Are Reported Beginning," *International Herald Tribune*, March 2, 1981; "Rozhovor s maršalom Sovietskeho zväzu Viktorom Kulikovom," *Pravda* (KSS), May 13, 1981.

[1492] *Rozbor velitelsko-štábního cvičení "Sojuz-81"*, 8. 4. 1981, FMO GŠ/OS, kr. 48, p. č. 180, č. j. 37032, VÚA, Prague.

[1493] In conversations with his military colleagues, Kulikov allegedly often stressed that there would be no restoration of order in Poland without foreign military action. Quoted in Ouimet, *The Rise and Fall of the Brezhnev Doctrine*, 200. The source of this information, however, was Anatoly Gribkov, the chief of Staff of the Unified Armed Forces. He would exaggerate his testimony after a change of the political situation in the 1990s, in order to emphasize a different position from which he had held during the Polish Crisis.

influenced by nationalism that they would not be willing to "shoot into their own people."[1494]

In the end, the planned imposition of martial law during the "Soyuz-81" drills failed as the situation in the PPR deteriorated rapidly after the so-called Bydgoszcz incident.[1495] Therefore, the Polish leadership just asked Moscow for a prolongation of the proceeding maneuvers. Kania and Jaruzelski thought that presence of the allied troops would prevent the opposition from other actions.[1496] Thus, not only the Kremlin, but also the Polish representatives tried to raise an impression that the Warsaw Treaty Organization was ready to intervene, if needed. Such an approach corresponded to Brezhnev's speech delivered at the 26th CPSU Congress in March, when the Soviet leader referenced the December 1980 meeting in Moscow as proof that Poland could fully rely on its Warsaw Pact allies.[1497] In fact, a factor of the presence of the alliance in the country should have provided the Polish leadership with psychological and political support and threatened the opposition at the same time.

The Western press immediately linked prolongation of the maneuvers to a deteriorating situation in Poland, reminding its readers that a similar scenario had preceded the 1968 invasion of Czechoslovakia.[1498] Western diplomats and observers paid special attention to whether a military framework for intervention, allegedly created by the Warsaw Pact drills, would be preserved. However, the Polish leadership stressed that

---

[1494] Quoted in Gaddis, John Lewis, *Studená válka* (Praha: Slovart, 2006), 195.

[1495] On March 16, 1981, the Polish opposition seized the Unified People's Party secretariat. Very brutal actions from the security forces followed, to which Solidarity reacted by calling for a nationwide strike. Paczkowski, *From Solidarity*, 19-20.

[1496] Mastny, "The Soviet Non-Invasion of Poland," 198.

[1497] *Sprawy polskie w obradach XXVI Zjazdu KPZR*, undated, March 1981, PZPR KC, XIA/1265, AAN, Warsaw. In the GDR, the "Soyuz-81" maneuvers were presented as a contribution to the peace initiatives of the 26th CPSU Congress and practicing of defense against "all the enemies of socialism." Such rhetoric could suggest that the drills were also a test of action against the Polish opposition. *Claris nr. 1105/210 z Berlina*, 7. 4. 1981, XIA/1278, ibid.

[1498] "Warsaw Pact maneuvers extended," *The Times*, April 1, 1981; "Troop maneuvers recall Czechoslovakia in 1968," *International Herald Tribune*, April 6, 1981.

the invasion had never been considered as a possible solution of the crisis. Nonetheless, in private talks with Western diplomats, this claim did not seem to be very persuasive. The press also warned that the incursion into Poland would have probably resulted in an outbreak of war.[1499] Despite these stories, Washington was given relatively accurate information from an unnamed Polish diplomat in West Germany about Jaruzelski's intention to suppress the opposition solely by the Polish armed and security forces. The rest of the Eastern Bloc were only supposed to give their political blessing to this action.[1500]

     Under the impetus of Jaruzelski and Kania, Moscow extended the "Soyuz-81" maneuvers to the second decade of April. The matter was not decided by the alliance Unified Command but by the Soviet defense minister Ustinov. Although the Polish leadership wanted to use the presence of the troops to put pressure on the opposition, Kania and Jaruzelski acted very carefully and never talked about the outside intervention frankly. A part of this psychological game was also participation of the Czechoslovak and East-German defense ministers at the following Soviet-Polish military exercise "Druzhba-81." At Polish request, Kulikov and the other representatives of the Unified Armed Forces remained in the country.[1501] The Supreme Commander used this situation and made his presence conditional on having the right to directly command the Polish units and their more active involvement in the following drills.[1502] Thus, the Soviet military activity in Poland continued also after the end of the "Soyuz-81" exercise. It included mostly numerous flights of supply planes to the Soviet military bases in the country.[1503]

[1499] "As Pact Maneuvers End, Poles Speculate," *International Herald Tribune*, April 9, 1981.

[1500] Mastny, "The Soviet Non-Invasion of Poland," 198-199.

[1501] East German Report of discussion with Marshal Viktor Kulikov, April 7, 1981, in *From Solidarity to Martial Law* (doc. 41): 246.

[1502] East German Report of the Military Council Meeting in Sofia, April 24, 1981, in *A Cardboard Castle?* (doc. 90b): 445.

[1503] The Soviet approach was facilitated by integration of military systems within the Warsaw Pact proceeding since the 1960s. The alliance's unified anti-aircraft system made it almost impossible for individual member-states to control their airspace independently. Unlike 1956, when the Polish military largely had a clear picture of the Soviet troops' moves and had informed its

Vojtech Mastny considers this as proof that Moscow still had not abandoned the idea of intervention altogether.[1504]

Presence of the Warsaw Pact armies in the country failed to threaten the Polish opposition. On March 27, Solidarity paralyzed Poland by a 24-hour warning strike and even threatened with a four-day general strike. Moscow responded by sending the KGB experts to Warsaw, who learned about the Polish plans of imposition of martial law. Afterwards, they urged some changes in the sense of a substantial tightening of repressions. However, despite these Soviet instructions, the Polish defense ministry let this issue peter out.[1505] On the contrary, the PUWP leadership tried to achieve some compromise with the opposition in this moment.[1506] It was becoming clear that Moscow meant its claims about a domestic solution of the crisis. In spite of the deteriorating situation in the PPR, the outside military intervention was not coming.

Helplessness of the Kremlin

Such an approach of the Polish officials did not meet with understanding by the Kremlin. On April 2, at the Soviet politburo session, Brezhnev complained about Kania, who formally agreed with the Soviet assessment of the situation and proposed strategy, but absolutely failed to follow them in praxis. However, Moscow chose a bilateral tactic, instead of bigger involvement of the entire Warsaw Treaty Organization. Another meeting of the alliance member-states' party leaders was seen as an

---

political leadership about it, in 1981 the Polish side knew only about increased activity of the Soviet air forces over its territory without an opportunity to get any other details. Wanner, *Brežněv*, 68-69.

[1504] Mastny, "The Soviet Non-Invasion of Poland," 200.

[1505] Ibid, 199.

[1506] At the same time, the top Polish officials also informed Oskar Fischer, the East-German minister of foreign affairs, during his visit to Warsaw, about the intention to solve the crisis primarily by political means, however they vaguely admitted "other variants of defense of socialism" too. *Návštěva ministra zahraničních věcí NDR O. Fischera v PLR*, 24. 3. 1981, DTO 1945-1989, i. č. 25, e. č. 5, č. j. 012.179/81, AMZV, Prague.

emergency solution.[1507] The politburo tasked the KGB chairman Andropov and defense minister Ustinov to deal with the existing situation personally, at a secret meeting with Kania and Jaruzelski.[1508] This aimed at escalation of political pressure on the Polish representatives.[1509] In this moment, a certain helplessness of the Soviet leadership was becoming apparent: during March, the Kremlin dispatched a network of its confidents in Poland, consisting of members of the State Planning Commission, the KGB, and the General Staff.[1510] In addition, Marshal Kulikov was present in the country since the beginning of the "Soyuz-81" drills and held important political talks under Ustinov's authorization. He even met Kania and Jaruzelski without witnesses and conveyed to them instructions from Moscow.[1511] However, Mark Kramer's claim that the supreme commander of the Warsaw Pact acted as an informal envoy of the Soviet politburo in the country[1512] is not accurate. Despite Kulikov presenting himself in this way to the Polish side, there was no direct informational exchange between the Marshal and the politburo. It was conveyed through Dmitri Ustinov. This fact, as explained later, resulted in some serious misunderstandings or even intentional shifts in the meaning of conveyed messages at crucial moments.

Moreover, during the meeting of the Ustinov-Andropov duo with Kania and Jaruzelski in the night from April 4 to 5, held in a train nearby Brest, the Polish and Soviet sides' different views on the issue of reliability

---

[1507] This opinion was presented by Brezhnev himself. Within the politburo, the main supporter of holding an extraordinary meeting of the Warsaw Pact was Dmitri Ustinov.

[1508] On April 9, the Soviet politburo even decided that the Warsaw Pact countries' leaders will not be informed about the meeting.

[1509] Transcript of CPSU CC Politburo Meeting, April 2, 1981, in *From Solidarity to Martial Law* (doc. 39): 239.

[1510] Kulikov told the East-German military officials that these persons conveyed Brezhnev's direct instructions to trustworthy Polish officials. In the context of the events, this is not very convincing. Moreover, they basically represented various institutions. Such a multiplicity of communication lines between Moscow and Warsaw rather suggests some splitting in the Soviet leadership.

[1511] East German Report of discussion with Marshal Viktor Kulikov, April 7, 1981, in *From Solidarity to Martial Law* (doc. 41): 246.

[1512] Kramer, "Colonel Kuklinski and Polish Crisis," 52.

of the PPA fully showed. The Polish representatives stated that the intervention by foreign troops was not acceptable, but at the same time, it was also impossible to impose martial law by domestic means, given the existing situation.[1513] Already after the 9th plenum of PUWP CC, held in March, Kania had declared that the Polish armed forces were able to follow orders with serious troubles only. But Kulikov still advocated the claim that the Polish army was reliable and there was no reason to fear using it against the "counterrevolution." Nevertheless, even those officers who had assured the supreme commander of the Warsaw Pact about the Polish army's loyalty in the past now stated that no more than 50-60% of the armed forces remained trustworthy. They suggested to the Marshal that in case of the allies' invasion, a rebellion of a large number of Polish units was likely.[1514] Even for this reason, Jaruzelski asserted at the domestic scene a thesis that the only possible solution of the existing social conflicts was a dialogue.[1515]

On the contrary, at the beginning of April, Kulikov recommended to the Polish representatives the following strategy: The government should declare a state of emergency according to Article 33 of the constitution. He did not rule out the possibility of military invasion in case of significant complications. The Marshal expected much more positive political impacts if the state and party leadership asked for the intervention, rather than if the action was launched solely by the Kremlin's decision.[1516]

[1513] Transcript of CPSU CC Politburo Meeting, April 9, 1981, in *From Solidarity to Martial Law* (doc. 43): 259.

[1514] Report regarding a confidential discussion with the Supreme Commander of the Combined Military Forces of the Warsaw Pact countries on 7 April 1981 in LEGNICA (PR Poland) following the evaluation meeting of the Joint Operative-Strategic Command Staff Exercise "SOYUZ-81," Cold War International History Project Bulletin, no, 11 (1998): 120.

[1515] *Notatka dotycząca nadesłanych do Wydziału Organizacyjenego i Biura Listów i Inspekcji KC PZPR uchwał, rezoluciji, apeli, listów, otwartych i indywidualńych organizacji i członków partii bezpośrednio przed i po IX Plenum KC PZPR*, 7. 4. 1981, PZPR KC, XII 3120, 945/11, AAN, Warsaw.

[1516] Report regarding a confidential discussion with the Supreme Commander of the Combined Military Forces of the Warsaw Pact countries on 7 April 1981 in LEGNICA (PR Poland) following the evaluation meeting of the Joint Operative-

The supreme commander most likely acted on his own desire, or just conveyed informal stances of individual persons in the Soviet leadership. In fact, Brezhnev actually obviously opposed the invasion at the time. During his visit to the 26th CPCz Congress, he noted that NATO core members had clearly showed that the military invasion of Poland would have not only definitely ended détente, but also resulted in harsh Western sanctions. Regarding these threats, the Soviet leader complained that the West, in fact, wanted the USSR to abandon some key elements of its security policy, including a provision of "fraternal assistance" to its satellites.[1517] Moreover, despite the propagandistic thesis, the Soviet politburo in reality did not admit that the unrests in Poland imminently jeopardized the country's membership in the Warsaw Treaty Organization. They thought that geographical location of the PPR did not allow the opposition to challenge membership in the alliance for two reasons: persisting concerns about the German threat and fears of outside intervention. Therefore, Moscow did not appreciate the activities of the PUWP left wing,[1518] which saw no way out of the crisis without the foreign troops' intervention. The Kremlin considered such a strategy counterproductive as it did not result in isolation of hardliners only, but also the "healthy forces" in the party and Polish society.[1519]

Unwillingness of key persons in the Soviet leadership to involve the Warsaw Treaty Organization in the solution of the crisis is documented by the fact that the official agenda of the 23rd meeting of the Military Council of the Unified Armed Forces, held on April 21-23 in Sofia, did

---

Strategic Command Staff Exercise "SOYUZ-81," Cold War International History Project Bulletin, no, 11 (1998): 120.

[1517] *Praga – Szyfrogram Nr 303/II*, 9. 4. 1981, PZPR KC, XIA/1278, AAN, Warsaw.

[1518] The PUWP left wing represented mostly Tadeusz Grabski, Andrzej Żabiński, and Stefan Olszowski, the politburo members, or Stanisław Kociołek, the head of party organization in Warsaw. A strong supporter of the Warsaw Pact's invasion was also Stanisław Ciosek, the Central Committee secretary. Kamiński, *Przed i po 13 grudnia*, XXXI.

[1519] Extract from Protocol No. 7 of CPSU CC Politburo Meeting, April 23, 1981, in *From Solidarity to Martial Law* (doc. 45): 267.

not reflect the situation in Poland at all.[1520] Instead, on Erich Honecker' initiative, a secret meeting of the party leaders of the USSR, GDR, and CSSR took place in Moscow on May 16.[1521] The Polish question was the only item to be discussed. Later, the Kremlin informed Janos Kadar, Todor Zhivkov, and Fidel Castro about the results. Nicolae Ceausescu and especially the Polish leadership were not acquainted about the session.[1522] Such an approach actually resembled a scheme of negotiations which had preceded the invasion of Czechoslovakia.

Honecker failed to get the Warsaw Treaty Organization more involved in a solution of the crisis. Despite formally dismissing outside military intervention, the East-German leader recommended installing these persons in the Polish leaderships who would not hesitate to take tough action against the opposition. Preparation for the imposition of martial law should have proceeded in cooperation with a special group established within the Warsaw Pact. Gustáv Husák tried to move negotiations even more to the level of the entire alliance. He suggested holding another meeting of the Warsaw Pact's officials, even though not necessarily the Political Consultative Committee, similar to December 1980. However, Moscow most likely already abandoned a possibility to solve the crisis on the multilateral basis at the time. Therefore, Brezhnev did not react to Husák's initiative, although he had recommended the same

---

[1520] *Výsledky 23. zasedání vojenské rady Spojených ozbrojených sil*, 23. 4. 1981, FMO GŠ/OS, kr. 49, p. č. 182, e. č. 37501, VÚA, Prague. In informal talks, Kulikov just informed the alliance member-states' military representatives about the background of a Polish request for prolongation of activity of the command post in Legnica. East German Report of the Military Council Meeting in Sofia, April 24, 1981, in *A Cardboard Castle?* (doc. 90b): 445.

[1521] The majority of the Soviet politburo attended the meeting as well. All of them received Brezhnev's instruction to deny holding such talks in case the West informed about it publicly.

[1522] On May 12, the question of the Polish side's information about opinions of the Eastern Bloc countries' leadership on development in Poland was raised at the Polish politburo meeting by Andrej Żabiński. He accused some top Polish officials of keeping reports in secret from the lower party collective. Kania denied this, claiming that the Central Committee secretaries had the same information as the politburo. *Protokół Nr. 92 z posiedzenia Biura Politycznego KC PZPR 12 maja 1981r.*, PZPR KC, V/165, mkf. 2991, AAN, Warsaw.

tactic to the politburo previously. The Soviet leader only vaguely informed that Kulikov had worked out various plans to suppress the Polish opposition.[1523]

The Moscow meeting brought no important result. Nevertheless, attention should be paid to Honecker's speech because of another reason: The SED CC General Secretary absurdly stated that military intervention in the Warsaw Pact member-states was possible by virtue of the alliance statutes. In fact, this absolutely groundless opinion was not unique among the orthodox communists. Just a few weeks before the outbreak of the Polish crisis, it was also presented at a military-theoretical conference, held in Moscow on the 25th alliance anniversary[1524] by General Jaroslav Klícha, the deputy chief of the CSPA's Main Political Administration. Referring to the events of 1968, he declared that the Warsaw Pact remained even in the future ready to use its armed forces "*against internal reactionists striving for restoration of capitalism*".[1525] Also, then analytics presented the opinion that indivisibility of internal and external security was a key principle for the Warsaw Pact countries as the West allegedly supported inside enemies and in this manner created a "counterrevolution atmosphere" in the alliance member-states.[1526] Similar thoughts were undoubtedly fueled, among other things, by some radical statements, for

---

[1523] Memorandum of Meeting between Leonid Brezhnev, Erich Honecker, Gustáv Husák et al, in Moscow, May 16, 1981, in *From Solidarity to Martial Law* (doc. 49): 280.

[1524] The conference was held in the scope of a regular session of the Military Council.

[1525] *Vystoupení náčelníka ideologické správy – zástupce náčelníka HPS ČSLA genmjr. J. Klíchy na vojensko-teoretické konferenci k 25. výročí podepsání Smlouvy o přátelství, spoluráci a vzájemné pomoci ve Varšavě konané v Moskvě 22.5. 1980,* FMO GŠ/OS, k. 47, p. č. 169, e. č. 37007, VÚA, Prague. There is the following passage in the draft of Klícha's speech: "*The principle of socialist internationalism consists also of a need to provide an assistance to progressive revolutionary forces of social and national liberation by adequate means, including military-ones, if results of national-liberation revolution are in jeopardy. Such a mission of the Warsaw Pact member-states' armed forces is based on the essence of socialism.*" However, it is not sure whether Klícha really delivered these words, as they are additionally crossed in the document.

[1526] Rubin, F., "The Theory and Concept of National Security in the Warsaw Pact Countries," *International Affairs,* no. 4 (1982): 650-651.

example Kulikov's warning published in the Western press that "the union of fraternal armies" was determined to protect socialist gains,[1527] or Honecker's information to Zenkō Suzuki, the Japanese prime minister, that the Warsaw Treaty Organization would not tolerate threats to socialism in Eastern Europe posed by the upheavals in the PPR.[1528]

Rise of anti-Soviet incidents in Poland

During spring 1981, the Polish party and state authorities noted a significant growth of incidents with an anti-Soviet subtext. On May 19, at the Polish politburo meeting, Prime Minister Wojciech Jaruzelski warned that escalation of anti-Soviet propaganda in media controlled by the opposition had reached a dangerous stage. Moreover, provocations towards Soviet soldiers had also occurred. Jaruzelski appealed to the Polish leadership's obligation to ensure their safety.[1529] The essence of the anti-Soviet campaign was mostly stressing some controversial historical events,[1530] blaming the USSR for Polish economic failures, but also referencing the unequal position of Warsaw in relation to Moscow, or questioning whether military spending was adequate to Polish capacities.[1531] This issue directly concerned the Warsaw Pact's existing mechanisms. After all, a negative impact of growing military budgets on the Polish economy was also admitted by Nikolai Baibakov, the Soviet deputy prime minister. But given the deteriorating international situation, he called it necessary.[1532]

[1527] "Pact's army chief says armies back Warsaw," *The Times*, June 22, 1981.

[1528] However, according to Husák's words, the GDR should have participated in the military operation to help Poland in case of an outside threat to the country. "Warsaw Pact 'would act over threat to Poland'," *The Times*, May 28, 1981.

[1529] *Protokół Nr. 93 z posiedzenia Biura Politycznego KC PZPR 19 maja 1981r.,* PZPR KC, V/165, mkf. 2991, AAN, Warsaw.

[1530] For example, the 1920 Poland-Russia war, the Rapallo treaty, the Molotov-Ribentropp Pact and the Katyn massacre.

[1531] *Założenia działalnsci propagandowej w kwestii stosunków polsko-radzieckich,* 22. 5. 1981, PZPR KC, V/166, mkf. 2992, AAN, Warsaw.

[1532] *Rozmowy delegacji rządowych PRL i ZSRR pod przewodnictwem Wicepremierów, Przewodniczących Komisji Planowania Z Madeja i N.K. Bajbakowa,* 1. 10. 1981, XI/471, ibid.

The Soviet ambassador to the PPR, Boris Aristov, raised an official protest to Jaruzelski against growing anti-Soviet feelings in Poland. The undermining of the "friendly atmosphere" allegedly negatively affected even military cooperation within the Warsaw Treaty Organization.[1533] He warned Jaruzelski that Solidarity was no longer just a trade union movement, but a political power which influenced Poland-Soviet Union relations and therefore the Polish leadership had to face it. At the same time, Aristov criticized the generally insufficient propagandistic work of the PUWP in the field of Polish allied ties and foreign policy.[1534] The fact that these fundamental issues were discussed with the Polish leadership by an official representative of the Soviet state, suggests some overestimation of the role played by Marshal Kulikov in negotiations between Moscow and Polish officials,[1535] as well as the generally unsystematic features of Soviet policy.

Large-scale actions against growing acts of anti-Sovietism and anti-communism should have been taken during preparations of the upcoming 9th PUWP extraordinary congress.[1536] In this regard, some callings from the party collective could be heard that new elected delegates

---

[1533] Aristov pointed to alleged cases of assault on the Soviet soldiers. Solidarity, on the opposite, claimed that victims of similar assaults were its members and sympathizers. He also noted that in Legnica voivodeship, the opposition had launched a campaign to return objects used by the Soviet troops and the Polish government had failed to protest.

[1534] *Protokół Nr. 94 z posiedzenia Biura Politycznego KC PZPR 21 maja 1981r.,* PZPR KC, V/165, mkf. 2991, AAN, Warsaw.

[1535] Despite the rise of anti-Soviet feelings in the PPR, Kulikov did not notice any significant problems in its approach towards Soviet soldiers on Polish territory. The exception was a meat supply. This was marginal trouble, as shortages were eliminated by air deliveries from the USSR. But he complained about the spreading opinion in the Polish population that it was time for the USSR to withdraw its troops from the country. At the time, the Warsaw Pact's supreme commander already frankly interfered with internal affairs of the Polish state. In his frequent talks with Jaruzelski, for example, the Marshal made clear he considered it sensible to remove Mieczysław Rakowski, the deputy prime minister. Report on Conversation between Marshal Kulikov and Senior East German Military Officials, June 13, 1981, in *A Cardboard Castle?* (doc. 91): 446.

[1536] *Wnioski z dyskusji w dniu 9.06. br. do wykorzystania przez komisje wnioskow,* 9. 6. 1981, PZPR KC, XII 3119, 945/37, AAN, Warsaw.

must have declared that Poland remain a Warsaw Pact member and would not change the line of its foreign and military policy.[1537] A similar statement was issued by the Polish ministry of foreign affairs for press departments of embassies in Warsaw.[1538] Despite these efforts, at the beginning of June, belief that the USSR was mostly responsible for the current troubles of the country strongly resonated in Polish society.[1539]

Escalation of Moscow's duress

Moscow decided to use the tumultuous period before the 9th PUWP extraordinary congress to escalate political pressure on the Polish side. At the beginning of June, in a phone call, Brezhnev warned the Polish first secretary that he was personally responsible for events in Poland.[1540] Just three days later, the PUWP CC received a harsh letter by its Soviet counterpart which called for mobilization of "healthy forces" for defense of socialism in Poland as well as joint security of the Warsaw Pact countries. Its diction noticeably resembled a letter of so-called Warsaw Fife to Dubček leadership in July 1968.[1541] An outgrowth of speculation in Polish society about the forthcoming Warsaw Pact's invasion followed. The Eastern Bloc countries' intelligence obtained information that some secret consultations were held within NATO on joint strategy in case the Warsaw Treaty Organization invaded Poland. Nevertheless, the top Western politicians mostly considered the Soviet letter as just a demonstrative act.[1542] However, to many supporters of the intervention

---

[1537] *Wojewódzka komisja przedzjazdowa PZPR w Gdańsku – Stanowisko na IX nadzwyczajny zjazd partii*, květen 1981, PZPR KC, XII 44, 957/158, AAN, Warsaw.

[1538] *Informacja bieżąca Nr 113*, 3. 6. 1981, XIA/1278, ibid.

[1539] *List Klubu partyjnej inteligencji twórczej "Warszawa 80"*, 3. 6. 1981, XII 890, 899/138, ibid.

[1540] *Protokół Nr. 94 z posiedzenia Biura Politycznego KC PZPR 2 czerwca 1981r.*, V/166, mkf. 2992, ibid.

[1541] CPSU CC Letter to the PUWP CC, June 5, 1981, in *From Solidarity to Martial Law* (doc. 50): 294.

[1542] Informacja wywiadu CSRS o aktualnej sytuacji w PRL, June 1981, in *Przed i po 13 grudnia*, tom II (doc. 199), eds. Łukasz Kamiński, Iskra Baeva and Petr Blažek (Warszawa: IPN, 2007): 84.

among the Eastern Bloc countries' representatives, the letter meant a welcomed incentive. In a direct reaction to it, the SED CC Politburo, for example, approved the NPA troops' participation in the upcoming alliance maneuvers. So, the East-German leadership still kept thoughts about a possible military solution of the Polish crisis.[1543]

The June 5 letter significantly influenced preparation of the 11th plenum of the PUWP CC. There, based on Moscow's instructions, the Polish leadership would declare its willingness to protect socialism by all available means and marked an alliance with the countries of socialist commonwealth and the Warsaw Pact membership as axioms. At the same time, the Soviet side demanded a statement that the PUWP would do everything to solve the crisis on their own.[1544] Kania's leadership did not object that. However, at the plenum, such a formulation was broadened: The crisis should have been solved independently, but in a peaceful way.[1545] This allowed hardliners, led by the PUWP CC secretaries Tadeusz Grabski and Stefan Olszowski, to attack Kania. But the first secretary kept his position, especially because of support by Jaruzelski and the other army generals on the Central Committee.[1546] This event resulted in substantial strengthening of the Polish army officials' influence on politics.[1547]

Despite the fact that the PPA command structures essentially contributed to Kania's persistence as the head of the party, the Warsaw Pact's supreme commander did not change his opinion on the reliability of the Polish army. In mid-June, during a secret meeting with the East-

---

[1543] *Politbüro des ZK Reinschriftenprotokoll nr. 7*, 9. 6. 1981, DY 30/ J IV 2/2/1895, BArch, Berlin.

[1544] *Notaka w spraše propózicji dokumentów XI Plenum KC PZPR*, undated 1981 PZPR KC, XII 3119, 945/37, AAN, Warsaw.

[1545] *XI plenum komitetu centralnego polskiej zjednoczonej pertii robotniczej*, 9. 6. 1981, XII 890, 899/138, ibid.

[1546] Paczkowski, *From Solidarity*, 25.

[1547] *Vystoupení Mieczyslawa Rakowského na IX. mimořádném sjezdu PSDS dne 15. července 1981*, DTO 1945-1989, i. č. 31, e. č. 53, č. j. 014.504/81, AMZV, Prague.

German military representatives in Dresden,[1548] he recommended deepening cooperation with the Polish armed forces in solving the crisis. For this purpose, the Marshal suggested intensifying personal contacts, exchange of delegations, and joint maneuvers of the Soviet, Polish, East-German, and Czechoslovak troops; at the same time, he also advocated activation of the entire Warsaw Pact. Kulikov actually acted in contradiction to Brezhnev who did not want to involve the alliance in a solution of the Polish question. The Marshal once again tried to act on behalf of Defense Minister Ustinov. This suggests that the Soviet leadership was not absolutely unanimous in its policy towards Poland, even though the situation was not apparent during official politburo meetings.[1549] The supreme commander announced an intention to use the Warsaw Pact drills "Shield-81" to practice crossing Polish borders by units of the North Group of the Soviet Troops in the GDR and the East-German NPA. Advancing divisions were supposed to avoid big cities in order not to raise the suspicions of the civilian population.[1550] Once again, the orders concerning the Unified Armed Forces' maneuvers on the Polish territory were not given by the Unified Command but by the chief of the Soviet general staff, Marshal Ogarkov.[1551] On the other hand, participation of the PPA in defining the territory of the drills is another proof of the Soviet top

---

[1548] The meeting was kept secret from the Polish side. The supreme commander actually complained to the East-German generals about being "watched by the Polish comrades." If Kulikov's claim was true, it meant totally unprecedented practice proving an absolute loss of trust between the Soviet and Polish leadership. By this behavior, the Polish representatives would have demonstrated their fears of the marshal's unclear activities.

[1549] Karel Durman claims, mostly in memoirs, that given the poor health of the Soviet leader at this time, key decisions of the Kremlin were no longer made in Brezhnev's office or at relatively short politburo meetings, but rather in the Hickory chamber, where triumvirate of Ustinov-Andropov-Gromyko regularly met. Durman, *Útěk od praporů*, 201.

[1550] Report on Conversation between Marshal Kulikov and Senior East German Military Officials, June 13, 1981, in *A Cardboard Castle?* (doc. 91): 446.

[1551] However, Kulikov emphasized ensuring a permanent connection between the troops in the GDR and representatives of the Staff of the Unified Armed Forces on Legnica base, including the supreme commander himself.

military's relatively careful approach and effort to coordinate actions with the Polish generality.[1552]

At the turn of May and June, in the countries outside the Eastern Bloc, speculations emerged again that the military intervention in Poland was just a matter of time. In May, the U.S. accused Moscow of a policy of intimidation that limited Poland's ability to solve its troubles without an outside invasion. This stance of Washington was also supported, for example, by members of the Yugoslavian leadership during private talks.[1553] Therefore, at the end of June, the West sent several warning signals to the Kremlin. The Polish diplomats detected some reports that the Western countries would try to use the upcoming stage of the CSCE summit in Madrid as an instrument to warn the USSR against potential intervention in Poland during the unstable period of reparation of the 9th PUWP extraordinary congress.[1554] Also Pierre Trudeau, the Canadian prime minister, during his confidential meeting with the Soviet ambassador to Canada, Alexander Yakovlev, noted that the poor international situation was not result of the Soviet policy in Afghanistan, but also of the crisis in Poland.[1555]

This was mostly a political game once again. In June, the CIA actually stated that even hardliners like Erich Honecker did not envisage the Warsaw Pact's intervention under existing conditions. In case of invasion, the U.S. intelligence expected a wide and bloody conflict. According to their estimation, Moscow would need to use at least 50

[1552] Notatka do akt dla ge. Fritza Streletza, July 29, 1981, in *Przed i po 13 grudnia*, tom II (doc. 229): 175.

[1553] *Biuletyn Ministerstwa Spraw Zagranicznych nr 5 B*, May 1981; *Belgrad – Szyfrogram nr 2258/II*, 29. 5. 1981, PZPR KC, XIA/1278, AAN, Warsaw.

[1554] *Informacja bieżąca Nr 104*, 22. 5. 1981, ibid; There were also some speculation within NATO that an emergency summit of the Warsaw Pact on the highest level could have taken place before the planned PUWP congress. The Western diplomatic sources saw Honecker's as well as Husák's regime as the initiators of such an alliance meeting. Allegedly, the issue of the potential pact's intervention in the PPR should have been discussed there. Collection of materials by the British diplomacy observing the Warsaw Pact's activities in 1981, FCO 28/4389, TNA, London.

[1555] *Ottawa – Szyfrogram nt 2059/II*, 26. 5. 1981, PZPR KC, XIA/1278, AAN, Warsaw.

divisions of both Soviet Army and Warsaw Treaty Organization to handle the operation successfully. The Polish leadership anticipated about 600-800 thousand victims on both sides in case of clashes, while the Soviet analytics allegedly expected even bigger bloodshed.[1556] Obviously, fears of such a scenario existed in the Soviet leadership. Mark Kramer notes that a term "the second Afghanistan" was used by both Andropov and Gromyko during politburo meetings.[1557] Regarding the potential armed conflict in the PPR, the Americans correctly pointed out that on one hand, the Polish army was the second strongest within the Warsaw Pact, but on the other hand, it was dispatched mostly on western borders and its operational plans were also oriented in this direction. That fact significantly limited the possibility of successful resistance against the Soviet or the Warsaw Pact's invasion.[1558]

Growth of speculations about the military intervention in the PPR at the turn of spring and summer was accompanied by an increased effort of Hungarian representatives to probe in unofficial ways via Polish diplomats how the PPA would have reacted if the Warsaw Pact's intervention proved to be necessary in the end.[1559] Actually, at least some officials of the Eastern Bloc states considered such reports on the forthcoming invasion beneficial. In May, for example, the East-German representatives claimed that fear of other socialist countries was the only reason why Solidarity did not speak frankly about seizing power.[1560]

The Czechoslovak intelligence also anticipated intensifying of the Western propagandistic campaign warning about the forthcoming invasion by the Warsaw Treaty Organization in the period ahead of the 9th PUWP Congress. In this regard, they stated that the West used a double-

---

[1556] CIA National Intelligence Daily, "USSR-Poland: Polish Military Attitudes", June 20, 1981, in *From Solidarity to Martial Law* (doc. 54): 310.

[1557] *Wejdą nie wejdą,* 160.

[1558] Memorandum from Roland I. Spiers to the Secretary of state, "Polish Resistance to Soviet Intervention", June 15, 1981, in *From Solidarity to Martial Law* (doc. 52): 304.

[1559] Budapeszt – Szyfrogram nr 2263/II z ambasády PRL do MSZ, May 29, 1981, in *Przed i po 13 grudnia*, tom II (doc. 184): 50.

[1560] *Hodnocení situace v PLR*, 19. 5. 1981, DTO 1945-1989, i. č. 25, e. č. 5, č. j. 013.384/81, AMZV, Prague.

track policy: On one hand, it firmly warned the USSR against the intervention in Poland; while at the same time, it would have welcomed such a move as it would have allowed them to escalate the campaign labeling the Soviet Union as the main violator of world peace. Within NATO, however, different perceptions existed regarding suppression of the Polish opposition by national forces or the Warsaw Pact troops. While the West-European governments allegedly did not consider the use of force by the Polish units an important international issue, a significant shift in the U.S. stance occurred. From this point, Washington made it clear that even this step would have been seen as an "order from Moscow," simply because of the fact that the Polish army remained fully loyal to the Warsaw Treaty Organization.[1561]

Paradoxically, warnings against the weakening loyalty of the PPR to the alliance were a significant element in the Eastern Bloc countries' propaganda. In May, the East-German regime defined continual membership of the country in the Warsaw Pact as one of the axioms in the Polish question. It was not a very successful move. An analysis by the SED politburo had to admit that this aspect was not challenged by the main factions of the Polish opposition.[1562] A more efficient approach was used by Bulgarian propagandists. Besides warnings against the potential withdrawal of Poland from the Warsaw Pact (in a sense of leaving the Soviet sphere of influence), they also noted the real military troubles caused by the crisis. They pointed out the real weakening of abilities of the PPA, the second largest army in the alliance, and the threat to transit on strategic communication connecting the USSR with its troops in the GDR.[1563] Concerns about a weakening of the Polish role in the socialist commonwealth's defense system, in fact the Warsaw Pact, were also

---

[1561] Informacja wywiadu CSRS o sytuacji w PRL po XI Plenum KC PZPR, July 1, 1981, in *Przed i po 13 grudnia*, tom II (doc. 219): 145.

[1562] Sytuacja i perspektiwy – analiza rozwoju wydarzeń w PRL przeznaczona dla członków i zastępców członków Biura Politycznego KC SED, May 1981, ibid (doc. 183): 45.

[1563] Biuletyn Informacji Wewnątrzpartyjnej BPK, nr 3, June 17, 1981, ibid (doc. 210): 114.

expressed by the CPCz CC in a letter to its Polish counterpart.[1564] The Hungarian party and state institutions talked intensively and constantly about the Polish opposition's efforts to weaken the Warsaw Pact's armed forces as well.[1565] The deteriorating situation in Poland, however, caused these initially strictly propagandistic claims to slowly transition into having some real basis. In mid-1981, the Soviet ambassador to the PPR, Aristov, warned the Soviet politburo that the situation had reached the stage where the international commitments of Poland could be really affected. The diplomat pointed out the spreading argumentation of some parts of the Polish opposition that the eventual democratization of the country was not compatible with the Warsaw Treaty Organization membership.[1566]

Even alteration of party cadres at the 9th PUWP extraordinary congress, which took place on June 14-19, 1981, did not significantly calm down the situation in Poland. In August, at the Central Committee plenum, the secretary of this body, Kazimierz Barcikowski,[1567] pointed out that whilst Solitary had called anti-Soviet incidents just acts of provocations in the past, some of its representatives had already taken open anti-Soviet positions in the moment. In the PUWP leadership, tendencies emerged to face this development by stressing the need to maintain the existing allied ties of Poland because of unforeseeable international events linked to the

---

[1564] *Návrh dopisu ústředního výboru KSČ ústřednímu výboru Polské sjednocené dělnické strany*, 18. 6. 1981, 1261/0/8, sv. P12/81, b. 8, NA, Prague. It is not clear whether the Czechoslovak side acted on the Soviet instruction or the letter was initiated by somebody within the CPCz leadership.

[1565] Budapeszt – Szyfrogram nr 327/IV z ambasády PRL do MSZ, October 8, 1981, in *Przed i po 13 grudnia,* tom II (doc. 291): 297.

[1566] Kramer, "Poland 1980–81," 118. Also, an analysis of the Czechoslovak minister of foreign affairs pointed to activities of the more radical opposition's faction, like the Confederation of Independent Poland. Its leader, Leszek Moczulski, did not hide that in the long-term perspective, the country should have withdrawn from the Warsaw Treaty Organization, as such a move would have given Poland a free hand for acting in the international arena. *Mezinárodní aspekty polské krize - J. Stich, ZPO,* 30. 9. 1981, DTO 1945-1989, i. č. 31, e. č. 53, AMZV, Prague.

[1567] At the 9th PUWP Congress, Barcikowski unsuccessfully ran for the post of first secretary of the party against Kania.

fall of détente. Some Polish communists thus replaced traditional rhetoric about "German revanchism" with an alleged threat to Poland posed by a general escalation of the Cold War. In this context, the PUWP CC secretary, Jan Grzelak, pointed out that given the international situation, the entire Eastern Bloc expected the PPR to guarantee long-term stability in both domestic and foreign policy.[1568]

These stances were summed up in the best way by Tadeusz Tuczapski, the deputy defense minister. According to his opinion, the crisis hit Poland at the beginning of the new phase of the Cold War: détente was falling; "hysteria" of the Soviet threat was occurring in the West as the public there were threatened by politicians referring to increasing armament of the Warsaw Treaty Organization. He pointed out the fact that Poland lay in the main direction of presumed combat operations in Europe, there were strategic communications on its territory, the country played an important role in the Warsaw Pact's defense system, and had a non-negligible weapons industry. Referring to historical experience and the impossibility to base a defense of the country on its own, Tuczapski advocated the alliance with Moscow. He stated that the existing situation already undoubtedly jeopardized the defense capabilities of the state, as the demonstrations and strikes caused troubles to weapons industry and mobilization plans.[1569]

The official Polish propaganda called for a strengthening of a political as well as defensive alliance with the USSR and the Eastern Bloc. It still presented the thesis that Warsaw Pact membership was the only guarantee of Polish independence and preserving of peace.[1570] However, some voices emerged in the PUWP demanding creation of its own, non-Russian model of socialism which would have been mainly based on the real geopolitical situation and historic experience of the Polish nation.[1571]

[1568] *II plenarne posiedzenie KC PZPR*, 11. 8. 1981, PZPR KC, XIA/1215, AAN, Warsaw.

[1569] Ibid.

[1570] *List 1. sekretarza partyjnej komisji koordynacynej srodowiska naukowego Warszawy Josefa Ludwickiego do Stanisława Kaniu*, 10. 6. 1981; *List KW PZPR Kielce dla KC PZPR*, 12. 6. 1981, XII 890, 899/138, ibid.

[1571] *List egzekutywy oop td przy hucie "Katowice" do KC PZPR* 11. 6. 1981, ibid.

Historical anti-Russian feelings came out even within the party during the crisis.

Withdrawal from Brezhnev doctrine

At the beginning of August, Erich Honecker tried once more to activate the Warsaw Treaty Organization. During his conversation with Brezhnev on Crimea, he appealed for holding a meeting, similar to the one in December 1980. But the aging Soviet leader tactfully dismissed Honecker's call, claiming he would wait on such a move for the outcomes of his upcoming talks with Kania and Jaruzelski. Brezhnev urged Honecker to be patient about the situation in Poland and let the Soviet side work on gradual improvement. He tried to calm Honecker's concerns by informing the SED CC General Secretary about the existence of a top-secret Suslov commission.[1572] The conversation suggests that Brezhnev probably did not want to deal with the complex situation on the multilateral basis of the Warsaw Treaty Organization, where radical statements of hardliners, like Honecker or Husák, could have been presented. As Moscow most likely abandoned military intervention altogether, the possibility of putting the Polish leadership under pressure through the alliance was the last, emergency option for Brezhnev.

Kania and Jaruzelski met Brezhnev in Crimea on August 14. The talks were characterized by vague and well-known phrases. The PUWP CC First Secretary only repeated an ideological assessment of the international consequences of the Polish Crisis, including claims that imperialistic and anti-Socialist forces in the West wanted to use the situation in the PPR to disrupt the Warsaw Treaty Organization as well.[1573] At the same time, he reassured Brezhnev that the Polish foreign policy still fully abided by the joint line of the alliance and was based on the last Political Consultative Committee resolution.[1574] Jaruzelski reported a

---

[1572] Record of Brezhnev-Honecker Meeting in the Crimea, August 3, 1981, in *From Solidarity to Martial Law* (doc. 58): 330.

[1573] During his talks with the Soviet satellites' leaders, Kania presented a thesis similar to the one before on Crimea. *Tezy do rozmow na Węgrzech*, 19. 3. 1981, PZPR KC, XIA/1255, AAN, Warsaw.

[1574] *Tezy do rozmow I sekretarza KC PZPR w ZSRR*, 14. 8. 1981, XIA/1257, ibid.

good situation in the Polish army. He suggested that plans for its action against the opposition were being prepared. Kania added that the Polish leadership was ready to defend socialism by all means. The weakening Brezhnev was able to only vaguely appeal for mobilization of the PUWP forces and urged the Polish officials to act on their own. Moscow had already helped enough, he stated.[1575] Despite its general features, the Soviet leader's statement must be considered as an expression of substantial change in the Kremlin's policy, especially signaling a withdrawal from the Brezhnev doctrine.

The Polish Crisis erupted in the moment when the Eastern Bloc faced several fundamental challenges. Their combination did not only weaken the possibility of military intervention to support the Polish communist regime, but also resulted in a forced withdrawal from the "socialist internationalism" proclaimed by Moscow. First of all, the increasing dependency of the USSR and its satellites on the West became apparent. Improvement in economic and science-technological cooperation with Western countries, which the Eastern Bloc expected from the Helsinki Act, was essentially disrupted by the Soviet invasion of Afghanistan. Moscow was well aware of its technological lagging and dependency on the import of Western food. Any eventual military action in Poland would have almost certainly resulted in a significant widening of economic sanctions applied by the West against Moscow after December 1979 as well as other cuts in technologies and food import.[1576] Moreover, the Reagan administration suggested a much tougher reaction than that of Carter to the Afghan events. Actually, during the Polish Crisis it was becoming clear that the Soviet army was unable to break a standoff in the Afghan conflict. The material costs and losses of lives were growing with no relevant saving strategy in sight.[1577] At the turn of the 1970s and 1980s, détente failed and the new threat of Cold War escalation emerged. Documents suggest that the Kremlin had enormous interest to prevent a new round of arms race. On one hand, massive Soviet armament programs

---

[1575] *Protokół Nr. 3 z posiedzenia Biura Politycznego KC PZPR 18 sierpnia 1981r.*, V/169, mkf. 2995, ibid.

[1576] Tejchman and Litera, *Moskva a socialistické země na Balkáně,* 101.

[1577] Gaddis, *Studená válka,* 194.

launched in the mid-1960s probably gave the Warsaw Treaty Organization little military advantage. On the other hand, these considerably exhausted the economy of the USSR. Therefore, especially after the Afghan conflict started, the Soviet political leadership tried to take into account its relations with the West, mainly because there were no resources for another round of massive armament.[1578] As noted before many times, concerns about the Western reaction strongly resonated even in the approach towards the Polish Crisis. Also, biological decay of the key persons in the aged Soviet leadership was a limiting factor. The time had passed when the politburo could act decisively and its sessions had been held almost permanently in critical moments. This time, meetings were short and largely formal.[1579] Various changes in attitude towards the Polish Crisis therefore largely reflected the helplessness of the Soviet leadership. However, given the absence of Soviet sources and features of the decision-making process in the Kremlin, to analyze it would be an extremely difficult task.

The Soviet politburo probably informed some Eastern Bloc countries' leadership about its intention not to intervene militarily in Poland. Along these lines, Czechoslovak officials in their talks with the Polish diplomats categorically ruled out the possibility of outside intervention, unless there was a risk that the PPR would leave the Warsaw Treaty Organization.[1580] A strategic aspect began to prevail over the ideological dogmas in the Soviet approach. Rather than using phrases about a need to preserve "dictatorship of proletariat," Brezhnev asked Kania a rhetorical question: Who would guarantee the inviolability of key strategic communications to the GDR, if Solidarity came to power?

[1578] This issue probably resulted in increasing tensions between the Soviet political and military elites. Since the beginning of the Afghan war, regarding the Western policy, the Soviet general staff, led by its chief Nikolai Ogarkov, recommended "belt-tightening" of the population in favor of armament programs, while the politburo supported a "consumer" line, which inevitably required keeping international tensions within acceptable limits. Durman, *Útěk od praporů*, 268, 294-295.

[1579] Ibid, 170.

[1580] Praga – Szyfrogram nr 1794/III z ambasády PRL do MSZ, August 19, 1981, in *Przed i po 13 grudnia*, tom II (doc. 242): 198.

Without them, he said, the Warsaw Pact would be unable to protect the results of World War II.[1581] Indeed, it was an exaggerated claim. However, Georgy Shakhnazarov the member of Suslov's commission, retrospectively noted that from Moscow's overall point of view, Polish withdrawal from the Warsaw Treaty Organization would have forced the USSR to revisit its entire defense conception that rooted back to the Yalta conference. In the 1990s, Valery Musatov, former deputy head of the CPSU CC International Department, also testified in the same spirit. He pointed out the rationalism of the Soviet leadership which was aware in 1981 that the intervention in Poland was not possible for various reasons. Moscow allegedly stopped thinking about the issue from an ideological point of view. Henceforward, it only cared about geopolitical ties with Warsaw. According to Musatov, the Kremlin was ready to negotiate on these matters even with an eventual new non-communist Polish government. This claim was also supported in the 1990s by a former chief of Staff of the Unified Armed Forces, General Gribkov.[1582]

Solution without assistance of the Warsaw Pact

Even the massive "West-81" drills, held since August in the Baltic Sea, failed to calm down the situation in Poland significantly.[1583] On invitation by Dmitri Ustinov,[1584] defense ministers of all the Warsaw Pact countries, except for Bulgaria, observed the exercises, accompanied by the Cuban vice-president, Raúl Castro,[1585] and defense ministers of Vietnam

---

[1581] Zapis wypowiedzi Leonida Breżnewa w rozmowie ze Stanisławem Kanią przekazany Jánosowi Kádárowi, September 11, 1981, ibid (doc. 251): 218.

[1582] *Wejdą nie wejdą,* 148, 169, 192.

[1583] Regarding the naval exercise, the U.S. media pointed out that the Soviet, East-German and Czechoslovak ground forces had been put on combat readiness. The Polish military officials later confirmed that the Polish troops had been symbolically joined by one Soviet tank and one Czechoslovak motor-rifle unit. Also, Marshal Kulikov took part in the maneuvers. *ČTK-Mezinárodní redakce, Zahraniční aktuality,* 10. 8. 1981, 506/0/45, k. 5968, NA, Prague.

[1584] *Pozvání na manévry SSSR,* 14. 8. 1981, FMO GŠ/OS, k. 50, p. č. 188, e. č. 37515, VÚA, Prague.

[1585] In September 1981, Honecker visited Cuba. Castro expressed his concerns about the imperialist's threats to "the liberty island", especially in case of the Warsaw Treaty Organization deciding to intervene in Poland. Honecker assured

and Mongolia. Nevertheless, the U.S. military officials officially stated they did not consider such an increase of military activity as a glimpse of a forthcoming invasion of Poland.[1586]

At the end of summer, no improvement of the situation, for which many blamed on the passivity of the Polish leadership, undermined positions of Kania as well as Jaruzelski. Honecker urged Moscow to install hardliner Stefan Olszowski as the head of the PUWP.[1587] Also, rumors about a conspiracy of radical Polish officers spread. On the initiative of Kulikov, they allegedly planned to topple Jaruzelski, unless the general committed to an immediate imposition of martial law.[1588] But stepping up the Kremlin's political pressure was the decisive factor. On September 14, the CPSU CC sent to its Polish counterpart a note protesting the constantly growing acts of anti-Sovietism in the PPR.[1589] One day later, Brezhnev, in

---

him that Cuba would not remain alone and was the "socialist camp" member. Castro wished Gromyko would send a clear message to the U.S. that the USSR would not leave a potential American attack on the island without response. In this regard, he complained that Cuba was not a Warsaw Pact member simply because of the negative Soviet stance. Honecker tried at least partially to satisfy Castro by reference to Raúl Castro's participation in "Brotherhood in Arms-81" and "West-81" in the role of observer, which the East-German leader absurdly compared to the alliance membership. Protokół rozmowy Ericha Honeckera z Fidelem Castro przeprowadzonej 13 września 1981 r. w Hawanie, September 15, 1981, in Przed i po 13 grudnia, tom II (doc. 253): 222.

[1586] "Cvičení Západ 81 pokračuje," Rudé právo, September 9, 1981; "Washington not worried by Baltic manoeuvres," The Times, August 15, 1981.

[1587] Paczkowski, From Solidarity, 27.

[1588] Existence of the conspiracy has never been proven and Jaruzelski himself strictly denied it retrospectively. Mastny supports him by claiming that during 14 years in the office, the Polish defense minister was able to establish a system of very close ties to his subordinates based on absolute loyalty. Mastny, "The Soviet Non-Invasion of Poland," 201. Also, Marceli Kosman, the Polish historian, points out the fact that Jaruzelski was well aware of the situation within the PPA, as he consistently directed cadre policy and appointed persons to important command posts. After all, since his youth, the army had been his second home. Kosman, Los Generała, 109.

[1589] CPSU CC Communication to the PUWP CC "Intensifying Anti-Soviet Feelings in Poland", September 14, 1981, in From Solidarity to Martial Law (doc. 63): 357.

a telephone conversation,[1590] accused Kania of failing to implement anything he had suggested to the Soviet leadership during previous negotiations. He also harshly criticized the decrease of combat capabilities of the Polish army.[1591] The Soviet leader's words that the Warsaw Pact members could not allow for a revision of the status quo after World War II and strategic communications to be jeopardized, the PUWP CC First Secretary understood as a direct threat of intervention.[1592] Probably because of the harshening of Soviet rhetoric, Kania, on the same day, openly talked about the imposition of martial law at the Polish politburo session.[1593] Moscow managed to threaten Kania without supporting its political duress by any real measures.

On September 13, a meeting of the Polish Committee of National Defense took place with Kania's participation. Jaruzelski announced there that preparation for the imposition of martial law had been finished. However, a decision to order the action was postponed.[1594] Regarding the Warsaw Treaty Organization, it is good to point out that none of the Polish officers who took part in the meeting held any senior post in the Unified Command.[1595]

Thus, the official military structures of the alliance did not formally participate in planning the imposition of martial law in Poland, despite their involvement in preparations for suppression of the Polish

---

[1590] Information about the phone conversation with Kania was sent by Brezhnev to the Soviet ambassadors to the Warsaw Pact countries, except for Romania, to be conveyed to the party leaders. Honecker, Husák, Kadar as well as Zhivkov fully identified with Brezhnev's message to Kania. Excerpt from CPSU CC Politburo Meeting Regarding Brezhnev-Kania Conversation, September 17, 1981, ibid (doc. 67): 368.

[1591] Information on the Bhreznev-Kania Telephone Conversation, September 15, 1981, ibid (doc. 64): 360.

[1592] Kamiński, *Przed i po 13 grudnia*, XXXVI.

[1593] *Protokół Nr. 7 z posiedzenia Biura Politycznego KC PZPR w dniu 15. IX. 1981r.*, PZPR KC, V/170, mkf. 2996, AAN, Warsaw.

[1594] Paczkowski, *From Solidarity*, xli.

[1595] Protocol No. 002/81 of Meeting of the Homeland Defense Committee, September 13, 1981, in *From Solidarity to Martial Law* (doc. 62): 350; *Seznam účastníků 22. zasedání vojenské rady Spojených ozbrojených sil*, 9. 10. 1980, FMO GŠ/OS, k. 48, p. č. 178, č. j. 37023, VÚA, Prague.

opposition being previously asserted by Honecker or Kulikov, for example. The probable reason must be seen as a reluctance of the top Soviet political circles to give their blessing to this approach.

Considerations of the Political Consultative Committee session

In spite of the aforementioned shift in preparations for the imposition of martial law, uninitiated communist functionaries still considered the situation in Poland critical. Therefore, during August, some appeals of lower Polish party representatives for an acceleration of a solution to the crisis in assistance with the Warsaw Pact's bodies could be heard. The head of the Warsaw Organization of the PUWP, Stanisław Kociołek, one of the prominent Polish hardliners, sought support in the GDR. In his conversation with Konrad Naumann, the East-German politburo member, Kociołek called for continued, visible pressure on the Polish leadership from the alliance. Without this, he claimed, the Polish leadership would not decide to impose martial law for another 2-3 years.[1596] In Berlin, Kociołek appealed for an immediate holding of the Political Consultative Committee session, where the Polish delegation would have been forced to sign a binding schedule of measures against the opposition. He appreciated Jaruzelski and his influence in the PPA, but at the same time demanded the Warsaw Pact supreme body guarantee a tough approach before the turn of October and November, when the rotation of soldiers on conscription would have taken place. The Polish official also made it clear that the Political Consultative Committee session had to be of a practical character, as the parties' leadership had already discussed the Polish question in detail bilaterally. Therefore, to only hold a consultative

---

[1596] The claim cannot be taken conclusively. Kociołek belonged to a group of Stanisław Kania opponents which were defeated on the 11th plenum of the PUWP CC. Thus, it is very likely that he tried to exaggerate the real state of things in order to picture the actual Polish political and military leadership as incompetent in front of the Eastern Bloc countries.

meeting of the Warsaw Treaty Organization, Kociołek considered pointless.[1597]

The fact that the Polish hardliners contacted the East-German officials instead of Moscow with their proposal, suggests that during the Polish Crisis, the Soviet leadership was not interested in cooperation on key issues with radicals within the PUWP. Unlike the Prague Spring, the top party representatives were not instigators of social change. Therefore, the Kremlin chose a totally different model; it focused on negotiations with the legitimate Polish leadership and did not seek "Polish Bil'ak," although some contacts between Moscow and radical Polish functionaries undoubtedly existed.[1598]

According to lower-party functionaries from the Warsaw region, Solidarity would have most likely reacted to the implementation of tougher measures approved by the Political Consultative Committee with a general strike. In this case, the Polish security forces were supposed to take firm action. Bloodshed would have been justified as a move preventing a much worse massacre in the future. On October 8, such an initiative was discussed by the representatives of the CPCz and SED CC International Departments. They recommended to deliberate an option to hold the Warsaw Pact supreme body session, at least. However, Kania should have been prevented from using the meeting to voice concerns about general strike and civil war, by which the Polish side had just threatened Moscow, according to the CPCz and SED officials. For this reason, Honecker's leadership considered a separate approach by the USSR, GDR, and CSSR more effective, despite it not dismissing negotiation on the basis of the entire Warsaw Treaty Organization altogether. In this case, the 1968 scenario, when the events in Czechoslovakia had been discussed separately and without any formal framework by selected alliance member-states only, would have largely recurred. The East-German side even contacted Moscow with this initiative. However, Konstantin Rusakov, the CPSU CC secretary,

---

[1597] Notatka z rozmowy Konrada Naumanna se Stanisławem Kociołkiem przeprowadzonej 28 września 1981 r., September 30, 1981, in *Przed i po 13 grudnia*, tom II (doc. 284): 279

[1598] Paczkowski, *From Solidarity*, 28; Durman, *Útěk od praporů*, 234.

politely, but firmly rejected such a proposal. He stated that the USSR intended to continue bilateral talks with the Polish leadership.[1599]

Some hardliners in low levels of the Polish party collective tried to use the Warsaw Treaty Organization mostly politically, as a tool of duress. An overall thoughtlessness of these conceptions is documented by the absence of any suggestions of what should have followed if the Polish leadership did not fulfill approved commitments. Content of the initiatives also shows that lower party representatives were just poorly informed about the real mechanism of the alliance's bodies. The Soviet leadership, Husák or Honecker, in their thoughts about involvement of the Warsaw Pact in a solution of the Polish Crisis never considered holding an official Political Consultative Committee session. They always talked about an extraordinary meeting of the party leaders of the alliance member-states, as largely settled; the very propagandistic scheme of the Warsaw Pact supreme body meetings did not seem to be an appropriate pressure forum. The situation is well illustrated by calls of some functionaries of the PUWP International Department for the Warsaw Pact Committee of Minister Affairs session, scheduled at the beginning of December, to deal with the crisis in Poland. However, the purpose of the meetings of this body was almost exclusively shaping a joint foreign policy in negotiations with the West. Therefore, key officials of the alliance member-states did not reflect this initiative at all.

Jaruzelski's attempt to get the Warsaw Pact's guarantee

A turning point for further development in the Polish Crisis became the 4th plenum of the PUWP CC, held on October 16-18, 1981. There, Kania offered his resignation from the post of First Secretary, which was accepted by the delegates.[1600] Subsequently, General Jaruzelski was elected to this position while still holding the post of prime minister as well as defense minister. Initially, this massive accumulation of

---

[1599] Record of a Meeting between Representatives of the CPCz CC and SED CC International Relations Department in East Germany, October 8, 1981, *Cold War International History Project Bulletin*, no. 11 (1998): 71.

[1600] Paczkowski, *From Solidarity*, xlii.

functions was not seen as a danger, even by the opposition, but rather a guarantee of stability. For a certain period, the general enjoyed the trust of a large part of Polish society.[1601] Jaruzelski became the PUWP CC First Secretary with backroom Soviet assistance. Moscow actually thought that Kania was one of the main obstacles to the imposition of martial law in Poland.[1602] Brezhnev himself welcomed installation of the general as the PUWP head very much and placed great hopes in the new party leader.[1603]

However, some tightening of reprisals, which the new party chief undertook,[1604] did not prevent an escalation of strikes at the end of October. Some of them even took place on the national level.[1605] At this moment, Jaruzelski most likely decided to impose the long-delayed martial law, although during his conversation with Kulikov he allegedly did not promise any concrete date.[1606] To its consternation, the Kremlin soon realized that the Polish leader returned to the idea of imposition of martial law in assistance of the Warsaw Treaty Organization.[1607] On October 29, during the discussion of the Soviet politburo about policy towards the new Polish leadership, Yuri Andropov stressed that there was no possibility to hear any request of "the fraternal countries for military assistance". To the contrary, the USSR should have consistently adhered to its existing approach and avoided sending the troops to the PPR under any circumstances. Such an attitude was also supported by Ustinov, referring mostly to the hostile positions of the Polish population.[1608] But

---

[1601] Wanner, *Brežněv*, 132.

[1602] Kramer, "Jaruzelski, the Soviet Union," 7.

[1603] Notes of Brezhnev-Jaruzelski Telephone Conversation, October 19, 1981, in *From Solidarity to Martial Law* (doc. 71): 392.

[1604] The Polish general prosecution, for example, began to sue people for public statements and publications attacking the allied ties of the country, based on the Article 133 of penal code - public acts against allied commitments of the PPR. *Informacja sekretariatu KC o przebiegu dotychczasowej realizacji uchwały IV plenum KC*, November 1981, PZPR KC, XII 884, 899/370, AAN, Warsaw.

[1605] Paczkowski, *From Solidarity*, xlii.

[1606] Mastny, "The Soviet Non-Invasion of Poland," 202. Mastny refers to a not very trustworthy source, Ryszard Kukliński's testimony.

[1607] Ibid.

[1608] Transcript of CPSU CC Politburo Meeting on Rusakov's Trip to Eastern Europe, October 29, 1981, in *From Solidarity to Martial Law* (doc. 72): 395.

only the political leadership was unanimous in its decision of nonintervention. Among the Soviet military officials, the politburo's approach allegedly caused a split.[1609]

Eugenius Molczyk, the deputy defense minister of the PPR, also tried to strengthen the hopes placed on the reconstructed Polish leadership within the Warsaw Pact's military structures. At the session of the Military Council, held in Budapest on October 27-30, he stated with a proper ideological accent that the deep crisis in Poland continued. However, he assured them that the PPA was ready to take action to protect socialism in the country on an order of the party leadership. Also, the final meeting of the Military Council members with the Hungarian leadership proceeded in a combative atmosphere.[1610] The crisis in Poland was in the center of discussion. Kulikov once more reiterated Brezhnev's statements that Poland was and would be a socialist country. Kadar, too, trusted in Jaruzelski. Via Molczyk, he urged the Polish leader to start the action against the opposition soon.[1611]

It was already noted that the official agenda of the Committee of Ministers of Foreign Affairs, held in December in Bucharest, did not include the Polish question. In his speech, Józef Czyrek just briefly informed his audience about the situation in the country.[1612] However, a

---

[1609] According to Gribkov's testimony (which must be seen critically of course), during talks at the Soviet defense ministry, the chief of the Soviet general staff Ogarkov presented a report of the Warsaw Pact's maneuvers, held in Poland in October, without mentioning the possibility of intervention. Then, an argument between the alliance's supreme commander Kulikov and chief of Staff of the Unified Armed Forces Gribkov took place. The former advocated the invasion of Poland, the latter was against. Sergei Akhromeyev, deputy chief of the Soviet general staff, took no clear stance. Mastny, "The Soviet Non-Invasion of Poland," 202.

[1610] We are reminded that this was an absolutely standard practice. Kulikov regularly met top party and state officials of countries hosting the Warsaw Pact's military bodies' meetings.

[1611] *24. zasedání vojenské rady Spojených ozbrojených sil*, 3. 11. 1981, FMO GŠ/OS, kr. 50, p. č. 186, e. č. 37512, VÚA, Prague.

[1612] *Informace o zasedání výboru ministrů zahraničnich věcí členských států Varšavské smlouvy v Bukurešti ve dnech 1. a 2. prosince 1981*, December 7, 1981, Accessed October 16, 2019. http://www.php.isn.ethz.ch/kms2.isn.ethz.ch/service

new propagandistic approach of Moscow appeared. Just ahead of this session, the Soviet ideologist Boris Ponomarev instructed the Soviet satellites at a special meeting[1613] to present the vague resolution of the Warsaw Pact Moscow meeting in December 1980 as their official statement on the Polish Crisis. In order to strengthen its significance, the gathering should have been called the Political Consultative Committee session, retrospectively.[1614] Along these lines, the Moscow meeting of the previous year was also mentioned by Gromyko at the aforementioned gathering of the alliance ministers of foreign affairs.[1615] The Soviet leadership itself, which in reality rejected every attempt to involve the Warsaw Treaty Organization in a solution of the crisis, initiated increasing the role of the alliance in propagandistic statements. This fact suggests that despite intensification of political cooperation within the Warsaw Pact during the 1970s, the alliance still remained largely a tool of Eastern propaganda.

A final decision to suppress the opposition by force was made at the end of October, 1981.[1616] But the Jaruzelski leadership appeared to be very shaky. They were not sure, whether they could manage to deal with the situation. Although Moscow insisted on a solely domestic solution, the Polish plans still mentioned resources for transport of the Warsaw Pact troops.[1617] A report by the Polish interior ministry evaluating possible scenarios of development after the imposition of martial law stated that

---

engine/Files/PHP/20107/ipublicationdocument_singledocument/26ac242c-cd31-4c09-8bfa-4b11db3809da/cs/011281_InformationReport.pdf.

[1613] Delegations of Cuba, Laos, and Vietnam took part in the Moscow ideological meeting too. This was a typical phenomenon of widening the circle of so called "socialist countries" at the time.

[1614] *Zpráva o průběhu a výsledcích porady tajemníků ústředních výborů komunistických a dělnických stran socialistických zemí pro mezinárodní a ideologické otázky,* 17. 11. 1981, 1261/0/8, sv. P24/81, b. 7, NA, Prague.

[1615] *Vystoupení ministra zahraničních věcí Svazu sovětských socialistických republik A. A. Gromyka, 1. prosince 1981,* December 1, 1981, Accessed October 16, 2019. http://www.php.isn.ethz.ch/kms2.isn.ethz.ch/serviceengine/Files/PHP/20104/ipublicationdocument_singledocument/8729f79e-d282-4d6b-99d1-1d380dd1e26c/cs/011281_SpeechSovietFM.pdf.

[1616] Kukliński, *Wojna z narodem,* 12.

[1617] Mastny, "The Soviet Non-Invasion of Poland," 202.

"the assistance of the Warsaw Pact forces is not ruled out" in case of massive violent resistance by the opposition. This document refutes the latter Jaruzelski's claim that the imposition of martial law prevented the imminent risk of outside military intervention.[1618] On the contrary, at the beginning of December, Jaruzelski himself tried to get a guarantee by the Warsaw Treaty Organization for his action, code-named "Operation X." For this purpose, he used a regular session of the Committee of Defense Ministers, which took place in the Soviet capital on December 1-4. At the meeting, the Polish question was discussed very intensively. Essentially, this was the first time when the Warsaw Pact's official structures seriously dealt with the crisis in the PPR.

In their speeches, the ministers of the USSR, GDR, and BLR, as well as the supreme commander Kulikov dealt with the situation in Poland in detail. Also, the Czechoslovak minister Martin Dzúr mentioned it marginally.[1619] Instead of Jaruzelski, who was busy with domestic problems, Florian Siwicki, the chief of general staff of the PPA, led the Polish delegation. He put forward the Polish leader's proposal: The Committee of Defense should have publicly declared concerns of the entire alliance about activities of antisocialist forces threatening fulfilling commitments of the Polish army to the Warsaw Treaty Organization and determination of the alliance to take proper steps to ensure joint security of "the socialist commonwealth in Europe."[1620] In this manner, Jaruzelski *de facto* tried to cover the "Operation X" by the Warsaw Pact. The statement should have been issued in the form of an amendment to the official communiqué of the meeting.

[1618] Machcewicz, Pawel, "The Assistance Of Warsaw Pact Forces Is Not Ruled Out," *Cold War Ineternationa History Project Bulletin*, no. 11 (1998): 40.

[1619] Dzúr repeated well-known claims about a potential threat to the interests of all the Warsaw Pact countries posed by the events in the PPR. His declaration of support to the Polish communists was unspecific and very general as well. *Vystoupení ministra národní obrany Československé socialistické republiky na 14. zasedání výboru ministrů obrany k otázce "Analýza stavu a tendence rozvoje ozbrojených sil agresívního bloku NATO"*, undated, 1981, FMO GŠ/OS, k. 49, p. č. 184, e. č. 37503, VÚA, Prague.

[1620] Report on the Committee of Ministers of Defense Meeting in Moscow, December 1-4, 1981, in *From Solidarity to Martial Law* (doc. 79): 420.

Individual party leaderships were immediately informed by telephone about Jaruzelski's proposal in Moscow. His request for guarantees of the Warsaw Treaty Organization was motivated, among other things, by an effort to refute the Western propagandistic claims that the PPR had already lost support of its allies. In this regard, Siwicki emphasized that this was mostly a "morally-political" shield for the Polish leadership, which had enough strength to solve the situation on its own, as he assured. The General tried to convince the present delegations that the Warsaw Treaty Organization had to supplement this statement with no concrete military measures. Such a declaration was supposed to mainly be a "cold shower for the counterrevolution in the PPR."[1621]

Approval of the amendment was blocked by Constantin Olteanu, the Romanian defense minister. The importance of the issue was underlined by the fact that because of Jaruzelski's proposal an imminent extraordinary meeting of the RCP Executive Committee was held. Nicolae Ceausescu stated that the matter was strictly political. He considered it necessary to prevent the military from discussing it, moreover at the regular session of the Warsaw Treaty Organization which was not linked to the Polish Crisis in any way. The Romanian leader made clear he would not approve any document that would have indicated any future commitment of Romania, the Warsaw Pact member-state, to the military intervention in Poland or connected Bucharest with such an action.[1622] Dmitri Ustinov, who chaired the Committee of Defense Ministers session under the regular rotation, unsuccessfully appealed to the Romanian delegation, noting that the PPR needed solely political support that allegedly was not connected with any potential military actions of the Warsaw Pact at all. But Romania blocked even an attempt to publish the

---

[1621] *Informace o průběhu a výsledcích 14. zasedání výboru ministrů obrany konaného ve dnech 1.-4. prosince 1981 v Moskvě*, FMO GŠ/OS, k. 49, p. č. 184, e. č. 37503, VÚA, Prague.

[1622] Stenogram z nadzwyczajnego posiedzenia Politycznego Komitetu Wykonawczego KC RPK, December 3, 1981, in *Przed i po 13 grudnia*, tom II (doc. 339): 384.

statement with an amendment that this was not the Romanian defense minister's stance.[1623]

During subsequent talks, also Lajos Czinege, the Hungarian defense minister, used the Romanian position. He declared that the HPR would support the amendment in case of unanimity only. The Romanian objection probably served the Hungarian leadership as a welcomed opportunity to avoid any, however vague commitments, to an eventual military involvement in Poland. On the contrary, Czinege advised the Poles to approach the Warsaw Pact member-states' party leaders with the request for support, as this was purely a political matter. At the same time, he harshly criticized the Polish side for its attitude so far. He asked the question of who actually should have been supported by the alliance statement, as the Polish communist had been constantly giving in to the opposition up to his point.[1624] Behind this statement, were Hungarian fears that the Polish leadership, in fact, would not be able to deal with the situation on their own. Despite his previous declarations that the PPR had to remain unconditionally a Warsaw Pact member, Janos Kadar was probably afraid of the continual weakness of the PUWP leadership. For this reason, he refused to issue any guarantee on behalf of the alliance, which would have potentially established any claim for the Hungarians troops' participation in the military intervention in Poland in case of resuming passivity of the Polish officials. The situation even resulted in an argument between the Soviet and Hungarian delegations. In trying to persuade the Hungarian ministers to support at least a separate appeal, the chief of the Soviet general staff Ogarkov snapped at Czinege, about whether he remembered the year 1956 and the bloodshed in Hungary. The Hungarian delegation considered his words an offense to its leadership and raised an official protest against the Marshal's behavior.[1625]

[1623] Report on the Committee of Ministers of Defense Meeting in Moscow, December 1-4, 1981, in *From Solidarity to Martial Law* (doc. 79): 420.

[1624] *Informace o průběhu a výsledcích 14. zasedání výboru ministrů obrany konaného ve dnech 1.-4. prosince 1981 v Moskvě,* 8. 12. 1981, FMO GŠ/OS, k. 49, p. č. 184, e. č. 37503, VÚA, Prague.

[1625] Report on the Committee of Ministers of Defense Meeting in Moscow, December 1-4, 1981, in *From Solidarity to Martial Law* (doc. 79): 420. The incident had a sequel during talks of J. Kadar with the Soviet Prime Minister, N.

Rejection of Jaruzelski's request ended all relevant considerations about at least the political involvement of the alliance in the imposition of martial law in Poland. Although Ustinov and Kulikov afterwards tried to link a solution of the crisis to holding a meeting of the Warsaw Pact countries' party leaders, they failed.[1626] Thus, tensions coming from disagreement in the Polish question could also be felt during the final meeting of Brezhnev with the Committee of Defense Ministers' members.[1627]

Siwicki's speech in Moscow deserves great attention. In his report on ideological subversion of NATO countries towards the Eastern Bloc states,[1628] stressing the actual crisis in the PPR, he expressed some concerns that the social as well as economic situation in Poland could significantly distort weapons supplies to the Warsaw Pact's armies in the upcoming years. He put a strong emphasis on political agitation in the army. Through its intensification, the Polish military command tried to prevent the armed forces from decay. However, their actual "morally-political" condition - the level of loyalty to the communist regime and army's hierarchy structure - was described by the chief of the Polish

---

Tikhonov, on December 9. In his efforts not to corrupt relations with Moscow, the HSWP CC First Secretary apologized for rejecting the stance towards the Polish request for support. He instrumentally pointed to an almost simultaneously-held meeting of the Committee of Ministers of Foreign Affairs, where the Polish problem had not been discussed. Moreover, according to Kadar, the Warsaw Pact military body's interference with Polish affairs could have elicited an unwanted Western response and launched a new media campaign. Notatka Mátyása Szürösa dla Jánosa Kádára i Györgya Lázára na temat stanoviska zajętego przes Lajosa Czinegego na 14. posiedzeniu Komitetu Ministrów Obrony państw stron Układu Warszawskiego w Moskwie, December 10, 1981, in *Przed i po 13 grudnia*, tom II (doc. 348): 412.

[1626] Report on the Committee of Ministers of Defense Meeting in Moscow, December 1-4, 1981, in *From Solidarity to Martial Law* (doc. 79): 420.

[1627] *Informace o průběhu a výsledcích 14. zasedání výboru ministrů obrany konaného ve dnech 1.-4. prosince 1981 v Moskvě*, 8. 12. 1981, FMO GŠ/OS, k. 49, p. č. 184, e. č. 37503, VÚA, Prague.

[1628] Materials and drafts of reports were prepared in advance, in cooperation with the Staff of the Unified Armed Forces. Documents reveal that in this regard the real interaction took place within the Unified Command. *14. zasedání výboru ministrů obrany členských států Varšavské smlouvy*, 7. 4. 1981, ibid.

general staff as sufficient.[1629] Thus, Siwicki reassured the Warsaw Pact member-states' defense ministers about the ability of the Polish armed forces to take action against the opposition. Paradoxically, Jaruzelski was absolutely not sure about it. That is why he tried to get the alliance's support. In reality, by imposing martial law on December 13, 1981, the Polish leader did not forestall the imminent risk of foreign military invasion, but to the contrary, he had tried to create a framework for the potential involvement of the Warsaw Treaty Organization in suppression of the Polish opposition.

Jaruzelski's actions were probably based on the Polish politburo's assumption that at least Solidarity's "hard core" did not anticipate the foreign military intervention any more. On December 5, Siwicki put forward his report for the politburo on the Committee of Defense Ministers' session. In the following discussion, the members of the PUWP leadership *de facto* identified with his evaluation of the situation in the country, which had been voiced in Moscow. They stressed the weakening of the entire alliance's defense capabilities due to the events in the PPR because of transit complications and risk to the Polish military supplies. But only Olszowski asserted other consultations with the alliance member-states' party leaders. It is surprising that the politburo did not deal with the tumultuous course of the Committee of Defense Ministers' session in detail.[1630] Therefore, it cannot be ruled out that the attempt to get the Warsaw Pact's guarantee for action against the opposition was the initiative by Jaruzelski and his closest coworkers, rather than the wider Polish leadership.

After the failure of the Committee of Defense Ministers' meeting, Jaruzelski could not understand why "*the allies do not want to shoulder any of the responsibility even though they have constantly asserted that the Polish problem is a problem for the whole Warsaw Pact, not just for Poland.*"[1631] Since December 7, Kulikov was present in Poland once

---

[1629] *Informace o průběhu a výsledcích 14. zasedání výboru ministrů obrany konaného ve dnech 1.-4. prosince 1981 v Moskvě,* 8. 12. 1981, ibid.

[1630] Protocol No. 18 of PUWP CC Politburo Meeting, December 5, 1981, in *From Solidarity to Martial Law* (doc. 80): 425.

[1631] Kramer, "Jaruzelski, the Soviet Union," 8.

again.[1632] Most likely, he was supposed to probe in which phase preparations for the imposition of martial law were. However, one week before the "Operation X" was launched, communication between Moscow and the Jaruzelski leadership had proceeded in a vague and unequivocal way. The Warsaw Pact's supreme commander, who again acted in an extremely suspicious and unclear role, was probably partially responsible for that. In his conversation with Rusakov, Jaruzelski reported that on December 8, Kulikov had personally promised him military assistance. Mark Kramer reminds us that already in April, the Marshal had informed the Polish officials that given the international situation, the invasion of Poland had not been realistic. However, he had not ruled out the possibility of a supporting outside intervention, if the action against the opposition by the Polish security forces failed and representatives of the PPR asked their allies for "fraternal assistance" afterwards. Therefore, it is plausible that a few months later, Jaruzelski got back to Kulikov's message and he relied on the outside help in case of complications during the imposition of martial law.[1633]

Jaruzelski's intention to coordinate the imposition of martial law with the Warsaw Pact allies was conveyed to the Soviet politburo through Nikolai Baibakov on December 10. In the scope of this tactic, the Polish leader referred to Kulikov's alleged guarantees. As a result, in the following discussion, some members of the Soviet leadership, especially Yuri Andropov, blamed Kulikov for disrespecting the politburo's official policy. But the Soviet vice prime minister advocated for the supreme commander. The Marshal was said to simply repeat Brezhnev's unspecific statement that the allies would not leave the PPR in a time of trouble. It remains a question of what exactly Kulikov had told Jaruzelski. In the effort to encourage the Polish side to start action against the opposition, it cannot be ruled out that he really had spoken about the intervention by the Soviet, or potentially even the other Warsaw Pact countries' troops. However, Ustinov vouched for Kulikov in the politburo. He gave assurances that the supreme commander had not exceeded Brezhnev's

---

[1632] Mastny, "The Soviet Non-Invasion of Poland," 203.
[1633] Kramer, "Jaruzelski, the Soviet Union," 8.

previously vague pledge of assistance.[1634] As already explained, the Marshal liked to use the Soviet leaders' words from the 26th CPSU Congress. Thus, in an extremely tense situation, Jaruzelski could have understood it in a different, deeper sense that corresponded to his actual strategy.[1635]

Regarding problems in communication through Kulikov, Gromyko proposed to instruct the Soviet ambassador to convey to Jaruzelski the official position of the Soviet politburo that the intervention was out of the question. Moreover, Andropov ruled out the invasion even if Solidarity actually seized power and did not jeopardize unproblematic transit between the USSR and GDR. Soon before Jaruzelski's action, Moscow actually defined a red line for free development in Poland - securing strategic communications on the Polish territory.[1636]

The Soviet leadership was not informed in detail about preparations of the action to suppress the Polish opposition. Andropov deduced from Baibakov's report that Jaruzelski probably had not made a final decision on the imposition of martial law yet. The Kremlin just knew that such a move had already been unanimously approved by the Polish politburo.[1637] So it is very unlikely that in a situation where Moscow had no concrete reports on Jaruzelski's intentions, any substantial exchange of information about planned moves of the Polish leadership with its Warsaw Pact's counterparts proceeded, as Jaruzelski had already indicated in his earlier talk with Baibakov. In fact, the Soviet politburo was not even sure about the exact date of the launch of "Operation X." The initial information was talked about in the night from December 11 to 12; later, a one-day postponement was reported. Afterwards, the Kremlin were given a brief message that the action would take place sometime around December

---

[1634] Transcript of CPSU CC Politburo Meeting, December 10, 1981, in *From Solidarity to Martial Law* (doc. 81): 446.

[1635] Jaruzelski himself admitted that in the period just ahead of the imposition of martial law, he had been in an extremely complicated personal state of mind. Jaruzelski, *Historia nie powinna dzielić*, 14.

[1636] Transcript of CPSU CC Politburo Meeting, December 10, 1981, in *From Solidarity to Martial Law* (doc. 81): 446.

[1637] Ibid.

20.[1638] It is interesting that when Jaruzelski had asked for an audience in Moscow in mid-November, the Soviet politburo had offered him a term of December 14-15.[1639] This either proves that the Soviet leadership lacked information about Jaruzelski's intentions for a longer time, or that in November, the General himself had not had a clear vision about the imposition of martial law yet. Thus, there was a significant contradiction in the Polish leader's approach: On one hand, the General tried to get the Warsaw Pact's, or at least Soviet guarantee for his action; on the other hand, he failed to inform the Kremlin about his plans in detail.

According to General Gribkov's memoirs, on December 10, Jaruzelski asked the Soviet politburo for these acts of support during the imposition of martial law: Moscow should have sent a politburo member to Poland, who would have supervised the action, promised bigger economic assistance, issued a public statement that the operation was in the Soviet interest, and pledged military assistance if the situation in Poland became critical. In this manner, Jaruzelski practically tried to get similar guarantees by Moscow, which he had failed to obtain few days earlier from the Warsaw Treaty Organization. But the Kremlin categorically rejected any military assistance. The Soviet politburo only vaguely promised to work out a declaration of support for Jaruzelski during the course of events. One day later, therefore, the General tried to affect Kulikov. He warned him against the risk of Polish withdrawal from the Warsaw Treaty Organization, if the "Operation X" went out of control.[1640] Based on Jaruzelski's urgings, Kulikov came to conclusion that the Polish leadership was trying to make the imposition of martial law conditional on a pledge of eventual outside intervention. However, after the Soviet politburo meeting on December 10, Ustinov gave the Warsaw Pact supreme commander a clear directive for talks with the Polish representatives: The PPR had to solve its problems on its own; the USSR

[1638] Mastny, "The Soviet Non-Invasion of Poland," 203.

[1639] Extract from Protocol No. 37 of CPSU CC Politburo Meeting, November 21, 1981, in *From Solidarity to Martial Law* (doc. 73): 400.

[1640] Mastny, "The Soviet Non-Invasion of Poland," 204.

was not preparing to send troops to the Polish territory.[1641] According to Gorbachev's memories, just ahead of the imposition of martial law, Jaruzelski allegedly phoned Suslov, who told the intrusive Polish leader that Moscow would not renounce its commitment of protecting Poland against foreign threats, but the Soviet troops would not intervene against an inside enemy.[1642] This phone conversation must be seen as the moment of abandonment of the Brezhnev doctrine, in fact.

International impacts of the imposition of martial law

From a military point of view, the imposition of martial law on December 13, 1981, was a brilliant operation which achieved almost immediate success with marginal losses only. The chaotic reaction by Solidarity shows that the movement did not anticipate such a massive repression. Problems with reliability of the PPA were minimized by using mostly the Motorized Reserves of the Citizens' Militia (ZOMO) special units.[1643] On the day the action was launched, the Polish politburo held its session. Siwicki informed the politburo about the course of "Operation X." He claimed that the action was proceeding in close communication with the Unified Command of the Warsaw Treaty Organization as well as the North Group of the Soviet Troops on the Polish territory.[1644] However, Jaruzelski's retrospective testimony shows that Moscow did not take any significant measures within the Warsaw Pact. For instance, information about the breakthrough in Poland caught Gribkov, the chief of Staff of the Unified Armed Forces, on holiday. He immediately contacted Ustinov and asked whether he was supposed to come back. But the Soviet defense minister left the nominally second-highest military official of the Warsaw Treaty Organization out of duty.[1645]

[1641] "The Anoshkin Notebook on the Polish Crisis," December 1981, *Cold War International History Project Bulletin*, no. 11 (1998): 19.

[1642] Kramer, "Jaruzelski, the Soviet Union," 6.

[1643] Wanner, *Brežněv*, 133.

[1644] Protocol No. 19 of PUWP CC Politburo Meeting, December 13, 1981, in *From Solidarity to Martial Law* (doc. 85): 461.

[1645] "Commentary by Wojciech Jaruzelski," *Cold War International History Project Bulletin*, no. 11 (1998): 33.

Moscow naturally welcomed the action. An official statement of the USSR emphasized that Jaruzelski not only confirmed the leading role of the PUWP, but also fully confirmed the commitments resulting from the Warsaw Pact. Afterwards, the CPSU Politburo instructed Soviet ambassadors to inform the socialist countries' leaders about the situation in Poland. Symbolically, this directive included Fidel Castro, Lê Duẩn and Kaysone Phomvihane,[1646] but not Nicolae Ceausescu.[1647] At the same time, Józef Czyrek, the Polish minister of foreign affairs, held an informative meeting with the socialist states' ambassadors to Poland.[1648]

The Romanian leadership, despite being ignored by Moscow, was very interested in the events in Poland. The RCP Executive Committee's extraordinary session was held as early as December 13. Ceausescu's leadership mostly appreciated the solution of the crisis on the domestic basis as it had considered the previous effort to incorporate the amendment concerning the PPR into the Committee of Defense Ministers communiqué as an attempt to get approval for outside intervention. In this regard, Ceausescu expressed an opinion that the USSR would have had to intervene in Poland alone, whereas solely Bulgaria would have symbolically supported it with a small unit. He probably underestimated the willingness of the CSSR and GDR to send their troops to Poland. In the discussion, the fact was also pointed out that despite Jaruzelski's speech announcing that Poland would solve the existing situation on its own as a sovereign state, an alliance with Moscow would remain a key element of Polish foreign policy.[1649]

[1646] Lê Duẩn was a leader of the Vietnam Socialist Republic in the years 1975-1986, Kaysone Phomvihane held the office of prime minister of the Laos People's Democratic Republic in the years 1975-1991.

[1647] Extract from Protocol No. 40 of CPSU CC Politburo Meeting, December 13, 1981, in *From Solidarity to Martial Law* (doc. 86): 473.

[1648] Protocol No. 19 of PUWP CC Politburo Meeting, December 13, 1981, in *From Solidarity to Martial Law* (doc. 85): 461.

[1649] *Shorthand Record of the Meeting of the Executive Bureau of the CC of the RCP on the Events in Poland*, December 13, 1981, Accessed October 16, 2019. http://www.php.isn.ethz.ch/kms2.isn.ethz.ch/serviceengine/Files/PHP/16628/ipu blicationdocument_singledocument/f2fe4d72-88f4-48f6-9e96-98704dc38981/en /811213_shorthand_record.pdf.

Thus, the imposition of martial law in Poland was appreciated by all the Warsaw Pact member-states, including Romania. In this regard, an analysis of the situation for the Polish leadership pointed out Ceausescu's criticism of the invasion of Czechoslovakia. His actual support for Jaruzelski's move was ascribed to Romanian internal troubles. However, there were different motives behind the Romanian leader's position. The SRR officially appreciated a domestic solution of the situation on the PPR without any outside interventions.[1650] Ceausescu had condemned the operation "Danube" because of violation of state sovereignty, a principle which had been emphasized as the core element of Bucharest's foreign policy at least since 1964. The Romanian leader did not regret the end of the reform process in the CSSR; he never had sympathy for the Prague Spring. It is almost certain that in case of a recurrence of the Czechoslovak scenario, Romania would have opposed such an action again. At the same time, taking into account the condition of its economy, Bucharest rejected Jaruzelski's request for material assistance which would have exceeded the scope of a standard trade exchange within the Eastern Bloc.[1651]

Right after "Operation X" was launched, Józef Czyrek expressed his concerns about medial interpretation in the West. The U.S. President Reagan immediately ended his holiday and the American state secretary, Alexander Haig, initiated emergency consultations within NATO. But the Western alliance decided not to react to the situation in Poland as it had no information about the deployment of Soviet troops.[1652] The fact that no foreign units were taking part in the repressions and Marshal Kulikov was just providing Soviet officials with reports on development, did not go

---

[1650] *Reakcje partii komunistycznych i robotniczych na wprowadezenie stanu wojennego w Polsce*, 6. 1. 1982, PZPR KC, XII 3121, 945/27, AAN, Warsaw.

[1651] *Stenographic Transcript of the Meeting of the Consultative Political Committee of the Central Committee of the Romanian Communist Party*, December 17, 1981, Accessed October 16, 2019. http://www.php.isn.ethz.ch/ kms2.isn.ethz.ch/serviceengine/Files/PHP/16627/ipublicationdocument_singled ocument / 02670c0a-92c0-480f-811a-29950004329c / en / 811217_stenographic _transcript.pdf.

[1652] Protocol No. 19 of PUWP CC Politburo Meeting, December 13, 1981, in *From Solidarity to Martial Law* (doc. 85): 461.

unnoticed by the CIA.[1653] In the end, the West stuck to warnings against any additional invasion by the Warsaw Treaty Organization. Indeed, Jaruzelski's move surprised many Western politicians as they mostly shared the opinion that the communist regime in Poland was able to survive in the case of outside military assistance only.[1654] For this reason, the West tried to protract the ongoing CSCE summit in Madrid. The Western countries were awaiting whether the confrontation in Poland would not result in foreign intervention after all. In that case, the meeting would have served as an important protest platform. The present delegations of NATO countries at least condemned Jaruzelski's action, referring to violations of human rights and inconsistency with the Final Act principles. On the contrary, the Warsaw Pact countries criticized NATO for interference with internal affairs of the PPR. Romania demonstrated its specific position by refraining from making any comments.[1655]

Risk of the Warsaw Pact's invasion as a justification for the "Operation X"

Until his death in 2014, Jaruzelski, whose action on December 13 prolonged the life of the communist regime in Poland by almost a decade, often called the imposition of martial law a lesser evil.[1656] In fact, in the period that imminently followed its realization, "Operation X" was not presented by Polish propaganda as salvation from outside intervention but rather a measure to stabilize a bad economic situation and terminate chaos

---

[1653] CIA National Intelligence Daily, "Poland: Test of Government's Measures", December 14, 1981, ibid (doc. 88): 476.

[1654] Valery Musatov pointed out this fact at the Jarchanka conference. He also noted memoirs of some Western politicians, Alexander Haig and Margaret Thatcher, for example. Both admitted that the way out of the Polish Crisis had caught them very much by surprise, as they had generally anticipated the Soviet invasion instead. *Wejdą nie wejdą,* 151.

[1655] *Informace o průběhu podzimní etapy madridské schůzky KBSE,* 28. 12. 1981, PK 1953-1989, kat. č. 824, k. č. 229, č. j. 010.019/82, AMZV, Prague.

[1656] Jaruzelski, *Historia nie povinna dzielić,* 11.; Jaruzelski, Wojciech, *Pod prąd. Refleksje rocznicowe* (Warszawa: Comandor, 2005), 13; Jaruzelski, Wojciech, *Stan wojenny – dlaczego* (Warszawa: Polska Oficyna Wydawnicza BGW, 1992).

and "counterrevolution." Jaruzelski's move allegedly prevented the risk of civil war and unspecified "serious international consequences" for Poland. However, propaganda stressed that the army had remained strong enough to solve the situation on its own, without foreign assistance. These messages, on the contrary, were supposed to dismiss speculations about military involvement of the USSR and the Warsaw Treaty Organization. On the other hand, Jaruzelski's leadership put strong emphasis on a campaign that should have infused the constant principles of Polish foreign policy with the population, convincing them of the benefits of cooperation with Moscow and the Eastern Bloc and appreciating their help in the critical period.[1657] The Hungarian development after 1956 served as the best example of a successful consolidation of a domestic situation after military suppression of the "counterrevolution."[1658]

Jaruzelski began to use in his defense the claim that "Operation X" averted the Warsaw Pact's invasion after the fall of the communist regime.[1659] He pointed to incalculable consequences which jeopardizing strategic communications in Poland could have caused in the context of Cold War. On the eve of the imposition of martial law, the country was allegedly in a situation where outside intervention was not only inevitable, but also already decided. He accused the historians claiming the opposite of an instrumental selection of archival materials. Jaruzelski retrospectively called the Soviet leadership's concerns about Polish

---

[1657] Report by the PUWP CC Ideological Department for the PUWP CC First Secretary, W. Jaruzelski, of tasks of upcoming political influence to the countryside population, December 22, 1981, PZPR KC, XII 3152, 934/80, l.dz. WR/11-76/81, AAN, Warsaw; *Wprowadzenie stanu wojennego v PRL, jego podstawy prawne i miedzynarodowo uwarunkowania*, 23. 12. 1981, XII 3121, 945/27, ibid; *Zadania instancji i organizacji paryjnych wynikające z referatu biura politycznego i uchwały VII plenum KC PZPR*, 6. 3. 1982, XII 866, 934/81, ibid.

[1658] Collection of materials of the Polish embassy in Budapest concerning detailed analysis of "consolidation" of the Hungarian leadership and society after suppression of the 1956 revolution, 1982, XII 3060, 980/26, ibid.

[1659] In J. Kopeć's book "Dossier Generala" published in 1991, based on interviews with Jaruzelski, a claim appeared that the Polish leader had preceded the Warsaw Pact's supreme commander Kulikov by only four days in the imposition of martial law. Kosman, *Los Generala*, 199.

resistance baseless and emphasized the Warsaw Pact's military potential. He doubted that the alliance was allegedly capable of seizure while Western Europe had been unable to organize a successful local intervention within the Soviet sphere of influence even despite potential resistance of the population.[1660] According to Jaruzelski, everybody who lived in Poland at the time was afraid that the slightest threat to the Warsaw Pact's interests could have become a pretext for invasion. As a typical example of encroachment on strategic cooperation within the Eastern Bloc, he considered the eventual disruption of military transports.[1661] He also referred to Zbigniew *Brzeziński's* book *Game Plan* from 1987. The former security adviser to the U.S. president called Poland the key to accessing the whole of Eastern Europe, as vital strategic communications were located in the country.[1662]

Jaruzelski also pointed out the long-standing combat readiness of the NPA and CSPA - in the former case until April; in the latter until even July of 1982.[1663] Indeed, the Czechoslovak units which had taken part in the "Giant Mountains" action remained in full combat readiness till the year 1982.[1664] A night before the imposition of martial law, the East-German NPA and border-patrol units were put in a state of combat readiness and the plans of December 1980 were adjusted. Allegedly, an order to prepare for incursion into Poland and take positions south from Warsaw was even issued. However, much more significant was Marshal Kulikov's order for the Soviet troops to remain absolutely calm.[1665] In fact, opinions of individual Soviet satellites' leadership of whether to solve the situation in Poland by outside military intervention were not that significant. A final decision depended on the Kremlin only.[1666]

[1660] Jaruzelski, *Historia nie povinna dzielić*, 25-38.

[1661] "Commentary by Wojciech Jaruzelski," *Cold War International History Project Bulletin*, no. 11 (1998): 37-38.

[1662] *Wejdą nie wejdą*, 185.

[1663] Jaruzelski, *Historia nie povinna dzielić*, 40.

[1664] Tůma, "The Czechoslovak Communist Regime and the Polish Crisis," 64; Kříž, *Krkonoše 80*, 40.

[1665] Mastny, "The Soviet Non-Invasion of Poland," 205.

[1666] Kamiński, *Przed i po 13 grudnia*, XXI.

In 1992, the former chief of Staff of the Armed Forces, Gribkov, who played an important role in the Polish events, was the first to speak against Jaruzelski's claim that the imposition of martial law prevented outside intervention. But Jaruzelski refused the Soviet general's statement. On the contrary, he marked him as a key person asserting the military intervention and putting pressure on the Polish leadership.[1667] Gribkov did not deny that plans for sending the Warsaw Pact's troops to Poland had really been worked out. However, he meant the tense period of December 1980.[1668] Existence of the plans for an invasion of the PPR was retrospectively confirmed even by Georgy Shakhnazarov, a member of the CPSU international department. According to his testimony, it was understood that Moscow prepared various scenarios for solving the situation, as development in Poland had the potential to distort the Soviet security perimeter built since World War II.[1669]

During his life, the former Warsaw Pact's supreme commander, Marshal Kulikov, strictly denied the existence of any real plan of the alliance for military intervention in the PPR. His statement was based on the fact that not a single document of the Warsaw Pact's military structures frankly mentioning an invasion of Poland was known so far. As for the reason, Kulikov described a lack of political will of both Polish and Soviet leadership to take such a step.[1670] Moreover, the Marshal declared that polish withdrawal from the Warsaw Treaty Organization would have been obviously troublesome, but not catastrophic. He assumed that in case of war with NATO, the Soviet troops would have enough time to pass through the territory of the PPR with no serious problems.[1671] Inconsistency of claims by two of the former highest-ranking military officials of the Unified Command is another proof that in researching the Warsaw Treaty Organization, a fully critical and cautious approach has to be applied to testimonies of former generals as well as defectors.

[1667] Kramer, "Jaruzelski, the Soviet Union," 5.
[1668] Kramer, "In Case Military Assistance is Provided," 102.
[1669] *Wejdą nie wejdą,* 192.
[1670] Ibid, 156.
[1671] Mastny, "The Soviet Non-Invasion of Poland," 204.

To weigh General Jaruzelski's behavior during the Polish Crisis is very complicated. General Siwicki testified that in case of need, if "Operation X" had not proceeded according to plans, the Soviet officials had intended to replace Jaruzelski by some of the real hardliners, like Generals Eugeniusz Molczyk and Wlodzimierz Sawczuk, or a civilian Tadeusz Grabski.[1672] Kukliński claimed that since November 1981, the preparations for suppression of the opposition had been in such an advanced stage that if Jaruzelski had further hesitated, he would have been toppled. Afterwards, martial law would have been imposed by the aforementioned Molczyk.[1673] Documents confirm that during the whole crisis, the Soviet Union really put pressure on the Polish leadership. Only its forms and intensity were changing. This psychological game undoubtedly would raise some concerns about the risk of outside military intervention. After all, this was most likely the aim of the Soviet approach in some period. Nevertheless, a historiography has relatively sizable evidence today about an extreme unwillingness of the Soviet leadership to initiate a military engagement in Poland. However, not all of them were available to Jaruzelski in 1980 and 1981. The Polish leadership, for example, lacked important information about discussions at the Soviet politburo sessions.[1674] The documents known so far clearly prove that the USSR, at least on the eve of December 13, 1981, did not plan any military intervention either on its own or through the Warsaw Treaty Organization. Jaruzelski verified this fact during his attempts to get guarantees for "Operation X" of the alliance, or at least Moscow alone. It is good to add that at the same time, unaware of today's available documents, the Polish leader could not be sure how the USSR would have reacted in the case of a substantial political breakthrough in the PPR or personal changes in the Kremlin.

[1672] Kramer, "Poland 1980–81," 123.
[1673] Kukliński, *Wojna z narodem*, 11.
[1674] Kramer, "Jaruzelski, the Soviet Union," 5.

Consequences of the Polish Crisis for the Warsaw Treaty Organization

The fact that the Polish Crisis did not result in an outside military invasion, despite the real risk of collapse of the communist regime in Poland and thus even the threat to integrity of the alliance, was seen by Vojtech Mastny as the most important consequence of the 1980-1981 events for the Warsaw Pact.[1675] He revisited his claim later. Referring to a retrospective statement of Marshal Kulikov, Mastny notes that the leaders of Solidarity, in fear of the possible military intervention, tried to avoid any declaration challenging the loyalty of Poland to the Warsaw Treaty Organization.[1676] This claim, however, is misleading. Despite those retrospective words of the former supreme commander, concerns for membership of the PPR in the alliance or disruption of military-strategic interests resonated very strongly within the Eastern Bloc countries during the years 1980 and 1981.

At the same time, Mastny considers the realization of "Operation X" the biggest achievement of the Warsaw Treaty Organization, as it was built after 1969. According to his opinion, the imposition of martial law by Polish officials proves that a generation of top-rank Eastern officers were brought up, for whom the Warsaw Pact's interests were more important than loyalty to their own population.[1677] He points out that since the beginning of the crisis, the Polish armed forces' command planned how to suppress Solidarity by force. But it was rather their own initiative than a directive from Moscow. In fact, stances of large part of the Polish officer corps corresponded to the Kremlin's interests.[1678] That thesis, based on the words of Ronald Reagan, who after the action against the Polish opposition talked about the Soviet officers in Polish uniforms, is not precise. The Polish army command consisted of a large number of orthodox communists. Therefore, it is not surprising they did not want to tolerate the opposition's activities, which from their point of view was

---

[1675] Mastny, *A Cardboard Castle?*, 50.
[1676] Mastny, "The Warsaw Pact," 152.
[1677] Ibid, 153.
[1678] Mastny, *A Cardboard Castle?*, 51.

dragging the country into anarchy. Also, military thinking in the scope of a bipolar world could clearly be seen, eloquently presented by General Tadeusz Tuczapski. Anticipating a new escalation of the Cold War, military officials considered it necessary to maintain order in Poland and the country to be ready for a potential conflict. Jaruzelski also talked about the need to ensure domestic stability in the period when massive maneuvers of NATO and the Warsaw Treaty Organization were getting more frequent in the region.[1679] However, during the entire crisis, the Polish generality preferred a domestic solution.

Consequences of the Polish Crisis for the Warsaw Treaty Organization can be divided into political and military. From the former point of view, it is important that the alliance's official institutions were not involved in the solution of the Polish question, even though attempts to implement such a scenario existed. It is symptomatic that these were mostly initiatives of the Soviet satellites, not Moscow itself. Through its approach, the USSR confirmed the political role of the Warsaw Treaty Organization as framed since 1969. Its primary task remained a unification of the Eastern Bloc countries' foreign policy, especially toward the West and international forum in general. The Kremlin did not deem it necessary to broaden the range of the organization to solving internal troubles in its European sphere of influence. On these issues, the Soviet leadership intended to keep its privileged decisive authority.

From the military perspective, Jaruzelski did not enjoy the full trust of the Eastern Bloc countries for a short period. Despite the General's reassurance about loyalty to the Warsaw Treaty Organization and alliance with the USSR, a military exercise was held in 1982 presuming an attack by China rather than NATO, whereas not only the Romanian, but also Polish armed forces failed to join combat operations.[1680] According to the scenario, the PPA had to be deployed to eliminate domestic unrest and efforts to withdraw from the Warsaw Pact.[1681] During the "Druzhba-82"

[1679] Jaruzelski, *Historia nie povinna dzielić*, 44.

[1680] Mastny, *A Cardboard Castle?*, 56.

[1681] Memorandum of conversation with Marshals Ustinov and Kulikov concerning Soviet War Games, June 14, 1982, in *A Cardboard Castle?* (doc. 95): 462-465.

drills, held in January on the Czechoslovak territory,[1682] the Western news agencies immediately began to speculate that given the participation of 25 thousand troops, the action was supposed to put more pressure on the PPR.[1683] However, mistrust was broken quickly and as early as in the second half of March, when Poland hosted the "Friendship-82" maneuvers including Soviet, Polish, and East-German troops. The exercise was even commanded by General Molczyk, from his position of deputy supreme commander of the Warsaw Pact.[1684]

A long-term military impact of the Warsaw Treaty Organization was mostly bad economic conditions for the PPR. In December 1982, during talks with Kulikov, the Polish general staff recommended some austerity measures, like reduction of the Polish air forces by decommissioning obsolete aircraft types. Poland also failed to implement the Unified Command's urging to modernize its navy. Given its own economic troubles, the USSR was not able to provide the required financial assistance and Warsaw alone did not have enough resources for the program.[1685] Moreover, in reaction to "Operation X," Reagan abolished the "most favored nation" statute for Poland. This confronted the Eastern Bloc, and especially Moscow, with the need to increase economic assistance, as no consolidation of the Polish communist regime was imaginable without it.[1686] At the Military Council meeting, held in October 1982 in Warsaw, Jaruzelski assured the council that the crisis had not affected the role of the PPA within the Unified Armed Forces. However, he suggested that stabilization of the economy would slow down

---

[1682] The Soviet, Czechoslovak, and Hungarian troops took part in the exercise and also Kulikov traveled to the country. "Spojenecké cvičení Družba 82 zahájeno," *Rudé právo*, January 26, 1982; "состоялась беседа," *Правда*, January 29, 1982.
[1683] The CSSR considered it necessary to insist that the maneuvers were not taking place nearby Polish, but East-German borders. "Kam mířila 'zmýlená'," *Rudé právo*, January 29, 1982.
[1684] "Spojenecké cvičení na území PLR," *Rudé právo*, March 13, 1982.
[1685] Jarząbek, *"Poland in the Warsaw Pact"*.
[1686] *Západní agentury o zasedání ministrů zahraničních věcí Varšavské smlouvy v Moskvě*, 21. 10. 1982, 506/0/45, k. 5969, NA, Prague.

rearmament plans.[1687] The Polish representatives argued along these lines two months later during the above-mentioned talks with Kulikov. They loyally agreed with the righteousness of demands by the Unified Command and the Soviet general staff for acceleration of armaments modernization. But at the same time, they pointed out that the condition of the Polish economy did not allow them to even fulfill the objectives of the military five-year plan of 1981-1985. Warsaw wanted to deal with the problem during bilateral negotiations with Moscow, at the level of expert groups.[1688]

[1687] *Informace o přijetí vojenské rady spojených ozbrojených sil dne 22. 10. 1982 prvním tajemníkem ÚV PSDS W. Jaruzelskim*, 24. 9. 1982, FMO GŠ/OS, k. 53, p. č. 204, e. č. 38013, VÚA, Prague.

[1688] *PROBLEMY do rozmów z Naczelnym Dowódcą Zjednoczonych Sił Zbrojnych Układu Warszawskiego Marszałkiem Związku Radzieckiego W. G. KULIKOWEM w dniu 15. 12. 1982 r.*, Accessed October 16, 2019. http://www.php.isn.ethz. ch/ kms2.isn.ethz.ch/serviceengine/Files/PHP/111557/ipublicationdocument_single document/b37c6821-5c2a-483e-a505-46ae172041ad/pl/1982_Talks_Kulikov. pdf.

The Warsaw Treaty Organization and "The Second Cold War"

The Soviet military intervention in Afghanistan, in practice, definitely terminated détente, at least in the form the Brezhnev leadership had been striving for during the entire 1970s. Contrary to this, the first half of the following decade was one of a so-called Second Cold War. It was characterized mostly by a cooling of the East-West relations, a new escalation of the arms race, and an overall growth of international tensions. Such a situation posed new challenges for the Warsaw Treaty Organization. As explained in this chapter, the alliance's political structures - very rigid and fully controlled by Moscow - were not able to effectively react to the Afghan problem, the Euromissiles Crisis, and the failure of disarmament talks.

The Soviet intervention in Afghanistan

The Soviet Union involved the Warsaw Treaty Organization in its actions in Afghanistan in neither military nor hardly political terms. However, it is necessary to at least basically explain how the Soviet leadership made a decision to send its troops to the peaky Asian country. The situation actually illustrates well how key international and security issues were discussed in the Kremlin at the turn of the 1970s and 1980s. Thus, it helps to understand why the Warsaw Pact's political activity stagnated significantly in the last years of Leonid Brezhnev's rule.

As mentioned before, simultaneously with the declining health of the CPSU CC General Secretary, the triumvirate of Yuri Andropov, Andrei Gromyko, and Dmitry Ustinov was increasing its power within the top Soviet leadership. In the second half of the 1970s, these persons more and more took part in defining key priorities of the Soviet foreign and security policy. In the issues related to the Third World countries, Mikhail

Suslov also held some influence.[1689] It is good to add that the cooperation of these officials took place in secret, behind the back of the aging leader, as they were well aware that if they raised Brezhnev's suspicions, the General Secretary still had effective levers to remove them quickly from top functions.[1690]

Andropov, Gromyko, and Ustinov initially supported the military intervention in Afghanistan which aimed to protect the local communist regime from serious troubles with a resistance of non-negligible parts of the traditional Islamic population. In December 1978, Afghanistan and the Soviet Union signed a treaty of friendship, neighborliness, and cooperation. Since this moment, Moscow saw the Asian state as a part of its sphere of interest. Key persons in the Soviet leadership considered a potential loss of the country from Moscow's influence unacceptable, from a geopolitical as well as ideological point of view. The physically wilting Brezhnev actually did not take part at all in the essential debates which proceeded in the Kremlin before the intervention. He was gathering strength at his dacha. Moreover, initially he tended to lean more and more against the military operation. The General Secretary's opinion was changed by the brutal killing of his personal friend Nur Muhammad Taraki, the Afghan president. In the scope of power struggle within the People's Democratic Party, his rival Hifizullah Amin had him assassinated.[1691] A final decision on the invasion of Afghanistan was made at the Soviet politburo meeting on December 12, 1979. In fact, only half of the members of this body were present. Brezhnev joined them one and a half hours after the beginning of the session. His physical condition was very poor. According to witnesses, for the first 3-4 minutes he just sat quietly. Then he called the new Afghan communist leader Amin an "indecent man" and left the room.[1692]

The whole situation reflected general features of the decision-making process within Brezhnev's leadership. In some moments, policy defined by the General Secretary was challenged. This, however, never

---

[1689] Webber, "'Out of Area' Operations," 126-127.
[1690] Zubok, A Failed Empire, 251.
[1691] Ibid, 260-262.
[1692] Bacon, "Reconsidering Brezhnev," 15.

resulted in a deeper crisis. Debates actually did not take place at proper forums, but in closed top-echelon circles of the Soviet power hierarchy instead. Because of that, key decisions were not questioned at first glance. Actual strategy always formally enjoyed unanimous support. Such tendencies were even increased by Brezhnev's deteriorating health. Basically, the fundamental issues, like deployment of new SS-20 missiles, or the intervention in Afghanistan, were subjects of minimal official discussion. The effort to avoid disputes in burning matters often resulted in the postponing of important decisions. Inevitably, this fact not only reduced the ability to act, but also the flexibility of the Soviet leadership. It also contributed to the genesis of a deep internal crisis in the USSR as well as to the fall of détente.[1693] At the turn of the 1970s and 1980s, a number of vital Soviet international initiatives declined. The stagnation of Brezhnev's system apparently hit even foreign policy.[1694]

There was no offensive in the expansive plans behind the Soviet invasion of Afghanistan. Carter's security adviser Zbigniew Brzezinski and other Western military analytics at the time wrongly assumed that Moscow was creating a bridgehead to the Persian Gulf. In fact, the USSR had just reacted to concerns about the new U.S.-China rapprochement and the rise of Islamic fundamentalism in the Middle East after the 1979 Iran revolution. This phenomenon posed a potential threat to stability in the Soviet central-Asian republics. "*If we leave Afghanistan today, then tomorrow we will have to defend our borders against Muslim hordes in Tajikistan and Uzbekistan*," Andropov commented on the situation.[1695] The whole issue was also complicated by the cooling of relations between both superpowers. After the coup in Iran, the Soviet general staff warned against the possible transfer of the U.S. military bases to neighboring Pakistan and the creation of a bridgehead to a future seizure of Afghanistan by American troops. Moreover, the KGB reports said that the Afghan communist leader, the People's Democratic Party CC General Secretary Amin had played a double game. Soviet intelligence suspected him of side-negotiations with Washington. Moscow monitored this approach with

---

[1693] Bowker, "Brezhnev and Superpowers Relations," 103.
[1694] Nowak, *Od hegemonii do agonii,* 50.
[1695] Durman, *Útěk od praporů,* 203.

great concern. The experience of loss of Sadat's Egypt was still fresh in its memory.[1696]

Yuri Andropov, the KGB chairman, played an important role in the decision on the intervention. Similar to the Prague Spring, he provided Brezhnev with distorted reports. In this manner, he tried to influence the Soviet leader in favor of military action. In this effort, Andropov was backed by Defense Minister Ustinov. The beginning Euromissile Crisis, among other things, gave them a welcome base for argumentation. Together, the duo warned the General Secretary against possible dislocation of the U.S. short-range missiles in Afghanistan. In this case, the Soviet military bases in Kazakhstan and Siberia would have been put in danger. Paradoxically, a non-negligible part of the top Soviet military, led by the chief of general staff Nikolai Ogarkov, opposed the intervention in Afghanistan. However, Ustinov, whose relations with the Marshal were strained in the long term, tried to silence the army command. He rejected any relevant objections of the general staff. Not only the defense minister, but also Andropov attacked Ogarkov. On December 10, at the Soviet politburo meeting, he snapped at the Marshal: "*Focus on military affairs! Leave policy-making to us, the party, and Leonid Ilyich!*"[1697] This, among other things, proves that the decision-making process in the USSR was still under tough control of the political power. The military elites had just very limited influence in essential matters.

Taking all this into account, it is not surprising that the question of potential military intervention in Afghanistan was not discussed at all in advance at the Warsaw Pact's military as well as political sessions. In May 1979, at the meeting of the Committee of Ministers of Foreign Affairs, Andrei Gromyko just declared support for Afghan "progressive forces."[1698] The forthcoming invasion was not even indicated by the next session of this body, held in Moscow only a few days before the operation was launched. To the contrary, the alliance loudly promoted Brezhnev's

---

[1696] Zubok, *A Failed Empire,* 262.

[1697] Ibid, 263-264.

[1698] *Stenografický záznam projevu ministra zahraničních věcí SSSR A.A. Gromyka na zasedání Výboru ministrů zahraničních věcí členských států Varšavské smlouvy, které se konalo ve dnech 14.-15. května 1979 v Budapešti,* 1261/0/7, sv. 108, a.j. 109/info8, NA, Prague.

peace initiatives.[1699] At the same time, Moscow failed to consult its intentions with the allies outside the Warsaw Pact's institutions. With tongue in cheek, the Eastern Bloc countries' top officials learned about the Soviet military intervention from newspapers. The situation is illustrated well by the Hungarian case. On December 28 - four days after the operation had started - the Soviet ambassador to Budapest, Vladimir Pavlov, conveyed to the HSWP leadership a secret report on the events in Afghanistan. Moscow considered it sufficient compensation for the total absence of preliminary consultation with the Warsaw Pact countries about its intentions.[1700] We are reminded that Article III of the founding charter stipulated that the alliance member-states would consult together on all the important international issues which concerned their common interests.[1701] Therefore, this was another in a long series of Soviet violations of the fundamental rules of the alliance.

Absence of consultations on the Soviet plans in Afghanistan totally reversed the practice that had been established within the Warsaw Treaty Organization during the Helsinki process, when the Bloc's joint foreign policy was shaped at the alliance meeting. This is another proof that at the time of the CSCE negotiations, Moscow ascribed its satellites the biggest role ever, although the Kremlin still streamlined their actions. Therefore, the importance of the alliance's political talks reached its peak in the first half of the 1970s. This experience gave the Eastern Bloc countries a reason to expect that in the case of such a serious step, as the intervention in Afghanistan undoubtedly was, there would be some preliminary consultations within the Warsaw treaty Organization. In fact, the Soviet politburo decided on military action 12 days before the combat

---

[1699] Wanner, *Brežněv*, 61.

[1700] Békés, Csába, "Why Was There No 'Second Cold War' in Europe? Hungary and the East-West Crisis Following the Soviet Invasion of Afghanistan," in *NATO and the Warsaw Pact. Intrabloc Conflicts*, eds. Mary Ann Heiss and S. Victor Papacosma (Kent: Kent State University Press, 2008): 219.

[1701] NA, f. 1261/0/44, i. č. 41, k. 20, *SMLOUVA o přátelství, spolupráci a vzájemné pomoci mezi Albánskou lidovou republikou, Bulharskou lidovou republikou, Maďarskou lidovou republikou, Německou demokratickou republikou, Polskou lidovou republikou, Rumunskou lidovou republikou, Svazem sovětských socialistických republik a Československou republikou*, 14. 5. 1955.

operations started. There was plenty of time for talks with the allies, or at least informing them. This prompted a certain parallel with the Caribbean Missile Crisis. The Soviet satellites once again had to face the consequences of the Kremlin's unpredictable decision; not only were they unable to influence it, but also were given no reports in advance. The fact that the Brezhnev leadership took the whole problem as a strictly bilateral issue with the Afghan communists was a sort of band-aid for the Warsaw Pact member-states' officials. Moscow did not try to involve the allies in the solution even subsequently.[1702]

The official Soviet statement was not delivered to the Warsaw Pact members until mid-January 1980.[1703] Then, Gromyko justified the intervention in Afghanistan with the existence of a mutual allied treaty and repeated requests by the Afghan government. The Kremlin added that in case of the fall of the communist regime in the country, a very dangerous military bridgehead would have been established on the southern borders of the Soviet Union.[1704] The issue was never discussed in detail at the Warsaw Pact meetings. In the Committee of Ministers of Foreign Affairs, for example, it was mentioned marginally in the era of Mikhail Gorbachev. Before, it was limited to complaints by the Soviet representatives that the West used the whole problem as a pretext for a "demagogic" attack on the socialist countries.[1705]

It must be added that no Eastern Bloc country was seriously interested in opening the Afghan question. An opinion prevailed that it would be better if the USSR paid the consequences alone and not engage the Warsaw Treaty Organization in its Greater Middle-East policy.[1706] Reaction of the Carter administration to the Soviet incursion into Afghanistan was actually surprisingly fast and uncompromising. The U.S.-Soviet relations soon reached the coldest point since the Caribbean Crisis in 1962. The United States imposed economic sanctions on the USSR and announced a boycott of the Olympic Games in Moscow. The

---

[1702] Békés, "Why Was There No 'Second Cold War' in Europe," 219-221.

[1703] Wanner, *Brežněv*, 61.

[1704] *Informace o vývoji událostí v Afghánistánu*, 26. 2. 1980, 1261/0/7, sv. P 132/80, b. 10, NA, Prague.

[1705] Locher, "*Shaping the Policies of the Alliance*".

[1706] Wanner, *Brežněv*, 61.

CIA was given permission to start military assistance to the Afghan Islamic rebels.[1707]

Futile attempts to restore détente

Despite a deteriorating international situation, the Eastern Bloc countries still considered military relaxation as their top priority, especially some sort of disarmament in Europe. The Warsaw Pact's aim was holding a conference that would have supplemented the Helsinki Act with a similar document for the military area. After all, before the Afghan war broke out, some Western countries showed their interest in such a scenario. The states of the Soviet sphere of influence planned to use the upcoming CSCE summit in Madrid to assert their goals. Their approach was naturally based on previous resolutions of the alliance political bodies. The situation was not only complicated by the Soviet actions in Afghanistan, but also by more and more emphasis put by the West on respect for human rights within the Eastern Bloc. The European Economic Community countries actually intended to focus on that issue in Madrid. The Warsaw Pact, on the contrary, desperately sought to distract attention from it through appeals for military relaxation.[1708]

Moscow came to the conclusion that given the gradually less favorable situation it would be useful to hold the Political Consultative Committee session after one and a half years. Initially, the meeting should have mostly been of celebrative features. It was envisaged that the gathering would take place on the 25th anniversary of the alliance; symbolically in the Polish capital, where the founding charter had been signed. However, the Western reaction to the Soviet approach in Afghanistan forced the Kremlin to revisit the scenario: in the new situation, the Warsaw Treaty Organization should have demonstrated its unity and strength, while at the same time sending a clear message that the

---

[1707] Westad, "The Fall of Détente," 23-24.
[1708] *Informace o stavu příprav madridské schůzky zástupců účastnických států Konference o bezpečnosti a spolupráci v Evropě, 9. 1. 1980, 1261/0/7, sv. P127/80, info6, NA, Prague.*

alliance was ready to continue in détente.[1709] During preparation of the meeting, Gromyko informed the Soviet satellites that a too-sharp tone of the resolution would not have been desirable. In these terms, Moscow was supported by the PPR which, as the hosting country, actively participated in the shaping of the document. Poland recommended stressing the alliance's willingness to continue in cooperation with "peaceful" and "realistic" forces in the West.[1710]

It is not surprising that works on the part of the resolution concerning Afghanistan took the longest time.[1711] In spring 1980, the Soviet satellites already clearly realized that the problem influenced the international events significantly and put their interests in danger too.[1712] After the West announced the boycott of the upcoming Olympic Games in Moscow at the end of January 1980, the Kremlin ordered the Warsaw Pact countries' officials to cancel their scheduled meetings with the West-European representatives. The move was supposed to put additional pressure on European NATO members to reject deployment of new U.S. missiles on their territories. However, the East-European states followed this Soviet directive only with great exasperation, as it directly contradicted their economic intentions. But only further research will

---

[1709] *Informace z porady tajemníků ÚV šesti bratrských stran konané v Moskvě 26. února 1980*, 4. 3. 1980, sv. P 132/80, b. 10, ibid; *Zasedání politického poradního výboru států Varšavské smlouvy*, 5. 3. 1980, b. 9, ibid; Report on six socialist states' party officials meeting in Moscow, February 28, 1980, PZPR KC, XIA/596, AAN, Warsaw.

[1710] Telegram by E. Wojtaszek, the Polish minister of foreign affairs, to E. Gierek, the PUWP CC First Secretary, April 18, 1980, XIA/590, ibid; Report by E. Wojtaszek, the Polish minister of foreign affairs, to E. Gierek, the PUWP CC First Secretary, on preparations of the Political Consultative Committee resolution, April 18, 1980, ibid.

[1711] Report by E. Wojtaszek, the Polish minister of foreign affairs, to E. Gierek, the PUWP CC First Secretary, on preparations of the Political Consultative Committee resolution, May 14, 1980, ibid.

[1712] The Czechoslovak report on the security situation, for example, correctly stated that the U.S. wanted to use the Soviet actions in Afghanistan as a pretext to isolate not only the USSR, but also the other socialist states on the international stage. *Průběžná informace o bezpečnostní situaci v ČSSR*, 22. 5. 1980, 1261/0/7, sv. P139/80, b. 1a, NA, Prague.

show how much the consequences of the Afghan war affected general relations between individual Warsaw Pact members and Moscow.[1713]

Given the unfavorable international situation, the Soviet Union intended to use the Warsaw Pact's supreme body session as a massive demonstrative action of the coherency of its sphere of influence. This plan was complicated by Romania. In the wake of the Euromissile Crisis and the Afghan war, the Ceausescu regime stepped up its criticism of the Soviet policy. In March 1980, for example, the Romanian delegation condemned the situation in Afghanistan at the UN General Assembly, calling it a threat to peace. In his conversation with Gromyko, Ceausescu also accused Moscow of irresponsible support for the "few revolutionaries estranged from the people" through the intervention.[1714] The Romanian dictator did not hesitate to criticize the Soviet approach in the Greater Middle East during negotiations with Western leaders. However, he strictly rejected their careful probes as to whether the SRR would have joined retaliatory measures in some way.[1715]

[1713] In the first few months of 1980, a visit of the West-German minister of foreign affairs, Hans Dietrich Genscher, to Prague as well as a meeting of Chancellor Helmut Schmidt with Erich Honecker were cancelled. Bulgaria, although the country did not plan such visits, received an instruction to not even think about these. The problem most affected Hungary, as its relation with the West had been developing significantly since the signing of the Final Act. The HUWP was given the Soviet directive to cancel the scheduled visit of its minister of foreign affairs to the FRG only one week before and to postpone a trip of its parliamentary delegation to the U.S. On January 29, a tumultuous discussion on the issue took place in the Hungarian politburo. It was probably the most dramatic session after 1956. Part of the Hungarian leadership recommended dismissing Moscow's demand and protecting the economic interests of the country. The backers of such an approach gained a majority, but Janos Kadar's personal intervention overturned the balance of power. Békés, "Why Was There No 'Second Cold War' in Europe," 225; Békés, Csába, "Hungary and the Warsaw Pact".

[1714] *Informace z porady tajemníků ÚV šesti bratrských stran konané v Moskvě 26. února 1980,* 4. 3. 1980, 1261/0/7, sv. P 132/80, b. 10, NA, Prague; *Rumunska Partia Komunistyczna – Charakteristika polityki międzynarodowej,* 31. 3. 1980, PZPR KC, XII 3056, 937/8, AAN, Warsaw.

[1715] On one hand, during a visit of Peter Carrington, the British minister of foreign affairs, to Romania, Ceausescu condemned the invasion of Afghanistan and supported demands to withdraw the Soviet troops. But at the same time, he did not want to join any counteraction, like the boycott of the Olympic Games in

The Political Consultative Committee, held in Warsaw, May 14-15, 1980, brought no breakthrough. At the meeting, the Warsaw Treaty Organization was supposed to formulate new tactics on how to face the fall of détente and the risk of a new round of the arms race. But the only result was a blessing to Brezhnev's initiative to hold a global top-level conference on military relaxation. It had to be clear that given the existing situation, such a proposal had minimal chance to succeed. Thus, the main aim was to show that the Eastern Bloc countries were willing to enter new forms of dialogue in order to prevent further escalation of international tensions.[1716]

Traditionally, at the center of the closed session was the Soviet leader's speech. Unlike the previous years, this one had a relatively sharp and offensive tone. The aging Brezhnev identified imperialism - especially from the U.S. - as the culprit for the fall of détente. He even made some reminiscences with the Korean, Berlin, and Caribbean crises. The General Secretary criticized the U.S. for its intensified armament, increasing cooperation with China,[1717] and strengthening its positions in Asia. Keeping with the Soviet security interests, he saw the possible deployment of new American accurate missiles in Europe as a fundamental complication for stability of the international situation. The Soviet leader stated that such an installation would have changed the existing balance of power to favor the West.[1718] He also warned the Warsaw Pact against the

---

Moscow or economic sanctions against the USSR. *Návštěva ministra zahraničních věcí V.Británie v SRR*, 10. 4. 1980, DTO 1945-1989, i. č. 34, e. č. 97, č. j. 012.820/80, AMZV, Prague.

[1716] *Zpráva o průběhu a výsledcích zasedání politického poradního výboru Varšavské smlouvy ve dnech 14.-15. května 1980 ve Varšavě*, 27. 5. 1980, 1261/0/7, sv. P139/80, b. 7, NA, Prague.

[1717] Within his raging against Beijing's policy, the CPSU CC General Secretary accused the Chinese communists of instigating anti-Soviet activities in Afghanistan, among other things.

[1718] Brezhnev expressed his suspicion that presence of Pershing-II rockets and cruise missiles capable of hitting strategic targets in the Soviet Union accurately in terms of minutes could have strengthened positions of U.S. hawks operating with a thesis of the first strike. This would have eliminated part of the strategic capacities of the USSR and at the same time exposed European countries of NATO to a retaliatory assault.

U.S. policy of approaching differently the individual countries of the Soviet sphere of influence. Brezhnev considered it an effort to harm relations between Moscow and its allies. He used this as a pretext to traditionally urge the Eastern Bloc countries' unified attitude on almost all international issues. In this regard, he even questioned an explanation of the Warsaw Pact's strictly European range.[1719]

The CPSU CC General Secretary had to mention - at least marginally - the burning problem of Afghanistan. Probably taking into account initial successes of the Soviet army, he promised the allies that the USSR would stay in the country only as long as necessary. Nevertheless, Moscow did not want to discuss this issue at the Political Consultative Committee in detail. Brezhnev talked about it mostly in connection with the Western anti-Soviet campaign.[1720] He conveyed a position taken by the Soviet politburo, which was based on ideological, rather than relevant arguments: Washington chose Afghanistan as a pretext to restart the arms race, strengthen its positions in the Persian Gulf, and continue an anti-Soviet crusade.[1721] Given the existing level of loyalty of the Soviet satellites, it is not surprising that the Afghan strategy was formally supported in the speeches of all the party leaders, except for Nicolae Ceausescu. He, on the contrary, appealed for withdrawal of the Soviet troops, calling such a move a necessary condition for a peaceful solution of the conflict.[1722] The Warsaw Pact's supreme body resolution actually called for a political settlement, most likely regardless of the real Soviet intentions.[1723]

Like in many previous cases, the Romanian delegation did not present its different stances publicly, but at closed talks only. This allowed

[1719] *Projev vedoucího sovětské delegace generálního tajemníka ÚV KSSS a předsedy prezidia Nejvyššího sovětu SSSR L.I. Brežněva na zasedání Politického poradního výboru Varšavské smlouvy konaném ve dnech 14.-15. 5. 1980 ve Varšavě*, 1261/0/7, sv. P139/80, b. 7, NA, Prague.
[1720] Ibid.
[1721] Durman, *Útěk od praporů*, 206.
[1722] *Zpráva o průběhu a výsledcích zasedání politického poradního výboru Varšavské smlouvy ve dnech 14.-15. května 1980 ve Varšavě*, 27. 5. 1980, 1261/0/7, sv. P139/80, b. 7, NA, Prague.
[1723] *Politbüro des ZK Reinschriftenprotokoll nr. 17*, 29. 4. 1980, DY 30/ J IV 2/2/1836, BArch, Berlin.

Ceausescu to act very frankly. He blamed not only the West, but also the Soviet Union and the other socialist countries for the fall of détente. He tried to mediate the tone of the final documents, demanding not to name NATO, the U.S., or the FRG.[1724] Such an approach, motivated by an effort not to complicate mutual relations with the Western countries, became a typical Romanian strategy at the Warsaw Pact sessions in the first half of the 1980s.

The main message of the Political Consultative Committee resolution was a pleasing phrase that the policy of relaxation had and even could have no rational alternative.[1725] A retrospective conversation of Erich Honecker and the Polish minister of foreign affairs, Emil Wojtaszek, reveals that the priority during the shaping of the document was to prevent the West from rejecting it categorically and to open a room for dialogue in this way. However, this aim was met only partially. The moderate diction as well as the absence of significant and publicly-presented different stances of Romania surprised the West.[1726] But the Warsaw Pact resolution still failed to calm down the international situation. Actually, the British

---

[1724] Also, the fact that the Warsaw Pact wartime statute was finally approved at the meeting contributed to the strained situation. As explained in the second chapter, the Romanian leader was so dissatisfied with the agenda and course of the session that he left the Polish capital right after it ended. The Six considered Ceausescu's stances as directly contradicting policy of Moscow as well as the entire Warsaw Treaty Organization. But they appreciated that the SRR did not seriously harm the outwardly presented unity of the alliance and after some modifications, its delegation signed the final documents. *Zpráva o průběhu a výsledcích zasedání politického poradního výboru Varšavské smlouvy ve dnech 14.-15. května 1980 ve Varšavě*, 27. 5. 1980, 1261/0/7, sv. P139/80, b. 7, NA, Prague.

[1725] *Politbüro des ZK Reinschriftenprotokoll nr. 17*, 29. 4. 1980, DY 30/ J IV 2/2/1836, BArch, Berlin; Draft of E. Gierek's toast on occasion of the Political Consultative Committee session, undated, most likely May 1980, PZPR KC, XIA/590, AAN, Warsaw.

[1726] *Ministr zahraničních věcí PLR v NDR – informace*, 19. 6. 1980, DTO 1945-1989, i. č. 25, e. č. 2, č. j. 014.452/80, AMZV, Prague; Actually, before the Political Consultative Committee session, speculations spread among the Western diplomats that Ceausescu might not take part in the meeting in the Polish capital at all. Cable from the British embassy in Belgrade regarding the upcoming 25th anniversary of the Warsaw Treaty Organization, April 22, 1981, FCO 28/4062, 061/1, TNA, London.

diplomacy labeled the declaration as a long-expected "peace" initiative by the USSR who was virtually losing ground since the outbreak of the war in Afghanistan. Given the Afghan conflict, London compared the effort to hold the global top-level conference on military relaxation to a burglar calling for an anti-crime convention. [1727]

Stagnation of the Warsaw Pact's foreign political activities

The 17th session of the Political Consultative Committee was the last Warsaw Pact action which Leonid Brezhnev actively took part in. Until his death in November 1982, the body did not meet anymore. The Soviet leader probably did not know that his speech would not only balance 25 years of the alliance's activities, but also his own contribution to its development. He praised the Warsaw Treaty Organization for becoming "*a forgery, where our coordinated international policy as well as the firm shield protecting the socialist gains of our nation has been forged.*" The Soviet leader actually linked establishing a strategic balance in the world to the alliance's activities so far. He also pointed out the significant improvement of its mechanisms since the 1960s.[1728] In fact, the Warsaw Treaty Organization had been losing the rest of its political ability to act. The alliance was unable to avoid the fall of détente, effectively influence the international situation, or solve the Eastern Bloc's internal troubles, as the crisis of the Polish communist regime soon proved in 1980 and 1981.

In the last months of Brezhnev's rule, the alliance meeting lost the rest of their dynamics. Sessions were limited to an exchange of opinions on various international issues, but the Soviet policy was neither substantially revisited nor criticized.[1729] In this regard, Romania was an

---

[1727] Report by the British Foreign Office on the Warsaw Pact's Political Consultative Committee declaration, wrongly dated 20. 5. 1979, probably May 20, 1980, FCO 28/4062, ibid.

[1728] *Projev vedoucího sovětské delegace generálního tajemníka ÚV KSSS a předsedy prezidia Nejvyššího sovětu SSSR L.I. Brežněva na zasedání Politického poradního výboru Varšavské smlouvy konaném ve dnech 14.-15. 5. 1980 ve Varšavě,* 1261/0/7, sv. P139/80, b. 7, NA, Prague.

[1729] Wanner, *Brežněv,* 64.

exception. In the wake of existing international complications, the Ceausescu regime was increasingly breaking with Moscow's course. Especially, the Committee of Ministers of Foreign Affairs was considered an appropriate platform where Bucharest could present its own initiative and oppose those of the USSR at the same time. The rest of the member-states tried to formulate their interests together with Moscow, not to challenge it. Having at least a basic conformity of stances within the Six was still literally a sacred rule.[1730] To some extent, absence of collective meetings of the party leaders was substituted by the talks of the central committee's secretaries. However, they served even more as just a forum to inform the satellites about the Soviet opinion on actual issues. Rebellious Romania was ordinarily not invited. On the contrary, the Soviet non-European satellites' representatives often took part in these meetings.[1731] Holding collective meetings of deputy foreign ministers under the formal auspices of Comecon, instead of the Warsaw Treaty Organization, was another innovation. Issues of foreign policy were mostly discussed there, but this framework also allowed for involving in consultations the non-European Soviet allies - Cuba, Vietnam, Mongolia, Laos, and North Korea.[1732]

The Warsaw Pact's political bodies' meetings, held before Brezhnev's death, were characterized by a desperate effort to save what remained of détente. Moscow revived the idea of signing a non-aggression

---

[1730] Locher, "*Shaping the Policies of the Alliance*".

[1731] *Zpráva o průběhu a výsledcích porady tajemníků ústředních výborů bratrských stran šesti socialistických zemí*, 15. 8. 1980, 1261/0/7, sv. P144/80, b. info2, NA, Prague; *Zpráva o průběhu a výsledcích porady tajemníků ústředních výborů komunistických a dělnických stran socialistických zemí pro mezinárodní a ideologické otázky*, 17. 11. 1981, 1261/0/8, sv. P24/81, b. 7, ibid. We are reminded that this trend was proven even during the Polish Crisis, when Moscow provided the non-European Soviet satellites' representatives with more information about the events in the PPR than the Ceausescu regime.

[1732] *Note on the Meeting of Deputy Foreign Ministers of the Socialist Countries (WP Member States, Mongolia, Vietnam (Ambassador), Cuba, and North Korea) With Responsibility for International Organizations*, September 9, 1980, Accessed October 16, 2019. http://www.php.isn.ethz.ch/kms2.isn.ethz.ch/service engine/Files/PHP/17390/ipublicationdocument_singledocument/c2691c16-68cb-4f2d-aa61-7bdd59cfc13c/en/800909_Note_E.pdf.

treaty between both alliances as well as a pledge not to use nuclear weapons first.[1733] The Soviet leadership most likely naively thought that in this way it would be able to reverse the international development without its own concessions on key matters. Instead of solving the real problems, the Kremlin continually used the Warsaw Treaty Organization as a tool to cover its peaceful, yet largely propagandistic initiatives. For example, in June 1980, fully along these lines, a meeting of the member-states' parliament representatives was held in Minsk, at the invitation of the Supreme Soviet of the USSR. Basically, it did not differ from the previous sessions of this platform. Individual delegations only manifested support for the last documents of the Political Consultative Committee. Moscow also intended to coordinate the approach of its satellites at the upcoming gathering of the Inter-Parliamentary Union, which would have been used to appeal for continuation of détente. Practically, the marginal importance of the Minsk meeting was illustrated by its absolutely vague final resolutions.[1734]

The Kremlin most likely underestimated the seriousness of the international situation. In August 1980, during the talks between representatives of the Six in Budapest, Boris Ponomarev, the CPSU CC secretary, informed the representatives about the actual priorities in foreign and domestic security policy. The USSR was trying to prevent a disruption of the existing balance of military forces in Europe. It wanted to stop the deployment of new U.S. missiles on the territories of West-European NATO countries, without the need to withdraw or reduce its own SS-20 rockets. In fact, the Soviet leadership saw in the Euro-missile Crisis an opportunity to foment discord between the United States and their allies on the Old Continent. Therefore, it focused on spreading propaganda that tried to refute the claims that a threat of war was posed by the existence of the Warsaw Pact and stressed its different features from

---

[1733] Mastny, *A Cardboard Castle?*, 55.

[1734] *Zpráva o setkání představitelů parlamentů členských států Varšavské smlouvy v Minsku*, 27. 6. 1980; Ibid, *Výzva parlamentům a poslancům států Evropy a světa přijatá v Minsku dne 17. června 1980*; *Komuniké o setkání představitelů parlamentů členských států Varšavské smlouvy v Minsku*, 17. 6. 1980, 1261/0/7, sv. P143/80, b. 16, NA, Prague.

NATO.[1735] Détente actually survived longer on the European level than in both superpowers relations. In general, the West-European countries showed more interest in the development of economic and cultural ties with the Eastern Bloc than the U.S. did. The Kremlin was aware of this fact. In 1980, at the Warsaw Pact's meeting, Gromyko stressed that there were differences between NATO members on the issue of Euromissiles. He stated that the U.S. policy enjoyed unreserved support by Great Britain, plus one or two countries only. According to the opinion of Moscow, the rest of the Western alliance was not interested in escalation of the confrontation.[1736]

The USSR placed great hopes on the aforementioned CSCE summit in Madrid. Its beginning was scheduled for November 1980. Moscow planned to prepare carefully for the gathering. Its success should have been ensured traditionally through close cooperation of the Eastern bloc countries. Scenarios of strategy were finalized at a series of meetings of the deputy heads of diplomatic resorts and the Committee of Ministers of Foreign Affairs of the Warsaw Treaty Organization. The main question was how to prevent the West from using the summit in Madrid as a platform to attack the socialist countries for violations of human rights and the war in Afghanistan, instead of talks about holding a conference on military relaxation. The following tactic was chosen: Basically, the Warsaw Pact countries were supposed to present their well-known peace initiatives once again, especially those which contained some points of contact with the stances of some Western governments.[1737] Such an approach was complicated by Romanian policy and also largely by the beginning Polish Crisis. The Committee of Ministers of Foreign Affairs' session, held in Warsaw in October 1980, was affected significantly by the

---

[1735] *Vystoupení s. B.N. Ponomarjova v Budapešti 15.7. 1980*; Ibid, *Zpráva o průběhu a výsledcích porady tajemníků ústředních výborů bratrských stran šesti socialistických zemí*, 15. 8. 1980, sv. P144/80, b. info2, ibid.
[1736] Locher, *"Shaping the Policies of the Alliance"*.
[1737] *Informace o poradě náměstků ministrů zahraničních věcí členských států Varšavské smlouvy k otázkám přípravy madridské schůzky signatářských států Závěrečného aktu konference o bezpečnosti a spolupráci v Evropě*, 6. 8. 1980, 1261/0/7, sv. P144/80, info8, NA, Prague.

latter factor. The domestic event in the PPR stifled the appeals concerning the Madrid summit.[1738]

In 1980, Romanian policy towards the Warsaw Treaty Organization became gradually unclear and less consistent than before. In practice, an approach of the Ceausescu regime at the international stage substantially differed from the unified course of the alliance. However, economic reasons[1739] and later also concerns about the development in Poland forced it to keep correct relations with the Eastern Bloc. Moreover, Bucharest was very interested in hosting the next CSCE summit.[1740] This intention was hardly realizable without political support by the alliance partners. Therefore, the Romanian representatives' behavior at the Warsaw Pact meetings as well as bilateral negotiations within the Eastern Bloc oscillated without any coherent conception between extreme confrontation and specious loyalty.[1741] If the USSR tolerated the deviation of Romanian

---

[1738] Gromyko basically just informed the Soviet satellites that the strategy for the Madrid summit had been clearly defined by the recent Political Consultative Committee resolution. At the time, no signatory of the Final Act openly opposed the key idea of holding a conference on military relaxation and disarmament. However, the USSR was concerned that during the meeting in Madrid, NATO would stipulate unacceptable preconditions. Therefore, the Warsaw Pact should have pushed forward holding the conference in two stages. Moscow also did not like the NATO countries' efforts to extend so-called mutual confidence building measures. The Kremlin feared a possible leaking of the Warsaw Pact's military secrets. For this reason, Moscow intended to promote holding two separate conferences - the first on disarmament and the second on the possibility to strengthen trust and control. *Vystoupení ministra zahraničních věcí Svazu sovětských socialistických republik A.A. Gromyka*, 19. 10. 1980; *Informace o zasedání výboru ministrů zahraničních věcí ve Varšavě*, 6. 11. 1980, sv. P153/80, info3, ibid.

[1739] Given the deteriorating economic situation of Romania, Ceausescu had to moderate his independent policy. He intended to boost the Romanian economy through better cooperation within the Comecon. Moreover, after the Iranian revolution, the SRR desperately needed oil supplies. *Návštěva delegace NDR v SRR*, 8. 7. 1980, DTO 1945-1989, i. č. 25, e. č. 2, č. j. 014.952/80, AMZV, Prague.

[1740] *Note on the Meeting of the Deputy Foreign Ministers*, October 18, 1980, Accessed October 16, 2019. http://www.php.isn.ethz.ch/lory1.ethz.ch/collections/colltopicf16e.html?lng=en&id=17393&navinfo=15700.

[1741] At the aforementioned meeting of the Warsaw Pact countries' parliament representatives in June, Nicolae Giosan surprisingly appreciated the importance of that forum. He even stated that the SRR was ready to intensify cooperation on

foreign policy and deliberately avoided taking harsh actions during the entire 1970s, in the existing situation Moscow had no other choice, in practice. Given the damaged East-West relations, the war in Afghanistan, and the crisis in Poland, no tough intervention against the Ceausescu regime was possible.[1742] In addition, the Soviet leadership desperately tried to keep an absolute illusion of unity of its sphere of influence. Therefore, at the Warsaw Pact meetings, the Six practically did not confront Romania. The Soviet delegations at least tried to achieve some

---

this level. He also praised the final documents of the Political Consultative Committee session in Warsaw. But only three months later, at a meeting of deputy foreign ministers, Bucharest informed its Warsaw Pact partners that regardless of the alliance's position, the country was about to propose at the UN the freezing, or even reduction of existing military budgets at best and the beginnings of a broad discussion on the economic and social impacts of the arms race. In October, at the Committee of Ministers of Foreign Affairs the SRR even tried to distance itself from the line defined by the Political Consultative Committee in order to get a free hand of acting at the CSCE summit in Madrid. In December, to the contrary, the Romanian Prime Minister, Ilie Verdeţ, assured his Czechoslovak counterpart about willingness of the SRR to act alongside the Warsaw Pact countries on the international stage, despite some different opinions. In the communiqué from this bilateral meeting, Bucharest supported the Political Consultative Committee documents once again. *Zpráva o setkání představitelů parlamentů členských států Varšavské smlouvy v Minsku*, 27. 6. 1980, 1261/0/7, sv. P143/80, b. 16, NA, Prague; *Note on the Meeting of Deputy Foreign Ministers of the Socialist Countries (WP Member States, Mongolia, Vietnam (Ambassador), Cuba, and North Korea) With Responsibility for International Organizations*, September 9, 1980, Accessed October 16, 2019. http://www.php.isn.ethz.ch/kms2.isn.ethz.ch/serviceengine/Files/PHP/17390/ipublicationdocument_singledocument/c2691c1 6-68cb-4f2d-aa61-7bdd59cfc13c/en/800909_Note_E.pdf; *Informace o zasedání výboru ministrů zahraničních věcí ve Varšavě*, 6. 11. 1980, 1261/0/7, sv. P153/80, info3, NA, Prague; *Zpráva o průběhu a výsledcích přátelské pracovní návštěvy předsedy vlády ČSSR s. L. Štrougala v Rumunské socialistické republice*, 15. 12. 1980, sv. P158/80, b. 10, ibid.

[1742] At the end of the 1970s, some Western analysts still considered the Soviet military intervention against Ceausescu's regime not completely out of the question. They thought, however, that it would have happened in a very specific situation, for example, if Romania were to leave the Warsaw Pact. Given the international impacts of such a move, they presented the Soviet invasion of the RSR as an emergency last resort in case Moscow would have felt a direct threat to its fundamental interests. *European Political Cooperation: Ministeral Meeting in Brussels: 4 December 1978*, December 1, 1978, FCO 28/3273, TNA, London.

compromise with the Romanian representatives in backrooms, in order to pass final documents unanimously.[1743] Over time, especially after the death of Brezhnev, such an approach became increasingly complicated.[1744]

From the Warsaw Pact's point of view, the international situation failed to improve even in 1981. The long-running CSCE summit in Madrid brought no significant results. The key Eastern initiative of holding a conference on military relaxation and disarmament, the West tried to connect with the issue of respect for human rights in the socialist states.[1745] The existing disarmament forums, especially the Vienna talks, were in a deadlock.[1746] In December 1980, the North Atlantic Council even suggested that NATO countries could have left the negotiations in the Austrian capital.[1747] Thus, in spring 1981, the Eastern Bloc states had to

---

[1743] *Informace o zasedání výboru ministrů zahraničních věcí ve Varšavě*, 6. 11. 1980, 1261/0/7, sv. P153/80, info3, NA, Prague.

[1744] Within the Soviet leadership, probably Dmitri Ustinov asserted the tougher approach towards Romania. In mid-1982, during a conversation with the Polish representatives, he said that at the next Political Consultative Committee session, the party leaders of the Six should have talks to Ceausescu completely freely, which essentially meant freedom to criticize his policy. He also expressed beliefs that the Romanian General Secretary was either not aware of the seriousness of the international situation or had totally lost his mind. The Soviet defense minister spitefully noted that if Ceausescu had been really such a great statesman as he liked to present himself, he would have been able to convince NATO to dissolve a long time ago. In that situation, the USSR allegedly would have disbanded the Warsaw Treaty Organization immediately. Memorandum of conversation with Marshals Ustinov and Kulikov concerning a Soviet War Games, June 14, 1982, in *A Cardboard Castle?* (doc. 95): 462-465.

[1745] *Informace o dosavadním jednání II. etapy madridské schůzky KBSE*, 17. 4. 1981, PK 1953-1989, kat. č. 816, k. č. 224, č. j. 012.662/81, AMZV, Prague.

[1746] Instead of constructive proposals, the Warsaw Treaty Organization emphasized the fact that in 1980, the USSR had finished a unilateral withdrawal of 20 thousand troops and one thousand tanks from the GDR. On the contrary, since the beginning of the Vienna talks, the U.S. military presence in the FRG had increased by 24.5 thousand persons. It is good to add that neither did the West show the will to make a real deal. *Informácia o priebehu a výsledkoch 23. kola viedenských rokovaní o vzájomnom znížení ozbrojených síl a výzbroje v sjednej Európe*, 15. 4. 1981, č. j. 012.661/81, ibid.

[1747] *Informácia o priebehu a výsledkoch 22. kola viedenských rokovaní o vzájomnom znížení ozbrojených síl a výzbroje v sjednej Európe*, 9. 1. 1981, kat. č. 812, k. č. 221, ibid.

admit that the only at least partially working platform for talks on reduction in military area remained the Committee for Disarmament in Geneva.[1748] There, unlike in Vienna, the talks did not proceed on the basis of blocs. This, in practice, made it impossible for the Kremlin to engage the Warsaw Treaty Organization in its strategy substantially.

The Soviet foreign political concessions

In this situation, the Soviet leadership slowly decided on certain corrections of its policy so far. The 26th CPSU Congress, held in late February 1981, put emphasis on the situation in Europe and Euro-Asia area respectively. Global ambitions, which Moscow operated with in the second half of the previous decade, took a back seat.[1749] There were probably some conceptual disputes in the Kremlin. These also hit also the ruling triumvirate. The part of the leadership represented mostly by Dmitry Ustinov asserted other military measures. On the opposite hand, diplomats led by Andrei Gromyko were increasingly aware that further armament was counterproductive. According to their opinion, competition with the West should have been limited to the political dimension.[1750] Others, like Yuri Andropov, for instance, resorted to catastrophic scenarios. The KGB chairman began to warn that the U.S. was trying to escalate the international situation and intentionally increase the risk of outbreak of war.[1751] However, it is necessary to add in his defense, that at the time, the United States' representatives, led by William Casey, the CIA director, and

---

[1748] *Informace o průběhu a výsledcích jarního kola zasedání Výboru pro odzbrojení v Ženevě v r. 1981*, 7. 5. 1981, kat. č. 816, k. č. 224, č. j. 013.077/81, ibid.

[1749] Litera, "Sovětský svaz, 'Eurorakety',": 259.

[1750] Mastny, *A Cardboard Castle?*, 55.

[1751] In May 1981, Andropov invited Brezhnev to a closed meeting of the Soviet intelligence. There he told the surprised participants that the U.S. was making preparations for a surprise nuclear attack on the USSR. Therefore, he asked for establishing a new early warning system based on cooperation between the KGB and the GRU, the Soviet military intelligence. The operation was named RJAN (raketno-jadernoje napadeniie). Paradoxically, an idea of sudden nuclear assault by the West was not resurrected by the Soviet military circles. They had considered such a scenario basically unreal since the early 1970s. It was most likely Andropov's own initiative.

Caspar Weiberger, the defense minister, actually approved several very provocative operations, including military exercises nearby Soviet borders and naval bases. The aim was to escalate pressure on the Kremlin.[1752]

At least some concessions began to emerge in the Soviet foreign policy. In 1981, in the UN General Assembly, the USSR presented a declaration on prevention of nuclear disaster. Its essence was to convince the nuclear powers to take the no-first-use pledge. One year later, on the soil of the international community, Moscow demonstratively made such a commitment unilaterally and called the other countries with atomic arsenals for the same move.[1753] In November 1981, during his visit in West Germany, Brezhnev also suggested some willingness to change the approach in the Euromissile Crisis. For the first time, he admitted a possibility of unilateral reduction of the Soviet mid-range rockets. Up to this point, Moscow had tried to deter NATO from the deployment of new U.S. missiles in Europe mostly by threats of strengthening its own arsenals. A concrete condition, however, should have been negotiated at appropriate talks between both superpowers, which began at the end of the month in Geneva.[1754]

On the issue of Euromissiles, the Soviet Union required maximum loyalty from its satellites. Thus, Moscow followed with great displeasure the Ceausescu regime, which took completely different stances. Any threat of countermeasures, which would have increased the number of destructive weapons on the Old Continent, Bucharest considered a mistake. On the contrary, it proposed unilateral reduction of military budgets and troop numbers.[1755] By this factual promotion of a return to Khrushchev's military policy of the second half of the 1950s, Romania continually undermined the already barely effective initiatives of the Warsaw Treaty Organization.[1756] The Romanian course climaxed at the Geneva talks in the frank accusation that the USSR was not seeking a necessary compromise with the American side.[1757] Romania tried to play

---

[1752] Zubok, *A Failed Empire,* 271.
[1753] Śluszarczyk, *Układ Warszawski,* 84.
[1754] Wanner, *Brežněv,* 65.
[1755] Mastny, *A Cardboard Castle?,* 55.
[1756] Mastny, "The Warsaw Pact," 153.
[1757] Locher, "*Shaping the Policies of the Alliance*".

a role of neutral mediator between the blocs. Along these lines, by the end of 1981, Ceausescu addressed Brezhnev and Reagan, and subsequently all the heads of the Final Act signatory states, with an appeal for stopping deployment of mid-range nuclear missiles in Europe, followed by their total withdrawal.[1758]

On the level of the Warsaw Treaty Organization, the Euromissile Crisis was discussed at the Committee of Ministers of Foreign Affairs session, held after more than a year on December 1-2, 1981, in Bucharest. Gromyko conveyed Moscow's opinion on the threat of mid-range missiles: The Soviet SS-20s were not capable of striking the United States. The deployment of new American missiles, on the contrary, posed a direct threat to the Soviet Union's territory. In this statement, the Soviet minister *de facto* admitted that Moscow looked at the entire situation from its point of view only. The security of West-European NATO countries as well as the satellites in the Warsaw Pact was secondary. At the starting Geneva talks, Moscow planned to use the fact that three nuclear powers were members of NATO, whereas the Soviet Union had a monopoly on nuclear weapons within the Warsaw Treaty Organization. The claims about achieving parity in atomic forces the Kremlin considered as an opportunity to split the Western nuclear power. Moscow wanted to present the issue as a struggle between both alliances rather than the superpowers only. In this scenario, the USSR would have achieved an advantage over the United States, as Washington would have to share its quotas with Great Britain and France.[1759] However, the general approach of Moscow towards nuclear disarmament remained indecisive. At the Bucharest meeting of the Committee of Ministers of Foreign Affairs, for example, Gromyko

---

[1758] *Vystoupení ministra zahraničních věcí Rumunské socialistické republiky Št. Andrei, Bukurešť*, December 2, 1981, Accessed October 16, 2019. http://www.php.isn.ethz.ch/kms2.isn.ethz.ch/serviceengine/Files/PHP/20103/ipu blicationdocument_singledocument/cf3abe65-6e5c-4a81-9be5-0ccbc33488e5/c s /021281_SpeechRomanianFM.pdf.

[1759] *Vystoupení ministra zahraničních věcí Svazu sovětských socialistických republik A. A. Gromyka*, December 1, 1981, Accessed October 16, 2019. http:// www. php.isn.ethz.ch/kms2.isn.ethz.ch/serviceengine/Files/PHP/20104/ipublic ationdocument_singledocument / 8729f79e-d282-4d6b-99d1-1d380dd1e26c/cs/0 11281_SpeechSovietFM.pdf.

vaguely supported an idea of establishing nuclear-free zones in the world. But this was nothing more than an empty, propagandistic statement without actual content. The Soviet minister just spoke in favor of a two-year-old Bulgarian proposal to create the Balkan nuclear-free zone. Given the existing situation, that initiative was obviously dead already. Instead of looking for more constructive paths, the Warsaw Pact Committee of Ministers of Foreign Affairs competed in the criticism of the fall of détente and policy of the Reagan administration. The old thesis that the U.S. moved mankind closer to an atomic war came to the forefront again.[1760]

A relatively important aspect of the Bucharest meeting of the Committee of Ministers of Foreign Affairs was the Soviet outline of the character of future cooperation within the Warsaw Pact's political structures. In backrooms, the USSR announced that it considered two years as an appropriate interval between the Political Consultative Committee sessions. The Committee of Ministers of Foreign Affairs, on the contrary, should have been held even two times per year. To a large extent, it was supposed to take the role of a key body of the alliance. Without unavailable Soviet documents, it is impossible to clearly identify what exactly stood behind this decision. One of the reasons could also have been Brezhnev's poor health.[1761] It was much easier for the 75-year-old politician to send a delegation to the Committee of Ministers of Foreign Affairs session which conveyed to the Warsaw Pact countries the actual foreign political strategy of Moscow. Moreover, as already explained, the Soviet minister's speech was always a key moment of those meetings. It was actually Gromyko, who clearly defined the overall diplomatic strategy of the USSR at the time. Therefore, in the first half of the 1980s, the Political Consultative Committee obviously took a back seat. In the period between May 1980 and October 1985, it met just once. This situation, of course, was also influenced by the crisis in the Soviet leadership caused

---

[1760] *Informace o zasedání výboru ministrů zahraničních věcí členských států Varšavské smlouvy v Bukurešti ve dnech 1. a 2. prosince 1981*, December 7, 1981, Accessed October 16, 2019. http://www.php.isn.ethz.ch/kms2.isn.ethz.ch/service engine/Files/PHP/20107/ipublicationdocument_singledocument/26ac242c-cd31-4c09-8bfa-4b11db3809da/cs/011281_InformationReport.pdf.
[1761] This thought was presented by the Czechoslovak diplomacy, for instance. Ibid.

by a quick series of deaths of three Kremlin leaders over a period of 28 months. In that period, the Committee of Ministers of Foreign Affairs became the main platform of the alliance's political cooperation and shaping of the Eastern Bloc's strategy.

The U.S.-Soviet relations further deteriorated in 1982. The way out of the Polish Crisis undoubtedly contributed to this fact. Reagan considered the imposition of martial law in Poland as a personal insult. Given Colonel Ryszard Kukliński's desertion, the American side had enough information about the pressure Moscow put on the Polish leadership during 1980 and 1981. Therefore, the head of the White House was determined to seek revenge for the events in the PPR and undertake measures maximizing the troubles of the Soviet economy.[1762] On the contrary, in the last months of Brezhnev's life, the vast majority of the Soviet politburo still hoped that Reagan would come back to "realism" and restore cooperation with Moscow. In the effort to help such a process through influencing public opinions in the West, Brezhnev actually delivered his speech in June 1982, in which he committed to a no-first-use policy. Soon thereafter, Defense Minister Ustinov demonstratively declared that the Soviet army did not count on achieving victory in a nuclear war. There are suggestions that the Soviet leader in his twilight moments took the fall of détente personally. He realized that the intervention in Afghanistan had been a gross error. According to testimony of his foreign policy adviser Andrei Alexander-Agentov, the General Secretary once complained to Andropov and Ustinov: "*You got me into*

[1762] The economic situation of the USSR and the entire Eastern Bloc at the end of the Brezhnev era was negatively affected by the costs of an ambitious global foreign policy in the 1970s and massive armament. Many Warsaw Pact member-states increasingly became economically dependent on the West. Two months before his death, the General Secretary returned from his holiday in the Crimea. He read to the Politburo a report on traditional meetings with the communist leaders. He criticized that mutual economic cooperation suffered by the inability of many Soviet resorts to supply the socialist states with promised goods. Brezhnev said that more and more disillusion from the Comecon was spreading among its states' leaders. In this way, he indirectly called for some reforms of the organization. Litera, "Sovětský svaz, 'Eurorakety'," 259; Zubok, *A Failed Empire,* 264; Volkogonov, *The Rise and Fall,* 321.

this mess!"[1763] Nevertheless, Brezhnev's interference with politics was mostly counterproductive at the time. But the Soviet leadership's members were not courageous enough to tell the party boss that his physical condition was limiting him in work and the time had come for his resignation. The situation was made more complicated by the absence of a tradition of honorable retirement from public functions in the USSR. Moreover, many thought that Brezhnev still played an important, at least stabilizing role, regardless of his poor health.[1764]

Two weeks before he passed away, Brezhnev met the participants of the Moscow meeting of the Committee of Ministers of Foreign Affairs. Removing the risk of nuclear war he once again called the most important actual task. However, the will to take any essential steps was still missing. At the aforementioned Warsaw Pact session, held on October 21-22, 1982, Gromyko conveyed to the member-states the Soviet leadership's vague opinion that it was necessary to defend and deepen especially the "political capital" resulting from détente. He blamed Washington for unwillingness to reach agreement on reduction of nuclear weapons until new American missiles were deployed in Europe in order to get better position for negotiations. On the contrary, he put to the forefront the Soviet declarative pledge not to use nuclear weapons first. In this moment, not only Romania tried to correct the obviously more and more unsuccessful policy of the Soviet Union.[1765] Some of Moscow's intentions were carefully and in

---

[1763] Zubok, *A Failed Empire,* 264; 271-272.

[1764] The future CPSU CC General Secretary, Mikhail Gorbachev, who was very critical to Brezhnev, denied retrospective claims that his predecessor had been just a funny cartoon at the end of his life. Gorbachev, for example, said to Gromyko that it was necessary to support the aging Brezhnev despite his condition. He considered the General Secretary as a guarantee of stability both in domestic issues as well as at the international stage. Thatcher, Ian D., "Brezhnev as Leader," in *Brezhnev Reconsidered,* eds. Erwin Bacon and Mark Sandle (New York: Palgrave, 2002): 28, 33.

[1765] At the meeting, the SRR provocatively stated that the no-first-use initiative had not been the original Soviet move, but China had actually taken such a step 18 years earlier. Through Stefan Andrei words, Bucharest also refused to support publicly the Soviet approach at the Geneva talks on reduction of strategic arms. On the contrary, it proposed the negotiations on atomic weapons in Europe not to proceed on the U.S-Soviet level only, but also with the other European states participation, as the issue concerned the whole continent.

much less confrontational tones also criticized by Hungary. Both states tried to moderate diction of the final communiqué, rejecting verbal assaults on the Reagan administration and labeling it as the main culprit of the strained international situation. On the opposite side, Frigyes Puja, the Hungarian minister of foreign affairs, asserted influencing world events through improving relations with selected Western countries. He noted that a non-negligible part of political specter in NATO European members, Canada, as well as Japan did not agree with the actual U.S. policy and there was a strong peace movement.[1766]

There are suggestions that in the eve of the death of Brezhnev, the Soviet leadership tried to involve the Warsaw Treaty Organization more in its effort to improve the international situation. Thus, for the first time, the Soviet side conveyed the communiqué of the Moscow meeting of the Committee of Ministers of Foreign Affairs to the UN as an official document[1767] and at the same time it was sent to all the governments of European countries, the U.S., and Canada. For the first time, it was also announced when the next alliance session would be held. After long years, ministers of foreign affairs were supposed to meet regularly and systematically. Furthermore, the Political Consultative Committee session was scheduled in December 1982.[1768] The agenda of the meeting in Prague should have included the struggle against nuclear war and defining new ways in the efforts to save and also to strengthen détente in future.[1769] But on November 10, the preparations were paused by the news that the CPSU CC General Secretary had died in his sleep. Without his participation, the Warsaw Pact's supreme body meeting was unimaginable.

[1766] *Informace o zasedání výboru ministrů zahraničních věcí - Moskva 21. a 22. října 1982*, 10. 11. 1982, 1261/0/8, sv. P52/82, info2, NA, Prague.

[1767] Solely the Political Consultative Committee declarations were sent to the UN before.

[1768] *Informace o zasedání výboru ministrů zahraničních věcí - Moskva 21. a 22. října 1982*, 10. 11. 1982, 1261/0/8, sv. P52/82, info2, NA, Prague.

[1769] As usual, the meeting and the agenda were proposed by the hosting country, the CSSR this time. In fact, the place and date as well as items to be discussed were defined by the Soviet politburo. *Politbüro des ZK Reinschriftenprotokoll nr. 21,* 25. 5. 1982, DY 30/ J IV 2/2/1948, BArch; *Anlage Nr. 14 zum Protokoll Nr. 48 vom 23. 11. 1982* DY30/ J IV 2/2/1979, ibid.

Andropov's changes

After the death of Brezhnev, the Soviet politburo almost immediately installed 68-year-old Yuri Andropov to the position of General Secretary. For the first time in history of the USSR, change of the party's head took place without power intrigues. Not only did Cold War tensions contribute to this outcome, but also the fact that the new leader enjoyed the support of the other key persons in the Soviet hierarchy - Gromyko and Ustinov.[1770] Actually, the new General Secretary had very much benefited from his ties with the defense minister.[1771] After he became the chief of the CPSU, Andropov suggested in his speeches that he did not intend to change the essence of the Soviet foreign policy. However, one could notice a willingness to act more flexibly and more rationally in international issues than Moscow had done in the previous era. The USSR and its satellites were supposed to strictly split their approach towards international problems from those of ideological matters.[1772]

At the time of the Andropov inauguration, the Soviet leadership considered the international situation serious. Therefore, the Political Consultative Committee session, originally scheduled for December 1982 in Prague, was postponed just by a couple of weeks.[1773] The Warsaw Pact's supreme body, attended by Andropov for the first time, met on January 4-5, 1983. The Soviet leader's speech, in fact, provided little in new ideas and instead developed on the previous strategy. It agreed with the old claims that there was a military balance between the blocs and dusted off the idea of simultaneous dissolution of NATO and the Warsaw Pact. Andropov ruled out, of course, the possibility of unilateral disbandment of the latter organization. Moreover, he did not hide that the peace initiatives were intended mostly as a tool of political pressure on Washington. In this

---

[1770] Zubok, *A Failed Empire,* 272.
[1771] Volkogonov, *The Rise and Fall,* 329, 331.
[1772] Litera, "Sovětský svaz, 'Eurorakety'," 259.
[1773] Letter by G. Husák, the CPCz CC General Secretary, to W. Jaruzelski, the PUWP CC First Secretary, 16. 11. 1982, PZPR KC, XIA/1414, AAN, Warsaw; *Příprava zasedání politického poradního výboru členských států Varšavské smlouvy v Praze,* 3. 12. 1982, 1261/0/8, sv. P54/82, b. 2, NA, Prague.

regard, he also asserted better cooperation with the West-European left wing. Some parts of Andropov's speech revealed that despite close connections to Defense Minister Ustinov, the new General Secretary was not a big advocate of a further arms race. He realized that the economic situation did not allow the Soviet Union to compete with the West in the field of recent sophisticated weapons. Therefore, he put a strong emphasis on the campaign against militarization of the outer space, which was mentioned increasingly in connection with the American SDI project. "*For Reagan, it is no problem to cancel dozens of billions in funding for social programs and allocate them to the military-industrial complex. But we cannot neglect the welfare of working people,*" Andropov said. It is good to remember that these claims totally opposed the then appeals of Marshal Kulikov at the alliance's military bodies' meetings. In front of the allies, the CPSU CC General Secretary condemned the Western doctrine of "rational, limited" nuclear war. He dismissed thoughts about a possibility to survive and win in a long-term nuclear conflict. Such ideas actually increasingly resonated in the military command of the Warsaw Treaty Organization.[1774] Conclusion of a non-aggression agreement between NATO and the Warsaw Pact was set as the main objective of the Eastern Bloc countries' foreign policy by the Soviet leader.[1775]

The Prague meeting of the Political Consultative Committee confirmed that in the existing international situation, some different positions began to emerge among the Warsaw Pact member-states. This was most obvious in the attitudes of Poland and Hungary towards the West. After Wojciech Jaruzelski came to power, up to this point a relatively accommodating policy of Warsaw towards the Western countries changed rapidly, especially in terms of relations with the United States. Along these lines, the Polish delegations tried to sharpen the tone of the resolution and speak out against the Reagan administration as a reaction to the U.S. sanctions. Hungary, which was interested in

---

[1774] *Projev generálního tajemníka ÚV Komunistické strany Sovětského svazu soudruha Jurije Vladimiroviče Andropova na zasedání politického poradního výboru členských zemí Varšavské smlouvy v Praze dne 4. ledna 1983,* sv. P57/83, b. 1, ibid.
[1775] *Výsledky zasedání politického poradního výboru členských států Varšavské smlouvy v Praze,* 13. 1. 1983, ibid.

maintaining good relations with the West due to economic reasons, tried to blunt the Warsaw Pact's declaration, in contrast. It instrumentally operated with the claim that placing too much emphasis on the risk of outbreak of armed conflict could have created a "war-psychosis" among the Eastern Bloc countries' population. The Polish and Hungarian delegations also presented different opinions on further armament. Jaruzelski repeated the claim that despite economic troubles, the PPR was ready to strengthen its armed forces, while Kadar cautiously criticized the military course so far. He talked about the necessity of a balanced use of resources for civil and military needs. However, he wanted to solve the problem mostly through a general strengthening of the economy, calling defense a priority.[1776]

Change on the post of the CPSU CC General Secretary did not result in any rapprochement of Romania and the Eastern Bloc. Andropov held bilateral talks with Ceausescu in December 1982. He had a strong antipathy for the Romanian dictator. Regarding the opposition behavior of Romanian delegations at the Warsaw Pact forums, he asked Ceausescu what would have Romania achieved through withdrawal from the alliance. The RCP leader immediately replied: *"We lose nothing."*[1777] In light of this, it is not surprising that at the Prague meeting of the Political Consultative Committee, Bucharest again strongly opposed the actual Soviet course. It even blamed the Eastern Bloc countries' policy thus far for the bad international situation. Although Romania agreed with the basic principles of the planned peace initiatives,[1778] it considered them too limited and less ambitious. Moreover, Ceausescu justified this by referring to an instrumentally selected Lenin thesis about the preference of economic development rather than armament. Along these lines, the SRR proposed reduction of the Warsaw Pact countries' military budgets by 20% till 1985, regardless of NATO. At the same time, it refused incorporating

---

[1776] Ibid.

[1777] Volkogonov, *The Rise and Fall*, 373.

[1778] These were mostly a proposal to freeze and later reduce armament spending of both NATO and the Warsaw Treaty Organization, an initiative by Czechoslovakia, East Germany, and Poland to create a chemical-weapons-free zone in Central Europe and the Soviet proposals to reduce troop numbers in that region.

into the Political Consultative Committee resolution any mention of a possible military reaction to the planned deployment of new U.S. missiles in Europe. In the end, Romanian intransigence resulted in a first-time situation where a risk existed that the document would be issued only on behalf of the Six, not the entire alliance.[1779]

  Soon after coming to power, Andropov announced that relations between the socialist countries had been negatively affected by certain mistakes in the past. He probably wanted to seek new methods on how to combine the interests of individual states and the whole bloc. He saw an outcome in improvement of political cooperation and deeper economic integration.[1780] Andropov also planned to modify slightly the existing forms of collaboration within the Warsaw Treaty Organization. He declared a will to hold the Political Consultative Committee at least once a year and make its sessions more effective. After all, during the Prague meeting, the final documents were already modified more collectively than in Brezhnev's era. The new Soviet leader also criticized the poor work of the Joint Secretariat. He even asked a question of whether it was worthy to strive for reform of the body which had been failing in its job for 7 years, or to think about a different solution instead.[1781] The task was assigned to ministers of foreign affairs. According to Andropov's vision, the Warsaw Treaty Organization should have also increased cooperation with Cuba,

---

[1779] The Six called the Romanian positions "blackmailing, destructive, and classless." In the end, after long negotiations, the delegation of the SRR accepted a compromise: The country surrendered its demand for unilateral reduction of military budgets in exchange for deleting the thesis about retaliatory measures in the Euromissiles Crisis. *Výsledky zasedání politického poradního výboru členských států Varšavské smlouvy v Praze*, 13. 1. 1983, 1261/0/8, sv. P57/83, b. 1, NA, Prague; *Anlage Nr. 1 zum Protokoll Nr. 1 vom 11. Jan. 1983,* DY 30/ J IV 2/2/1983, BArch, Berlin.

[1780] Litera, "Sovětský svaz, 'Eurorakety'," 259.

[1781] Seriously ill Nikolai Firyubin, the Warsaw Pact's General Secretary, did not take part in the Prague meeting. At the last minute, he was replaced by Dušan Spáčil, the Czechoslovak diplomat. This is another proof of the marginal importance of this post. *Výsledky zasedání politického poradního výboru členských států Varšavské smlouvy v Praze*, 13. 1. 1983, 1261/0/8, sv. P57/83, b. 1, NA, Prague.

Vietnam, Laos, Mongolia, as well as with Yugoslavia and North Korea, despite their often different stances on international events.[1782]

The resolution of the Prague meeting of the Political Consultative Committee of January 1983 became the basic guideline for the Eastern Bloc countries' foreign policy activity in the next two and a half years. In the document, the Warsaw Treaty Organization put the strongest emphasis on disarmament in its history so far. At the same time, some willingness to shift in the Euromissile Crisis was suggested. The situation should have been solved through withdrawal of all tactical nuclear weapons and their mid-range carriers from Europe.[1783] The document was translated to the major world languages. Then, the Warsaw Pact member-states promoted it massively at all the international forums, much more intensively than ever before.[1784] The initial Western reaction to the resolution was cold. It was criticized for its propagandistic targeting of a peace movement. However, Reagan quite surprisingly appreciated it. Subsequently, the document began to be seen as proof of a possibility to continue a dialogue between the blocs. Ministers of foreign affairs of Great Britain and Germany even started to admit that the firm U.S. approach at the Geneva talks on the Euromissiles was not necessarily the right one.[1785] Some proposals by the Political Consultative Committee were publicly supported even by Javier Péréz de Cuellar, the UN General Secretary. The width, concreteness, and relatively moderate language of the East caught

---

[1782] *Projev generálního tajemníka ÚV Komunistické strany Sovětského svazu soudruha Jurije Vladimiroviče Andropova na zasedání politického poradního výboru členských zemí Varšavské smlouvy v Praze dne 4. ledna 1983*, sv. P57/83, b. 1, ibid.

[1783] *POLITICKÁ DEKLARACE ČLENSKÝCH STÁTŮ VARŠAVSKÉ SMLOUVY*, undated, 1983, ibid.

[1784] The Czechoslovak diplomacy was especially very agile in spreading the document. The allies said with pleasure that in the Warsaw Pact's history so far, no document had been promoted in this way. Only Albania rejected accepting it. But the country didn't participate in the CSCE either and dismissed the Warsaw Pact's materials in the long term. *Příprava zasedání výboru ministrů zahraničních věcí členských států Varšavské smlouvy v Praze ve dnech 6.-7. dubna 1983*, 15. 3. 1983, sv. P63/83, b. 5, ibid.

[1785] *Výsledky zasedání politického poradního výboru členských států Varšavské smlouvy v Praze*, 13. 1. 1983, sv. P57/83, b. 1, ibid.

the West by surprise. Initial efforts by right-wing politicians to negate the appeal as recurrence of the previous Soviet and Warsaw Pact's proposals and a part of communist propaganda faded.[1786]

The success of the document caused the Committee of Ministers of Foreign Affairs, held on April 6-7, 1983, in Prague, to issue just a brief communiqué. It was again of a very moderate tone. On the Soviet cue, it should have avoided any direct insults, including to the Reagan administration. The criticism was limited to unspecified "imperialistic circles." The Warsaw Treaty Organization suggested its willingness to continue in bilateral talks with NATO countries and build the ground for future multilateral negotiations in this manner. It also appealed for starting a discussion on reduction of military spending.[1787]

Above all, the Committee of Ministers of Foreign Affairs' meeting confirmed that Andropov had seriously meant his previous appeals for changes in cooperation within the Warsaw Pact's political structures. The heads of diplomatic resorts were supposed to consider further strategy on the issues of the non-aggression agreement between NATO and the Warsaw Treaty Organization and removal of chemical weapons from Europe particularly. However, they were permitted to deal with other topics in their speeches in order to make the meeting more flexible.[1788] There were also shifts in work on final documents. For the basis of the text a draft was used by the host Czechoslovakia, which was afterwards collectively modified by diplomats of all the member-states. Above all, special expert groups were established and assigned with elaboration of specific tasks: The first was supposed to be to check the Joint Secretariat activities and propose how to make the alliance's political

---

[1786] *Příprava zasedání výboru ministrů zahraničních věcí členských států Varšavské smlouvy v Praze ve dnech 6.-7. dubna 1983,* 15. 3. 1983, sv. P63/83, b. 5, ibid.

[1787] *KOMUNIKÉ ze zasedání ministrů zahraničních věcí členských států Varšavské smlouvy,* 7. 4. 1983, sv. P67/83, info3, ibid.

[1788] In this manner, the USSR complied with the request of the Romanian side. At the same time, it showed its willingness to make talks within the Warsaw Treaty Organization more flexible. Thus, Romania used a situation and through Stefan Andrei urged starting direct negotiations between NATO and the Warsaw Pact on all actual international issues. Bucharest admitted that some joint alliance's initiatives could have been presented there.

cooperation more effective in general. The second focused on defining strategy for negotiations with NATO on reduction of military budgets. The third worked on details of the initiative to remove chemical weapons from Europe. The Kremlin's effort to treat the Warsaw Pact countries more as partners was underlined by the fact that the expert groups did not work solely in Moscow, but also in Bucharest and Sofia.[1789]

A moderate tone, seeking the paths of new dialogue between the blocs and, in comparison to the past, more constructive proposals in the Warsaw Pact's resolutions resulted in a certain rise of the organization's prestige. Responses to the communiqué of the Prague session of the Committee of Ministers of Foreign Affairs in the Western media as well as by some political representatives were not negligible. The West-German vice-chancellor, Hans-Dietrich Genscher, for example, stated that the Warsaw Pact's proposals on the issue of disarmament had to be analyzed in detail.[1790] However, putting forward any catchy initiatives and appeals within the so-called "peace offensive of the socialist countries" was not able to compensate for the absence of any breakthrough in the Euromissile Crisis. It is good to add that in the first half of 1983, more intransigence was shown by the U.S. side. The USSR, in contrast, declared its will to reduce the numbers of launch pads for mid-range missiles and warheads to the level of 1976, before the dislocation of the SS-20 started. Such a move, which would have been seen as a goodwill gesture at the end of the 1970s and probably would have stopped escalation of the crisis, the Reagan administration did not accept. This assured Moscow that political representation in Washington was not seeking compromise and insisted on the deployment of new missiles in Europe at any cost.[1791] Moreover, on

---

[1789] *Informace o řádném zasedání výboru ministrů zahraničních věcí členských států Varšavské smlouvy v Praze ve dnech 6.-7. dubna 1983*, 20. 4. 1983, 1261/0/8, sv. P67/83, info3, NA, Prague.

[1790] The West-German press did not miss that the document was formulated also with aim of attracting the West-European peace movement. Therefore, newspapers tried to confront it with some militant statements by Kulikov and Ustinov. Ibid.

[1791] In practice, the Geneva talks were blocked by the U.S. side's unwillingness to include its air and submarine weapons as well as British and French nuclear arsenals in a potential agreement. Washington insisted on reduction of missiles, which the USSR much more relied on. *O výsledcích nedávno skončeného kola*

March 8, 1983, Reagan publicly called the Soviet Union "an evil empire." In an unprecedented way, he sharpened the rhetoric of the previous U.S. governments which had never challenged the Soviet regime's legitimacy publicly. Thirteen days later, the head of the White House announced the beginning of the "Star Wars" project.[1792] These steps of the American president unambiguously deepened the atmosphere of mistrust and provided a welcomed opportunity to justify escalation of the arms race and militant rhetoric of the top Warsaw Pact military officials.

Despite its public noting Reagan's calls for a "crusade" against communism, Moscow, in fact, considered neither deployment of the U.S. missiles in Europe nor the SDI project as an imminent threat to the balance of military forces. In February 1983, Vadim Zaglanin, the Soviet ideologist, informed representatives of the Six about the unreliability of Pershing-II missiles so far. Only one of seven pieces hit a target accurately.[1793] An expert group, led by physicist Evgeni Velikhov, came to the conclusion that the SDI did not require imminent Soviet counteraction. Nevertheless, the discussion within the military circles did not end. The Soviet generality thought that the consideration of weapons in outer space would inevitably lead to the development of new even more destructive technologies in the long-term.[1794] But, the Eastern Bloc lacked the necessary technical and economic capacities for such programs.

Given this situation, in June 1983, the USSR convened an extraordinary meeting of the Warsaw Pact member-states' party leaders to Moscow. The agenda concerned mostly the Euromissile Crisis. The question of why the Soviet leadership did not want to discuss the issue on the level of the Political Consultative Committee can be hardly

---

sovětsko-amerických jednání o omezení jaderných zbraní v Evropě /informace ÚV KSSS/, 6. 5. 1983, 1261/0/8, sv. P68/83, info2, NA, Prague.

[1792] Zubok, *A Failed Empire,* 272-273.

[1793] *Informace o pracovním setkání zástupců vedoucích oddělení ústředních výborů bratrských stran socialistických zemí /Moskva, 16. února 1983/,* 4. 3. 1983, 1261/0/8, sv. P61/83, b. 8, NA, Prague.

[1794] Zubok, *A Failed Empire,* 273. After all, development of the Polyus military space station, which was supposed to destroy enemy satellites by powerful lasers, began in the Soviet Union.

answered.[1795] Andropov had to admit with displeasure that despite the Warsaw Pact's intense activity, the international situation had deteriorated since the beginning of the year, in fact. He expressed concerns that the West was trying to strengthen its military position to the point that would have allowed it to dictate its own course in the international arena. We are reminded that this strategy was pursued in the first half of the 1970s by Andropov's predecessor, Leonid Brezhnev. The Soviet leader also complained that the U.S. did not want to make any constructive deal on the issue of Euromissiles and instrumentally put forward those proposals which, in fact, would have weakened the USSR significantly. He once more talked about the need to conclude agreements preventing militarization of outer space. The Soviet General Secretary still preferred a defensive approach.[1796] He assured the Warsaw Pact members that Moscow would present new initiatives on a solution of the Euromissile Crisis, including withdrawal of some SS-20s from the European part of the USSR. But their numbers were supposed to stay at least on the level of British and French arsenals. At the center of the Eastern Bloc's proposal remained the non-aggression agreement between NATO and the Warsaw

---

[1795] For the Soviet leadership, it was probably easier to convene an extraordinary meeting which required no long preparations. The purpose was actually to react to the G7 summit in Williamsburg, where the most developed countries of the world had supported the U.S. approach on the issue of mid-range nuclear weapons. Some historians, like Bohuslav Litera, wrongly considered the meeting in Moscow as the Political Consultative Committee session. Litera, "Sovětský svaz, 'Eurorakety'," 262. However, the extraordinary meeting in Moscow and the Political Consultative Committee session in Prague were put on the same level of ideological and propagandistic dimension. *Informativní zpráva o poradě tajemníků ÚV bratrských stran socialistických zemí pro mezinárodní a ideologické otázky v Moskvě*, 15. 12. 1983, 1261/0/8, sv. P92/83, b. 5, NA, Prague.
[1796] The defensive approach of the Soviet Union is also proven by its actions at the CSCE summit in Madrid. Moscow was content with the Final Act signatories' willingness to hold a conference on measures to improve trust and security and disarmament in Europe. It supported without reservation a draft of final documents put forward by a group of neutral countries. *PODKLADY pro jednání se soudruhem ANDROPOVEM*, undated, 1983, FMO GŠ/OS, k. 24, p. č. 79, č. j. 41102, VÚA, Prague.

Treaty Organization.[1797] However, obstinacy of the U.S. side, undoubtedly fueled by calls for a firm approach from the orthodox part of the Soviet leadership, inevitably led to another sharpening of Moscow's rhetoric. Supported by its allies, the USSR warned that in case of deployment of new American weapons in Europe, appropriate military countermeasures would be taken, including dislocation of new missiles on the territories of East Germany and Czechoslovakia. Erich Honecker loyally assured that the GDR was ready to take such a step.[1798]

It would be odd to say that Andropov preferred this scenario. Although he reluctantly stated that "*the Soviet people would understand an increase of defense spending, for sure,*" in fact, he was seriously concerned about economic complications. Therefore, he paid close attention to the Western effort to put the East not only under political duress, but also to reduce existing economic cooperation. After all, the Polish leader, Wojciech Jaruzelski, took his word for it when calculating damages to the economy of the PPR caused by the U.S. sanctions at six billion USD.[1799] Moscow placed some hopes on peace and left-wing movements in the USA and West Europe to overturn the situation. In this regard, the PUWP CC First Secretary urged intensification of the Eastern propaganda on the West-European population. He noted that in this field, the Eastern Bloc countries lagged behind the West so much.[1800]

[1797] *Wystąpienie Sekretarza Generalnego KC KPZR, Przewodniczącego Prezydium Rady Najwyższej ZDRR J. W. Andropowa na naradzie przywódców Państw-Stron Układu Warszawskiego w dniu 28 czerwca 1983 rok*, PZPR KC, XIA/1481, AAN, Warsaw.

[1798] *Zpráva o setkání vedoucích představitelů stranických a státních činitelů BLR, ČSSR, MLR, NDR, PLR, SRR a SSSR v Moskvě dne 28. června 1983*, 30. 6. 1983, 1261/0/8, sv. P75/83, b. 8, NA, Prague.

[1799] Given the growing economic troubles of the Eastern Bloc states, it is almost laughable that at the meeting in Moscow, the Polish leader did not miss an opportunity to present obligate phrases about "the deep internal crisis of capitalism."

[1800] One of the most important actions against deployment of the U.S. missiles in Europe was the World Peace Convention which took place in Prague, June 21-26, 1983, with participation of hundreds of representatives of the Western left-wing and peace movement. *Zpráva o stavu příprav Světového shromáždění za mír a život proti jaderné válce a návrh na účast vedoucích československých*

The USSR used the Moscow meeting to get public support of the Warsaw Pact member-states for its strategy in negotiations with the U.S. on reduction of nuclear weapons. But in reality, the alliance was not unanimous. This did not just include the traditionally opposing Romania.[1801] The GDR and CSSR approved the dislocation of additional Soviet nuclear forces on their territories only with great reluctance.[1802] The operation was supposed to start in 1984. Especially the East-German regime showed significant unwillingness to make such a move. Its position began to correspond to that of Hungary. A long-term interest of Budapest was to cooperate with the West-European countries regardless of ideological differences. At the beginning of the 1980s, Hungary was followed by the GDR which substantially increased mutual economic collaboration with West Germany. At the beginning of the Euromissiles Crisis, the HPR and GDR had the most developed economic interaction with the West of all the Warsaw Pact members. The SED leadership was aware that escalation of tensions due to American, but also Soviet, missiles

---

představitelů na některých akcích, 1. 7. 1983, 1261/0/8, sv. P75/83, b. 9, NA, Prague.

[1801] At the meeting in Moscow, the SRR argued with the Six about all the essential issues. Romania rejected the formulation that the U.S. and its allies strove for military superiority. On the question of Euromissiles, Bucharest took a totally unclear position. It refused to directly criticize the FRG, Great Britain, and Italy for allowing the deployment of new U.S. weapons on their territories as well as the statement that these missiles were capable of hitting the socialist countries deep in their hinterlands. On the contrary, the Romanians as usual emphasized freezing of military budgets. They tried to push forward some formulation into the final resolution which would have allowed for a critical view on the Soviet policy. Ceausescu demanded the document include an appeal for stopping dislocation of new missiles not only on the West-German territory, but also in the GDR and CSSR. Furthermore, he recommended withdrawing the Soviet SS-20 missiles to at least 300-400 kilometers from the western borders of the USSR. In the end, the Romanian leader was streamlined by Gromyko and Ustinov. They probably succeeded even because of the fact that Ceausescu was eminently interested in strengthening cooperation within the Comecon which he hoped could result in economic recovery in his country. *Zpráva o setkání vedoucích představitelů stranických a státních činitelů BLR, ČSSR, MLR, NDR, PLR, SRR a SSSR v Moskvě dne 28. června 1983*, 30. 6. 1983, sv. P75/83, b. 8, ibid.

[1802] Up to this point, only old Soviet tactical missiles Scud and Frog were located on both countries' territories.

in Europe complicated its intentions significantly. Moreover, the situation had an important security aspect. The East-German and Czechoslovak representatives clearly realized that their countries could have become a target of a NATO preemptive strike, or be hit by the West right after an outbreak of war as the Soviet SS-21 and SS-23 missiles were capable of hitting the enemy territories in terms of a few minutes. Therefore, it was obviously an absolute priority of NATO to eliminate them soon. Nevertheless, till the turn of October and November 1983, the GDR and CSSR failed to show their reluctance to deploy the additional Soviet nuclear weapons. Up to this point, they probably counted on either a possibility of breakthrough at the Geneva talks or disapproval of dispatching the U.S. missiles by parliaments of the West-European members of NATO.[1803]

The effort to prevent the dislocation of new U.S. weapons in Europe suffered a big blow on September 1, 1983. After a series of fatal mistakes, a Soviet fighter-jet shot down a South Korean civilian Boeing 747 over Sakhalin. All 269 people onboard were killed. The international outrage as well as the initial reaction by Moscow and the other Warsaw Pact countries' representatives[1804] negated all the previous Eastern Bloc initiatives. Moscow failed to restore the credibility of its "peace offensive."[1805] Thus, for example, a document proposed by Andropov

---

[1803] Mastny, *A Cardboard Castle?*, 56-57; Litera, "Sovětský svaz, 'Eurorakety'," 262-263.

[1804] In October 1983, the incident was discussed in general at an extraordinary meeting of the Committee of Defense Ministers. In this regard, Ustinov reminded the Committee of Reagan's words about a "crusade" against communism, which should have just masked the anti-Soviet campaign and the effort to destroy socialism, according to the Soviet minister. He called the shooting down of the South Korean jetliner a U.S. provocation the American hawks used to ask for another raise in their military budget and to escalate their anti-Soviet campaign. Also, Ustinov's Czechoslovak counterpart Dzúr spoke in a similar style. He talked about "cunning provocation" by the U.S. intelligence. *VYSTOUPENÍ člena předsednictva ÚV KSSS, ministra obrany SSSR, maršála Sovětského svazu D. F. USTINOVA na mimořádném zasedání výboru ministrů obrany členských států Varšavské smlouvy 20. října 1983; Vystoupení ministra národní obrany ČSSR na mimořádném zasedání výboru ministrů obrany*, undated, 1983, FMO GŠ/OS, k. 55, p. č. 215, č. j. 38108, VÚA, Prague.

[1805] Litera, "Sovětský svaz, 'Eurorakety'," 262.

called "An initiative of the Soviet state in connection with the forthcoming entry of mankind into the third millennium," put forward in the UN, was totally forgotten. Moscow had in fact proposed there a very complex program for the years 1985-2000, which included reducing the risk of an outbreak of nuclear war and fighting against poverty, famine, disease, and the devastation of environment. Given the circumstances, the international community did not accept the plan, of course. Such an ambitious proposal by a state that waged a bloody war in Afghanistan, shot down over its territory an aircraft with hundreds of people onboard, and flagrantly violated human rights was not taken seriously at all by a large part of the world.[1806]

In autumn 1983, the forthcoming deployment of Pershing-II and BGM-109G cruise missiles was the main item on the agenda of the Warsaw Pact meetings. The Committee of Foreign Ministers met in Sofia, October 13-14. Gromyko warned that the level of the USSR-U.S. tensions was the highest it had been for a long-time. He still saw a solution in the non-aggression agreement between NATO and the Warsaw Treaty Organization; however, he considered it a long-term goal. The head of the Soviet diplomacy urged the Warsaw Pact to formulate constructive initiatives which NATO could not just simply reject.[1807] Nevertheless, given the American side's intransigence, the initially declared effort to make a deal on the issue of Euromissiles was being gradually replaced with the threat of retaliation.[1808] Just one week later, on October 20, an extraordinary meeting of the Committee of Defense Ministers took place in East Berlin at Kulikov's formal initiative. It was strictly of a political and demonstrative character. For the first time, it was opened to media. The body publicly stated that the deployment of new U.S. missiles in Europe was changing the existing military balance in favor of NATO and dangerously escalated the situation on the continent. In this regard, an acceleration of modernization of the Warsaw Pact armies' armaments in

---

[1806] Volkogonov, *The Rise and Fall*, 358.
[1807] *Informace o řádném zasedání výboru ministrů zahraničních věcí členských států Varšavské smlouvy v Sofii ve dnech 13.-14. října 1983*, 26. 10. 1983, 1261/0/8, sv. P86/83, info6, NA, Prague.
[1808] Litera, "Sovětský svaz, 'Eurorakety'," 261.

the ongoing as well as the upcoming five-year plan was announced. In addition, the Soviet defense minister threatened that in case of an implementation of the U.S. intention, the USSR would resort to deployment of other mid-range missiles nearby its western borders, strengthen missile units, and immediately rearm the units belonging to the Unified Armed Forces. Above all, Ustinov and Honecker signed a treaty on dislocation of new Soviet missiles in the GDR. A similar document was being prepared for Czechoslovakia.[1809] On October 24, the USSR and the governments of these two Central-European satellites declared that preparations for the dispatching of weapons had begun.[1810] In this regard, the East-German leader publicly talked about a dangerous rise in the threat of outbreak of World War III.[1811]

Despite these moves, the USSR still considered it necessary to maintain the impression of an "open door" policy. On December 9-10, 1983, it convened the Warsaw Pact member-states' parliament representatives meeting in Sofia. The aim was to appeal to the legislatures of the CSCE countries on the issue of deployment of U.S. missiles in Europe. Although the Soviet, Bulgarian, and Czechoslovak delegations supported the installation of Soviet rockets in the GDR and CSSR, the final resolution stated that preventing destructive weapons from dislocation in Europe was still possible if the political will existed.[1812]

[1809] *MATERIÁLY pro mimořádné zasedání (poradu) výboru ministrů obrany členských států Varšavské smlouvy*, undated 1983; *Mimořádné zasedání výboru ministrů obrany členských států Varšavské smlouvy*, October 1983; *INFORMACE o průběhu a výsledcích mimořádného zasedání výboru ministrů obrany konaného dne 20. října 1983 v Berlíně*, FMO GŠ/OS, k. 55, p. č. 215, č. j. 38108, VÚA, Prague.
[1810] Litera, "Sovětský svaz, 'Eurorakety'," 262.
[1811] *INFORMACE o průběhu a výsledcích mimořádného zasedání výboru ministrů obrany konaného dne 20. října 1983 v Berlíně*, FMO GŠ/OS, kr. 55, p. č. 215, č. j. 38108, VÚA, Prague.
[1812] *Zpráva o konzultativním setkání představitelů parlamentů členských států Varšavské smlouvy v Sofii*, 14. 11. 1983; *Výzva parlamentům účastnických států Konference o bezpečnosti a spolupráci v Evropě*, 10. 11. 1983, 1261/0/8, sv. P89/83, b. 12, NA, Prague.

At all the aforementioned meetings, Romania acted in a very confrontational manner.[1813] In November 1983, Ceausescu once more put forward his own proposals of a solution to the Euromissile Crisis, which differed significantly from the Warsaw Pact's initiatives. Andropov swept them off the table quickly. He accused Bucharest of a very wrong evaluation of the existing correlation of nuclear forces in Europe. He also easily disproved the Romanian leader's doubtful argument that dislocation of the U.S. and Soviet rockets violated the non-proliferation treaty.[1814] One month later, the Romanian stances were attacked harshly by Andropov's future successor, Konstantin Chernenko, in front of representatives of the Six gathered in Moscow. He stated that regardless of the SRR, necessary measures to maintain military-strategic balance would be implemented within the Warsaw Treaty Organization.[1815]

Thus, the Soviet Union definitely left détente. After the deployment of U.S. weapons on the territories of NATO's European members was approved, Moscow withdrew from the Geneva talks on mid-range nuclear weapons and strategic forces in November 1983. One month later, at the Vienna disarmament talks, the Warsaw Pact countries refused

---

[1813] Above all, Romania rejected all the Warsaw Pact's military countermeasures to deploy the U.S. missiles. On the contrary, it put in the forefront freezing the military budget, which the country unilaterally resorted to. In practice, Bucharest ascribed the same responsibility for rise of armament to the USSR and the United States. For escalation of the Euromissile Crisis, it blamed not only NATO, but also these Warsaw Pact countries where new nuclear weapons would be deployed. In a possible solution, Romania tended towards U.S. proposals, in fact. According to the Ceausescu regime, the issue of British and French nuclear forces should have been solved later, within global disarmament. *Informace o řádném zasedání výboru ministrů zahraničních věcí členských států Varšavské smlouvy v Sofii ve dnech 13.-14. října 1983*, 26. 10. 1983, sv. P86/83, info6, ibid; *Zpráva o konzultativním setkání představitelů parlamentů členských států Varšavské smlouvy v Sofii*, 14. 11. 1983, sv. P89/83, b. 12, ibid; *INFORMACE o průběhu a výsledcích mimořádného zasedání výboru ministrů obrany konaného dne 20. října 1983 v Berlíně*, FMO GŠ/OS, k. 55, p. č. 215, č. j. 38108, VÚA, Prague.

[1814] *Odpověď J. V. Andropova na list N. Ceausesca týkaúca se otázky jadrových zbraní v Evrópe*, 25. 11. 1983, 1261/0/8, sv. P90/83, info6, NA, Prague.

[1815] *Informativní zpráva o poradě tajemníků ÚV bratrských stran socialistických zemí pro mezinárodní a ideologické otázky v Moskvě*, 15. 12. 1983, sv. P92/83, b. 5, ibid.

to set a date for another round of negotiation.[1816] However, the Soviet leadership avoided total suspension of contacts with NATO states. The Kremlin was aware that some willingness towards economic cooperation with the Eastern Bloc persisted in the West.[1817] Furthermore, at the end of 1983, Ustinov stated that the preparation for deployment of new nuclear weapons in the CSSR and GDR would be accelerated. At the same time, additional rockets would have been aimed at the U.S. territory. According to the Soviet defense minister, the government of the United States should have felt an impact of its moves in Europe.[1818]

During the short rule of Andropov, the Warsaw Treaty Organization changed neither its basic essence nor its role in the Eastern Bloc mechanism. Despite this fact, some part-shifts and an increase of cooperation in the alliance's political framework apparently took place. First of all, there was an effort to meet regularly, especially on the level of the Committee of Ministers of Foreign Affairs as well as their deputies. Unlike before, consultations were held not only at a moment when Moscow needed to have its new strategy formally blessed by its satellites. This was reflected also by the fact, that communiqués usually repeated previous initiatives of the Warsaw Pact only.[1819] The Soviet diplomacy did not work on foreign political appeals and proposals alone any longer. During 1983, a collective working method on the level of the previously mentioned expert groups was relatively established. In October, Gromyko informed them that Moscow appreciated this practice and envisaged it also in future.[1820]

Between March and June 1983, one of the expert groups worked in the Bulgarian capital on alteration of the Warsaw Pact's political structures mechanism. Hungary praised the fact that, for the first time in

---

[1816] Durman, *Útěk od praporů*, 260; Litera, "Sovětský svaz, 'Eurorakety'," 263.

[1817] Locher, "*Shaping the Policies of the Alliance*".

[1818] Statement by Marshal Ustinov at the Committee of Ministers of Defense Meeting in Sofia, December 5-7, 1983, in *A Cardboard Castle?* (doc. 102): 490-491.

[1819] *KOMUNIKÉ ze zasedání výboru ministrů zahraničních věcí Varšavské smlouvy v Sofii 13.-14. října 1983*, 1261/0/8, sv. P86/83, info6, NA, Prague.

[1820] *Informace o řádném zasedání výboru ministrů zahraničních věcí členských států Varšavské smlouvy v Sofii ve dnech 13.-14. října 1983*, 26. 10. 1983, ibid.

the alliance's history, such a discussion proceeded openly and had practical features. However, the Soviet representatives showed that Moscow was not interested in any radical changes in the organization. But it did not oppose those modifications which would have resulted in the strengthening of allied ties and help solving actual problems. The USSR therefore proposed to intensify meetings of deputy foreign ministers (or prospectively even hold them regularly), bigger activity of the expert groups, and rotation on the post of general secretary of the alliance. The post should have been held by a diplomat of the country where a meeting took place. There was unanimous agreement on the last aspect. The death of Nikolai Firyubin in February 1983 contributed to this fact. Ever since then, the post had remained empty. The Joint Secretariat's general role was subjected to a long discussion. Poland and Czechoslovakia, supported by East Germany and Bulgaria, wanted to transform it into a permanent political institution. The Soviet Union, Hungary, and Romania,[1821] in contrast, asserted just improving its technical and organization role. Some modification of the Soviet approach towards the Warsaw Pact member-states was illustrated by the fact that neither Czechoslovakia, nor Poland, nor East Germany consulted their proposals with Moscow in advance. Although the Soviet delegation tried to influence the stances of these countries during negotiations in backrooms, it admitted some concessions on the other hand: Moscow did not oppose establishing a forum that would have met every two or three months to deal with actual matters concerning the Warsaw Treaty Organization.[1822]

[1821] It is interesting that Romania, unlike the previous talks about possible reform of the Warsaw Treaty Organization, remained largely passive this time. The country avoided any obstructions and did not actively participate in discussion. Bucharest just stated that it was satisfied with the last adjustment of the Warsaw Pact's mechanisms of 1976, but at the same time did not oppose making the existing rules more effective. Establishing of expert groups on the reduction of military budgets was a much more burning issue for the Ceausescu regime, as the Soviet Union strictly dismissed such a step.

[1822] *Report on Session of the Working Group for the Reform of the Political Cooperation Mechanisms within the WP in Sofia*, June 7, 1983, Accessed October 16, 2019. http://www.php.isn.ethz.ch/kms2.isn.ethz.ch/serviceengine/Files/PHP/16991/ipublicationdocument_singledocument/bd0351d5-067e-4073-9512-f4d57470fb82/en/830607_report.pdf.

Some modifications were also expected in preparation of final documents. The member-states should have been given more of an opportunity to influence these texts and put forward their own proposals. The Political Consultative Committee and the Committee of Ministers of Foreign Affairs were supposed to automatically deal with all incentives sent at least ten and seven days, respectively, before a meeting. The expert groups were also tasked with modification of documents during the alliance's sessions. Original documents should have been archived in the PPR, as the founding charter was also in its depository. The end of chairmanship rotation at the alliance political meetings was also envisaged. In the future, a single representative of the hosting country was supposed to chair them.[1823] Not only the regular holdings of the Committee of Ministers of Foreign Affairs and the establishing of the expert groups were appreciated by the Warsaw Pact states. They commended more intensive talks on the level of deputy foreign ministers as well.[1824] After all, this had been happening since the beginning of the 1980s already.[1825]

[1823] *Bericht der Expertenarbeitgruppe über die Ergebnise ihrer Arbeit zur Prüfing von Vorslägen für die Aufgaben und die Arbeitswiese des Vereinten Sekretariats sowie zu anderen Fragen des Arbeitsmechanismus der Zusamenabrbeit in Rahmen des Warschauer Vertrages*, 22. 8. 1983, DY 30/ J IV 2/2/2093, BArch, Berlin.

[1824] In 1983 alone, deputy foreign ministers held their talks three times at least. First, they met ahead of the final stage of the CSCE summit in Madrid; second, in order to develop further strategy on the issue of a non-aggression treaty between NATO and the Warsaw Pact. The third meeting was scheduled before the Stockholm Conference on Confidence- and Security-Building Measures and Disarmament in Europe, which was supposed to start in January 1984. *Informace o řádném zasedání výboru ministrů zahraničních věcí členských států Varšavské smlouvy v Sofii ve dnech 13.-14. října 1983*, 26. 10. 1983; *Projev ministra zahraničních věcí Svazu sovětských socialistických republik A. A. Gromyka*, 13. 10. 1983, 1261/0/8, sv. P86/83, info6, NA, Prague.

[1825] After 1980, deputy foreign ministers met annually before the UN General Assembly session and tried to coordinate a joint position of the Warsaw Pact member-states. Representatives of Laos, Cuba, Mongolia, Vietnam, and North Korea, as well as those of two Soviet republics, the UN members Belarus and Ukraine, attended these talks. If held in Moscow, Gromyko usually met the participants separately and informed them in detail about actual Soviet foreign policy. Thus, these meetings were an important communication line between the Kremlin and its satellites. Békés, "*Records of the Meetings*".

It must be emphasized that the aforementioned changes were not intended to harm the Soviet hegemony in any way. They were rather Moscow's sops and most likely an effort to make the largely rigid mechanisms of the alliance's political structures - which since the second-half of the 1970s till Brezhnev's death had gradually stagnated - more effective. However, the changes made in Andropov's rule made room for future shifts in the alliance cooperation implemented in Gorbachev's era.

It is good to add that during Andropov's rule, multilateral mechanisms of the Warsaw Treaty Organization were strengthened not only in political, but also in the military framework of the alliance. Since 1984, regular talks between the supreme commander and his deputies from the member-states' armies took place. This measure was supposed to improve the practical work of the Unified Command. The scheme of these meetings was not defined in detail. They should have been held according to actual needs, but at least once per six months. These multilateral consultations took place in all the alliance member-states, usually following the Military Council session. A very narrow circle of top military officials participated. They evaluated the course of implementation of tasks given to the Unified Armed Forces by both military and political bodies of the alliance and analyzed the actual international situation.[1826]

Chernenko's intermezzo

Since October 1983, Andropov's health was deteriorating rapidly. From this point, mostly Gromyko and Ustinov played key roles in shaping the foreign and security policy of the USSR.[1827] The last important question the CPSU CC General Secretary decided was the dislocation of Soviet missiles in the GDR and CSSR. He did that from a hospital bed. On February 9, 1984, Andropov died. The politburo chose Konstantin Chernenko as his successor. Straightforward acting was typical of this 72-year-old man without a proper education. He lacked the long-term vision

---

[1826] *Periodické porady hlavního velitele SOS se zástupci hlavního velitele SOS spojeneckých států*, 10. 2. 1984, FMO GŠ/OS, k. 57, p. č. 218, č. j. 38117, VÚA, Prague.
[1827] Litera, "Sovětský svaz, 'Eurorakety'," 267.

needed to lead the country and party in the increasingly complicated
conditions at that time. Mostly, Dmitri Ustinov's support had contributed
to his election.[1828] The defense minister most likely anticipated that old
Chernenko, himself in a poor state of health, would not jeopardize his
power positions. At the time of taking the function, the new General
Secretary was called a "walking mummy." He also took high doses of
sedatives.[1829]

       Chernenko's election strengthened the positions of Gromyko and
Ustinov only temporarily. In fact, the new party leader's opinion on
domestic as well as foreign policy contradicted those of his
predecessor.[1830] Moreover, it was obvious that the Soviet strategy at the
international stage had so far failed and some revision was necessary.
Already during Andropov's funeral in Moscow, the Warsaw Pact member-
states' party leaders agreed to hold a top-level meeting soon. They also
expressed their interest to organize the Political Consultative Committee
session quickly. Death of the Soviet general secretary also should not have
threatened the Committee of Ministers of Foreign Affairs meeting
scheduled in April.[1831]

       The session took place in Budapest, April 19-20, 1984. At the
time, however, the new Soviet leadership took a few steps which put the
USSR in deep international isolation. East-West relations continued to
cool down. Slow rapprochement with China, started during Andropov's
rule, stopped.[1832] Furthermore, Moscow revisited its policy towards Bonn.

[1828] The defense minister saw in Chernenko a much better candidate than Grigory
Romanov or Victor Grishin, the politburo members, or Victor Chebrikov, the head
of the KGB.
Volkogonov, *The Rise and Fall*, 379, 384, 387.
[1829] Zubok, *A Failed Empire,* 276.
[1830] Litera, "Sovětský svaz, 'Eurorakety'," 267.
[1831] *Protokoll Nr. 6/84 der Sitzung des Politbüros des Zentrakomitees vom 15.
Februar 1984,* DY 30/ J IV 2/2/2041, BArch, Berlin.
[1832] Andropov was aware that the USSR could not afford a conflict with China,
not while its economy was being squeezed by the arms race with the United States,
waging an endless war in Afghanistan, and keeping enormous military
contingents in Eastern Europe. Therefore, he tried to find some *modus vivendi*
with the Asian communist power. However, it soon became clear that disputes
were too deep to be overcome quickly. The Soviet General Secretary even came

West Germany began to be labeled as the enemy on the same level as the U.S., which should have been harshly punished for the dislocation of new American weapons on its territory.[1833] In his speech at the Committee of Ministers of Foreign Affairs, Gromyko stubbornly made restoration of negotiations on nuclear forces in Europe conditional on removal of deployed missiles. Nevertheless, he noted that the USSR preferred an "open door" approach; it did not plan to take such radical measures like severing relations with West Germany. To the contrary, he emphasized an effort to assert withdrawal of the weapons by political means. In this regard, the Committee of Ministers of Foreign Affairs stated in its communiqué that the Eastern retaliatory measures - dislocation of the Soviet missiles in the GDR and CSSR - would be cancelled after the U.S. rockets were removed from Europe. The Warsaw Treaty Organization also proclaimed it did not strive for military superiority, but maintaining parity only.[1834] The alliance also desperately urged the West to start talks on the non-aggression agreement. In such a document, both pacts were supposed to take a pledge of not using force against third-world countries and implement measures preventing a risk of a sudden surprise attack.[1835] The issue as well as the Soviet effort to instigate negotiations between NATO and the Warsaw Treaty Organization on refraining from increasing military budgets and their reduction, presented in March 1984, basically faded out.[1836] The ruling garniture in Washington actually increasingly realized that the USSR was collapsing under the burden of the arms race and its centrally-directed economy was not able to solve this problem. Therefore, the Reagan administration showed no will to compromise. This situation was even deepened during Chernenko's intermezzo. Whereas in

---

up with an idea to involve China in the campaign against U.S. missiles in Europe. But its implementation failed. Volkogonov, *The Rise and Fall*, 360, 376.

[1833] Litera, "Sovětský svaz, 'Eurorakety'," 271.

[1834] *Informace o řádném zasedání výboru ministrů zahraničních věcí členských států Varšavské smlouvy v Budapešti ve dnech 19.-20. dubna 1984*, 7. 5. 1984, 1261/0/8, sv. P106/84, info8, NA, Prague.

[1835] *VÝZVA členských států Varšavské smlouvy členským státům Severoatlantického paktu týkající se uzavření Smlouvy o vzájemném nepoužití vojenské síly a zachování mírových vztahů*, 20. 4. 1984, 1261/0/8, sv. P106/84, info8, NA, Prague.

[1836] Śluszarczyk, *Układ Warszawski*, 87.

the case of Andropov some Western politicians thought about mutual talks, they largely refused any dialogue with his successor.[1837]

Thus, repeating the old initiatives did not meet with any success. This was proven during the first stage of the Stockholm conference on Confidence- and Security-Building Measures and Disarmament in Europe. Its start in January 1984 was the first political achievement of the Warsaw Treaty Organization after a long time.[1838] Nevertheless, in the Swedish capital, the West rejected the well-known proposals of the Eastern Bloc, calling them demonstrative, non-binding, and uncontrollable, or already contained in documents like the UN Charter of the CSCE Final Act. To the Warsaw Pact's displeasure, the neutral countries' positions were close to those of NATO. In addition, the effort by the Soviet sphere of influence was also undermined by Romania.[1839]

At the Warsaw Pact's political bodies meetings, the Soviet Union was still able to reach agreement within the Six without serious complications. In reality, under this mask, the Eastern Bloc was becoming more and more divided. The Soviet satellites were not in a position to implement openly a more independent policy like Romania. During the alliance meetings, their ministers of foreign affairs usually just agreed with Gromyko's evaluations. They took none of their own initiatives. However, through some actions, the Eastern Bloc countries showed their discontent to the Soviet course. In the existing situation, even loyal Bulgaria considered it necessary that Moscow officially accept a nuclear-free statute of the country. On the Zhivkov regime's appeals, the USSR

[1837] Volkogonov, *The Rise and Fall*, 389-390.

[1838] The conference was convened by decision of the CSCE summit in Madrid. All the signatories of the Final Act took part in it. We are reminded that holding such a conference was one of the Warsaw Pact's priorities since the late 1970s.

[1839] The SRR acted totally separately and put forward its own proposals. Afterwards, the Western states referred to selected parts of these. The Six had to admit that preparation for the Stockholm conference had not been sufficient and improvisation had been needed in certain moments. Thus, other consultations within the alliance's framework were expected to aim at setting an approach on how to influence Western countries in favor of Eastern proposals. *Informace o jednání prvního kola I. etapy Konference o opatřeních k posílení důvěry a bezpečnosti a o odzbrojení v Evropě, konaného ve Stockholmu ve dnech 17. ledna - 16. března 1984*, 4. 5. 1984, 1261/0/8, sv. P105/84, info2, NA, Prague.

publicly stated that there were no atomic weapons on the Bulgarian territory. Soon after the death of Andropov, the HSWP International Department announced that Hungary was open to act as an intermediary in restoration of the superpowers' dialogue. Also, the East-German leader Honecker imposed himself in the role of a "moderate voice", a possible mediator between Moscow and Washington and an advocate of disarmament. Nevertheless, the USSR insisted that the Bloc's unity was a top priority at the time. It only had a little sympathy for the Hungarian and East-German activities. It must be added that in this regard, the Kremlin had to rely mostly on the Czechoslovak regime which remained absolutely loyal.[1840]

The East-German activity, especially the SED leadership's effort to continue in cooperation with West Germany, was seen as the main threat to cohesion of the Six by the Kremlin. Honecker rejected any criticism, claiming he was just trying to "reduce damages" caused by the Euromissile Crisis. Since mid-1984, Moscow sent the Honecker regime clear messages that the GDR-FRG relations had to be cooled down. At the same time, this was a warning to the other countries of the Bloc which were interested in development of a collaboration with the West. Discrepancy between Soviet and East-German policies climaxed in August, when Moscow banned the SED CC General Secretary from his trip to West Germany.[1841] Some cracks had begun to appear in the previously near-ultimate loyalty of the GDR to the USSR when attempts for a conciliatory solution to the Euromissile issue had failed. East Berlin reacted to the West-German approval of dislocation of new U.S. weapons in an unusually moderate way. Honecker stated that such a move negatively affected relations between both German states. But at the same time, he stressed that the

---

[1840] At the end of March 1984, the article "National and International in the CPCz Politics" was published in Rudé právo newspaper. In the text, conservative members of the party's international department, close to the Presidium member Vasil Biľak, harshly attacked the efforts of unnamed countries not to follow strictly a joint foreign policy line of the Warsaw Pact, which allegedly played into the hands of the capitalist states. The text was not just anti-Romanian, but also anti-Hungarian. Litera, "Sovětský svaz, 'Eurorakety'," 269-272; Schaefer, "*The GDR in the Warsaw Pact*".

[1841] Ibid.

GDR would do its best to minimize the resulting damages. His behavior probably was not motivated by concerns for reduction of trade exchange with the western neighbor only. A possibility of construction of new Soviet missile bases on the East-German territory hung in the air and Berlin would have been obligated to participate financially.[1842]

In August 1984, Chernenko and the other Soviet officials attacked the SED leadership during talks in Moscow. The CPSU CC General Secretary complained that at the last meeting two months earlier, Honecker had verbally fully agreed with the Soviet views on the international situation as well as proposed strategy. But despite this fact, the GDR had surprisingly announced its willingness to start negotiations with the FRG on relaxation of rules for West-German citizens' visit to the East-German territory. Moscow suspected the Honecker regime of pursuing economic profits at the expense of security risks in this way. The formal chief of the Kremlin raged that while the Soviet press spouted stories about the West-German revanchism, the East-German newspapers remained silent. Moreover, these facts were not missed by the Western media which began to speculate about a possible rift between intentions of the GDR and USSR. Also, Mikhail Gorbachev, the Soviet politburo member, noted that the East had tried to discourage Bonn from the approval of deployment of U.S. missiles, among other things, through the threat of significant deterioration of both German states' relations. But in reality, the exact opposite happened: There was a growth of contacts, economic cooperation, and loans. This undermined the Warsaw Pact's previous joint resolutions. Hermann Axen, the SED CC secretary and politburo member, tried to calm down the Soviet leadership by noting that through this approach, the GDR was just reducing tensions on borders. But he was harshly attacked by Dmitri Ustinov. The Soviet defense minister accused East Berlin of leaving a firm attitude towards the West-German state, which provided 50% of ground forces and 30% of air forces to NATO combat capacities in Europe. In addition, he warned against the potential arrival of West-German spies under a pretext of family visits.[1843]

[1842] Litera, "Sovětský svaz, 'Eurorakety'," 263.

[1843] Transcript of Honecker-Chernenko Meeting in Moscow, August 17, 1984, in *A Cardboard Castle?* (doc. 104): 496-499.

The relatively weak Chernenko was unable to settle these disputes. Thus, relations between Berlin and Moscow continued to fall. The East-German actions also concerned the Warsaw Pact allies. The GDR unilaterally expanded its territorial waters and began to block the mouth of Oder River nearby Świnoujście, obstructing Polish vessels from free sailing.[1844]

Not only the Eastern Bloc's unity crumpled. Fragmentation of opinions also took place within the Soviet leadership. General Secretary Chernenko and his protégé Mikhail Gorbachev gradually strengthened their power positions against the duo Gromyko-Ustinov. The latter group rigorously insisted on previously stipulated conditions for a return to the negotiating table with the United States. Despite that such a policy was losing support in the politburo, it remained backed by the vast majority of military circles. On the contrary, in late June 1984, Gorbachev suggested a possibility of starting an "honest dialogue" with Washington. The General Secretary began to make similar statements in the media. At the turn of August and September, the Soviet policy was obviously turning its course so far. Moscow announced that Gromyko would meet the U.S. president during his trip to the UN General Assembly in New York.[1845] Subsequently, on September 12, the USSR ended its boycott of disarmament talks.[1846] Afterwards, the Soviet minister of foreign affairs bilaterally informed the Warsaw Pact allies that Moscow intended to negotiate with the United States especially on strategic nuclear weapons, Euromissiles, and risk of militarization of outer space.[1847] These issues also dominated the agenda of the Committee of Ministers of Foreign Affairs, held in East Berlin December 3-4, 1984. Restoration of the Soviet-American dialogue should have been preceded by the long-postponed

---

[1844] In this regard, Vojtech Mastny notes that there was a risk of military confrontation between both countries because of this situation. This claim seems to be too exaggerated. Mastny, *A Cardboard Castle?*, 57.

[1845] Litera, "Sovětský svaz, 'Eurorakety'," 267, 274.

[1846] Durman, *Útěk od praporů*, 294.

[1847] *Niederschrift über das Gespräch des Genossen Erich Honecker mit Genossen Andrej Andrejewitsch Gromyko am 8. Oktober 1984 im Hauze des Zentralkomitees der SED,* DY 30/2380, BArch, Berlin.

session of the Political Consultative Committee that was finally scheduled for January of the upcoming year.[1848]

The Warsaw Pact's supreme body meeting was cancelled in the end as Konstantin Chernenko's health worsened rapidly at the beginning of 1985. On January 9, the Soviet politburo agreed that the General Secretary was not capable of traveling. Somebody else was supposed to lead the delegation of the USSR.[1849] However, this emergency plan was never implemented. The Political Consultative Committee never met again during Chernenko's life. The sixth Soviet leader passed away on March 10. Thus, he became the only post-Stalin Kremlin leader during his rule in which no top summit of the Warsaw Treaty Organization was held. There was basically no reason for such a meeting. The East-West relations were on ice. The Soviet diplomatic activity stagnated at the beginning of Chernenko's rule.[1850] In the moment when the Kremlin's foreign political strategy began to change, the Soviet leadership decided to wait with the Political Consultative Committee session on results of the presidential elections in the United States scheduled for November 1984.[1851]

Prolongation of the Warsaw Treaty

Chernenko's short intermezzo did not influence the inside development of the Warsaw Treaty Organization at all. The most important question in this period was prolongation of the alliance's validity which would end on May 14, 1985. At the end of 1983, the USSR showed its satisfaction with the existence of the multilateral alliance between the

---

[1848] *Informace o řádném zasedání výboru ministrů zahraničních věcí členských států Varšavské smlouvy v Berlíně ve dnech 3. a 4. prosince 1984*, 10. 12. 1984, 1261/0/8, sv. P123/85, info12, NA, Prague.

[1849] Volkogonov, *The Rise and Fall*, 428.

[1850] Christian Nünlist speculates that the reason why the Political Consultative Committee session was not held in 1984 was the Soviet effort to prevent the GDR, SRR, and HPR from presenting their different foreign political stances at that forum. But this seems to be unlikely. Actually, two Committee of Ministers of Foreign Affairs' meetings took place in that period and the Six acted unanimously at first glance. Nünlist, "*Cold War Generals*".

[1851] *Niederschrift über das Treffen zwischen Genossen K. U. Tschernenko und Genossen E. Honecker am 14. Juni 1984 in Moskau*, 15. 6. 1984, DY 30/2380, BArch, Berlin.

countries of its European sphere of influence; Moscow was interested in its continuation. Initially, the Kremlin considered two options: to prolong the Warsaw Treaty or to establish a brand-new alliance. The former one was preferred, of course. In the reality of the Eastern Bloc, the entire matter would have been just a formality. But some complications by the Romanian side were expected. Moscow was unambiguously interested in continual membership of the SRR in the alliance. The Soviet leadership thought that the Ceausescu regime would use the forthcoming end of the Warsaw Treaty's validity to launch a new campaign on dissolution of military-political blocs. On the other hand, the Kremlin doubted Bucharest's will to withdraw from the Warsaw Pact.[1852] Perhaps also for this reason, an argument was raised in the discussion on prolongation of the alliance, that such a move did not change at all an offer to disband the Pact, if NATO was dissolved. Despite all the issues being directed by the Soviet side, Poland became the main formal actor. At the beginning of 1984, the country put forward a proposal to hold talks on this question at the level of an expert group based in Warsaw. Naturally, everything would have proceeded in secret.[1853]

In April 1984, at the Committee of Minister of Foreign Affairs' session, the Romanian delegation blocked this approach. Nevertheless, eight months later, at the next meeting of the body, the SRR stated that the 13th RCP Congress had approved prolongation of the Warsaw Treaty. But the issue should have been discussed by the Political Consultative Committee or states' top representatives only.[1854] During 1984, the Six actually actively influenced Romania in this matter. Chernenko, in person, talked about the prolongation of the Warsaw Treaty Organization with

---

[1852] *Notatka z rozmowy tow. Rusakowa z gen. W. Jaruzelskim*, 10. 11. 1983, PZPR KC, XIA/1396, AAN, Warsaw.

[1853] *Pismo tow. gen. W. Jaruzelskiego, I Sekretarza KC PZPR do I Sekretarzy Komitetów Centralnych państw socjalistycnych stron Układu Warszawskiego w sprawie jego przedłużenia*, February 1, 1984, XIA/1397, ibid.

[1854] *Informace o řádném zasedání výboru ministrů zahraničních věcí členských států Varšavské smlouvy v Budapešti ve dnech 19.-20. dubna 1984*, 7. 5. 1984, 1261/0/8, sv. P106/84, info8, NA, Prague; *Informace o řádném zasedání výboru ministrů zahraničních věcí členských států Varšavské smlouvy v Berlíně ve dnech 3. a 4. prosince 1984*, 10. 12. 1984, sv. P123/85, info12, ibid.

Ceausescu. He tried to convince him to avoid any complications in the procedure.[1855] It is good to add that the rest of the Six just followed a line set by Moscow on this issue.[1856] In the end, the protocol prolonging the alliance's validity by 20 years with the possibility of another prolongation of an extra 10 years was worked out by the expert group during the first few months of 1985. Then, Wojciech Jaruzelski formally proposed the date of its signing in Warsaw on April 24. In reality, it was done two weeks later. Moscow did not consider it necessary to hold the Political Consultative Committee on this occasion.[1857]

At the time, Mikhail Gorbachev already held the position of the CPSU CC General Secretary. Biological decay of the rest of the Brezhnev era's top officials undoubtedly made his appointment easier. Dmitri Ustinov died on December 20, 1984. The last man alive of the formerly ruling "troika," Andrei Gromyko, pragmatically supported the election of 54-year-old Gorbachev, eventually. Given his age, he was moved from top of the Soviet diplomacy to the position of Chairman of the Supreme Soviet. This largely ceremonial function suited the age of the man, who had directed the Soviet ministry of foreign affairs for a long 27 years.[1858]

[1855] *Niederschrift über das Treffen zwischen Genossen K. U. Tschernenko und Genossen E. Honecker am 14. Juni 1984 in Moskau*, 15. 6. 1984, DY 30/2380, BArch, Berlin.

[1856] In October 1984, the USSR formulated an official letter to the Romanian leadership on the issue of prolongation of the alliance. Afterwards, the rest of the Warsaw Pact members wrote notes of the same spirit and handed them to the Romanian diplomats. In bilateral talks, the representatives of the Six tried to influence them in favor of prolongation of the Warsaw Treaty. The East-German politburo even suggested inviting Ceausescu to the GDR and discussing the question with him in person. *Usnesení 120. schůze předsednictva ÚV KSČ ze dne 24. října 1984*, 1261/0/8, sv. P120/84, b. 11, NA, Prague; *Protokoll Nr. 42 der Sitzung des Politbüros des ZK der SED vom 23. Okt. 1984*, DY 30/ J IV 2/2/2083, BArch, Berlin; *Protokoll Nr. 6 der Sitzung des Politbüros des Zentrakomitees vom 12. Februar 1985*, DY 30/ J IV 2/2/2098, ibid.

[1857] *Pismo tow. gen. W. Jaruzelskiego, I Sekretarza KC PZPR do I Sekretarzy Komitetów Centralnych państw socjalistycnych stron Układu Warszawskiego w/s jego przedłużenia*, 2. 4. 1985, PZPR KC, XIA/1398, AAN, Warsaw; Letter by W. Jaruzelski, the PUWP CC First Secretary, to G. Husák, the CPCz CC General Secretary, April 2, 1985; *Důvodová zpráva*, 10. 4. 1985, 1261/0/8, sv. P130/85, b. 6, NA, Prague.

[1858] Zubok, *A Failed Empire*, 277-278.

The first top alliance meeting under the new CPSU CC General Secretary took place in October 1985, in Sofia. In that moment, Gorbachev had already turned the Eastern Bloc to the path which resulted in its collapse at the end of the 1980s. Together with the Soviet sphere of influence in Europe, the Warsaw Treaty Organization itself was also destined to dissolution, of course.

Epilogue

Military-political alliances have various natures. Despite political statements and propagandistic proclamation, absolute equality of their member-states often remains merely an illusion in praxis. This is particularly the case when an alliance includes a number of countries whose real military, economic, and political power differs. In a historical context, it would be naive to believe that a stronger power, not to mention a superpower, in certain moments would not try to promote its interests which are not necessarily compatible with those of the junior allies. It is a level of dominance as well as methods used that matter.

As far as the Warsaw Pact is concerned, for most of the time of this organization's existence the Soviet Union acted as an effective hegemon, with slights shifts only. The fact that in the period examined in this book it did so more tactfully and with at least illusionary respect for its "allies," compared to the 1950s and 1960s, was far from sufficient to make it a real partner and guardian, a "senior brother" in the eyes of non-communist segments of the member-states' societies. On the contrary, during the first three decades of its existence, the Warsaw Pact became a symbol of submission to a foreign power. During the big crises of the Eastern Bloc in 1956, 1968, and 1980-1981, therefore, criticism of such an alliance - of at least the organization's real mechanisms - occurred at every moment.

Such a perception of the Warsaw Pact was well proven during the Polish events in the early 1980s. With a bit of exaggeration, the threat of the Warsaw Pact's tanks storming the streets of major Polish cities was seen as more eminent than the danger of "imperialistic aggression" that the organization should have deterred, according to official and propagandistic proclamations. Historical experience was too strong. If relatively few people accepted the possibility of the Warsaw Pact's military intervention against the reform process in Czechoslovakia in 1968, twelve years later, concerns of such a scenario were present since the outbreak of unrest in Poland. The fact that the suppression of the Prague Spring was not an official action by the organization was absolutely irrelevant in the context of events. Paradoxically, there was the same time interval between the Prague Spring and the Polish events as between the Hungarian Uprising and the attempt to liberalize real socialism in Czechoslovakia. And so, the reason might be also seen in the fact that Moscow intervened without the direct military assistance of its allies in 1956. Moreover, the 1950s and its methods were considered virtually a closed chapter in the second half of the 1960s.

During Brezhnev's era, the Warsaw Pact did not become a driver of events in the Eastern Bloc. Its structures, in fact, had little initiative. Agendas of the meetings were set elsewhere, in Moscow most often. Instead of being a starting point for development of the Soviet sphere of influence in Europe, activities in the alliance's framework were of an exactly opposite nature. The processes within the Warsaw Pact rather reflected a general situation in the Eastern Bloc, changes in the Kremlin's policy and shifts in relations between the Soviet Union and its European satellites. The result for a historian dealing with the Eastern Bloc history is that the analysis of the situation within the alliance would be a "litmus test" at best and that any transformations of its framework indicates wider changes in the geopolitical area. Their roots and impetuses, however, must be seen elsewhere.

In the era of Nikita Khrushchev, the Warsaw Pact reflected the style of his overall policy. A set of separate, poorly deliberated, and incoherent campaigns actually corresponded with irregular, *ad-hoc* talks held within the pact. Brezhnev's changes were linked with general efforts to stabilize the Eastern Bloc; a largely chaotic cooperation should have become more systematic in some ways. It is probably not just coincidental that the Warsaw Pact's political pinnacle coincided with the Brezhnev leadership's strongest activity in foreign policy. Similarly, there is definitely a connection between the later Brezhnevist stagnation and the Warsaw Pact's switch to some passivity. The attempts to alter the alliance's mechanisms and functioning in the short period of Brezhnev's successors just reflected Andropov's observation that both the domestic and foreign policy of the Soviet Union was unsustainable and that certain modifications were inevitable.

Given these facts, the Warsaw Pact remained completely linked with the Cold War reality and namely the Eastern Bloc's existence. It was widely seen as a tool of the Soviet political and military will - correctly, let us add. Therefore, no attempts to set new mechanisms and agenda for the alliance in Gorbachev's era seemed credible. So, it is no surprise that a formal dissolution of the Warsaw Pact became one of the most urgent issues when the East-European communist regimes collapsed in 1989 and the Soviet power empire disintegrated. Moreover, the alliance was already effectively dead at that time. The new garnitures in power in its member-states showed little interest in effort to reform deeply the organization whose roots and fundamental features of cooperation dated back to Brezhnev's era. In addition, the lesson learned was that in many cases, ways were found to circumvent the formal rules. It is no wonder that breaking such ties to Moscow was seen as an essential and logical move. However, to fully explain and analyze the course and nature of that process, whose main causes were formed in the years examined in this book, is a task for future historical research.

Sources and literature

Unpublished sources

Archive of Ministry of Foreign Affairs, Czech Republic (Archiv ministerstva zahraničních věcí České republiky - AMZV), Prague
Collections:
- Documentation of Teritorial Depratments (Dokumentace teritoriálních odborů - DTO) 1945–1989
- Teritorial Departments (secret) [Teritoriální odbory (tajné) – TO(t)] 1970–1974
- Meetings of Collegium (Porady kolegia - PK) 1953–1989

Archive of Modern Documents (Archiwum Akt Nowych - AAN), Warsaw
Collections:
- Polish United Worker's Party Central Committee (Polska Zjednoczona Partia Robotnicza. Komitet Centralny - PZPR KC)
- Bureau of Council of Ministers (Urząd Rady Ministrów - URM-KT)

Federal Archive (Bundesarchiv - BArch), Berlin
Collections:
- Socialist Unity Party of Germany, Departments and Secretariat (Sozialistische Einheitspartei Deutschlands, Büros und Sekretariate - DY 30/)

National Archive (Národní archiv - NA), Prague
Collections:
- CPCz CC Presidium 1966–1971 (Předsednictvo ÚV KSČ 1966–1971 - 1261/0/5)
- CPCz CC Presidium 1971–1976 (Předsednictvo ÚV KSČ 1971–1976 - 1261/0/6)
- CPCz CC Presidium 1976–1981 (Předsednictvo ÚV KSČ 1976–1981 - 1261/0/7)

- CPCz CC Presidium 1981–1986 (Předsednictvo ÚV KSČ 1981–1986 - 1261/0/8)
- Ministry of Foreign Affairs – press cuttings 1945–1987 (Ministerstvo zahraničních věcí – výstřižkový archiv 1945–1987 - 506/0/45)

The National Archives (TNA), London
Collections:
- Foreign and Commonwealth Office (FCO)

Central Military Archive (Vojenský ústřední archiv - VÚA), Prague
Collections:
- Federal Ministry of Defense, General Staff/Operational Department (Federální ministerstvo obrany Generální štáb/Operační správa - FMO GŠ/OS)

Published sources

Electronic archives and sources

- Parallel History Project on Cooperative Security – http://www.php.isn.ethz.ch/
- Wilson Center - http://digitalarchive.wilsoncenter.org/

Editions

- *A Cardboard Castle? An Inside History of the Warsaw Pact*, eds. Vojtech Mastny and Malcolm Byrne (New York: CEU Press, 2005).
- *Tajne dokumenty Biura Politycznego. Grudzień 1970*, ed. Paweł Domański (London: Aneks, 1991).
- *The Euromissiles Crisis and the End of the Cold War: 1977-1987*, vol. II, ed. Tomothy McDonnell (Washington: Wilson Center, 2009).
- *Przed i po 13 grudnia*, tom I, eds. Łukasz Kamiński, Iskra Baeva and Petr Blažek (Warszawa: IPN, 2006).
- *Przed i po 13 grudnia*, tom II, eds. Łukasz Kamiński, Iskra Baeva and Petr Blažek (Warszawa: IPN, 2007).
- *Plánování nemyslitelného*, ed. Petr Luňák (Praha: Ústav pro soudobé dějiny – Dokořán, 2007).

- *From Solidarity to Martial Law: The Polish Crisis of 1980-1981: A Documentary History*, eds. Andrzej Paczkowski and Malcolm Byrne (New York: CEU Press, 2007).
- *Selected Soviet Military Writings 1970-1975*, ed. William F. Scott (Washington: U.S. Government Print Office, 1976).

Memoirs, testimonies

- "Commentary by Wojciech Jaruzelski," *Cold War International History Project Bulletin*, no. 11 (1998): 32–39.
- Грибков, Анатолий Иванович, *Судьба Варшавского Договора. Воспоминания, документы, факты*, (Москва: Русская книга 1998).
- Jaruzelski, Wojciech, *Stan wojenny – dlaczego* (Warszawa: Polska Oficyna Wydawnicza BGW, 1992).
- Jaruzelski, Wojciech, *Pod prąd. Refleksje rocznicowe* (Warszawa: Comandor, 2005),
- Jaruzelski, Wojciech, *Historia nie povinna dzielić* (Toruń : Adma Marszałek, 2006).
- Kukliński, Ryszard, *Wojna z narodem. Rozmowa z byłym płk. dypl. Ryszardem J. Kuklińskim* (Warszawa: Kurs, 1987).
- Nowak, Jerzy, *Od hegemonii do agonii: upadek Układu Warszawskiego – polska perspektywa* (Warszawa: Bellona, 2011).
- *Wejdą nie wejdą: Polska 1980-1982 konferencja w Jachrance listopad 1997* (London: ANEKS, 1997).

Literature

- Bacon, Erwin, "Reconsidering Brezhnev," in *Brezhnev Reconsidered*, eds. Erwin Bacon and Mark Sandle (New York: Palgrave, 2002): 1–21.
- Baev, Jordan, *"The "Crimean Meetings" of the Warsaw Pact Countries' Leaders,"* Accessed October 16, 2019. 2003. http://www.php.isn.ethz.ch/lory1.ethz.ch/collections/crimea_mee tings4a8c.html?navinfo=16037.
- Baev, Jordan, "The Warsaw pact and Southern Tier Conflicts, 1959-1969," in *NATO and the Warsaw Pact. Intrabloc Conflicts*, eds. Mary Ann Heiss and S. Victor Papacosma (Kent: Kent State University Press, 2008): 193–205.
- Bange, Oliver and Kieninger, Stephan, "Negotiating one's own demise? The GDR's Foreign Ministry and the CSCE negotiations:

Plans, preparations, tactics and presumptions," *Cold War International History Project e-Dossier*, no. 17 (2007): 1–17.

- Bange, Oliver, "Comments on and Contextualization of Polish Documents related to SOYUZ 75 and SHCHIT 88," *Cold War International History Project e-Dossier*, no. 20 (2010).

- Benčík, Antonín, *Vojenské otázky československé reformy 1967-1970*; vol. 1, (Praha: Doplněk, 1996).

- Békés, Csába, "*Hungary and the Warsaw Pact, 1954-1989: Documents on the Impact of a Small State within the Eastern Bloc*," Accessed October 16, 2019. http://www.php.isn.ethz.ch /lory1.ethz.ch/collections/coll_hun/intro07f3.html?navinfo=15711.

- Békés, Csába, "*Records of the Meetings of the Warsaw Pact Deputy Foreign Ministers*," Accessed October 16, 2019. http://www.php.isn.ethz.ch/lory1.ethz.ch/collections/coll_defomi n/intro_bekes72dc.html?navinfo=15700.

- Békés, Csába, "Why Was There No 'Second Cold War' in Europe? Hungary and the East-West Crisis Following the Soviet Invasion of Afghanistan," in *NATO and the Warsaw Pact. Intrabloc Conflicts*, eds. Mary Ann Heiss and S. Victor Papacosma (Kent: Kent State University Press, 2008): 219–232.

- Békés, Csába, "Studená válka, détente a sovětský blok. Vývoj koordinace zahraniční politiky sovětského bloku (1953– 1975)," *Soudobé dějiny,* no. 1-2 (2011): 53–85.

- Bílek, Jiří and Koller, Martin, "Československá armáda v rámci sovětského bloku. 1945-1989," in *Od Velké Moravy k NATO. Český stát a střední Evropa*, ed. Josef Tomeš (Praha: Evropský literární klub 2002):196–207.

- Bílý, Matěj, "*Varšavská smlouva v 50. a 60. letech*," (PhDr diss., Charles University in Prague, 2011).

- Bílý, Matěj, "Počátky pokusu o reformu Varšavské smlouvy," *The Twentieth Century – Dvacáté století*, no. 1 (2011): 159–172.

- Bílý, Matěj, "Bratrská pomoc Polsku? Role Varšavské smlouvy v první fázi polské krize: červenec–prosinec 1980," *The Twentieth Century – Dvacáté století*, no. 1 (2012): 159–182.

- Bílý, Matěj, "Prosinec 1980: účast ČSSR v pokusu o multilaterální řešení polské krize," in *České, slovenské a československé dějiny 20. století VII*, eds. Tomáš Hradečný and Pavel Horák (Ústí nad Orlicí: Oftis, 2012), 207–217.

- Bílý, Matěj, "Rumunsko a Varšavská smlouva v první polovině 70. let," in *VI. ročník studentské konference o Balkáně – sborník vystoupení,* ed. Hana Suchardová (Praha: Rada pro mezinárodní vztahy, 2012) 18–25.

- Bílý, Matěj, "1981: Role Varšavské smlouvy ve druhé fázi polské krize," *Historie a vojenství,* no. 2 (2013): 20–40.
- Bílý, Matěj, "ČSSR v organizaci Varšavské smlouvy na přelomu 60. a 70. let," in *České, slovenské a československé dějiny 20. století VIII,* eds. Tomáš Hradečný and Pavel Horák (Ústí nad Orlicí: Oftis, 2013): 273–283.
- Bílý, Matěj, "Gierekovo vedení a prosazování polských zájmů v organizaci Varšavské smlouvy," *Historie a vojenství,* no, 3 (2014): 4–16.
- Bílý, Matěj, "Romania in the political structures of the Warsaw Treaty Organization at the turn of 1960s and 1970," *Oriens Aliter,* no 2 (2014): 44–69.
- Bílý, Matěj, "Od pražského jara k helsinskému procesu. Vývoj politické spolupráce v organizaci Varšavské smlouvy na přelomu 60. a 70. let," *Východočeské listy historické,* no. 33 (2015): 89–124.
- Bílý, Matěj, "Statut Varšavské smlouvy pro období války. Kontext – schválení – dopady," *Historie a vojenství,* no. 2 (2015): 15-27.
- Bílý, Matěj, "ČSSR a krach détente. Nástin československé politiky v rámci Varšavské smlouvy v druhé polovině 70. let," in *České, slovenské a československé dějiny 20. století IX,* eds. Tomáš Hradecký, Pavel Horák and Pavel Boštík (Ústí nad Orlicí: Oftis, 2015): 417-426.
- Bowker, Mike, "Brezhnev and Superpowers Relations," in *Brezhnev Reconsidered,* eds. Erwin Bacon and Mark Sandle (New York: Palgrave, 2002): 90–109.
- Burr, William, "'Is this the best they can do?' Henry Kissinger and the US quest for limited nuclear operation, 1969-75," in *War Plans and Alliances in the Cold War,* eds. Vojtech Mastny, Sven Holtsmark and Andreas Wenger (London: Routledge, 2006): 118–140.
- Caciagli, Federica, "The GDR's target in the early CSCE process. Another missed opportunity to freeze the division of German, 1969-73," *Origins of the European security system: the Helsinky process revisited, 1965–75,* eds. Andreas Wenger, Vojtech Mastny and Christian Nuenlist (Abingdon: Routledge, 2008): 107–123.

- Crump, Laurien, *The Warsaw Pact Reconsidered. International Relations in Eastern Europe, 1955–69* (London/New York: Routledge, 2015).
- Deletant, Dennis, "*Romania and the Warsaw Pact : Documents Highlighting Romania's Gradual Emancipation from the Warsaw Pact, 1956-1989*," Accessed October 16, 2019. http://www.php.isn.ethz.ch/lory1.ethz.ch/collections/coll_romani a/introduction0445.html?navinfo=15342.
- Dumitru, Laurentiu Cristian, *Romania nad the Warsaw Pact, 1955–1968. From Obedience to Defiance* (Rome: Italian Academic Publishing, 2014).
- Durman, Karel, *Útěk od praporů. Kreml a krize impéria 1964–1991* (Praha: Karolinum, 1998).
- Eisler, Jerzy, *Grudzień 1970: geneza-przebieg-konsekwencje* (Warszawa: Sensacje XX Wieku, 2000).
- Firth, Noel E. and Noren, James H., *Soviet Military Spending* (Texas: A&M University Press, 1998).
- Friszke, Andrzej, *Polska Gierka* (Warszawa: WSiP, 1995).
- Fučík, Josef, "Rozpočtové výdaje na obranu v zemích Varšavské smlouvy v letech 1955–1989," in: *Vojenské výdaje v letech studené války a po jejím skončení*, ed. Miroslav Krč (Praha: Ústav mezinárodních vztahů, 2000): 64–92.
- Gaddis, John Lewis, *Studená válka* (Praha: Slovart, 2006).
- Garthoff, Raymond L., "New Thinking and Soviet Military Doctrine," in *Soviet Military Doctrine from Lenin to Gorbachev 1915–1991,* eds. Willard C. Frank and Philip S. Gillette (Westport: Preager, 1992): 195–210.
- Giurescu, Dinu Constantin and Fisher-Galati, Stephen, *Romania. A Historic Perspective* (New York: Boulder, 1998).
- Gizicki, Wojciech, *Od Układu do Paktu: (r)ewolucyjna zmiana w polityce bezpieczeństwa Polski* (Lublin: Institut Sądecko-Lubelski, 2011).
- Griggs, Les, "*Origins and Significance of the Warsaw Pact Wartime Statute Documents*," Accessed October 16, 2019. https://www.cia.gov/library/publications/cold-war/wartime-statutes/wartime-statutes.pdf.
- Harrison, Mark, "Economic Growth and Slowdown," in *Brezhnev Reconsidered*, eds. Erwin Bacon and Mark Sandle (New York: Palgrave, 2002): 38–67.

- Heuser, Beatrice, "Warsaw Pact Military Doctrine in the 1970s and 1980s: Findings in the East German Archives," *Comparative Strategy*, no. 12 (1993): 437–451.
- Hoffenaar, Jan and Findlay, Christopher, eds., *Military planning for European Theatre Conflict During the Cold War. An Oral History Roundtable, Stockholm, 24-25 April 2006* (Zürich: Center for Security Studies, 2007).
- Holloway, David, "Jaderné zbraně a studená válka v Evropě," *Soudobé dějiny*, no 1-2 (2011): 32–52.
- Ionescu, Mihai E., "Rumunsko a vojenská reforma Varšavské smlouvy," *Historie a vojenství*, no 3-4 (2003): 699–705.
- Jarząbek, Wanda, *PRL w politycznych strukturach Układu Warszawskiego w latach 1955–1980* (Warszawa: PAN, 2008).
- Jarząbek, Wanda, "*Poland in the Warsaw Pact 1955–1991: An Appraisal of the Role of Poland in the Political Structures of the Warsaw Pact*," Accessed October 16, 2019. http://www.php.isn.ethz.ch/lory1.ethz.ch/collections/coll_poland /Introductionb85a.html?navinfo=111216.
- Jensen, Frade P., "The Warsaw Pact's Special Target. Planning the Seizure of Denmark," in *War Plans and Alliances in the Cold War*, eds. Vojtech Mastny, Sven Holtsmark and Andreas Wenger (London: Routledge, 2006): 95–117.
- Kamiński, Łukasz, *Przed i po 13 grudnia. Państwa bloku wschodniego wobec kryzysu w PRL, 1980 - 1982*, tom I, (Warszawa: IPN, 2006).
- Kopstein, Jeffrey, "Ulbricht Embattled. The Quest for Socialist Modernity in the Light of New Sources," *Europe-Asia Studies*, no. 4 (1994): 597–615.
- Kramer, Mark, "Warsaw Pact Military Planning in Central Europe: Revelations From the East German Archives," *Cold War International Project Bulletin*, no. 2 (1992): 1, 13–19.
- Kramer, Mark, "Comentary on Soyuz-75 and Shchit-88 Military Excercise Documents," *Cold War International Project e-Dossier*, no. 20 (2011).
- Kramer, Mark, "The Warsaw Pact and the Polish Crisis of 1980–81: Honecker's Call for Military Intervention," *Cold War International History Project Bulletin*, no. 5 (1995): 124.
- Kramer, Mark, "Poland 1980–81: Soviet Policy during the Polish Cisis," *Cold War International History Project Bulletin*, no. 5 (1995): 1, 116–126.

- Kramer, Mark, "In Case Military Assistance is Provided to Poland: Soviet Preparation for Military Contingencies, August 1980,". *Cold War International History Project Bulletin*, no. 11 (1998): 102–109.
- Kramer, Mark, "Jaruzelski, the Soviet Union, and the Imposition of Martial Law in Poland: New Light on the Mystery of December 1981," *Cold War International History Project Bulletin*, no. 11 (1998): 5–14.
- Kramer, Mark, "Colonel Kuklinski and Polish Crisis, 1980–81," *Cold War International History Project Bulletin*, no. 11 (1998): 48–59.
- Kramer, Mark, "Die Sowjetunion, der Warschauer Pakt und blockinterne Krisen während der Brežněv-Ära," in *Der Warschauer Pakt. Von der Gründung bis zum Zusammenbruch 1955 bis 1991,* eds. Torsten Diedrich, Winfried Heinemann, and Christian Ostermann (Berlin: Christoph Links Verlag, 2009): 273–336.
- Krč, Miroslav, "Vojenské výdaje v letech studené války a jejich determinanty s důrazem na NATO a Varšavskou smlouvu," in: *Vojenské výdaje v letech studené války a po jejím skončení*, ed. Miroslav Krč (Praha: Ústav mezinárodních vztahů, 2000): 35–49.
- Kříž, Antonín, *Krkonoše 80. Cvičení nebo...?* (s.n., after 1995).
- Kořalková, Květa, *Vytváření systému dvoustranných spojeneckých smluv mezi evropskými socialistickými zeměmi (1943-1949)* (Praha: Academia, 1966).
- Kosman, Marceli, *Los Generała. Wokoł medialnego wizerunku Wojciecha Jaruzelskiego* (Torun: Adam Marszałek, 2008).
- Litera, Bohuslav, "Unifikace sovětského bloku v 70. letech," *Slovanský přehled,* no. 1 (1994): 81–96.
- Litera, Bohuslav: "Sovětský svaz, 'Eurorakety' a eroze sovětského bloku," *Slovanský přehle,* no. 3 (1994): 59–276.
- Locher, Anna, "*Shaping the Policies of the Alliance – The Committee of Ministers of Foreign Affairs of the Warsaw Pact, 1976-1990*," Accessed October 16, 2019. http://www.php.isn.ethz.ch/lory1.ethz.ch/collections/coll_cmfa/c mfa_intro7e2f.html?navinfo=15699.
- Luňák, Petr, "War Plans from Stalin to Brezhnev. The Czechoslovak Pivot," in *War Plans and Alliances in the Cold War*, eds. Vojtech Mastny, Sven Holtsmark and Andreas Wenger (London: Routledge, 2006): 72–94.

- Luňák, Petr, *Plánování nemyslitelného* (Praha: Ústav pro soudobé dějiny – Dokořán, 2007).
- Madry, Jindřich, "Sovětské zájmy v pojetí obrany Československa (1965-1970)," *Historie a vojenství*, no. 5 (1992): 126–140.
- Machcewicz, Pawel, "The Assistance Of Warsaw Pact Forces Is Not Ruled Out," *Cold War Ineternationa History Project Bulletin*, no. 11 (1998): 40.
- Mastny, Vojtech, "We are in a Bind: Polish and Czechoslovak Attempts at Reforming the Warsaw Pact, 1956–1969," *Cold War International Project Bulletin*, no. 11 (1998): 230–235.
- Mastny, Vojtech, "The Soviet Non-Invasion of Poland in 1980–1981 and the End of the Cold War," *Europe-Asia Studies*, no. 51 (1999): 189–211.
- Mastny, Vojtech, *Learning from the Enemy: NATO as a Model for the Warsaw Pact* (Zürich: Forschungsstelle für Sicherheitspolitik und Konfliktanalyse der ETH, 2001).
- Mastny, Vojtech, "*Warsaw Pact Generals in Polish Uniforms,* " Accessed October 16, 2019. http://www.php.isn.ethz.ch/lory1. ethz.ch/collections/coll_polgen/introduction8f94.html?navinfo=1 5708.
- Mastny, Vojtech, "Superpower Détente: US-Soviet Relations, 1969-1972," *Bulletin of the German Historical Institute*, no 1 (2004): 19–25.
- Mastny, Vojtech and Byrne, Malcolm, eds., *A Cardboard Castle? An Inside History of the Warsaw Pact*, (New York: CEU Press, 2005).
- Mastny, Vojtech, "Imagining War in Europe. Soviet Strategic Planning," in *War Plans and Alliances in the Cold War*, eds. Vojtech Mastny, Sven Holtsmark and Andreas Wenger (London: Routledge, 2006): 15–45.
- Mastny, Vojtech, "The Warsaw Pact. An Alliance in Search of a Purpose," in *NATO and the Warsaw Pact. Intrabloc Conflicts*, eds. Mary Ann Heiss and S. Victor Papacosma (Kent: Kent State University Press, 2008): 141–160.
- Munteanu, Mircea, "The Beginning of the End for Détente: The Warsaw Pact Political Consultative Committee," Cold War International History Project e-Dossier, no. 24 (2011).
- Morgan, Michael Fotry, "North America, Atlanticism, and the making of the Helsinki Final Act," in *Origins of the European security system: the Helsinky process revisited, 1965–75*, eds.

Andreas Wenger, Vojtech Mastny and Christian Nuenlist (Abingdon: Routledge, 2008): 25–45.

-   Nünlist, Christian, "*Cold War Generals: The Warsaw Pact Committee of Defense Ministers, 1969-90,* " Accessed October 16, 2019.http://www.php.isn.ethz.ch/lory1.ethz.ch/collections/ coll_cmd/introductiond6c9.html?navinfo=14565.

-   Ochrana, František, "Pojetí vojenské síly států Varšavské smlouvy," in: *Vojenské výdaje v letech studené války a po jejím skončení,* ed. Miroslav Krč (Praha: Ústav mezinárodních vztahů, 2000): 50–63.

-   Opriş, Petre, "Die rumänische Armee und die gemeinsemen Manöver des Warschauer Paktes," in *Der Warschauer Pakt. Von der Gründung bis zum Zusammenbruch 1955 bis 1991,* eds. Torsten Diedrich, Winfried Heinemann, and Christian Ostermann (Berlin: Christoph Links Verlag, 2009): 185–208.

-   Ouimet, Matthew J, *The Rise and Fall of the Brezhnev Doctrine in Soviet Foreign Policy* (Chapel Hill: The University od North Carolina Press, 2003).

-   Paczkowski, Andrzej and Byrne, Malcolm, eds., *From Solidarity to Martial Law: The Polish Crisis of 1980-1981: A Documentary History* (New York: CEU Press, 2007).

-   Paczkowski, Andrzej, "Die Polnische Volksarmee im Warschauer Pakt, " in *Der Warschauer Pakt. Von der Gründung bis zum Zusammenbruch 1955 bis 1991,* eds. Torsten Diedrich, Winfried Heinemann, and Christian Ostermann (Berlin: Christoph Links Verlag, 2009): 119–132.

-   Pech, Radek, "Rumunsko let sedmdesátých – od liberalismu k represi," *Slovanský přehled,* no. 3 (1992): 268–279.

-   Retagan, Mihai, *Ve stínu pražského jara* (Praha: Argo, 2002).

-   Rey, Marie-Pierre, "The USSR and the Helsinki process, 1969-75," in *Origins of the European security system: the Helsinky process revisited, 1965–75,* eds. Andreas Wenger, Vojtech Mastny and Christian Nuenlist (Abingdon: Routledge, 2008): 65–81.

-   Rinoveanu, Carmen, "Rumänien und die Militärrefrom des Warschauer Paktes," in *Der Warschauer Pakt. Von der Gründung bis zum Zusammenbruch 1955 bis 1991,* eds. Torsten Diedrich, Winfried Heinemann, and Christian Ostermann (Berlin: Christoph Links Verlag, 2009): 209–224.

-   Ross Johnson, Artur, "*Soviet Control of East European Military Forces: New Evidence on Imposition of the 1980 "Wartime*

Statute"," Accessed October 16, 2019. https://www.cia.gov/
library/publications/cold-war/wartime-statutes/wartime-statutes.
pdf.

- Rubin, F., "The Theory and Concept of National Security in the
Warsaw Pact Countries," *International Affairs,* no. 4 (1982): 648–
657.

- Scott, William F., *Selected Soviet Military Writings 1970-1975*
(Washington: U.S. Government Print Office, 1976).

- Schaefer, Bernd, *"The GDR in the Warsaw Pact,"* Accessed
October 16, 2019. http://www.php.isn.ethz.ch/lory1.ethz.ch/
collections/coll_gdr/intro2644.html?navinfo=44755.

- Schaefer, Bernd, "The Sino-Soviet Conflict and the Warsaw Pact,
1969–1980," in *NATO and the Warsaw Pact. Intrabloc Conflicts,*
eds. Mary Ann Heiss and S. Victor Papacosma (Kent: Kent State
University Press, 2008): 206–218.

- Sebestyen, Victor, *1989. Pád východního bloku* (Brno: Computer
Press, 2011).

- Selvage, Douglas, "The Treaty of Warsaw (1970): The Warsaw
Pact Context," *Bulletin of the German Historical Institute,* no. 1
(2004): 67–79.

- Selvage, Douglas, "The Warsaw Pact and the European security
conference, 1964–69," in *Origins of the European security
system: the Helsinky process revisited, 1965–75,* eds. Andreas
Wenger, Vojtech Mastny and Christian Nuenlist (Abingdon:
Routledge, 2008): 85–106.

- Selvage, Douglas, "The Warsaw Pact and the German Question
1955–1970: Conflict and Consensus," in *NATO and the Warsaw
Pact. Intrabloc Conflicts,* eds. Mary Ann Heiss and S. Victor
Papacosma (Kent: Kent State University Press, 2008): 178–192.

- Ślusarczyk, Jaczek, *Układ Warszawski: Działalność polityczna
1955-1991* (Warszawa: Instytut Studiów Politycznych Polskiej
Akademii Nauk, 1992).

- Suri, Jeremy, "Henry Kissinger and the reconceptuailization of
European security, 1969-75," in *Origins of the European security
system: the Helsinky process revisited, 1965–75,* eds. Andreas
Wenger, Vojtech Mastny and Christian Nuenlist (Abingdon:
Routledge, 2008): 46–64.

- Tejchman, Miroslav, *Nicolae Ceausescu* (Praha: Nakladatelství
Lidové noviny, 2004).

- Tejchman, Miroslav and Litera, Bohuslav, *Moskva a socialistické
země na Balkáně 1964–1989* (Praha: Historický ústav, 2009).

- Tischler, János, "The Hungarian Party Leadership and the Polish Crisis of 1980-1981," *Cold War International History Project Bulletin*, no. 11 (1998): 77–79.
- Thatcher, Ian D., "Brezhnev as Leader," in *Brezhnev Reconsidered*, eds. Erwin Bacon and Mark Sandle (New York: Palgrave, 2002): 22–37.
- Thee, Marek, "The Polish Drama: Its Meaning and International Impact " *Journal of Peace Research,* no. 1 (1982): 1–10.
- Tůma, Oldřich, "The Czechoslovak Communist Regime and the Polish Crisis 1980-1981," *Cold War International History Project Bulletin*, no. 11 (1998): 60–76.
- Uhl, Mattias, "Storming on to Paris. The 1961 Buria Exercise and the Planned Solution of the Berlin Crisis," in *War Plans and Alliances in the Cold War*, eds. Vojtech Mastny, Sven Holtsmark and Andreas Wenger (London: Routledge, 2006): 46–71.
- Umbach, Frank, *Die Rotte Bündnis. Entwicklung und Zerfall des Warschauer Paktes 1955-1991* (Berlin: Links Verlag, 1996).
- Volkogonov, Dmitrij, *The Rise and Fall of the Soviet Empire. Political Leader from Lenin to Gorbachev* (London: HarperCollins Publishers, 1999).
- Vykoukal, Jiří and Litera, Bohuslav and Tejchman, Miroslav, *Východ. Vznik, vývoj a rozpad sovětského bloku 1944–1989* (Praha: Libri, 2000).
- Walsh, David M., *The Military Balance in the Cold War. US perception and policy, 1976-1985* (Abingdon: Routledge 2008).
- Wanner, Jan, "Sovětská hierarchie a politika SSSR 1968–1982," *Slovanský přehled*, no. 2 (1994): 165–186.
- Wanner, Jan, *Brežněv a východní Evropa* (Praha: Karolinum, 1995).
- Watts, Larry "The Soviet-Romanian Clash over History, Identity and Dominion," *Cold War International History Project e-Dossier*, no. 29 (2010).
- Webber, Mark, "'Out of Area' Operations: The Third World," in *Brezhnev Reconsidered*, eds. Erwin Bacon and Mark Sandle (New York: Palgrave, 2002): 110–134.
- Wenger, Andreas and Mastny, Vojtech, "New perspectives of the origin of the CSCE process," in *Origins of the European security system: the Helsinky process revisited, 1965–75*, eds. Andreas Wenger, Vojtech Mastny and Christian Nuenlist (Abingdon: Routledge, 2008): 3–22.

- Westad, Odd Arne, "The Fall of Détente and the Turning Tides of History," in: The Fall of Détente. Soviet-American Relations during Carter Years, ed. Odd Arne Westad (Boston: Scandinavian University Press, 1997): 3–33.
- Wilke, Manfred, Erier, Peter, Kubina, Michael, Laude, Horst and Muller, Hans-Peter, "The SED Politburo and the Polish Crisis, 1980/1982," Cold War International History Project Bulletin, no 5 (1995): 121, 127.
- Задорожнюк, Элла Григорьевна, Бласть - общество – реформы : Центральная и Юго-Восточная Европа. Вторая половина XX века, Москва: Наука, 2006).
- Zubok, Vladislav, A Failed Empire: The Soviet Union in the Cold War from Stalin to Gorbachev (Chapel Hill: The University od North Carolina Press, 2007).

Abbreviations

AAN	Archiwum Akt Nowych (Archive of Modern Documents, Poland)
AMZV	Archiv ministerstva zahraničních věcí České republiky (Archive of Ministry of Foreign Affairs, Czech Republic)
BArch	Bundesarchiv (Federal Archive, Germany)
BCP	Bulgarian Communist Party
BLR	Bulharská lidová republika (People's Republic of Bulgaria)
CC	Central Committee
CSSR	Czechoslovak Socialist Republic
CSCE	Conference on Security and Cooperation in Europe
CPCz	Communist Party of Czechoslovakia
CPSU	Communist Party of the Soviet Union
DTO	Dokumentace teritoriálních odborů (Documentation of Teritorial Depratments)
FCO	Foreign and Commonwealth Office
FMO GŠ/OS	Federální ministerstvo obrany Generální štáb/Operační správa (Federal Ministry of Defense, General Staff/Operational Department)
FRG	Federal Republic of Germany
GDR	German Democratic Republic
HPR	Hungarian People's Republic
HSWP	Hungarian Socialist Worker's Party
KBSE	Konference o bezpečnosti a spolupráci v Evropě (Conference on Security and Cooperation in Europe)
KC	Komitet Centralny (Central Committee)
KSČ	Komunistická strana Československa (Communist Party of Czechoslovakia)
MSDS	Maďarská socialistická dělnická strana (Hungarian Socialist Worker's Party)
NATO	North Atlantic Treaty Organization

NA	National Archive, Czech Republic
NDR	Německá demokratická republika (German Democratic Republic)
PK	Porady kolegia (Meetings of Collegium)
PLR	Polská lidová republika (Polish People's Republic)
PPR	Polish People's Republic
PRB	People's Republic of Bulgaria
PUWP	Polish United Workers Party
PSDS	Polská sjednocená strana dělnická (Polish United Workers Party)
PZPR	Polska Zjednoczona Partia Robotnicza (Polish United Workers Party).
RCP	Romanian Communist Party
RKS	Rumunská komunistická strana (Romanian Communist Party)
RSR	Rumunská socialistická republika (Socialist Republic of Romania)
SRR	Socialist Republic of Romania
SED	Sozialistische Einheitspartei Deutschlands (Socialist Unity Party of Germany)
SRN	Spolková republika Německo (Federal Republic of Germany)
SSSR	Svaz sovětských socialistických republik (Union of Soviet Socialist Republics)
TNA	The National Archive, United Kingdom
TO(t)	Teritoriální odbory (tajné) [Teritorial Departments (secret)]
URM	Urząd Rady Ministrów (Bureau of Council of Ministers)
USA	United States of America
USSR	Union of Soviet Socialist Republics
ÚV	Ústřední výbor (Central Committee)
VÚA	Vojenský ústřední archiv (Central Military Archive, Prague)
a. j.	archivní jednotka (archival unit)
e. č.	evidenční šíslo (evidence number)
i. č.	inventární číslo (inventory number)
k.	karton (box)
p. č.	pořadové číslo (seriál number)
sv.	svazek (folder)

Name Index